ANNALS OF THE NEW YORK ACADEMY OF SCIENCES

Volume 440

EDITORIAL STAFF

Executive Editor
BILL BOLAND

Managing Editor
JOYCE HITCHCOCK

Associate Editor
RANDI E. SCHOLNICK

The New York Academy of Sciences
2 East 63rd Street
New York, New York 10021

DISCRETE GEOMETRY AND CONVEXITY

ANNALS OF THE NEW YORK ACADEMY OF SCIENCES

Volume 440

DISCRETE GEOMETRY AND CONVEXITY

Edited by Jacob E. Goodman, Erwin Lutwak,
Joseph Malkevitch, and Richard Pollack

The New York Academy of Sciences
New York, New York
1985

Library of Congress Cataloging in Publication Data

Main entry under title:

Discrete geometry and convexity.

(Annals of the New York Academy of Sciences ISSN 0077-8923 ; v. 440)
Proceedings of a conference held Apr. 2-3, 1982, sponsored by the New York Academy of Sciences and other organizations.
Includes bibliographies and indexes.
1. Geometry—Addresses, essays, lectures. 2. Convex domains—Addresses, essays, lectures. I. Goodman, Jacob E. II. New York Academy of Sciences. III. Series.
Q11.N5 vol. 440 500 s 85-8856
[QA446] [516]
ISBN 0-89766-274-1
ISBN 0-89766-275-X (pbk.)

Cover: Figure courtesy of J. M. Wills (see page 225).

S/PCP
Printed in the United States of America
ISBN 0-89766-274-1 (cloth)
ISBN 0-89766-275-X (paper)
ISSN 0077-8923

ANNALS OF THE NEW YORK ACADEMY OF SCIENCES

VOLUME 440

May 30, 1985

DISCRETE GEOMETRY AND CONVEXITY*

Editors and Conference Organizers

JACOB E. GOODMAN, ERWIN LUTWAK, JOSEPH MALKEVITCH, AND RICHARD POLLACK

CONTENTS

* This volume is the result of a conference sponsored by The New York Academy of Sciences entitled Discrete Geometry and Convexity Days. It was held April 2–3, 1982 in New York City and was cosponsored by City College of the City University of New York, Courant Institute of New York University, Polytechnic Institute of New York, and York College of the City University of New York.

Part IV. Polyhedral Geometry

Part V. Tiling, Packing, Covering, and Weaving

Part VI. Computational Aspects

PREFACE

This volume is an outgrowth of a two-day meeting devoted to Discrete Geometry and Convexity which was held on April 2nd and 3rd, 1982 at the New York Academy of Sciences and the Courant Institute of Mathematical Sciences of New York University. The meeting was held under the auspices of the Mathematics Section of the New York Academy of Sciences, and financial support was provided by City College (CUNY), the Courant Institute of Mathematical Sciences, the Polytechnic Institute of New York, and York College (CUNY).

Twelve talks were given at the meeting, and expanded versions of nine of these appear in this volume. The editors decided to solicit additional papers in convexity and discrete geometry in the hope of compiling a volume that would reveal the breadth and depth of recent work in these areas, on both new and old problems.

Many people contributed directly or indirectly to the meeting and to this volume, and to all of them—as well as to the sponsoring institutions—we extend our gratitude. In particular, we would like to single out the numerous referees, for their generous efforts in substantially improving the volume. We also wish to thank the staff of the New York Academy of Sciences for its cooperation. Special thanks go to Ms. Randi Scholnick who prepared all the articles for publication with great diligence and dedication.

J. E. GOODMAN
E. LUTWAK
J. MALKEVITCH
R. POLLACK

INTRODUCTION

The area of mathematics treated in the present volume ranges over a wide spectrum which includes purely combinatorial questions involving the geometry of finite sets of points at one extreme and integral geometry at the other, and which branches off at various points into such applied fields as computational geometry and operations research. The editors have made an attempt to include representative papers in many, but by no means all, of the areas which could be subsumed under the heading *Discrete Geometry and Convexity*. Thus the reader will find no papers on valuation or matroid theory per se, or on algebro-geometric questions dealing with configurations of points in C^n, to mention but a few areas that were omitted. But exigencies of space aside, we hope the volume does succeed in what was our goal at the outset—to collect under one cover some representative current work in these areas of geometry so that the reader may be in a position to judge for himself the present state of these fields and their place in contemporary mathematics.

The volume begins with eight papers on discrete geometry. The first, by **P. Erdös**, discusses a variety of combinatorial problems, both old and new, including some that have been solved recently, and contains a number of challenging conjectures on bounds and asymptotic estimates for extremal problems. The paper of **J. E. Goodman** and **R. Pollack** that follows presents a discrete result analogous to the isotopy problem for planar configurations of points: the authors prove that one configuration can be moved to another of the same *order type* by small steps, defined in terms of *allowable sequences of permutations* which encode a refinement of this order type. **R. L. Graham**'s paper surveys the field of *Euclidean Ramsey theory*, which asks which configurations of points, up to some Euclidean motion, must lie in a single class whenever Euclidean n-space is partitioned into r classes; several new results on *sphere-Ramsey* sets are also presented. The paper of **H. Harborth** extends the previously known three infinite families of simple pseudoline arrangements containing as many as $[n(n-2)/3]$ (respectively, $[n(n-1)/3]$) triangles for n odd (respectively, even) to arbitrarily many such families by means of a new construction. The *slope problem*, finding the minimum number of slopes determined by n noncollinear points in the plane, is reviewed by **R. E. Jamison,** from its enunciation by P. R. Scott to its solution by P. Ungar, and beyond—to the study of configurations minimizing the number of slopes and to generalizations to C^2 and to higher dimensions. **W. O. J. Moser**'s paper, like that of Erdös, deals largely with extremal problems of finite configurations, and constitutes an update on the status of nineteen problems from his 1981 edition of "Research Problems in Discrete Geometry." The paper by **G. B. Purdy** uses the recently proven Dirac conjecture to find a lower bound for the size of the largest antichain in the set of regions determined by n lines in the Euclidean plane, with respect to a natural partial ordering on that set. **T. Zaslavsky**'s paper, on extremal hyperplane arrangements, indicates how matroid theory can be used to find sharp bounds on the number of flats and cells of Euclidean or projective arrangements.

The next set of papers, on quantitative convexity, begins with a paper by **G. D. Chakerian** and **M. A. Ghandehari**; they establish two theorems (regarding *midpoint polyhedra*) which when combined show that given r points on the unit sphere (of a normed linear space) with the property that their convex hull contains the origin, the sum of the distances determined by pairs of these points is $2r-2$. Weil's measure of support flats is the subject of **W. J. Firey**'s paper; a motion-invariant integral-geometric density for m-dimensional flats tangent to a sufficiently smooth convex body in Euclidean n-space (for all $m < n$) is described and an explicit representation involving the principal radii of curvature of the body is obtained. The paper by **J. Linhart** characterizes regular simplicial d-polytopes among all simplicial d-polytopes with the same number of facets and circumradius as those having maximum volume. **E. Lutwak** obtains a geometric inequality which acts as an intermediary between the Blaschke–Santaló inequality and a general version of the affine isoperimetric inequality. **C. M. Petty**'s paper, on affine isoperimetric problems, shows that regularity assumptions heretofore necessary in concluding equality in the affine isoperimetric inequality, the Blaschke–Santaló inequality, and a related inequality of Petty involving geominimal surface area, can actually be removed. **L. A. Santaló** finds a set of similitude-invariant functionals in Euclidean n-space which includes a functional previously considered by Franklin for $n=3$; a lower bound is obtained for these functionals which is close to the conjectured best lower bound. This section concludes with a paper by **R. Schneider** which contains a survey of known results regarding the case of equality in the Aleksandrov–Fenchel inequality; a conjecture regarding the general case of equality and several new results to support the conjecture are presented.

Five papers on qualitative convexity follow, beginning with **M. Breen**'s paper on theorems of Krasnosel'skii type, in which she discusses her recent results on the dimension and size of the kernel of a starshaped set in a Euclidean space. **R. Fourneau**'s paper is concerned with Choquet simplices in R^d; he gives a number of their properties in dimensions 2 and 3, which can then be used to give a complete enumeration of all Choquet simplices (compact or not) in those dimensions. A survey of properties that hold for *most* convex bodies in Euclidean d-space is given in **P. M. Gruber**'s paper, where *most* means all but a set of first Baire category in the topology induced by the Hausdorff metric; he also presents two new Baire category results, one on the number of intersection points of a convex body with its minimal ellipsoid and the other on the number of neighbors of a convex body in a lattice packing. **E. Heil** uses Morse theory to prove that any convex body in Euclidean n-space for $n \geq 3$ has six normals which meet at an interior point, improving an earlier result which required a smoothness assumption. **D. C. Kay**'s paper extends an earlier generalization of Tietze's theorem that a locally convex, closed, and connected subset S of R^d is convex, from the case where S contains only finitely many points of local nonconvexity whose removal leaves a connected set to the case where the set of locally nonconvex points is permitted to be infinite but required to satisfy certain additional conditions; the conclusion is a bound on the smallest number of convex subsets whose union gives S.

The section that follows, on polyhedral geometry, begins with a proof by **D. W. Barnette** of the somewhat surprising result that—even though there are toroidal maps which require as many as seven colors—every toroidal *polytope* is six-colorable. In the next paper, **G. Ewald** discusses two connections between algebraic geometry and the theory of convex bodies: Kouchirenko's characterization of the Milnor number of an isolated critical point of a complex polynomial via geometric properties of Newton polytopes, and a recent proof of the Aleksandrov–Fenchel inequality for mixed volumes by means of the Hodge index theorem. **C. W. Lee** takes up the problem of finding a minimal triangulation of the d-cube without using extra vertices, and gives a new proof of the minimality of a triangulation of the 4-cube due to Maria; he also discusses several methods of triangulation. **R. P. Stanley** discusses the problem of characterizing the face-vectors of simplicial polytopes and simplicial spheres from an historical point of view, and details the role that algebraic geometry played in his proof of the necessity of the McMullen conditions in the polytopal case. **J. M. Wills** defines and gives a number of examples of *semi-Platonic* polyhedral manifolds, which generalize (to higher genus) the combinatorial regularity properties of the bounding surfaces of the five Platonic solids.

Tiling, packing, covering, and weaving are the topics of the next group of papers, which begins with **H. E. Debrunner**'s work on simplicial tilings: this introduces new methods of finding simplices, congruent copies of which tile Euclidean d-space, and considerably enlarges the set of known simplices of this type for $d > 3$. **H. Groemer**'s survey collects a number of results on covering and packing problems by convex bodies in Euclidean n-space, and considers in particular both isometric and translative coverings (and packings) by sequences of bodies of bounded diameter and by sequences of strips and slabs. A computer search was used to obtain the complete list of isonemal fabrics of period ≤ 13 described in the paper of **B. Grünbaum** and **G. C. Shephard** (*isonemal* means having strand-transitive symmetry group); a number of these fabrics are illustrated. **J. Malkevitch** considers a variant of the question of which convex n-gons can be tiled edge-to-edge by unit squares and equilateral triangles, namely: Whenever such a tiling exists, what is the possible number (v_0) of non-extreme boundary points? He gives a complete enumeration of the minimum value of v_0 for each possible choice of exterior angle set. In the final paper in this section, **W. T. Trotter, Jr.** examines the problem of tiling bounded open sets S in R^n with (possibly infinitely many) rectangular tiles that touch the boundary and have edges parallel to the coordinate axes, and proves that this can be done (a) with squares if $n = 2$ and S is a homeomorph of a disk or annulus, but not in general; (b) with rectangular parallelepipeds for all $n \geq 2$; (c) not with cubes for $n = 3$ even if S is homeomorphic to a ball.

Five papers on the computational side of geometry conclude the volume. First, **D. Avis** and **J. Horton** consider a graph defined by Toussaint and known as the *sphere of influence graph* of a finite set S in the plane, in which two points of S are joined by an edge whenever the maximum open disks centered at them and containing no other point of S have nonempty intersection; they prove that if S consists of n points, then this graph has at most $29n$ edges. The problem of finding how much one can save on the total length of a planar network joining given points by inserting new vertices is considered by **F. R. K. Chung** and **R. L. Graham**, who use intricate computer-assisted algebraic computations to improve a previous lower bound and prove that for any finite set X the ratio of the length of the minimal Steiner tree through X to the length of the minimal tree *spanned* by X is $\geq 0.824167\ldots$, the conjectured value being $\sqrt{3}/2$. The second paper of **J. E. Goodman** and **R. Pollack,** on geometric sorting theory, extends the idea of sorting a set of points on the line to a higher-dimensional generalization, based on a theorem which asserts that the order type of a set of points in Euclidean d-space is determined by the number of points on the positive side of each oriented hyperplane spanned by $d+1$ of the points, and discusses algorithms for implementing this sorting process, as well as possible applications. **A. C. Hayes** and **D. G. Larman** discuss several variants of the problem of fitting a hyperplane to a set of points in R^{d+1}, including fitting k hyperplanes and fitting a hyperplane *bent* along a $(d-1)$-flat; algorithms polynomial in the number of points are given, which extend previous planar algorithms. The volume concludes with a paper by **P. F. Pickel** on the ellipsoid algorithm (Khachian's algorithm), with particular reference to its applicability to nonlinear and to discrete optimization problems; recent results on comparisons between the ellipsoid method and other nonlinear programming methods are discussed, as well as recent extensions of the algorithm by Karmarkar.

J. E. GOODMAN
E. LUTWAK
J. MALKEVITCH
R. POLLACK

PROBLEMS AND RESULTS IN COMBINATORIAL GEOMETRY[a]

P. Erdös

Hungarian Academy of Sciences
Budapest, Hungary

I have published during my very long life many papers with similar titles. In this note, I report on progress made on some of the old problems and discuss also some new ones, and I repeat some old problems which seem to have been forgotten by everybody (sometimes including myself). First, a very incomplete list is given of some papers dealing with these and related questions.

I have published three survey papers on this subject [1–3].

I want to call attention to the excellent book of Hadwiger and Debrunner which originally appeared in German and French. This book was translated by Klee who also added much original material and brought the book up-to-date (up to about 1964) [4]. This book contains a great deal of interesting material and can be used as a textbook to learn the subject, but contains few unsolved problems. (Unfortunately, it is out of print.)

A very fruitful source of problems and results is the much expanded 1981 edition of W. Moser's collection, "Research Problems in Discrete Geometry" (McGill University). The nucleus of this collection was the old problems of the late Leo Moser.

In *Elemente der Mathematik* there is a section entitled "Ungelöste Probleme" which was edited by Hadwiger and in the *American Mathematical Monthly* there is a similar section edited by Guy; both often contain geometrical problems. See also the papers by Fejes-Toth, Grünbaum, and Klee that are quoted in [1].

G. Purdy and I plan to write a book on various problems in combinatorial geometry. We hope the book will appear in this decade.

I

Let there be given n distinct points $x_1, \ldots, x_n$ in k-dimensional Euclidean space E_k. Denote by $d(x_i, x_j)$ the distance between x_i and x_j. $D_k(x_1, \ldots, x_n)$ denotes the number of distinct distances among the x_i's. Put

$$f_k(n) = \min_{x_1, \ldots, x_n} D_k(x_1, \ldots, x_n).$$

Denote further by $P_k(n)$ the largest integer for which there are $P_k(n)$ pairs x_i and x_j for which $d(x_i, x_j) = 1$. Both these functions have been extensively studied in [1] (with many references).

The best results until recently were

$$P_2(n) = o(n^{3/2}), \qquad f_2(n) > cn^{2/3}. \tag{1}$$

[a] Dedicated to the memory of my friend Hugo Hadwiger.

The first result in (1) is due to Szemerédi, the second to L. Moser. Recently, J. Beck and J. Spencer proved

$$P_2(n) < n^{3/2-c} \qquad \text{for some } c > 0 \tag{2}$$

and Fan Chung proved

$$f_2(n) > cn^{5/7}. \tag{3}$$

I still believe that my old conjectures

$$f_2(n) > cn/(\log n)^{1/2}, \qquad P_2(n) < n^{1+c/\log\log n} \tag{4}$$

are true and I offer \$500 for a proof or disproof. I offer \$250 for $P_2(n) < n^{1+\varepsilon}$.

Observe that L. Moser's proof in fact gives that there is a point x_i so that the number of distinct distances $d(x_i, x_j)$, $i \leq j \leq n$, $j \neq i$, is greater than $cn^{2/3}$. The proof of Fan Chung does not seem to give $cn^{5/7}$, but perhaps it is true. In fact, perhaps there always is a point x_1 so that the number of distinct distances $d(x_1, x_i)$, $2 \leq i \leq n$, is greater than $cn/(\log n)^{1/2}$. This and several other conjectures are discussed in [1].

For $k \geq 4$, $P_k(n)$ is much easier to handle than it is for $k = 2$.

Let $x_1, \ldots, x_n$ be n points on the surface of a two-dimensional sphere. Denote by $P'_2(n)$ the largest integer for which there are $P'_2(n)$ pairs x_i, x_j for which all the distances $d(x_i, x_j)$ are the same. I could not prove that

$$\lim_{n\to\infty} P'_2(n)/n = \infty.$$

Ulam recently asked me the following question: Let $x_1, \ldots, x_n$ be n points in the plane. Does one get interesting combinatorial and geometric questions if one modifies the metric and asks how often we can have $d(x_i, x_j) = 1$? E.g., he asked: What if we define the distance of two points as the sum of the absolute values of the differences of their coordinates? In this case I proved that if $n > 4$, $n \equiv 0 \pmod 4$, then $P_2(n) = (n^2 + n)/4$. I hope to return to these questions later.

Reference for Section I: [5].

II

Let there be given n points in the plane not all on a line. Join every two of them. Sylvester conjectured and Gallai proved in 1933 that there always is a line that passes through precisely two of these points. This problem and various extensions and generalizations are of course discussed in [1]. Motzkin conjectured that for $n > 13$, the number of ordinary lines (i.e., the number of lines going through precisely two of the points) is $\geq n/2$. Motzkin observed that for every even n, this conjecture—if true—is best possible. Hansen recently proved this conjecture; his proof, which is not yet published, is quite complicated [6].

Denote by $L_1, \ldots, L_m$ the lines determined by our points $x_1, \ldots, x_n$. Denote by $|L_i|$ the number of points on L_i. Put $|L_i| = y_i$, $y_1 \geq \cdots \geq y_m$. It easily follows from Gallai's theorem and also from a theorem of de Bruijn and myself that if $m > 1$, then $m \geq n$. Grünbaum asked, What are the possible values of m? I showed that for

$m > cn^{3/2}$ all possible values of m can occur except $\binom{m}{2} - 1$ and $\binom{m}{2} - 3$. The analogous problems in higher dimensions can probably be settled without great difficulty (lines are replaced by hyperplanes).

The following problem is probably quite difficult: Denote by $F(n)$ the number of possible choices of the set $\{y_1, \ldots, y_m\}$. We must of course have

$$\sum_{i=1}^{m} \binom{y_i}{2} = \binom{n}{2}. \tag{5}$$

(5) is by no means sufficient for the existence of a set $x_1, \ldots, x_n$ for which $|L_i| = y_i$ and perhaps it is not reasonable to expect a good necessary and sufficient condition. I conjectured several years ago that

$$F(n) < \exp cn^{1/2}. \tag{6}$$

It is easy to see that (6) if true is best possible, but I do not believe that to decide (6) will be easy and I offer \$250 for a proof or disproof of (6).

Denote by $h_k(x_1, \ldots, x_n)$ the number of indices i for which $|L_i| = k$. What is the possible range of values of $h_k(x_1, \ldots, x_n)$? As far as I know, this question has not yet been investigated.

Gallai's theorem can be stated as follows: Unless all the x_i are on a line, $h_2(x_1, \ldots, x_n) \geq 1$. Hansen's theorem states that for $n > 13$, $h_2(x_1, \ldots, x_n) \geq n/2$. Put

$$\max_{x_1, \ldots, x_n} h_k(x_1, \ldots, x_n) = t_k(n).$$

Trivially $t_2(n) = \binom{n}{2}$. $t_3(n)$ was first studied by Sylvester and was investigated in a very nice paper by Burr, Grünbaum, and Sloane [7].

Croft and I observed that for every k and $n > n_0(k)$, $t_k(n) > c(k)n^2$. We could not determine the largest possible value of $c(k)$. Put

$$\lim_{n \to \infty} t_k(n)/n^2 = c_k.$$

Trivially $c_k \leq 1/k(k-1)$ and for $k = 2$ and 3, equality is possible. Croft and I conjectured that

$$\lim_{k \to \infty} k^2 c_k = 0, \tag{7}$$

but we could not even prove $c_k < 1/k(k-1)$ for $k > 3$. (Recently, Szemerédi and Trotter proved $c_k < (\log k)^c/k^3$, where c is an absolute constant.) Put

$$H(n) = \max_{x_1, \ldots, x_n} \sum_{k \geq \sqrt{n}} h_k(x_1, \ldots, x_n).$$

In other words, $H(n)$ is the largest number of lines that contain at least $\sqrt{n}$ points. I conjectured that $H(n) < cn^{1/2}$, but could not even prove $H(n) = o(n)$. Recently, Szemerédi and Trotter proved $H(n) < cn^{1/2}$. This result has not yet been published.

I thought that perhaps $H(n^2) = 2n + 2$, but Szemerédi informed me that recently someone showed $H(n) > 3n^{1/2}$.

Purdy and I conjectured that

$$\max \sum_{i=1}^{n} |L_i| < cn^{4/3}. \tag{8}$$

The lattice points show that **(8)** if true is best possible. It is possible that the method of Szemerédi and Trotter will give **(8)**. More generally, perhaps the number of indices i for which $|L_i| > n^{1/2}2^{-u}$ is less than $c2^{2u}n^{1/2}$ and if true the lattice points show that it is best possible. Szemerédi and Trotter have proven

$$\max \sum_{i=1}^{n} |L_i| < c_1 n^{4/3}(\log n)^{c_2}.$$

Croft and I conjectured that to every $\varepsilon > 0$ there is a $k_0 = k_0(\varepsilon)$ so that if $|L_i| < \varepsilon n$ for all i, then

$$\sum_{|L_i| > k_0} \binom{L_i}{2} < \varepsilon \binom{n}{2}.$$

In other words, the number of pairs of points situated on "large lines" is small.

Assume that $x_1, \ldots, x_n$ is such that $|L_i| \leq k$; i.e., no line contains more than k points. Denote by $T_k(n)$ the maximum number of lines that contain k points. I conjectured that for $k > 3$

$$T_k(n) = o(n^2). \qquad \textbf{(9)}$$

For $k = 3$, Sylvester and Burr, Grünbaum, and Sloane [7] proved that $T_3(n) = (1 + o(1))(n^2/c)$, but the exact value of $T_3(n)$ is not yet known.

Grünbaum proved

$$T_k(n) > cn^{1+1/k};$$

perhaps the exponent $1 + 1/k$ is best possible.

Purdy observed that

$$\frac{1}{m} \sum_{i=1}^{m} |L_i| < 3$$

and that 3 is best possible.

Additional references for Section II: [8, 9].

III

Hadwiger and Nelson posed the following problem: Let G_k be a graph whose vertices are the points of E_k (the k-dimensional Euclidean space). Two points are joined if their distance is 1. Determine or estimate the chromatic number $h(k)$ of G_k. It is known that $4 \leq h(2) \leq 7$. I am almost sure that $h(2) > 4$. In this connection, L. Moser asked the following interesting question: Let R be large and S a measurable set in the circle of radius R so that no two points of S have distance 1. Denote by $m(S)$ the measure of S. Determine

$$\lim_{R \to \infty} \max m(S)/R^2. \qquad \textbf{(10)}$$

It seems very likely that the limit in **(10)** is less than $\frac{1}{4}$.

I asked the following question: Let S be a subset of the plane. Join two points of S if their distance is 1. This gives the graph $G(S)$. Assume that the girth (shortest circuit) of $G(S)$ is k. Can its chromatic number be greater than 3? Wormald proved that such a graph exists for $k \leq 5$. The problem is open for $k > 5$. Wormald suggested that his method may work for $k = 6$, but probably a new idea would be needed for $k > 6$.

A related (perhaps identical) question is: Does $G(S)$ have a subgraph that has girth k and chromatic number 4?

Let $0 < r_1 < \cdots < r_k$. $G_2(r_1, \ldots, r_k)$ is the graph whose vertices are the points of the plane and two points are joined if their distance is one of the numbers $r_1, \ldots, r_k$. Put ($h(G)$ is the chromatic number of G)

$$h_r(2) = \max_{r_1, \ldots, r_k} h(G_2(r_1, \ldots, r_k)).$$

It is easy to see that $\lim_{r \to \infty} h_r(2)/r = \infty$, but I do not know if $h_r(2)$ grows polynomially.

Larman and Rogers proved

$$ck^2 < h(k) < (3 + o(1))^k.$$

I conjectured that $h(k) > (1 + c)^k$. This conjecture was recently proved by Frankl and Wilson. The value of $\lim_{k \to \infty} h(k)^{1/k}$ is not known and in fact it is not even known if the limit exists.

V. T. Sós and I proved that if there are $n + 1$ triples in a set S of n elements, then there are always two of them whose intersection is a singleton. The proof is simple. We conjectured that if $|S| = n$, $n > n_0(k)$ and $A_i \subset S$, $|A_i| = k$, $1 \leq i \leq t_k$, $t_k > \binom{n-2}{k-2}$ is a family of distinct subsets of S, then there are always two A's whose intersection is a singleton. It was immediate that if true, this conjecture is best possible.

Katona proved the conjecture for $k = 4$ and Frankl proved it in the general case.

I further conjectured: Let $|S| = n$, $0 < \eta < \frac{1}{2}$, $A_i \subset S$, $1 \leq i \leq T_{n,\eta}$. Assume that for some r,

$$\eta n < r < (\tfrac{1}{2} - \eta)n, \qquad (A_i \cap A_j) \neq r \text{ for every } 1 \leq i < j \leq T_{n,\eta}.$$

Then

$$T_{n,\eta} < (2 - \varepsilon)^n, \qquad \varepsilon = \varepsilon(\eta). \tag{11}$$

It is easy to see that (**11**) implies $h(k) < (1 + c)^k$. (**11**) is still open; the proof of Frankl and Wilson did not use (**11**). [June 25, 1982—I just had a letter from Frankl: He and Füredi have proved (**11**).]

In a recent paper Simonovits and I investigated $h^*(k)$, the so-called essential chromatic number of E_k. $h^*(k)$ is the largest integer for which there is a finite $x_1, \ldots, x_n$ in E_k so that if we omit $o(n^2)$ edges of G_k in all possible ways, we are always left with a graph of chromatic number $\geq h^*(k)$. We prove that $h^*(4) = 2$ and conjecture that $h^*(k)$ tends to infinity exponentially. In fact, we only proved $h^*(k) \geq k - 2$. Several further problems (which I think are interesting and challenging) are discussed in our paper.

A well-known theorem of de Bruijn and myself states that every k-chromatic graph contains a finite subgraph of chromatic number k. Thus all the problems considered in this section are problems about finite sets and finite graphs.

Perhaps the following modification of the Hadwiger–Nelson problem is of some interest: Join two points in E_2 if their distance is between α and β, $\alpha < 1 < \beta$. Determine or estimate the chromatic number $h(2, \alpha, \beta)$ of this graph.

References for Section III: [10–16].

IV

An old problem of Heilbronn states as follows: Let $z_1, \ldots, z_n$ be n points in the unit circle. Denote by $A_k(z_1, \ldots, z_n)$ the smallest area of all polygons $\{z_{i_1}, \ldots, z_{i_k}\}$, $1 \le i_1 < \cdots < i_k \le n$. Put

$$g_k(n) = \max_{z_1, \ldots, z_n} A_k(z_1, \ldots, z_n).$$

Heilbronn asked for the determination or estimation of $g_3(n)$. He of course observed that trivially $g_3(n) < c/n$ and suspected that the order of magnitude of $g_3(n)$ is $1/n^2$. I observed that indeed $g_3(n) > c/n^2$. The first nontrivial result was due to K. F. Roth who proved $g_3(n) < 1/n(\log \log n)^{1/2}$. This was improved by W. Schmidt to $c/n(\log n)^{1/2}$ and later by Roth to $1/n^{1+c_1}$.

In a recent paper, Komlós, Pintz, and Szemerédi [17] improved the value of c_1, but their really surprising result was their proof of

$$g_3(n) > \frac{c \log n}{n^2}. \tag{12}$$

The proof of **(12)** uses a novel combination of combinatorial and probabilistic arguments, which will have many further applications. Szemerédi believes that **(12)** perhaps is best possible.

As far as I know, the first nontrivial results for $k > 3$ are due to Schmidt. He proved

$$g_k(n) > \frac{c_k}{n^{1+1/k}} \tag{13}$$

and observed that the proof of

$$g_4(n) = o\left(\frac{1}{n}\right) \tag{14}$$

presents considerable difficulties. As far as I know **(14)** is still open and seems to be a fundamental problem.

Szemerédi and I posed the following problem: Denote by $D(z_1, \ldots, z_n)$ the smallest distance between two of our z's and $\alpha(z_1, \ldots, z_n)$ is the smallest angle determined by three of our points (if three z's are on a line, then $\alpha(z_1, \ldots, z_n) = 0$). Is it true that

$$D(z_1, \ldots, z_n)\alpha(z_1, \ldots, z_n) = o(1/n^{3/2})? \tag{15}$$

It is almost trivial that **(15)** holds if $c/n^{3/2}$ replaces $o(1/n^{3/2})$. Perhaps in fact $o(1/n^{3/2})$ can be replaced by c/n^2. This conjecture is perhaps too optimistic. The regular polygon shows that if true, it is best possible. **(15)** if true may throw some light on Heilbronn's problem.

V. T. Sós, E. Straus, and I slightly modified Heilbronn's problem as follows: Denote by $L(z_1, \ldots, z_n)$ the largest integer for which there are $L(z_1, \ldots, z_n)$ z's on a line. Assume $L(z_1, \ldots, z_n) = o(n^{1/2})$. Is it then true that there are three z's, $z_{i_1}, z_{i_2}, z_{i_3}$, not on a line for which the area of the triangle $(z_{i_1}, z_{i_2}, z_{i_3})$ is $o(1/n)$? Perhaps this conjecture is too optimistic. Pintz proved that if we assume $L(z_1, \ldots, z_n) < cn^\alpha$ for some fixed positive α then the conjecture indeed holds. The lattice points show that $o(n^{1/2})$ can certainly not be replaced by $O(n^{1/2})$.

V

Euclidean Ramsey Problems

A finite subset C of E_n is called r-Ramsey for E_n if for any partition of E_n into r sets S_i, $\bigcup_{i=1}^{r} S_i = E_n$, some S_i always contains a subset C' which is congruent to C. If C is r-Ramsey for every r for some E_n, then it is called Ramsey.

The study of these problems was started by "us" (Graham, Montgomery, Rothschild, Spencer, Straus, and myself) a few years ago. Many very interesting and challenging problems remain on this subject. We prove that if C is Ramsey, then C must lie on the surface of some sphere. Further, we prove that any subset of the vertices of a rectangular parallelepiped is Ramsey. We do not know which (if any) of these alternatives characterize Ramsey sets, and I offer \$500 for an answer to this question.

Is it true that every nonequilateral triangle is 2-Ramsey in the plane? I offer \$250 for a proof or disproof. L. Shader proved that all right triangles are 2-Ramsey (in E_2).

Gurevich asked: Put $E_n = \bigcup_{i=1}^{r} S_i$. Is it then true that at least one S_i contains the vertices of a k-dimensional simplex of volume 1? Graham proved and generalized this. His paper refers to nearly all the relevant literature on Euclidean Ramsey problems. (See Graham's paper in this volume.)

We posed the following problem: Let $S \subset E_2$ be such that no two points of S have distance 1. Is it then true that $\bar{S}$ (the complement of S) contains the vertices of a unit square? We could not settle this problem. Juhász proved that the answer is affirmative. In fact, she proved that our theorem remains true if "unit square" is replaced by "arbitrary four point configuration." She further proved that the theorem certainly fails if four is replaced by 12. Many further problems and conjectures are discussed in our papers.

References for Section V: [18, 19].

VI

In this final section, I discuss miscellaneous problems.

1. Let $x_1, \ldots, x_n$ be n points in the plane not all on a line. Denote by $A(x_1, \ldots, x_n)$ and $a(x_1, \ldots, x_n)$ the area of the largest and the smallest nonzero, respectively, of the triangles (x_i, x_j, x_l), $1 \le i < j < l \le n$. Purdy, Straus, and I proved that

$$A(x_1, \ldots, x_n)/a(x_1, \ldots, x_n) \ge [n/2] \qquad \textbf{(16)}$$

and we determined all cases when there is equality in **(16)**. Straus extended **(16)** to higher dimensions. In **(16)**, one could perhaps replace triangles by polygons with r vertices, but as far as I know, this has not yet been investigated.

Our paper on this subject will soon appear in *Discrete Mathematics.*

2. Corrádi, Hajnal, and I asked: Is it true that n points in the plane not all on a line determine at least $n-2$ different angles? This is trivial if no three of the points are on a line, but seems to present curious difficulties in the general case and all we proved was that the number of distinct angles is $> cn^{1/2}$.

A related question is: Is it true that the smallest (nonzero) angle determined by our points is $\leq \pi/n$? We have equality for the regular polygon. This is again trivial if no three of the points are on a line, but in the general case we only get $c/n^{1/2}$.

An old problem of G. Dirac states: Is it true that if we join any two of our points, then there always is a point that is joined to at least $n/2 - c$ distinct lines? As far as I know, here also only $cn^{1/2}$ has been proved so far.
[**Editors' note:** A version of this problem has recently been solved by Szemerédi and Trotter, and independently by Beck.]

Scott asked: Is it true that our points determine at least $2[n/2]$ distinct directions? We have equality for the regular polygon. Scott proved $cn^{1/2}$. Here the problem was completely solved. First Burton and Purdy [20] proved $n/2$ and recently P. Ungár has proven $2[n/2]$, thus brilliantly settling the conjecture of Scott. It does not happen too often that a problem in this subject gets a complete solution.
[**Editors' note:** Ungár's paper has appeared in the *Journal of Combinatorial Theory, Series A* **33:** 343–347 (1982); see Jamison's paper in this volume for a summary of Ungár's proof as well as a history of the problem.]

The three problems discussed here are really unrelated.

3. A few years ago, I asked the following question: Let $f(n)$ be the largest integer for which there are n points $x_1, \ldots, x_n$ in E_2 so that there should be $f(n)$ distinct circles of unit radius passing through three of them. I observed

$$3n/2 < f(n) < n(n-1). \qquad \textbf{(17)}$$

I believe that $f(n)/n \to \infty$ but $f(n)/n^2 \to 0$, but have made no progress on this problem.

It might be difficult to give an exact (or even an asymptotic) formula for $f(n)$. Several related questions can be posed.

Straus and I observed that if $n_k \geq k + 2\binom{k-1}{2}\binom{k-1}{3}$ and $x_1, \ldots, x_{n_k}$ are in general position (i.e., no three on a line and no four on a circle), then there are always k of them so that the $\binom{k}{3}$ radii of the circumscribed circles are all different. Probably our estimate for n_k is far from being best possible. See [21].

4. Denote by n_k the smallest integer (if it exists) for which every set of n_k points (no three on a line) contains k, which form a convex polygon that contains none of the other points in its interior. It is easy to see that $n_4 = 5$. Ehrenfeucht and Harborth proved that n_5 is finite. Harborth in fact proved $n_5 = 10$ [22]. It is not yet known if n_6 exists. J. D. Horton just informed me that he showed that n_7 does not exist. His proof will be published in the *Canadian Mathematical Bulletin.*

I raised this problem as a sharpening of an old problem of E. Klein (Mrs. Szekeres): Let $f(k)$ be the smallest interior for which every set of $f(k)$ points, no three on a line, contains the vertices of a convex k-gon. Klein observed 50 years ago that $f(4) = 5$ and Szekeres conjectured $f(k) = 2^{k-2} + 1$. Makai and Turán proved this for $k = 5$. Szekeres and I proved

$$2^{k-2} + 1 \leq f(k) \leq \binom{2k-4}{k-2}$$

This problem is discussed and referenced in [1].

5. Let $x_1, \ldots, x_n$ be n points in k-dimensional space. Assume that $d(x_i, x_j) \geq 1$ and that either two distances are equal or they differ by at least 1, i.e.,

$$d(x_i, x_j) = d(x_k, x_l) \quad \text{or} \quad |d(x_i, x_j) - d(x_k, x_l)| \geq 1. \qquad \textbf{(18)}$$

Let

$$D_k(n) = \min_{x_1, \ldots, x_n} D(x_1, \ldots, x_n),$$

where $D(x_1, \ldots, x_n)$ is the diameter of $x_1, \ldots, x_n$ and the minimum is to be taken over all sets $x_1, \ldots, x_n$ in E_k satisfying (**18**). Trivially $D_1(n) = n - 1$; I expect that $D_2(n) > cn$, but I am very far from being able to prove this. Kanold proved $D_2(n) > cn^{3/4}$ (*Elemente der Mathematatik*, 1982, Manh) and the conjecture (**4**) would imply $D_2(n) > cn/(\log n)^{1/2}$: I find the conjecture $D_2(n) > cn$ a challenging and interesting problem and offer \$100 for a proof or disproof.

The lattice points in E_3 (respectively, E_k) easily show that

$$D_3(n) < c_3 n^{2/3}, \qquad D_k(n) < c_k n^{(k-1)/k}. \tag{19}$$

I would expect (without any real evidence) that the order of magnitude in (**19**) is best possible.

6. Let $x_1, \ldots, x_n$ be n points in E_2, at most k on a line. It is easy to see that if $k\binom{r}{2} < n$, then there are always r of them no three of which are on a line; i.e., there are $(1 + o(1))(2n/k)^{1/2}$ points no three of them on a line. Probably this can be very much improved, but I have no further results on this problem. The best upper bound is the trivial $2n/k$.

7. It is not difficult to prove that if S is a set in the plane of infinite two-dimensional measure, then it contains for every c the vertices of a triangle of area c. I could never prove that there is an absolute constant α so that every planar set of measure $>\alpha$ contains the vertices of a triangle of area 1. In fact, perhaps $\alpha = \pi/\sqrt{3}$ (the equilateral triangle inscribed in the circle of area $\pi/\sqrt{3}$ has area 1). This conjecture is perhaps again a bit too optimistic.

Let R be large and S a set of plane measure $>cR^2$ in a circle of radius R. Is it true that S contains the vertices of an equilateral triangle of side >1? Straus thought that this result perhaps already holds if we only assume that the area of S is greater than C, where C is a sufficiently large absolute constant.

An old conjecture of mine states: Let E be an infinite set of real numbers; then there always is a set S of positive linear measure that contains no subset similar (in the sense of elementary geometry) to E. I offer \$100 for a proof or disproof.

8. Two-distance sets and points in general position.

Denote by $g_k(n)$ the smallest integer for which every set of $g_k(n)$ points in E_n determines at least $k + 1$ distinct distances. Trivially $g_1(n) = n + 2$ and Blockhuis recently proved (sharpening a previous result of Delsarte, Goedels, and Seidel) that

$$g_2(n) \leq \frac{(n+1)(n+2)}{2}.$$

Denote by $G_k(n)$ the smallest integer for which every set of $G_k(n)$ points in E_n contains a subset of $k + 1$ points any two distances of which are distinct; i.e., any set of $G_k(n)$ points contains a subset of $k + 1$ points that determines $\binom{k+1}{2}$ distinct distances. I proved long ago that $G_2(2) = 7$ and Croft proved $G_2(3) = 9$. Bárány and Füredi proved

$$G_2(n) < n^{\log n}.$$

Recently Blockhuis proved

$$G_2(n) < c_1 n^2.$$

It would be interesting to determine the exact values of $g_2(n)$ and $G_2(n)$ and more generally $g_k(n)$ and $G_k(n)$, for $k > 2$. This is not easy even for E_2.

The 2^n vertices of an n-dimensional cube determine $n + 1$ distinct distances. Is it true that every set of 2^n points in E_n determines at least cn distinct distances, where c is an absolute constant independent of n?

9. To end the paper, I state two problems from the collection of W. Moser. Problem 60 (due to Sierpinski) states: Do there exist two point sets in the plane such that no matter how they are placed in the plane, their intersection contains exactly one point? I proved the existence of two such sets by transfinite induction.

Problem 59 due to Steinhaus states: Does there exist a point set (in the plane) such that no matter how it is placed on the plane, it covers exactly one lattice point? I found this old problem of Steinhaus very challenging and got nowhere with it. More generally, let S be a set of positive numbers. Is there a set P in the plane such that no two points of P have distance in S and no matter how P is placed in the plane it covers exactly one lattice point? In the problem of Steinhaus, S is the set of numbers $(u^2 + v^2)^{1/2}$.

Notes and References

1. Erdös, P. 1974. On some problems of elementary and combinatorial geometry. Annali di Mat. Ser. 4 **103:** 99–108.
2. Erdös, P. 1980. Some combinatorial problems in geometry. *In* Lecture Notes in Mathematics No. 792 (proceedings of a conference held in Haifa, Israel, 1979): 46–53. Springer-Verlag. West Berlin, FRG.
3. Erdös, P. 1981. Some applications of graph theory and combinatorial methods to number theory and geometry. Algebraic methods in graph theory. Coll. Math. Soc. J. Bolyai **25:** 131–148.
4. Hadwiger, H., H. Debrunner & V. Klee. 1964. Combinatorial Geometry in the Plane. Holt, Rinehart & Winston. New York, N.Y.
5. Erdös, P. 1967. On some applications of graph theory to geometry. Canad. J. Math. **19:** 968–971.
6. Hansen, S. Contributions to the Sylvester–Gallai theory. Preprint.
7. Burr, S. A., B. Grünbaum & N. J. A. Sloane. 1979. The orchard problem. Geom. Dedicata **2:** 397–424. This paper contains an extensive bibliography and many interesting historical and other remarks. Many further references can be found in Reference 1.
8. Erdös, P. 1972. On a problem of Grünbaum. Canad. Math. Bull. **15:** 23–25.
9. Erdös, P. & G. Purdy. 1978. Some combinatorial problems in the plane. J. Combin. Theory Ser. A **25:** 205–210.
10. de Bruijn, N. G. & P. Erdös. 1951. A colour problem for infinite graphs and a problem in the theory of relations. Indag. Math. **13:** 371–373.
11. Erdös, P. & M. Simonovits. 1980. On the chromatic number of geometric graphs. Ars Combin. **9:** 229–246.
12. Frankl, P. 1977. On families of finite sets no two of which intersect in a singleton. Bull. Austral. Math. Soc. **17:** 125–134.
13. Frankl, P. & R. M. Wilson. 1981. Intersection theorems with geometric consequences. Combinatorica **1:** 357–368.
14. Larman, D. G. & C. A. Rogers. 1972. The realization of distances within sets in Euclidean space. Mathematica **19:** 1–24.
15. Woodall, D. R. 1973. Distances realised by sets covering the plane. J. Combin. Theory Ser. A **14:** 187–200.
16. Wormald, N. 1970. A 4-chromatic graph with a special plane drawing. J. Austral. Math. Soc. Ser. A **28:** 1–8.

17. Komlós, J., T. Pintz & E. Szemerédi. 1982. A lower bound for Heilbronn's problem. J. London Math. Soc. (2) **25:** 13–24.
18. Graham, R. L. 1980. On partition of E_n. J. Combin. Theory Ser. A **28:** 89–94. This paper contains very extensive references to the literature.
19. Juhász, R. 1979. Ramsey type theorems in the plane. J. Combin. Theory Ser. A **27:** 152–160.
20. Burton, G. R. & G. Purdy. 1979. The directions determined by *n* points in the plane. J. London Math. Soc. (2) **20:** 109–114.
21. Erdös, P. Some problems on elementary geometry, I and II. Austral. Math. Soc. Gaz. **2** (1975): 2–3 and **5** (1978): 52–54.
22. Harborth, H. 1979. Konvexe Fünfecke in ebenen Punktmengen. Elem. Math. **33:** 116–118.

A COMBINATORIAL VERSION OF THE ISOTOPY CONJECTURE

Jacob E. Goodman[a]

City College
City University of New York
New York, New York 10031

Richard Pollack[b]

Courant Institute
New York University
New York, New York 10012

0. Introduction

$\{i_1, i_2, \ldots, i_k\}$ is a semispace of the numbered configuration $C = \{P_1, P_2, \ldots, P_n\}$ in E^2 if $\{P_{i_1}, P_{i_2}, \ldots, P_{i_k}\}$ is the intersection of C with an open halfplane. Two numbered configurations are *semiequivalent* if they have the same semispaces, which comes down to saying that they have the same convexity properties. The isotopy conjecture asserts that if C and C' are semiequivalent, then we can continuously move C' to C always staying semiequivalent to C.

To a numbered configuration C we can associate, in a natural way, an allowable sequence of permutations; the concept of semiequivalence of configurations gives rise to a similar notion for allowable sequences. We prove that if the allowable sequences S and S' are semiequivalent, then S' can be transformed to S by a sequence of "small" steps.

We discuss the relation of this theorem to the isotopy conjecture and other isotopy-type conjectures.

1. The Geometric Setting

Let $C = \{P_1, P_2, \ldots, P_n\}$ be a numbered configuration of n points in E^2, not all collinear. The convex properties of C are entirely determined by the knowledge of all possible intersections of C with halfplanes. We shall call $\{i_1, i_2, \ldots, i_k\}$ a semispace of C iff $\{P_{i_1}, P_{i_2}, \ldots, P_{i_k}\}$ is the intersection of C with a halfplane. Two configurations C and $\bar{C}$ which have the same family of semispaces we call semiequivalent and write $C \sim_{\text{sem}} \bar{C}$. Semiequivalent configurations are identical as far as their convexity properties are concerned. We call them semiequivalent rather than convex equivalent because the definition focuses on the semispace aspect of convexity. If we focus on separation properties (see [6]) we say that C and $\bar{C}$ are separation equivalent if whenever the line $\overline{P_i P_j}$ separates points P_k and P_l of C the corresponding line separates the corresponding points of $\bar{C}$ and we write, $C \sim_{\text{sep}} \bar{C}$. Since the

[a] Research supported in part by National Science Foundation Grant MCS-8201831.
[b] Research supported in part by National Science Foundation Grant MCS-8201342.

line $\overline{P_i P_j}$ separates P_k and P_l iff $\varnothing$, $\{k\}$, $\{l\}$, $\{i, k\}$, $\{i, l\}$ and their complements are semispaces of $\{P_i, P_j, P_k, P_l\}$, we see that C and $\bar{C}$ are separation equivalent if they are semiequivalent.

We might also focus on the order type (see [1, 4]) of the configuration. The order type of C is determined by the orientations of all the triples of points P_i, P_j, P_k. Two configurations have the same order type if corresponding triples always have the same orientation (or are collinear). Finally we say that C and $\bar{C}$ are order equivalent if C and $\bar{C}$ have the same order type or if C and $\bar{C}$ have opposite order type (corresponding triples all have opposite orientation), and we write $C \sim_{\text{ord}} \bar{C}$. Since P_i, P_j, P_k and P_i, P_j, P_l have the same or opposite orientation according to whether or not the line $\overline{P_i P_j}$ fails to separate the points P_k and P_l, it is easy to see that C and $\bar{C}$ are order equivalent if they are separation equivalent.

A directed line through a point P_i rotating counterclockwise about P_i will meet the remaining points of C in a periodic sequence of ordered sets, $S_{i_1}, S_{i_2}, \ldots, S_{i_k}$, called the local i-sequence of C. For example, the configuration of FIGURE 1 has the local sequences:

1. $\ldots (P_1, P_2, P_3), (P_1, P_4), (P_1, P_5), (P_3, P_2, P_1), (P_4, P_1), (P_5, P_1), \ldots$
2. $\ldots (P_2, P_4, P_5), (P_3, P_2, P_1), (P_5, P_4, P_2), (P_1, P_2, P_3), \ldots$
3. $\ldots (P_3, P_5), (P_3, P_4), (P_3, P_2, P_1), (P_5, P_3), (P_4, P_3), (P_1, P_2, P_3), \ldots$
4. $\ldots (P_2, P_4, P_5), (P_3, P_5), (P_5, P_1), (P_5, P_4, P_2), (P_5, P_3), (P_1, P_5), \ldots$

Clearly the (unordered) sets S_{i_j} are determined by the order type of C: They are simply the collinear sets containing P_i. Moreover, the order in which they appear is determined by the orientations of the triples P_i, P_{i_j}, P_{i_m}, where $P_{i_j} \in S_{i_j}$. Finally, if $S_{i_j} = (P_{i_1}, P_{i_2}, \ldots, P_{i_k})$, then this order is determined by the fact that the triples P, P_{i_1}, P_{i_2}; $\ldots$; P, $P_{i_{k-1}}$, P_{i_k}, where P is a point of C not in S_i, all have the same orientation; i.e., the local sequences are all determined by the order type of the configuration. Thus, configurations that are order equivalent either have the "same" local sequences or they are all reversed; i.e.,

$$\ldots S_{i_k}, S_{i_{k-1}}, \ldots, S_{i_1}, \ldots$$

instead of

$$\ldots S_{i_1}, S_{i_2}, \ldots, S_{i_k}, \ldots .$$

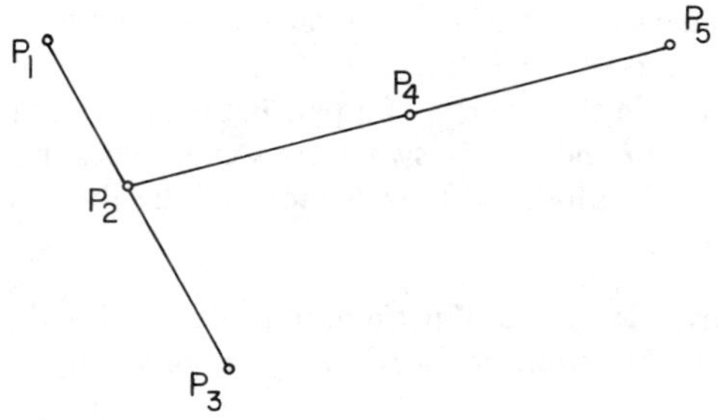

FIGURE 1

If C and $\bar{C}$ have the same or all reversed local sequences we say that C and $\bar{C}$ are locally equivalent and write $C \sim_{\text{loc}} \bar{C}$. Thus, configurations that are order equivalent must be locally equivalent.[c]

If $C(t)$ is a continuous family of configurations which stays in one semi-equivalence class, then the $C(t)$ are in fact all of the same order type since $P_i(t)$, $P_j(t)$, $P_k(t)$ cannot change orientation without becoming collinear. A configuration C and its reflection C^r in a line are semiequivalent even though they have opposite order types and opposite local types (corresponding triples have opposite orientation and corresponding local sequences are reversed). The isotopy problem that we address is: If $C \sim_{\text{sem}} \bar{C}$, must there exist a family $C(t)$ of configurations all semiequivalent to C such that $C(0) = C$ and either $C(1) = \bar{C}$ or $C(1) = \bar{C}^r$ (i.e., are C and $\bar{C}$ isotopic)? This question was first proposed, in a somewhat different setting, by Ringel in [5]. We shall prove a combinatorial version of the statement that if $C \sim_{\text{loc}} \bar{C}$, then C and $\bar{C}$ are isotopic; the truth of this statement would, of course, imply the isotopy conjecture, namely, semiequivalent configurations are isotopic.

The preceding definitions easily generalize to configurations in E^d: simply replace halfplane by halfspace, line by hyperplane, and triple by $d + 1$-tuple. Local sequences are a bit more complicated but they too can be defined inductively. Moreover, the question of whether semiequivalent configurations are isotopic is equally interesting in E^d. Unfortunately the combinatorial setting described in the next section only applies to the plane.

2. The Combinatorial Setting

Suppose C is the numbered configuration $\{P_1, \ldots, P_n\}$ in $\mathbb{E}^2$. Let L be a directed line not orthogonal to any line determined by two members of C, and project C orthogonally onto L. If we suppress the P's, we get a permutation of $[1, n]$. If we then allow L to turn in the counterclockwise direction, the permutation changes whenever the direction of projection passes through that given by two or more points of C, and we get a periodic sequence of permutations called *the circular sequence of permutations associated to C*.

An example is given in Figure 2 where we have indicated the move from each term of the sequence to the next.

Note that such a sequence Σ always has the following properties:

(1) Σ is periodic;

(2) the move from each term of Σ to the next consists of reversing one or more nonoverlapping substrings;

(3) if a move results in the reversal of a pair ij, then every other pair is reversed subsequently by the time i and j again switch; this guarantees that each period of Σ breaks up into two half-periods, with each move of the first half reversed in the second.

Definition 1. Any sequence Σ of permutations of $[1, n]$ with properties (1), (2), and (3) is called an *allowable sequence of permutations*, or simply an *n-sequence*. Each

[c] In fact, the four types of equivalence discussed so far are shown in [3] to be identical.

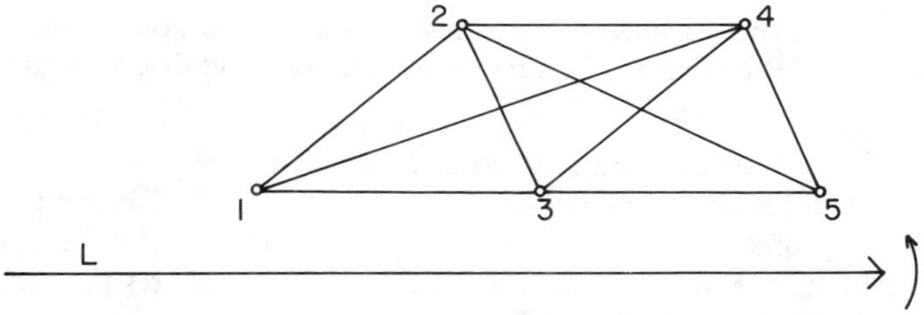

$$\ldots 12345 \xrightarrow{23,45} 13254 \xrightarrow{25} 13524 \xrightarrow{135,24} 53142 \xrightarrow{14} 53412 \xrightarrow{34,12}$$

$$54321 \xrightarrow{54,32} 45231 \xrightarrow{52} 42531 \xrightarrow{42,531} 24135 \xrightarrow{41} 21435 \xrightarrow{21,43} 12345 \ldots$$

FIGURE 2

move of Σ, from one term, Π, to the next, Π', consists of one or more substring reversals, or *switches*,

$$i_1 \cdots i_k,$$

where $i_1, \ldots, i_k$ is a complete substring of Π which is being reversed.

Observe that the allowable sequence associated to the configuration in FIGURE 1 is

$$\ldots 12345 \xrightarrow{123} 32145 \xrightarrow{14} 32415 \xrightarrow{15} 32451 \xrightarrow{245} 35421 \xrightarrow{35} 53421 \xrightarrow{34} 54321$$

$$\xrightarrow{321} 54123 \xrightarrow{41} 51423 \xrightarrow{51} 15423 \xrightarrow{542} 12453 \xrightarrow{53} 12435 \xrightarrow{43} 12345 \ldots.$$

Note also that if we suppress the P's in the local i-sequence of this configuration we get the sequence of switches involving i. In general, for any allowable sequence the local sequence of ordered switches of an index i is the periodic sequence of substring reversals that involve i. The reverse $\tilde{\Sigma}$ of a sequence Σ will be the sequence consisting of the same terms listed in the reverse cyclic order; for example, if Σ is the sequence of FIGURE 1 then $\tilde{\Sigma}$ is the sequence

$$\ldots 12345 \xrightarrow{34} 12435 \xrightarrow{35} 12453 \xrightarrow{245} 15423 \xrightarrow{15} 51423 \xrightarrow{14} 54123 \xrightarrow{123} 54321$$

$$\xrightarrow{43} 53421 \xrightarrow{53} 35421 \xrightarrow{542} 32451 \xrightarrow{51} 32415 \xrightarrow{41} 32145 \xrightarrow{321} 12345 \ldots,$$

which is the sequence associated to the reflection of this configuration.

We will use the notation $(i_1 j_1 \prec \cdots \prec i_r j_r)$ to mean that *within the half-period of* Σ *following the move in which the indices* i_1 *and* j_1 (*in that order*) *switch*, the indices i_2 and j_2 (in that order) switch in the same or in a subsequent move, then i_3 and j_3, and so on. We emphasize that all of these switches take place within a single half-period. For the sequence of FIGURE 2, for example, we have $(43 \prec 23 \prec 35 \prec 24 \prec 15)$. Note that for any three indices i, j, k, we always have

$$(ij \prec jk) \Rightarrow (ij \prec ik \prec jk).$$

Finally, we call a sequence *trivial* if all of the indices in it switch simultaneously. (Such a sequence is clearly associated to a collinear configuration of points.)

Remark 1. The n-sequence Σ associated to a configuration C encodes, in combinatorial terms, a number of geometric features of C (let us abbreviate P_i simply as "i"):

(a) $i_1, \ldots, i_k$ are collinear iff they switch simultaneously;

(b) i is in the convex hull conv $(i_1, \ldots, i_k)$ of $i_1, \ldots, i_k$ iff in every term of Σ, i is preceded by one of $i_1, \ldots, i_k$ (hence, "surrounded" by them), and $i_1, \ldots, i_k$ are convexly independent (i.e., none is in the convex hull of the rest) iff each of $i_1, \ldots, i_k$ precedes the rest in some term of Σ;

(c) i is an extreme point of C iff some term of Σ begins with i, or—equivalently—iff the local sequence of ordered switches belonging to i has the form

$$\ldots, iS_1, \ldots, iS_k, \bar{S}_1 i, \ldots, \bar{S}_k i, iS_1, \ldots;$$

(d) $\overline{ij}$ is parallel to $\overline{kl}$ iff i, j switch in the same move (but in disjoint substrings) as k, l;

(e) ijk has positive (i.e., counterclockwise) orientation iff $(ij \prec ik)$;

(f) line $\overline{ij}$ separates k from l iff when i and j switch, k and l are on opposite sides of the substring containing i and j which reverses; point i separates points j and k (collinear with i) iff when all three switch, i is between j and k;

(g) $\{i_1, \ldots, i_k\}$ constitute a semispace of C (i.e., all the points of C lying on one side of a line) iff some term of Σ has $i_1, \ldots, i_k$ (in some order) as an initial segment (hence, some other term has them as a terminal segment);

(h) a directed line L rotating in a counterclockwise direction through point i passes through the points of C in the cyclic order $\ldots, S_1, \ldots, S_k, \bar{S}_1, \ldots, \bar{S}_k, S_1, \ldots$, where each S_j is a *string* of indices containing i and $\bar{S}_j$ is the reverse of S_i, listed in the order induced by the direction on L, iff the local sequence of ordered switches belonging to i is precisely $\ldots, S_1, \ldots, S_k, \bar{S}_1, \ldots, \bar{S}_k, S_1, \ldots$.

Not every allowable sequence arises from a configuration of points (see [2]). Properties (a) through (h) of Remark 1 all make sense, however, for an arbitrary allowable sequence in which the indices $1, \ldots, n$ are regarded as the "points," and so we use them to define the corresponding properties of allowable sequences. Thus, when we speak of a semispace of an n-sequence Σ we shall mean an initial segment in some term of Σ, and so on. With this understanding, the definitions in the first section of semiequivalence, separation equivalence, order equivalence, and local equivalence apply equally well to allowable sequences. Moreover, the arguments given there that semiequivalence implies separation equivalence implies order equivalence implies local equivalence are equally valid for allowable sequences. What remains to complete the combinatorial analogy to the geometric setting is to have a combinatorial notion of "small" change in an allowable sequence comparable to infinitesimal change in a geometric configuration. Since we can identify a sequence with its sequence of moves and we do not wish to disturb the semiequivalence class of the sequence nor therefore its local sequences, a minimal change accomplishing this is what we call an elementary transformation:

An *elementary transformation* of one n-sequence into another consists either of amalgamating two successive moves whose substrings undergoing reversal are disjoint, or of the reverse, i.e., separating a move into two successive moves; in each case of course the corresponding amalgamation or separation must take place in every half-period of the sequence.

For example, we may apply an elementary transformation to the 6-sequence

$$\ldots 123456 \xrightarrow{34} 124356 \xrightarrow{124} 421356 \xrightarrow{56} 421365 \xrightarrow{136} 426315 \xrightarrow{26,\,15} 462351$$
$$\xrightarrow{46,\,235} 645321 \xrightarrow{45} 654321 \xrightarrow{43} 653421 \xrightarrow{421} 653124 \xrightarrow{65} 563124 \xrightarrow{631} \ldots$$

to get the 6-sequence

$$\ldots 123456 \xrightarrow{34} 124356 \xrightarrow{124,\,56} 421365 \xrightarrow{136} 426315 \xrightarrow{26,\,15} 462351$$
$$\xrightarrow{46,\,235} 645321 \xrightarrow{45} 654321 \xrightarrow{43} 653421 \xrightarrow{65,\,421} 563124 \xrightarrow{631} \ldots .$$

By a succession of two elementary transformations we can interchange two successive moves involving disjoint sets of indices.

Thus, we shall prove:

THEOREM. *If the n-sequences Σ and Σ' are locally equivalent, then we can transform Σ either into Σ' or into $\tilde{\Sigma}'$ by a sequence of elementary transformations.*

3. The Proof of the Theorem

We shall prove first that if Σ and Σ' have the same local sequences of ordered switches, then a succession of elementary transformations can be applied to each so that the resulting sequences have a common term. [Let us remark first that it is possible for two sequences satisfying the hypotheses to have *no* terms in common, even in the "simple" case (i.e., no multiple or simultaneous switches):

$$\ldots 123456 \xrightarrow{12} 213456 \xrightarrow{34} 214356 \xrightarrow{14} 241356 \xrightarrow{24} 421356 \xrightarrow{35} 421536$$
$$\xrightarrow{36} 421563 \xrightarrow{15} 425163 \xrightarrow{16} 425613 \xrightarrow{25} 452613 \xrightarrow{26} 456213 \xrightarrow{45} 546213$$
$$\xrightarrow{46} 564213 \xrightarrow{13} 564231 \xrightarrow{56} 654231 \xrightarrow{23} 654321 \ldots$$

and

$$\ldots 124356 \xrightarrow{35} 124536 \xrightarrow{36} 124563 \xrightarrow{12} 214563 \xrightarrow{14} 241563 \xrightarrow{15} 245163$$
$$\xrightarrow{16} 245613 \xrightarrow{13} 245631 \xrightarrow{24} 425631 \xrightarrow{25} 452631 \xrightarrow{26} 456231 \xrightarrow{23} 456321$$
$$\xrightarrow{45} 546321 \xrightarrow{46} 564321 \xrightarrow{43} 563421 \xrightarrow{56} 653421 \ldots$$

are two such sequences.]

Now suppose Σ and Σ' have the same local sequences of ordered switches; by applying elementary transformations, if necessary, we may assume each move of Σ and Σ' is a switch of a single string. Say 1 is an extreme point in Σ. Then its local sequence has the form

$$\ldots,\ 1S_1,\ 1S_2,\ \ldots,\ 1S_p,\ \bar{S}_1 1,\ \bar{S}_2 1,\ \ldots,\ \bar{S}_p 1,\ 1S_1,\ \ldots,$$

where $S_1, \ldots, S_p$ are strings whose disjoint union is $\{2, \ldots, n\}$, and $\bar{S}_i$ is the reverse of S_i. Hence, this is the local sequence of 1 in Σ' so that, in particular, 1 is also an extreme point of Σ'. By a succession of elementary transformations, the moves in Σ not involving 1 which intervene among $1S_1, \ldots, 1S_p$ can be moved either to the left of $1S_1$ or to the right of $1S_p$ as follows: Any switch involving only indices $\in S_1 \cup \cdots \cup S_k$ which occurs after S_k can be moved to the *right*, past $1S_p$, while any switch involving only indices $\in S_m \cup \cdots \cup S_p$ which occurs before S_m can be moved to the *left*, past $1S_1$. How can we get stuck? Only if we have a switch involving indices i and j, with $i \in S_k$ and $j \in S_m$, which occurs *between* $1S_k$ and $1S_m$. But then we would have, supposing $k < m$,

$$(1i \prec ij \prec 1j) \qquad \text{or} \qquad (1i \prec ji \prec 1j),$$

both of which are impossible. Hence, by a succession of elementary transformations, Σ (respectively, Σ' or $\tilde{\Sigma}'$) can be changed into a new sequence Σ_1 (respectively, Σ'_1) in which the moves

$$1S_1, 1S_2, \ldots, 1S_p$$

occur in that order, with no other moves intervening. But then in each of Σ_1 and Σ'_1 the term immediately preceding these moves must be $1S_1 S_2 \cdots S_p$, so that we have produced a common term.

We now proceed, by elementary transformations, to create more and more common terms without losing any of the old ones. Suppose at some stage Σ (respectively, Σ' or $\tilde{\Sigma}'$) has been transformed into a sequence Σ_r (respectively, Σ'_r) such that Σ_r and Σ'_r each contain the terms Π^* and Π^{**}, with Π^{**} occurring no more than a half-period later than Π^* in each. (Of course Σ_r and Σ'_r still have the same local sequences of ordered switches, since an elementary transformation has no effect on the set of local sequences.) Let the intervening moves and terms be as follows:

$$\Sigma_r = \cdots \Pi^* = \Pi_1 \xrightarrow{M_1} \Pi_2 \xrightarrow{M_2} \cdots \xrightarrow{M_{p-1}} \Pi_p = \Pi^{**} \cdots,$$

$$\Sigma'_r = \cdots \Pi^* = \Pi'_1 \xrightarrow{M'_1} \Pi'_2 \xrightarrow{M'_2} \cdots \xrightarrow{M'_{q-1}} \Pi'_q = \Pi^{**} \cdots.$$

We claim first that the sets $\{M_1, \ldots, M_{p-1}\}$ and $\{M'_1, \ldots, M'_{q-1}\}$ are equal: This is clear, since—as a result of the assumed equality of the local sequences in Σ_r and in E'_r—the set of *all* moves in Σ_r equals that in Σ'_r, and since precisely the same pairs i, j are switched as a result of the moves $M_1, \ldots, M_{p-1}$ as are switched as a result of the moves $M'_1, \ldots, M'_{q-1}$—namely, each pair whose position in Π^{**} is the reverse of its position in Π^*. Now consider M_1. If this is the same as M'_1, then necessarily $\Pi_2 = \Pi'_2$. If not, say $M_1 = M'_t$ for some $1 < t < q$. We claim that no index i appearing in M'_t can occur in any M'_s for $1 \leq s < t$: The reason is that otherwise i would be involved in two distinct moves (M'_s and M'_t) which occurred in Σ_r and Σ'_r *in the opposite order* and in each case within a portion of the sequence *no more than a half-period in length*, and this would violate the supposed equality of the local sequences belonging to i in Σ_r and Σ'_r. Hence, by a succession of elementary transformations starting with Σ'_r, M'_t can be moved to the left, to yield the sequence

$$\Sigma'_{r+1} = \cdots \Pi^* = \Pi'_1 \xrightarrow{M'_t} \Pi_2 = \Pi''_2 \xrightarrow{M'_1} \Pi'_3 \xrightarrow{M'_2} \cdots$$

$$\xrightarrow{M'_{t-1}} \Pi'_{t+1} \xrightarrow{M'_{t+1}} \cdots \xrightarrow{M'_{q-1}} \Pi'_q = \Pi^{**} \cdots,$$

and induction (on the number of common terms) then yields the result.

4. Concluding Remarks

What bearing does the theorem of the last section have on the geometric isotopy problem? One might hope to effect a single elementary transformation continuously; i.e., if C is a configuration with allowable sequence Σ and Σ' is obtained from Σ by an elementary transformation, we would like to move C continuously to a configuration C' with allowable sequence Σ' without disturbing its order type. Then, given semiequivalent configurations C and C', we could isotope C to C'', where C and C'' are combinatorially equivalent (i.e., they have the same associated allowable sequence). We would then have reduced the isotopy problem to what we call the combinatorial isotopy problem: If C and C' are combinatorially equivalent can we continuously transform C to C' without changing the associated allowable sequence along the way? Unfortunately, this program breaks down at the first step. It may happen that Σ' is not realizable, i.e., there is no configuration C' that has Σ' as its associated allowable sequence. This objection would be removed if we could strengthen our theorem to prove that if Σ and Σ' are both realizable, then there exists a sequence of elementary transformations taking us from Σ to Σ' that passes only through realizable sequences. Then the isotopy problem would be reduced to the combinatorial isotopy problem.

A closer look at the proof of the theorem reveals that the terms of the sequence are irrelevant. All that is being used is the sequence of moves. As a result of this observation we have answered another kind of realizability question: When is a sequence of "moves" the sequence of moves of an allowable sequence, i.e., when is it realizable by an allowable sequence? Answer: when it is "locally equivalent" to an allowable sequence!

References

1. Dreiding, A. S., A. Dress & H. R. Haegi. Chirotopes, a combinatorial theory of orientation. Preprint.
2. Goodman, J. E. & R. Pollack. 1980. On the combinatorial classification of nondegenerate configurations in the plane. J. Combin. Theory Ser. A. **29:** 220–235.
3. Goodman, J. E. & R. Pollack. 1984. Semispaces of configurations, cell complexes of arrangements. J. Combin. Theory Ser. A. In press.
4. Goodman, J. E. & R. Pollack. Geometric sorting theory. Ann. N.Y. Acad. Sci. (this volume).
5. Ringel, G. 1956. Teilungen der projektiven Ebene durch Geraden oder topologische Geraden. Math. Z. **64:** 79–102.
6. Zaslavsky, T. Combinatorial ordered geometry. I. Bilateral geometry. Preprint.

OLD AND NEW EUCLIDEAN RAMSEY THEOREMS

R. L. Graham
Bell Laboratories
Murray Hill, New Jersey 07974

INTRODUCTION

Mathematics has sometimes been called the science of order. From this point of view, the guiding principle of Ramsey theory is given by the statement, "Complete disorder is impossible." In a nutshell, Ramsey theory is the study of structure invariant under partitions. Typical examples of results in this subject are:

(i) (Adam [0].) Any rearrangement of the integers $\{1, 2, \ldots, n^2 + 1\}$ always contains a monotone subsequence of length $n + 1$.

(ii) (van der Waerden [18].) For any partition of the set $\mathbb{Z}^+$ of positive integers into finitely many classes, one of the classes must contain arbitrarily long arithmetic progressions.

(iii) (Ramsey [12].) For any partition of the k-element subsets of $\mathbb{Z}^+$ into finitely many classes, some class must contain all the k-element subsets of some infinite subset $X \subseteq \mathbb{Z}^+$.

(iv) (Hindman [9].) For any partition of $\mathbb{Z}^+$ into finitely many classes, some class must contain all the (finite) subset sums of some infinite subset $X \subseteq \mathbb{Z}^+$.

(v) (Graham *et al.* [5]; also see [16].) For any integers k, l, and r and any finite field F, there exists an integer $N = N(k, l, r, F)$ such that if all the k-dimensional subspaces of an n-dimensional vector space G over F are partitioned into r classes, then some class must contain all the k-dimensional subspaces of some l-dimensional subspace of G.

A few remarks may be in order here. First, we should point out that despite their similar appearance, (iv) is significantly more difficult to prove than (iii). The precursor to (iv) was the theorem of I. Schur [13], who in 1916 showed that if the integers $\{1, 2, \ldots, [er!]\}$ are partitioned into r classes, then some class must contain a solution to the equation $x + y = z$. This theme was picked up by Schur's illustrious student R. Rado, who, beginning with his dissertation [11] in 1933, developed a beautiful theory for the "partition regularity" of systems of linear equations. In more recent times, great progress in these (and other) directions has been made by Deuber, Nešetřil, Rödl, and many others. A description of much of this work can be found in [6].

In contrast to (iii) and (iv) there is no infinite analogue to (ii). This can be seen, for example, by the partition

$$\mathbb{Z}^+ = \{x: 2^{2k} \leq x < 2^{2k+1}: k \geq 0\} \cup \{x: 2^{2k+1} \leq x < 2^{2k+2}: k \geq 0\}.$$

Finally, we should remark that in some sense the first Ramsey theorem actually was due to Hilbert [8], who in 1892 proved that for any partition of $\mathbb{Z}^+$ into finitely many classes, some class contains for every m and suitable choices of positive integers $a, a_1, \ldots, a_m$ all 2^m sums

$$a + \sum_{k=1}^{m} \varepsilon_k a_k, \qquad \text{where } \varepsilon_k = 0 \text{ or } 1.$$

[This is a special case of (iv).]

Euclidean Ramsey Theory

For Euclidean Ramsey theory, the fundamental problem is this: Which configurations of points C must always occur, up to some Euclidean motion, in a single class, whenever Euclidean n-space $\mathbb{E}^n$ is partitioned into r classes?

Definition. For such C we will say that $R(C, n, r)$ holds.

For example, it is easy to see that if C_0 is the set of three vertices of some unit equilateral triangle, then $R(C_0, 4, 2)$ holds (by considering the five vertices of a unit simplex in $\mathbb{E}^4$) and $R(C_0, 2, 2)$ does not hold (by partitioning $\mathbb{E}^2$ into two classes of alternating half-open strips of width $\sqrt{3}/2$).

The following is typical of some of the many results available along these lines.

Fact. *$R(C, 6, 2)$ holds when C is the set of four vertices of some unit square.*

Proof. Consider the set $S \subseteq \mathbb{E}^6$ defined by $S = \{(x_1, \ldots, x_6)\colon x_i = 1/\sqrt{2}$ for exactly two values of i, and $x_i = 0$ for all other values of $i\}$. Any partition of $\mathbb{E}^6$ into two classes, say $\chi\colon \mathbb{E}^6 \to \{0, 1\}$, also partitions S into two classes. To each point $s = (s_1, \ldots, s_6) \in S$ we can associate a pair $\{i, j\}$ by letting i and j denote the two indices for which $s_k = 1/\sqrt{2}$. Thus, χ induces a partition of the edges of the complete graph on six points into two classes. By a standard result in (Ramsey) graph theory, in any such partition there must be a 4-cycle, say

$$\circlearrowright c_1 \to c_2 \to c_3 \to c_4,$$

in a single class. It is straightforward to check that this 4-cycle corresponds to the four vertices of a unit square in S, which proves our claim. ∎

It is no accident that in the examples presented thus far, a proof that $R(C, n, r)$ holds for a particular C was accomplished in fact by selecting a suitable *finite* subset of $\mathbb{E}^n$ and partitioning it (instead of requiring all of $\mathbb{E}^n$). A standard compactness argument (see [6]) shows that this must always be the case.

Before proceeding to more general considerations, we mention a tantalizing question that, besides being among the most fundamental in the theory, illustrates more than adequately how little we really know about what is going on in this area.

For this example we take for C the set C^* consisting of two points separated by a distance of 1. It is easy to see that $R(C^*, 2, 2)$ holds by considering the three vertices of a unit equilateral triangle. It is somewhat less obvious (but equally true) that $R(C, 2, 3)$ holds. In the graph shown in Figure 1, all edges represent unit distances. This graph has the property that it has chromatic number 4, i.e., any 3-coloring of the vertices must assign the same color to the two endpoints of *some* edge of the graph.

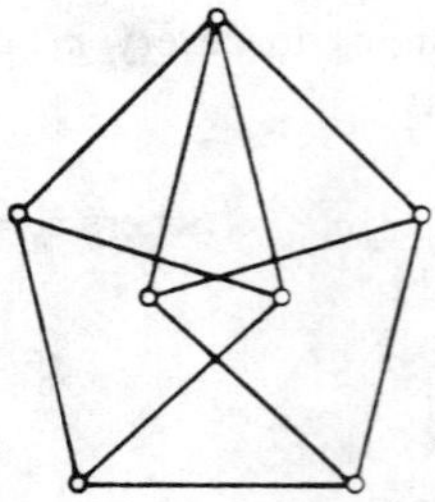

FIGURE 1. The Moser graph.

In the other direction, suppose $\mathbb{E}^2$ is tiled by regular hexagons of diameter 9/10. It is easy to partition these hexagons into seven classes so that no class contains two points at a distance of 1. Thus, $R(C^*, 2, 7)$ does not hold. At present, no one knows the *least* value of r such that $R(C^*, 2, r)$ does not hold. This is also known as the chromatic number of $\mathbb{E}^2$, denoted by $\chi(\mathbb{E}^2)$, since it is the chromatic number of the (uncountable) graph formed by taking each point of $\mathbb{E}^2$ as a vertex and each pair $\{x, y \in \mathbb{E}^2$: distance $(x, y) = 1\}$ as an edge. Thus, the best current bounds are

$$4 \leq \chi(\mathbb{E}^2) \leq 7.$$

More generally, it has recently been shown by Frankl and Wilson [4] that $\chi(\mathbb{E}^n)$, the chromatic number of Euclidean n-space, grows exponentially in n, verifying an earlier conjecture of Erdös. The best available bounds for the general case now are

$$(1 + o(1))(1.2)^n \leq \chi(\mathbb{E}^n) \leq (3 + o(1))^n.$$

RAMSEY SETS

A basic concept in Euclidean Ramsey theory is that of a Ramsey set.

DEFINITION. A configuration C is said to be *Ramsey* if for all r there exists an $N = N(C, r)$ such that $R(C, N, r)$ holds. The most general result for constructing Ramsey sets is given by the following.

THEOREM [1]. *If C_1 and C_2 are Ramsey sets, then the Cartesian product $C_1 \times C_2$ is a Ramsey set.*

Since any two-point set is a Ramsey set, any subset of the vertices of a rectangular parallelepiped (= "brick") is a Ramsey set. The determination of which n-simplexes are subsets of bricks is an interesting open problem. It is certainly necessary that any angle formed by three of the points be less than or equal to 90°. This also turns out to be sufficient for $n = 2$ and 3 but *not* for $n = 4$.

On the other side of the coin, one might well ask if there are any non-Ramsey sets. The simplest example of such a set is given by the following result.

THEOREM. *Let $C = \{\bar{x}, \bar{y}, \bar{z}\}$ be a set of three equally spaced collinear points (see* FIGURE 2*). Then C is not a Ramsey set.*

Proof. To each $\bar{u} \in \mathbb{E}^2$ assign the color $\chi(u) = [\bar{u} \cdot \bar{u}] \pmod 4$, where $[\alpha]$ denotes the greatest integer not exceeding α.

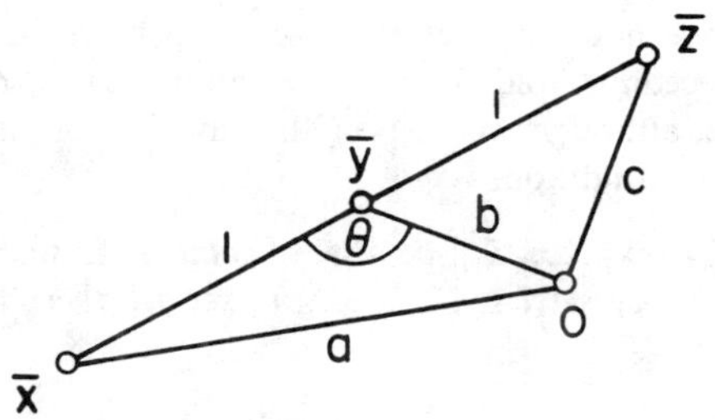

FIGURE 2

Suppose $C = \{\bar{x}, \bar{y}, \bar{z}\}$ occurs "monochromatically," i.e., for some i, there is a copy of C in $\chi^{-1}(i)$. Since

$$a^2 = b^2 + 1 - 2b \cos \theta,$$
$$c^2 = b^2 + 1 + 2b \cos \theta,$$

then

$$a^2 + c^2 = 2b^2 + 2. \quad \textbf{(1)}$$

Since $\chi(\bar{x}) = \chi(\bar{y}) = \chi(\bar{z}) = i$, then

$$a^2 = 4k_a + i + \varepsilon_a, \qquad 0 \le \varepsilon_a < 1,$$
$$b^2 = 4k_b + i + \varepsilon_b, \qquad 0 \le \varepsilon_b < 1,$$
$$c^2 = 4k_c + i + \varepsilon_c, \qquad 0 \le \varepsilon_c < 1,$$

for suitable integers k_a, k_b, k_c. Thus, by **(1)**

$$4(k_a - 2k_b + k_c) - 2 = -\varepsilon_a + 2\varepsilon_b - \varepsilon_c$$

which is, however, (barely) impossible because of the constraints on the ε's. ▮

This argument contains the kernel of an idea which when more fully developed can be used to prove the following result.

THEOREM [1]. *If C is a Ramsey set, then C must lie on the surface of some sphere.*

This is still the strongest restriction known for Ramsey sets.

A fair number of results in this general spirit appear in [1–3, 14, 17]. Rather than repeat these, we will conclude the next section with several very recent results which have not yet appeared in the literature.

EUCLIDEAN RAMSEY THEORY ON THE n-SPHERE

In this section we examine the corresponding questions when the underlying spaces are unit n-spheres

$$S^n = \left\{(x_0, \ldots, x_n) : \sum_{k=0}^{n} x_k^2 = 1\right\} \subseteq \mathbb{E}^{n+1}$$

and the allowed motions are orthogonal transformations of S^n onto itself. The corresponding Ramsey sets will be called "sphere-Ramsey." It will turn out that for sets $X \subseteq S^n$ which are not too large (in a sense to be made precise later), results similar to those preceding hold. For the remaining cases, only very preliminary results are available, although we suspect that much more is very likely true. We begin with a necessary condition.

THEOREM. *Let* $X = \{\bar{x}_1, \ldots, \bar{x}_m\}$ *be a set of points in* $\mathbb{E}^n$ *such that:*

(i) *for some nonempty set* $I \subseteq \{1, 2, \ldots, m\} \equiv [m]$, *there exist nonzero* α_i, $i \in I$, *such that*

$$\sum_{i \in I} \alpha_i \bar{x}_i = \bar{0};$$

(ii) *for all nonempty* $J \subseteq I$,

$$\sum_{j \in J} \alpha_j \neq 0.$$

Then there exists $r = r(X)$ *such that for any* N, *there is a partition* $S^N = \bigcup_{k=1}^{r} C_k$ *such that no* C_i *contains a copy of* X.

Proof. Consider the homogeneous linear equation

$$\sum_{i \in I} \alpha_i z_i = 0. \tag{2}$$

By (ii), Rado's theorem for the partition regularity of this equation over $\mathbb{R}^+$ (see [6] or [11]) implies that it is *not* regular, i.e., for some r there is an r-coloring $\chi\colon \mathbb{R}^+ \to [r]$ such that **(1)** has no monochromatic solution. Color the points of $S_+^N = \{(x_0, \ldots, x_N) \in S^N\colon x_0 > 0\}$ by

$$\chi^*(\bar{x}) = \chi(\bar{u} \cdot \bar{x}),$$

where $\bar{u}$ denotes the unit vector $(1, 0, 0, \ldots, 0)$. Thus, the color of $\bar{x} \in S_+^N$ just depends on its distance from the "north pole" of S^N.

For each nonempty subset $J \subseteq I$, consider the equation

$$\sum_{j \in J} \alpha_j z_j = 0. \tag{2$_J$}$$

Of course, by (ii) this also fails to satisfy the (necessary and sufficient) condition of Rado for partition regularity. Hence, there is a coloring χ_J of $\mathbb{R}^+$ (using r_J colors) so that **(2)$_J$** has no monochromatic (under χ_J) solution. As before, we can color S_+^N by giving $\bar{x} \in S_+^N$ the color

$$\chi_J^*(\bar{x}) = \chi_J(\bar{x} \cdot \bar{u}).$$

Now, form the *product* coloring $\hat{\chi}$ of S_+^N by defining for $\bar{x} \in S_+^N$,

$$\hat{\chi}(\bar{x}) = (\chi_I(\bar{x}), \ldots, \chi_J(\bar{x}), \ldots),$$

where the sequence has length $2^{|I|} - 1$ and the indices of the χ_J range over all nonempty subsets $J \subseteq I$. The number of colors required by the coloring $\hat{\chi}$ is at most $\prod_{\varnothing \neq J \subseteq I} r_J \equiv R$.

An important property of $\hat{\chi}$ is this. Suppose we extend $\hat{\chi}$ to $S_0^N \equiv \{(x_0, \ldots, x_N) \in S^N\colon x_0 \geq 0\}$ by assigning *all* R colors to any point in $S_0^N \backslash S_+^N$, i.e., having $x_0 = 0$. Then the *only* monochromatic solution to **(2)** in $\mathbb{R}^+ \cup \{0\}$ is $z_i = 0$ for all $i \in I$.

Next, construct a similar coloring $\check{\chi}$ on $S_-^N = \{-\bar{x}: \bar{x} \in S_+^N\}$, but use R completely different colors. This assures that any set X that hits both hemispheres S_+^N and S_-^N cannot be monochromatic. Finally, we have to color the equator

$$S^{N-1} = \{\bar{x} \in S^N: x_0 = 0\}.$$

By our construction, any copy of X that is not contained entirely in S^{N-1} cannot be monochromatic. Hence, it suffices to color S^{N-1} avoiding monochromatic copies of X where we may use any of the $2R$ colors previously used in the coloring of $S_+^N \cup S_-^N$. By induction, this can be done provided we can so color S^1. However, since $m > 1$, S^1 can in fact always be 3-colored without a monochromatic copy of X (in fact, of any 2-element subset of X since the corresponding graph has maximum degree 2). This proves the theorem. ▮

Note that if X is a constant distance $d \neq 90°$ from some point $\bar{t} \in S^n$, then X cannot satisfy both (i) and (ii). For

$$\sum_{i \in I} \alpha_i \bar{x}_i = \bar{0}$$

implies

$$\begin{aligned} 0 = \bar{t} \cdot \left(\sum_{i \in I} \alpha_i \bar{x}_i \right) &= \sum_{i \in I} \alpha_i \bar{t} \cdot \bar{x}_i \\ &= (\cos d) \cdot \sum_{i \in I} \alpha_i, \end{aligned}$$

i.e.,

$$\sum_{i \in I} \alpha_i = 0$$

since $\cos d \neq 0$.

However, these are not the only sets not ruled out from being possible Ramsey sets by the theorem. Another such example is given by the three-point set

$$T = \left\{(1, 0), \left(\frac{1}{2}, \frac{\sqrt{3}}{2}\right), \left(\frac{1}{2}, -\frac{\sqrt{3}}{2}\right)\right\} = \{t_1, t_2, t_3\}.$$

Their linear dependence is given by

$$t_1 - t_2 - t_3 = \bar{0}$$

which does not satisfy (ii).

We restate the theorem in its positive form.

THEOREM′. *If X is sphere-Ramsey, then for any linear dependence $\sum_{i \in I} \alpha_i \bar{x}_i = \bar{0}$ there must exist a nonempty $J \subseteq I$ such that $\sum_{j \in J} \alpha_j = 0$.*

The next result gives a sufficient condition for a set to be a sphere-Ramsey set. Let us call an m-dimensional brick with edge lengths $\lambda_1, \lambda_2, \ldots, \lambda_m$ *small* if

$$\sum_{i=1}^{m} \lambda_m^2 \leq 2. \tag{3}$$

THEOREM. *Every small brick is sphere-Ramsey.*

Proof. We sketch the proof (which has the same basic structure as that of the Hales–Jewett theorem given in [7]). Let a fixed number r of colors be given. For $m = 1$, the theorem is immediate: We simply consider the $r + 1$ points

$$\overbrace{\begin{array}{c}(\beta_1, 0, 0, \ldots, 0, \gamma)\\ (0, \beta_1, 0, \ldots, 0, \gamma)\\ (0, 0, \beta_1, \ldots, 0, \gamma)\\ \vdots \\ (0, 0, 0, \ \ldots, \beta_1, \gamma),\end{array}}^{r+1}$$

where $\beta_1 = \lambda_1/\sqrt{2} \leq 1$ and $\gamma^2 + \beta_1^2 = 1$. These $r + 1$ points are on S^{r+1}. Since they are r-colored, some pair must have the same color. This pair has distance $\beta_1\sqrt{2} = \lambda_1$, which is the desired conclusion.

In general, for a $\lambda_1 \times \cdots \times \lambda_m$ brick B, the set S of points we consider is of the form

$$(\overbrace{0, \ldots, \beta_m, \ldots, 0}^{N_m}, \overbrace{0, \ldots, \beta_{m-1}, \ldots, 0}^{N_{m-1}}, \overset{\cdots\cdots\cdots}{0, \ldots\ldots\ldots 0}, \overbrace{0, \ldots, \beta_1, \ldots, 0}^{N_1}, \gamma).$$

That is, S consists of $(N_m + N_{m-1} + \cdots + N_1 + 1)$-tuples in which exactly one of the entries in the jth block (of length N_j) is $\beta_j = \lambda_j/\sqrt{2}$ and all other entries are 0, with the exception of the last entry

$$\gamma = \left(1 - \sum_{j=1}^{m} \beta_j^2\right)^{1/2},$$

chosen so that all points are a unit N-sphere with $N = N_m + N_{m-1} + \cdots + N_1$. The hypothesis (**3**) guarantees that γ is real. The key to this construction is (as usual) in the choice of the N_j's. Needless to say, for the proof to work they must grow very rapidly.

As an example, we consider the case $m = 2$. Choose $N_1 = r + 1$, $N_2 = r^{r+1} + 1$. An r-coloring χ of S induces an r^{r+1}-coloring χ' of the set

$$S' = \{(\overbrace{0, \ldots, \beta_2, \ldots, 0}^{N_2}, \gamma)\}$$

by

$$\chi'(s_1') = \chi'(s_i'), \qquad s_1', s_2' \in S',$$

iff

$$\chi(s_1'\, t) = \chi(s_2'\, t)$$

for all

$$t \in \{(\overbrace{0, \ldots, \beta_1, \ldots, 0}^{N_1}, \gamma)\} = T_1,$$

where the concatenation $s'_1 t$ has the obvious interpretation of being an element of S. Since $|S'| = N_2 = r^{r+1} + 1$ and S' is r^{r+1}-colored, some pair of points $s'_1, s'_2 \in S'$ has $\chi'(s'_1) = \chi'(s'_2)$, i.e., $\chi(s'_1 t) = \chi(s'_2 t)$ for all $t \in T_1$. Since χ is an r-coloring and $|T_1| = N_1 = r + 1$, some pair of points $t, t' \in T_1$ has

$$\chi(s'_1 t) = \chi(s'_1 t').$$

Of course, this implies

$$\chi(s'_1 t) = \chi(s'_1 t') = \chi(s'_2 t) = \chi(s'_2 t').$$

But

$$d(s'_1 t, s'_1 t') = \beta_1 \sqrt{2} = \lambda_1 = d(s'_2 t, s'_2 t),$$
$$d(s'_1 t, s'_2 t) = \beta_2 \sqrt{2} = \lambda_2 = d(s'_1 t', s'_2 t')$$

so that these four points form the desired monochromatic $\lambda_1 \times \lambda_2$ brick.

The general result follows by the same techniques where, in general, we choose $N_1 = r + 1$ and $N_{j+1} = 1 + r^{N_1 N_2 \cdots N_j}$ for $j \geq 1$. Specifically, we think of S as $S(m) \times T(m)$, where $S(m)$ consists of the N_m N_m-tuples $(0, \ldots, \beta_m, \ldots, 0)$ and $T(m)$ consists of the $N_1 N_2 \cdots N_{m-1}$ complementary $(N_1 + \cdots + N_{m-1} + 1)$-tuples

$$(\overbrace{0, \ldots, \beta_{m-1},}^{N_{m-1}} \overbrace{\ldots, \beta_{m-2}, \ldots,}^{N_{m-2}} \ldots\ldots, \overbrace{\ldots, \beta_1, \ldots,}^{N_1} \gamma).$$

The initial r-coloring χ of S induces an $r^{N_1 \cdots N_{m-1}}$-coloring χ' of $S(m)$ by

$$\chi'(s'_1) = \chi'(s'_2), \qquad s'_1, s'_2 \in S(m),$$

iff

$$\chi(s'_1 t) = \chi(s'_2 t) \qquad \text{for all } t \in T(m).$$

Since

$$|S(m)| = N_m = 1 + r^{N_1 \cdots N_{m-1}},$$

there exists a pair of points, say $s_1, s_2 \in S(m)$, such that

$$\chi'(s'_1) = \chi'(s'_2).$$

Also, there is induced r-coloring $\hat{\chi}$ of $T(m)$ by

$$\hat{\chi}(t) = \chi(s'_1 t), \qquad t \in T(m).$$

By induction, there is a monochromatic $\lambda_1 \times \cdots \times \lambda_m$ brick under the coloring $\hat{\chi}$ of $T(m)$. By the definition of $\hat{\chi}$ and χ', this extends to a monochromatic $\lambda_1 \times \cdots \times \lambda_m$ brick in the original coloring of S. ▮

By suitable manipulations, it can be shown that the N_m satisfy

$$N_m \leq \overbrace{(r+2)^{(r+2)^{\cdot^{\cdot^{\cdot^{(r+2)}}}}}}^{m} = (r+2)\uparrow\uparrow m.$$

Bricks that have a main diagonal of length exceeding 2 seem much less tractable, although we expect that any $\lambda_1 \times \cdots \times \lambda_m$ brick with

$$\lambda_1^2 + \cdots + \lambda_m^2 < 4$$

is sphere-Ramsey. We can only prove this in the case $m = 1$.

THEOREM. *Let B be the set* $\{-\lambda/2, \lambda/2\}$, *where* $0 < \lambda < 2$. *Then B is sphere-Ramsey.*

Proof. It is enough to show that the graph G_n with vertex set S^n and edge set $\{\{\bar{x}, \bar{y}\}: d(\bar{x}, \bar{y}) = \lambda\}$ has chromatic number tending to infinity as n tends to infinity (where d denotes Euclidean distance). To prove this, we use the following recent result of Frankl and Wilson (suggested by Bárány, Füredi, and Pach).

THEOREM [4]. *Let* $\mathcal{F}$ *be a family of k-sets of* $[n]$ *such that for some prime power q,*

$$|F \cap F'| \not\equiv k \pmod q$$

for all $F \neq F'$ *in* $\mathcal{F}$. *Then*

$$|\mathcal{F}| \leq \binom{n}{q-1}.$$

For a fixed r, choose a prime power q so that

$$\binom{2(1+\varepsilon)q}{(1+\varepsilon)q} > r\binom{2(1+\varepsilon)q}{q-1},$$

where $\lambda = 2\beta\sqrt{2q}$, and choose $\varepsilon > 0$ and α so that

$$\alpha^2 + 2(1+\varepsilon)q\beta^2 = 1$$

and $N = (1+\varepsilon)q$ is an integer. Consider the set

$$S = \left\{(s_0, \ldots, s_{2N}): s_0 = \alpha,\ s_i = \pm\beta,\ \sum_{i=1}^{2N} s_i = 0\right\}.$$

To each $s \in S$ associate the subset

$$F(s) = \{i \in [2N]: s_i = \beta\}.$$

Thus, the family

$$\mathcal{F} = \{F(s): s \in S\}$$

consists of the $\binom{2N}{N}$ N-element subsets of $[2N]$. If $F, F' \in \mathcal{F}$, $F \neq F'$, then

$$|F \cap F'| \equiv N \pmod q$$

iff

$$|F \cap F'| = N - q = \varepsilon q.$$

If the elements of $\mathcal{F}$ are r-colored, then some color class must contain at least

$$\frac{1}{r}|\mathcal{F}| = \frac{1}{r}\binom{2N}{N} > \binom{2N}{q-1}$$

elements of $\mathcal{F}$. However, by Frankl and Wilson, if $|F \cap F'| = \varepsilon q$ *never* occurs, then the number of N-sets must be at most $\binom{2N}{q-1}$, which is a contradiction. Thus, some monochromatic pair F, F' must have

$$|F \cap F'| = \varepsilon q.$$

This means that the corresponding points $s, s' \in S$ must (up to a permutation of coordinate positions) look like

$$\begin{aligned} s &= (\alpha, \overbrace{\beta, \ldots, \beta}^{\varepsilon q}, \overbrace{\beta, \ldots, \beta}^{q}, \overbrace{-\beta, \ldots, -\beta}^{\varepsilon q}, \overbrace{-\beta, \ldots, -\beta}^{q}), \\ s' &= (\alpha, \underbrace{\beta, \ldots, \beta}_{\varepsilon q}, \underbrace{-\beta, \ldots, -\beta}_{q}, \underbrace{\beta, \ldots, \beta}_{\varepsilon q}, \underbrace{-\beta, \ldots, -\beta}_{q}). \end{aligned}$$

Note that

$$d(s, s') = \sqrt{8q\beta^2} = \lambda$$

and

$$d(s, 0) = d(s', 0) = \alpha^2 + 2(1 + \varepsilon)q\beta^2 = 1,$$

i.e., $s, s' \in S^{2N}$. This proves the theorem. ▮

As remarked previously, one would expect that the corresponding result should hold for any $\lambda_1 \times \cdots \times \lambda_m$ brick provided $\lambda_1^2 + \cdots + \lambda_m^2 < 4$. However, we are unable to prove this for even the case $m = 2$.

We conclude with a final observation. Many of the results in (Euclidean) Ramsey theory assert that the desired structure will occur whenever the dimension N of the space is sufficiently large. The proofs of these results typically end up constructing N's that are quite large. In this domain, bounds such as $N_m \leq (r + 2)\uparrow\uparrow m$ given before are considered very small. For the finite form of van der Waerden's theorem, which asserts for a suitable integer $W(k)$, any 2-coloring of $[W(k)]$ must contain a monochromatic arithmetic progression of length k, the only known upper bounds on $W(k)$ are not even primitive recursive (see [6]) (they grow like the Ackermann function). The best lower bound grows like $k \cdot 2^k$. Whether these gigantic upper bounds are a reflection of the truth or just of our inability to find the right proofs of these results is not currently known. Some evidence for the former is given by recent results of Paris, Harrington, and others (e.g., see [6, 10]) who give examples of combinatorial theorems of this general type which have *lower* bounds that grow this fast (and even much faster; see [15] for an amusing account).

References

0. Folklore.
1. Erdös, P., R. L. Graham, P. Montgomery, B. L. Rothschild, J. H. Spencer & E. G. Straus. 1973. Euclidean Ramsey theorems, I. J. Combin. Theory Ser. A **14:** 341–363.
2. Erdös, P., R. L. Graham, P. Montgomery, B. L. Rothschild, J. H. Spencer & E. G. Straus. 1973. Euclidean Ramsey theorems. II. Infinite and finite sets. Colloq. Math. Soc. Janos Bolyai **10:** 529–557.

3. ERDÖS, P., R. L. GRAHAM, P. MONTGOMERY, B. L. ROTHSCHILD, J. H. SPENCER & E. G. STRAUS. 1973. Euclidean Ramsey theorems, III. Colloq. Math. Soc. Janos Bolyai **10:** 559–583.
4. FRANKL, P. & R. M. WILSON. 1981. Intersection theorems with geometric consequences. Combinatorica **1:** 357–368.
5. GRAHAM, R. L., K. LEEB & B. L. ROTHSCHILD. Ramsey's theorem for a class of categories. Adv. in Math. **8** (1972): 417–433; Errata **10** (1973): 326–327.
6. GRAHAM, R. L., B. L. ROTHSCHILD & J. H. SPENCER. 1980. Ramsey Theory. John Wiley & Sons, Inc. New York, N.Y.
7. GRAHAM, R. L. 1981. Rudiments of Ramsey Theory. American Mathematical Society. Providence, R.I.
8. HILBERT, D. 1982. Über die irreduzibilität ganzer rationaler Functionen mit ganzzahlegen Koeffizienten. J. Reine Angew. Math. **110:** 104–129.
9. HINDMAN, N. 1974. Finite sums from sequences within cells of a partition of $\mathbb{N}$. J. Combin. Theory Ser. A **17:** 1–11.
10. PARIS, J. & L. HARRINGTON. 1977. A mathematical incompleteness in Peano Arithmetic. *In* Handbook of Mathematical Logic. J. Barwise, Ed.: 1133–1142. North-Holland. Amsterdam, the Netherlands.
11. RADO, R. 1933. Studien zur Kombinatorik. Math. Z. **36:** 424–480.
12. RAMSEY, F. P. 1930. On a problem of formal logic. Proc. London Math. Soc. **30:** 264–285.
13. SCHUR, I. 1916. Über die Kongruenz $x^m + y^m \equiv z^m (\text{mod } p)$. Jber. Deutsche Math.-Verein. **25:** 114–116.
14. SHADER, L. 1976. All right triangles are Ramsey in $\mathbb{E}^2$! J. Combin. Theory Ser. A **20:** 385–389.
15. SMORYNSKI, C. 1980. Some rapidly growing functions. Math. Intelligencer **2:** 149–154.
16. SPENCER, J. H. 1979. Ramsey's theorem for spaces. Trans. Amer. Math. Soc. **249:** 363–371.
17. STRAUS, E. G. 1975. A combinatorial theorem in group theory. Math. Comp. **29:** 303–309.
18. VAN DER WAERDEN, B. L. 1927. Beweis einer Baudetschen Vermutung. Nieuw Arch. Wisk. **15:** 212–216.

SOME SIMPLE ARRANGEMENTS OF PSEUDOLINES WITH A MAXIMUM NUMBER OF TRIANGLES

Heiko Harborth

Technische Universität
D-3300 Braunschweig, Federal Republic of Germany

A collection of simple closed curves in the projective plane is called a simple arrangement of pseudolines if each pair of curves has precisely one point in common, which is a point of intersection, and no point of the plane belongs to more than two curves.

A simple arrangement of n lines partitions the projective plane into $1 + \binom{n}{2}$ regions, and the appropriate Euclidean plane into $1 + \binom{n+1}{2}$ regions, $2n$ of them unbounded. What is the maximum number of these regions that can be triangles? Is it possible that no quadrangles occur? For references see [1, 3–5].

It is easy to prove (compare [1])

$$p_3 \leq \left[\frac{n(n-2)}{3}\right] = M_3(n) \qquad \text{for odd } n,$$

$$p_3 \leq \left[\frac{n(n-1)}{3}\right] = M_3(n) \qquad \text{for even } n,$$

if p_3 denotes the number of triangles of a simple arrangement in the projective plane. Only for $n \leq 10$, $n = 15$, and $n = 16$ were arrangements known with $M_3(n)$ triangles (compare the tables in [1, p. 56; 5, p. 319]). Recently in [2] a method was described which for odd n guarantees $M_3(2n-1)$ triangles in simple arrangements of $2n-1$ pseudolines if n pseudolines with $M_3(n)$ triangles are known. Thus, for $n = 2^t + 1$, $n = 3 \cdot 2^t + 1$, $n = 7 \cdot 2^t + 1$ $(t = 1, 2, 3, \ldots)$ and, together with the line at infinity, for the appropriate even numbers, simple arrangements exist with $M_3(n)$ triangles. For $n = 2^t + 2$ $(t \neq 1)$ and $n = 7 \cdot 2^t + 2$, no quadrangles occur in these arrangements (compare Conjectures 2.9 and 2.10 in [1]).

Here another method is presented which for odd n constructs arrangements of pseudolines with $M_3(3n)$ triangles if n pseudolines with $M_3(n)$ triangles are given.

THEOREM. *If in the Euclidean plane for* $n \equiv 3$ *or* $5 \pmod 6$ *a simple arrangement of* n *pseudolines exists, which has* $M_3(n)$ *triangles and* n *two-sided regions, then there exists a corresponding arrangement of* $3n$ *pseudolines with* $M_3(3n)$ *triangles and* $3n$ *two-sided regions.*

Proof. It follows from the assumptions, that in the arrangement of n lines each finite line segment is the side of exactly one triangle, that the n two-sided regions on the outside alternate with unbounded regions having at least four sides ($n \neq 3$), and that no bounded four-sided regions occur.

Now $2n$ new lines are added as follows: To the right and to the left of each of the n lines a nearly parallel line is drawn close by; that means, no point of intersection of the n lines lies between both lines. Then each finite line segment of the n lines is intersected by exactly two of these nearly parallel lines. This can be done in such a

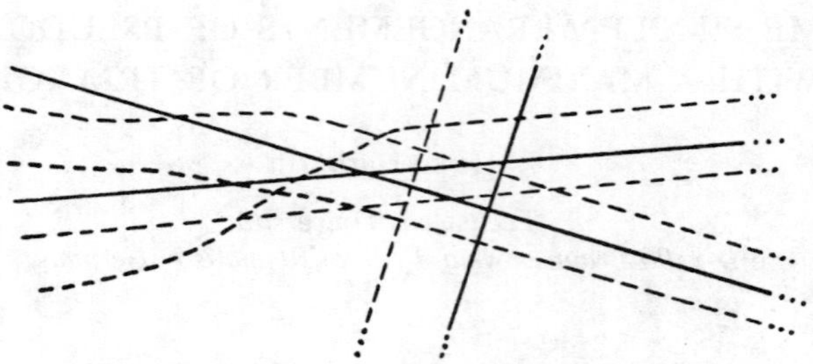

FIGURE 1

way that the point of intersection of these two lines lies between the line segment and that line nearly parallel to it that does not traverse the triangle neighboring the line segment. The four new lines are twisted around each two-sided region as shown in FIGURE 1.

Then each finite r-gon ($r > 4$) of the n lines is partitioned by the $2n$ lines into a central r-gon, r hexagons, and $2r$ triangles; each triangle of the n lines into one hexagon, and three triangles; each two-sided region of the n lines into one triangle, one unbounded five-sided, and two two-sided regions; and each unbounded r-sided region ($r \geq 4$) of the n lines into $2r - 1$ triangles, $r - 2$ hexagons, one pentagon, one two-sided, one four-sided, and one $(r + 2)$-sided unbounded region. Altogether there are $3n$ lines that form an arrangement having the described properties. ∎

By using the constructions of [2] and of the above theorem, arbitrarily many infinite sequences of odd values n are now known, which in the projective plane guarantee simple arrangements of n pseudolines with $M_3(n)$ triangles. Adding the

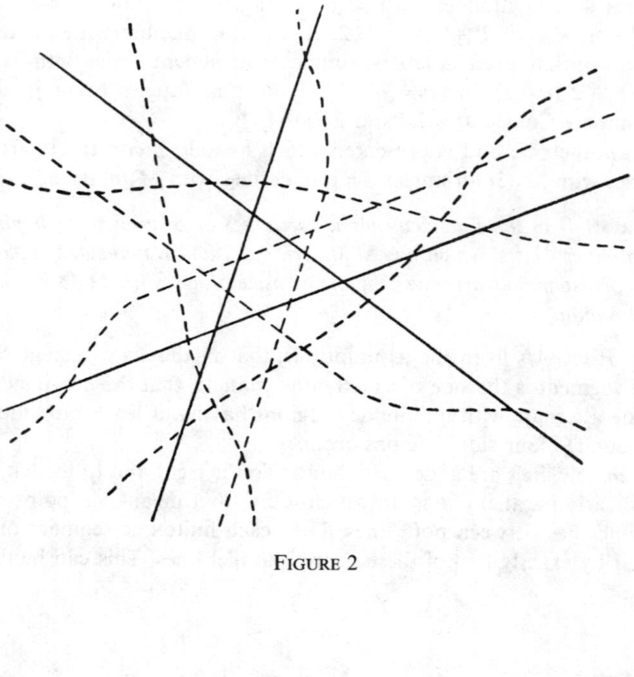

FIGURE 2

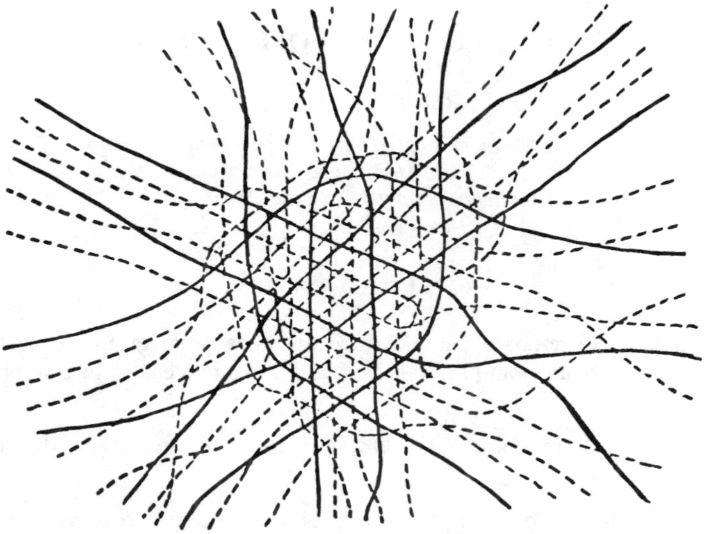

FIGURE 3

line at infinity, the next following even values n have the same property. Moreover, all arrangements of these even values n, $n \neq 4$, and $n - 1$ not generated by 7, do not contain quadrangles.

The table in [1, p. 56] can now be completed: Up to 50 for $3 \leq n \leq 10$, and $n = 13, 15, 16, 17, 18, 25, 27, 28, 29, 30, 33, 34, 45, 46, 49, 50$ arrangements of n pseudolines with $M_3(n)$ triangles are known. A few of these arrangements are known to be stretchable (possible in straight lines); however, in general the question of stretchability remains an open problem. FIGURES 2 and 3 illustrate the first two steps of the sequence $n = 3^t$ and, with the line at infinity, for $n = 3^t + 1$. Is the arrangement of 27 lines in FIGURE 3 stretchable?

REFERENCES

1. GRÜNBAUM, B. 1972. Arrangements and Spreads. American Mathematical Society. Providence, R.I.
2. HARBORTH, H. Two-colorings of simple arrangements. *In* Proceedings of the Sixth Hungarian Colloquium on Combinatorics (Eger, 1981). To appear.
3. PURDY, G. 1979. Triangles in arrangements of lines. Discrete Math. **25:** 157–163.
4. PURDY, G. 1980. Triangles in arrangements, II. Proc. Amer. Math. Soc. **79:** 77–81.
5. STROMMER, T. O. 1977. Triangles in arrangements of lines. J. Combin. Theory Ser. A **23:** 314–320.

A SURVEY OF THE SLOPE PROBLEM

Robert E. Jamison[a]

Mathematical Sciences Department
Clemson University
Clemson, South Carolina 29631

1. Introduction

When Erdös [8] re-posed the celebrated Sylvester Problem on the existence of ordinary lines (cf. Grünbaum [15, p. 17] for additional references), he also posed the following:

Line Problem. Show that n noncollinear points in the real plane determine at least n connecting lines.

Clearly, this is a problem in projective geometry since it is concerned only with point-line incidences. Its affine analogue asks for the minimum number of parallel classes (i.e., slopes) of the connecting lines:

Slope Problem. Show that n noncollinear points in the real affine plane determine at least $2[n/2]$ distinct slopes.

This problem was first posed by Scott [26] and finally solved by Ungar [28], using the allowable sequences of Goodman and Pollack [12]. Our purpose here is to trace the history of this problem and review some of the research it has spawned. A comparison with the development of the Erdös Line Problem will help to highlight the ensuing discussion. There are four main aspects of these problems which will be considered:

(1.1) the determination of the minimum,
(1.2) the classification of the configurations for which the minimum is achieved,
(1.3) extensions to higher dimensions and other fields,
(1.4) the determination of the spectrum—i.e., the possible numbers of lines and slopes that n points may generate.

Convention. In what follows, X will denote a generic configuration of n noncollinear points that determine m slopes.

For the Line Problem, (1.1)–(1.3) have been studied and have satisfactory solutions. Using the existence of an ordinary line (a line passing through exactly two points of X), one can easily obtain a solution to the Line Problem by induction starting from the *near-pencil* ($n - 1$ points on a line, the nth point off the line). In fact, it is not difficult to show by the same method that near-pencils are the only configurations determining the minimum number of n lines. As noted by de Bruijn and Erdös [3], the solution to the Line Problem does not actually depend on the existence of ordinary lines but can be derived purely combinatorially. In fact, it

[a] Partially supported by National Science Foundation (NSF) Grant MCS-80-02543 and NSF EPSCOR Grant IPS-80-11451.

depends only on the fact that the points and lines of a configuration determine a geometric lattice (matroid). It was this connection (and the higher-dimensional analogues) that motivated Basterfield and Kelly [1] to show that in a general matroid, the number of hyperplanes is at least as large as the number of points. This yields a solution to the Line Problem which is valid over any field and which has a satisfactory extension in higher dimensions. The classification of the minimal configurations is also straightforward, being given in terms of subgeometries (modular sublattices), which, over both the real and complex planes, correspond to near-pencils.

Scott's Slope Problem seems to be more involved. Ungar's solution, although not as complicated as those working on the problem had expected, is by no means trivial. It depends heavily on order and, as we shall see in Section 7, fails over the complex field.

In contrast to the simple classification of line-minimal configurations, the classification of slope-minimal configurations appears to be quite complicated and is not yet complete. Somewhat surprisingly, there is an overwhelming diversity of examples. For the Slope Problem, there is a natural distinction between the case of n odd and that of n even. If $m = n - 1$, then X is (*slope*)-*critical*. By Ungar's solution to the Slope Problem, this is possible only if n is odd. There are 4 infinite classes of critical configurations and 102 "sporadic" examples known. These were catalogued by Jamison and Hill [20] and will be discussed in Sections 3 and 4.

If $m = n$, then X is *near-critical*. The deletion of any point from a critical configuration results in an even near-critical configuration, but not all even near-critical configurations arise in this way. There are also several infinite families of odd near-critical configurations. Thus, the diversity of the near-criticals is much greater than that of the criticals. No comprehensive study has been made of the near-criticals, so our attention will be focused on the critical case.

Little is presently known about higher-dimensional analogues of the Slope Problem. Scott [26] mentions a special case in R^3 and we will consider some general examples in R^d in Section 6.

Curiously, the spectrum problem (1.4) has a simple solution for slopes, but it remains unsolved and is seemingly quite difficult for lines. This will be discussed with other variants of the Slope Problem in Section 8.

2. The Solution of the Slope Problem

When Scott [26] posed the Slope Problem, he noted that his conjectured bound was best possible for $n > 3$. By counting slopes on the connecting lines through a vertex, he obtained a lower bound on the order of $\sqrt{2n}$.

Independently, Shannon (private communication, 1973) also conjectured $m \geq 2[n/2]$ in connection with his work on arrangements of pseudohyperplanes [27]. Although he was unable to improve the trivial bound $m \geq \sqrt{n}$ (obtained by considering the maximum number of points on a line), he did discover several interesting critical configurations.

The first significant breakthrough was made by Burton and Purdy [2], who proved $m \geq (n - 1)/2$. They augmented X by adding the point at infinity (i.e., the direction) on each connecting line of X. To this augmented configuration X^* in the projective plane, they associated, in the manner of Kelly and Moser [21], a zonotope in R^3. By combining arguments on the geometry of zonohedra with manipulation of the Euler equation, they were able to estimate the number of points on the line at infinity and hence the number of slopes.

Cordovil [6] generalized the arguments of Burton and Purdy to oriented matroids of rank 3 and slightly improved the bound to $m \geq (n + 3)/2$. He also conjectured that Scott's conjecture should hold for all rank 3 oriented matroids.

The conjecture in this stronger form was settled by Ungar [28]. His proof reduces the problem via the allowable sequences of Goodman and Pollack [11] to an elegant, but short, combinatorial counting argument. The reduction to allowable sequences, as carried out by Goodman and Pollack [12], will be used in further discussion and thus is described briefly here.

Given a set X of n points in R^2, project X onto a directed line L. Label the points of X as 1 through n in the order of their projections onto L. Now allow L to rotate through some fixed center. Each time L passes through a position perpendicular to some connecting line—say K—in X, the order of the points of K in their projections onto L will reverse. Continuing in this way through 180°, we obtain a sequence of permutations Π_i corresponding to the projections of X onto the various positions of L. These satisfy the following properties:

(2.1) $\Pi_0 = 123 \ldots n$ and $\Pi_m = n\, n-1 \ldots 3\,2\,1$,

(2.2) Π_i is obtained from Π_{i-1} by reversing the order of some disjoint collection of increasing substrings of Π_{i-1}.

The disjoint substrings reversed at the ith move correspond to parallel connecting lines in X. The number m of moves required to bring Π_0 into reverse order is thus the same as the number of slopes determined by X. A sequence of permutations satisfying (2.1) and (2.2) will be called a *fundamental n-sequence*.

Rotating L further, we obtain the reversals of Π_1 through Π_m with Π_0 reached again at 360°, where the sequence begins to repeat. This full periodic sequence is called the *allowable sequence* of X, and m (the number of moves = slopes) is its *half-period*. (Note that a configuration will have several fundamental sequences, depending on the initial position of L.)

Ungar [28] showed that any $2k$-sequence requires at least $2k$ moves to reverse the initial permutation (unless the initial permutation is reversed in a single move—this case corresponds to collinear points). He did this by visualizing a barrier between the first k and the last k terms of each permutation and then keeping track of the moves in which terms cross this barrier.

THEOREM 2.3 (Ungar [28]). *Any n noncollinear points in the plane determine at least n − 1 slopes, with equality possible only for n odd.*

As remarked above, Ungar's result on allowable sequences actually settles the Slope Problem for oriented matroids of rank 3. Folkman and Lawrence [10] showed that every oriented matroid may be represented by an arrangement of "pseudohemispheres." In particular, a rank 3 oriented matroid may be represented by an arrangement of pseudolines. Using the Levi enlargement lemma and the Schoenflies theorem, Goodman and Pollack [13] have shown that every pseudoline arrangement may be sufficiently straightened to admit a representation by an allowable sequence. Conversely, Goodman and Pollack [12] have also shown how any allowable sequence may be associated via a "wiring diagram" with an arrangement of pseudolines. The connections among arrangements, oriented matroids, and allowable sequences have also been discussed by Cordovil [5].

Since Ungar's result applies to all allowable sequences and since these correspond to "generalized configurations," it is natural to ask about those sequences with the minimum number of moves. An n-sequence with half-period $m = n - 1$ will be called a *critical sequence*. An extension of Ungar's argument (cf. Jamison [19]) establishes the following fact:

THEOREM 2.4. *In any critical sequence, the middle term never leaves the middle position.*

This middle term (or in the case of configurations, the point corresponding to it) is called the *centrex*. As we shall see in Sections 4 and 5, it plays an important role in the study of critical configurations and critical sequences.

3. Infinite Families of Slope-Critical Configurations

The *centered polygon* $P_0(k)$ consists of the vertices and center of a regular k-gon. Scott [26], Shannon, and others observed that $P_0(k)$ is critical for all even $k > 2$.

A *bipencil* is a centrally symmetric configuration of $2k + 1$ points with $2k - 1$ of them on a line. This was observed to be critical by Ungar [28] (who attributes it to Scott) and others.

For nonnegative integer parameters s, t, and h, the *tricolumnar array* $TC(s, t; h)$ consists of the following $2(s + t + h) + 3$ points:

$$\begin{array}{lll} (\pm 1, k), & 0 \le k \le s, & k \text{ integral}, \\ (0, (s/2) \pm k), & 0 \le k \le t, & k \text{ integral}, \\ (0, (s \pm k)/2), & 0 \le k \le 2h, & k \text{ odd integral}. \end{array}$$

$TC(0, t; h)$ is a bipencil, $TC(1, 0; 0)$ is the quincunx $P_0(4)$, and $TC(1, 1; 0)$ is $P_0(6)$. The remaining TC-arrays are all new and, except for the equivalence of $TC(1, 0; 1)$ with $TC(2, 0; 0)$, are all distinct. Some special cases of these were discovered by Pollack (private communication, 1981) during his investigations of allowable sequences.

The last infinite family is, in a sense, the simplest. It also has the most natural analogues in higher dimensions and over the complex field. Curiously, it seems to have been the last discovered. For positive integers s, t and any real $\lambda > 1$, the *exponential cross* $EX_\lambda(s, t)$ consists of the following $2(s + t + 2) + 1$ points:

$$(0, 0), (\pm\lambda^i, 0), (0, \pm\lambda^j), \qquad 0 \le i \le s, 0 \le j \le t.$$

These determine the $2(s + t + 2)$ slopes of 0, ∞, and $\pm\lambda^k$, where $-s \le k \le t$.

It is natural in our context to regard X and any of its affine images as *equivalent* configurations since affine maps preserve collinearity and parallelism. More generally, two sets are *slope-isomorphic* if there is a bijection between them that preserves collinearity and parallelism of *their* points. Such a bijection need not be induced by an affinity of the whole plane. Indeed, any two bipencils of n points are slope-isomorphic but need not be affinely equivalent. Similarly, the affine equivalence class of a cross $EX_\lambda(s, t)$ depends on the parameter λ, whereas the slope-isomorphism type does not.

Since affine maps may drastically alter the visual appearance of a configuration, it is valuable, in studying critical sets, to have simple geometric criteria by which the members of the above four families may be recognized. Such criteria were given in [16], where the tricolumnar arrays and exponential crosses were first described.

THEOREM 3.1. *Any critical configuration contained in the union of two lines is either a bipencil or affinely equivalent to some exponential cross.*

THEOREM 3.2. *Any critical configuration contained in the union of three parallel lines is either a bipencil or affinely equivalent to some tricolumnar array.*

THEOREM 3.3. *Any critical configuration with all but two points collinear is a bipencil.*

The conditions yielding the centered polygons are somewhat more complicated. Following Scott [26], we say X is *convex* if it lies on the boundary of a convex polygon. Scott showed that if X is convex, then $m \geq n$. The convex configurations for which equality holds are classified in [16] but are too diverse to list here. This classification leads, however, to the following result:

THEOREM 3.4. *Suppose X is critical and obtained from a convex configuration by the addition of a single point. Then X is affinely equivalent to one of the following:*

(1) *a bipencil,*

(2) $P_0(k)$ *for some even* $k > 2$,

(3) $TC(s, 0; 0)$ *for some* $s > 0$,

(4) $TC(2, 1; 0)$, *the Fano Triangle, or the Golden Kite* (*see* FIGURE 1).

4. SPORADIC SLOPE-CRITICAL CONFIGURATIONS

In addition to the four infinite families, numerous sporadic criticals have also come to light. Some were discovered "by hand" by Shannon, by Grünbaum, and by the author. However, most were found by computer searches conducted for the author by Glenn Forney and by Dale Hill. (A catalogue of all known criticals, known to be complete for $n \leq 9$, was compiled by Jamison and Hill [20]. The names applied to configurations in the figures are taken from that catalogue.) Since there are 102 known sporadics, it is natural to wonder if there might not be a missing infinite family. However, the present evidence tends to indicate the contrary. Namely, from empirical observation it appears that

(a) even when recognizable patterns among sporadic examples suggest a series, the larger members in the series fail to be critical, and

(b) critical configurations seem to have coordinates which, if rational, involve small prime factors, and if irrational, satisfy very simple polynomial equations.

Moreover, the results of Jamison [17] show that

(c) there are some fairly strong theoretical restrictions on the possible structure of critical configurations.

Conjecture 4.1. Any slope-critical configuration with more than 49 points is (up to affine equivalence) in one of the four infinite families.

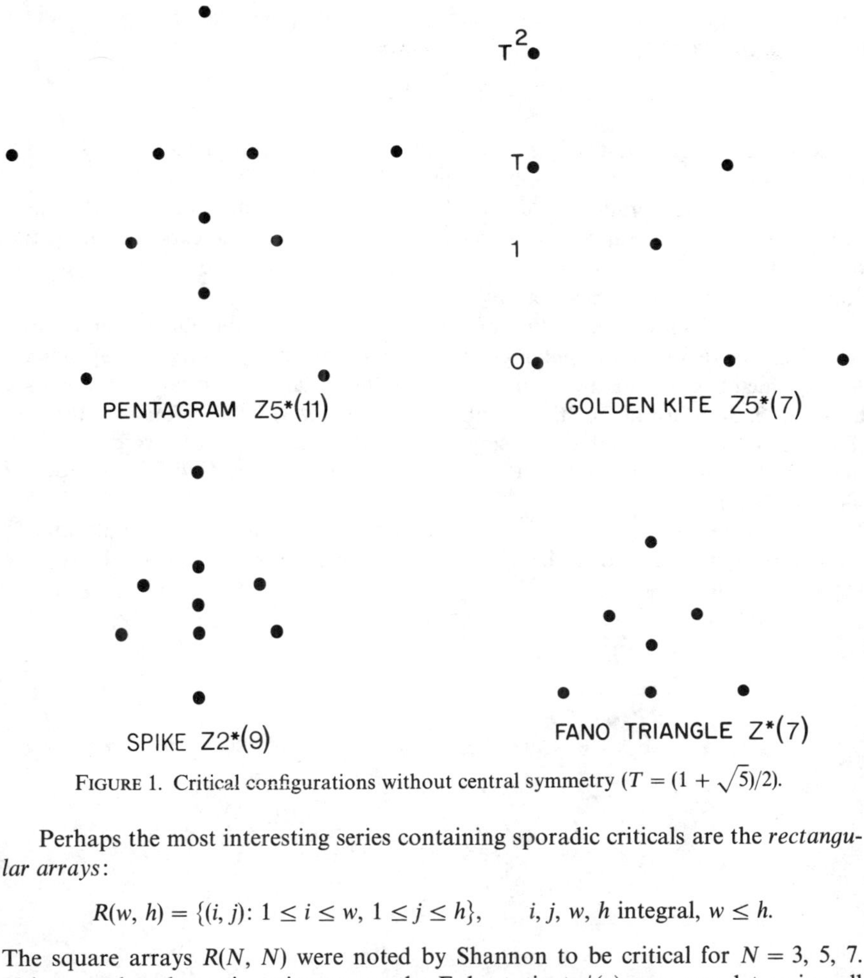

FIGURE 1. Critical configurations without central symmetry ($T = (1 + \sqrt{5})/2$).

Perhaps the most interesting series containing sporadic criticals are the *rectangular arrays*:

$$R(w, h) = \{(i, j): 1 \leq i \leq w,\ 1 \leq j \leq h\}, \qquad i, j, w, h \text{ integral, } w \leq h.$$

The square arrays $R(N, N)$ were noted by Shannon to be critical for $N = 3, 5, 7$. Using number-theoretic estimates on the Euler totient $\phi(n)$, one can determine all critical and near-critical $R(w, h)$.

THEOREM 4.2. *The following are the only arrays $R(w, h)$ with $m = n - 1$:*
(i) $R(3, N)$, *N odd,*
(ii) $R(5, 5)$, $R(5, 7)$, *and* $R(7, 7)$.
The following are the only arrays $R(w, h)$ with $m = n$:
(iii) $R(2, N)$,
(iv) $R(3, N)$, *N even,*
(v) $R(4, 4)$, $R(4, 5)$, *and* $R(4, 7)$.

Adding the center of symmetry to arrays (iii)–(v) makes them critical. Note, however, that $R(3, N)$ (N odd), $R_0(3, N)$ (N even), and $R_0(2, N)$ are all tricolumnar arrays. Thus we obtain, from (ii) and (v), only six new criticals.

The proof of THEOREM 4.2 will be given elsewhere. It is along much the same lines as the estimation of the number of totient points given by D. N. Lehmer [23].

His results imply that as w and h both approach infinity, the ratio of slopes to points in $R(w, h)$ approaches

$$2/\zeta(2) = 12/\pi^2 = 1.21585 \ldots .$$

When w, say, is fixed and only h is allowed to grow, the ratio is closer to 1, and a modification of Lehmer's argument is required.

Other series of sporadics include the "diamond strings" and "stretched lozenges" of FIGURE 2. There are also series obtained from TC-arrays and exponential crosses by the addition of one or two pairs of extra points. There are, however, only finitely many criticals in each of these series.

A less well defined series is that of the "double polygons" in which each member is a union of two centered polygons, with the same center and axes of symmetry. Known members of this series include the Fano Triangle (FIGURE 1), the square array $R(3, 3)$ (FIGURE 2), the Pentagram (FIGURE 1), the Double Hexagon (FIGURE 3), and the Double Octagon (FIGURE 3). The first three were found by Shannon (private communication, 1973) and the last two by Grünbaum (letter of October 1981).

The two examples in FIGURE 4 are of interest since they are the only known critical configurations other than the centered polygons whose convex hulls have more than eight vertices.

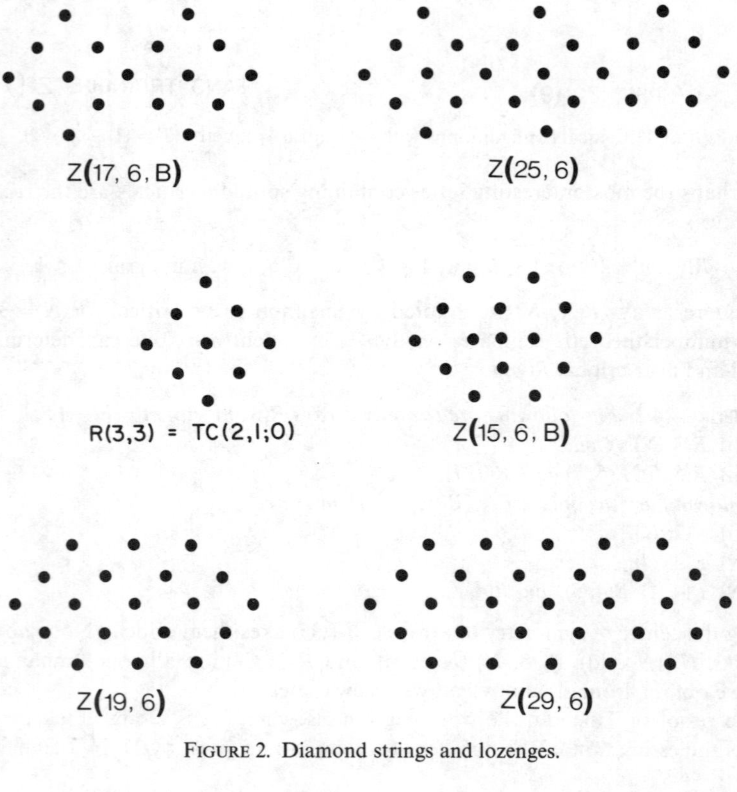

FIGURE 2. Diamond strings and lozenges.

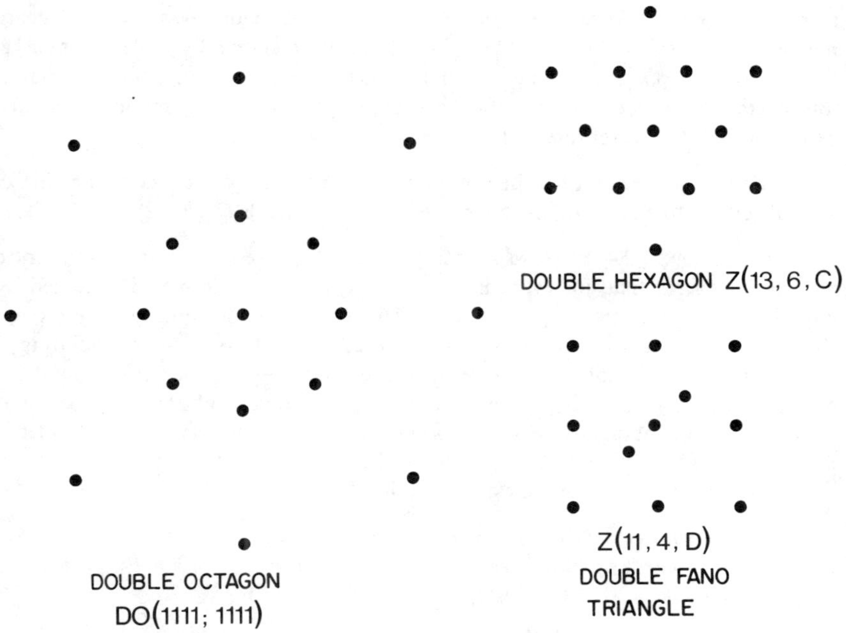

FIGURE 3. Double critical configurations.

It follows from the completeness of the first-order theory of real-closed fields (cf. Robinson [25]) that any finite configuration in the real plane is slope-isomorphic to a configuration whose points have real algebraic coordinates. Hence, any finite configuration may be realized in some finite extension of the rationals. In fact, most criticals seem to occur in the integer lattice. The known sporadic examples are distributed as follows over the fields below:

$$Q\text{: } 70, \qquad Q(\sqrt{2})\text{: } 16, \qquad Q(\sqrt{3})\text{: } 1, \qquad Q(\sqrt{5})\text{: } 15.$$

The tricolumnar arrays are defined in the integer lattice, and all bipencils and exponential crosses have slope-isomorphic representations with integer coordinates. The centered polygons, however, are exceptional.

It follows from the theory of affinely regular polygons (see Coxeter [4] and Nizette [24]) that any regular polygon is determined up to affine equivalence by its

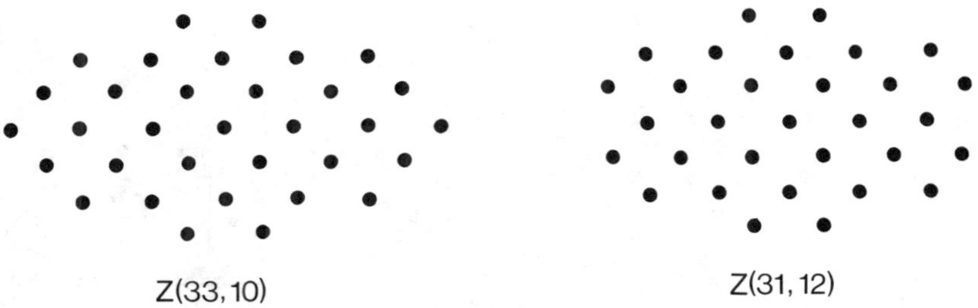

FIGURE 4. Sporadic critical sets with 10 and 12 vertices.

slope-isomorphism type. The ratio of a side of the regular k-gon to its nearest neighboring chord is $1 + 2\cos(2\pi/k)$, which has been shown by D. H. Lehmer [22] to have degree $\phi(k)/2$ over the rationals. Since this ratio is an affine invariant, it follows that as k goes to ∞, the critical configurations $P_0(2k)$ require ever larger extensions of Q to be represented.

Conjecture 4.3. Except for the centered polygons, every critical configuration can be realized, up to slope-isomorphism, over one of the fields Q, $Q(\sqrt{2})$, $Q(\sqrt{3})$, $Q(\sqrt{5})$.

From THEOREM 2.4 it follows that any critical X possesses a point z (the centrex) with the property that, for every line through z, the two half-lines determined by z contain equal numbers of points of X. The number d of connecting lines of X through z is the *central degree* of X and the $2d$ connecting half-lines radiating out from z are the *spokes* of X. Suppose these are listed counterclockwise around X as $S_1, S_2, \ldots, S_d, S_{d+1}, \ldots, S_{2d}$. Then S_i and S_{i+d} (addition mod $2d$) are opposite and contain the same number $v_i = v_{i+d}$ of points of $X \backslash z$. The number v_i will be called the *length* of S_i. The d-tuple $(v_1, v_2, \ldots, v_d)$ is the *central signature* of S. The following limitations on spoke length are established in [17]:

THEOREM 4.4. *Let X be critical with central degree $d > 2$.*
(i) *If for some i, $v_{i-1} > 1$, $v_i = 1$, and $v_{i+1} > 1$, then $v_{i-1} \leq 3$ and $v_{i+1} \leq 3$.*
(ii) *If for some i, $v_i > 1$ and $v_{i+1} > 1$, then $v_i \leq 7$ and $v_{i+1} \leq 7$.*

Conjecture 4.5. Any critical configuration with a connecting line of eight or more points is (up to affine equivalence) in one of the four infinite families.

With four exceptions (shown in FIGURE 1), the known critical configurations are centrally symmetric. Also with four exceptions (shown in FIGURE 5), the known

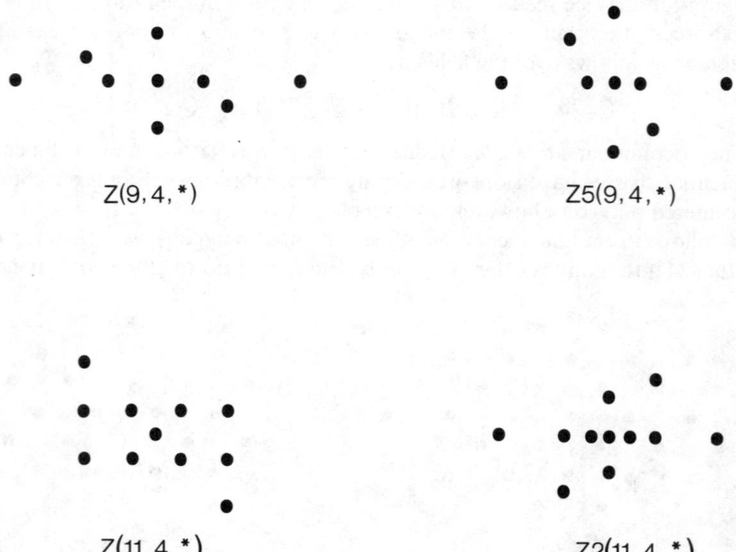

FIGURE 5. Critical configurations without bilateral symmetry.

criticals have bilaterally symmetric representations. The following symmetry results are also established in [17]:

THEOREM 4.6. *Let X be critical with central degree d.*
(i) *If d is even, then X is centrally symmetric.*
(ii) *If X has a spoke of length* 1 *and a spoke of length greater than* 1, *then X is either centrally symmetric or affinely equivalent to the Spike Z2*(9)* (*see* FIGURE 1).

Unlike central symmetry, bilateral symmetry is not an affine invariant and no analogue of THEOREM 4.6 is known.

Question 4.7. Are there other criticals lacking either central or bilateral symmetry?

This question is of additional interest since the configurations in FIGURE 1 together with the quincunx $P_0(4)$ are *minimal critical* in that they do not properly contain any critical subsets. All other known critical sets are centrally symmetric and hence contain a copy of the quincunx $P_0(4)$. In the other direction, the Fano Triangle is a subset of the Double Fano Triangle (FIGURE 3) and hence of the critical array $R(7, 7)$. The Golden Kite lies in the two configurations in FIGURE 6. The Spike and the Pentagram do not appear to be contained in any larger critical configurations.

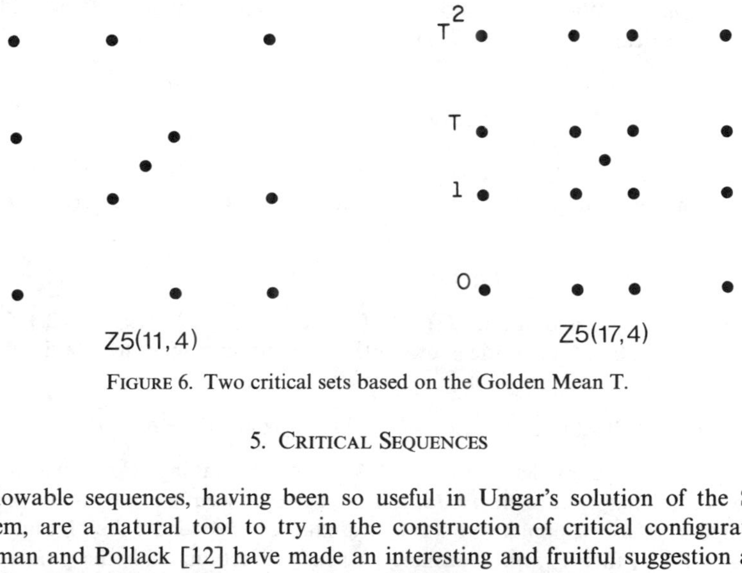

FIGURE 6. Two critical sets based on the Golden Mean T.

5. CRITICAL SEQUENCES

Allowable sequences, having been so useful in Ungar's solution of the Slope Problem, are a natural tool to try in the construction of critical configurations. Goodman and Pollack [12] have made an interesting and fruitful suggestion along these lines. A fundamental *n*-sequence is *greedy* iff in every move after the first, every maximal increasing substring is reversed. Such a process would seem to get one as quickly as possible to the reverse position. In fact, Ungar [28] also proved:

THEOREM 5.1. *Any greedy fundamental n-sequence reaches the reverse position n* ... 21 *in at most n moves.*

By THEOREM 2.3 for *n* even, this is best possible; for *n* odd there is a possible discrepancy of 1. This is accounted for, however, by the following result from [19].

THEOREM 5.2. *A greedy fundamental n-sequence is critical iff n is odd and the middle term is never moved.*

This provides a means to construct large classes of critical sequences. In [19] it is shown that, for each n, there are exponentially many greedy critical sequences which are not realizable by point configurations. Thus the critical sequences form a much more extensive class than the critical configurations. In fact, in contrast to the restrictions in THEOREM 4.4, it is shown in [19] that there are critical sequences of every possible signature.

THEOREM 5.3. *For any $d > 1$ and any sequence of positive integers $v_1, v_2, \ldots, v_d$, there is a critical sequence with signature $(v_1, v_2, \ldots, v_d)$.*

Such sequences cannot in general be greedy since it is easily seen that a greedy sequence can have at most one spoke of length $v_i > 1$. Thus even the greedy sequences account for only a small portion of all possible critical sequences. (Among the configurations in the Catalogue [20], the only ones which have greedy allowable sequences are the centered polygons $P_0(2k)$, the bipencils, and two types of TC-arrays: $TC(s, 0; 0)$ and $TC(1, t; 0)$.) It appears that the structure of critical sequences is quite flexible, perhaps so much so as to allow every sequence to be embedded in a critical sequence.

Problem 5.4. Which allowable sequences may be extended by the addition of points to critical sequences?

6. DIRECTIONS IN HIGHER DIMENSIONS

For $n \geq d + 1 \geq 3$, let $k_d(n)$ denote the minimum number of directions that can be determined in R^d by n points that affinely span R^d. Then our critical examples show

$$k_2(n) \leq n - \varepsilon(n) \qquad \text{for } n \geq 4, \tag{6.1}$$

where $\varepsilon(n) = 1$ if n is odd and $\varepsilon(n) = 0$ if n is even. Ungar's THEOREM 2.3 shows the reverse inequality. The critical examples may be extended to obtain a general inequality.

THEOREM 6.2. $k_d(n) \leq (d - 1)n - d^2 + 2d - \varepsilon(n)(d - 1)$ *for* $n \geq 2d$.

Proof. For brevity, let us call a set X, affinely spanning R^d, for which equality holds in THEOREM 6.2 with n odd, a *d-critical configuration.* Let us think of R^{d-1} situated in R^d as the points with dth coordinate 0. Let e_i denote the ith unit vector. If X is in R^{d-1}, then e_d will determine $|X|$ new directions with the points of X. If $X = -X$ and $0 \in X$, then $-e_d$ will determine the same directions to X that e_d does. Letting $X' = X \cup \{e_d, -e_d\}$, we then have

$$m(X') = m(X) + |X|,$$

where $m(X)$ is the number of directions determined by X. From this it is easily verified that if X is $(d - 1)$-critical, then X' is d-critical. Note that since X' again satisfies the central symmetry condition, we may also form X'', X''', and so on, each of which will be referred to as a *trivial extension* of X. By trivially extending, say, the bipencils, inequality in THEOREM 6.2 can be established for odd $n \geq 2d$ in any dimension d. The inequality in THEOREM 6.2 for even n follows by removing the center of symmetry. ∎

Conjecture 6.3. Equality holds in THEOREM 6.2.

In a certain sense, these trivial extensions may be regarded as analogues of the bipencils—which are themselves trivial extensions of symmetric "1-critical" configurations. The exponential crosses also have an easy generalization to higher dimensions.

Given positive integers $s_1, s_2, \ldots, s_d$, and real $\lambda > 1$, let $EX_\lambda(s_1, s_2, \ldots, s_d)$ consist of 0 and all points in R^d of the form

$$\pm\lambda^k e_i\colon\ k = 0, 1, \ldots, s_i;\ i = 1, \ldots, d.$$

THEOREM 6.4. *$EX_\lambda(s_1, \ldots, s_d)$ is d-critical.*

Proof. Let $S = \sum (s_i + 1)$. Then $X = EX_\lambda(s_1, \ldots, s_d)$ has $|X| = 2S + 1$ points. If we slice X with the i, j-coordinate plane, we obtain the two-dimensional cross $EX_\lambda(s_i, s_j)$ with $2(s_i + s_j + 2) + 1$ points and $2(s_i + s_j + 2)$ slopes. Since any pair of points in X lies in such a slice, every direction determined by X is counted in the sum

$$\sum_{i<j} 2(s_i + s_j + 2) = 2(d - 1)S.$$

However, the direction of each of the d coordinate axes gets counted $d - 1$ times in this sum. Hence, the total number of directions in X is

$$\begin{aligned} m &= 2(d-1)S - (d-2)d \\ &= (d-1)(2S+1) - d^2 + 2d - (d-1). \end{aligned}$$ ∎

In d-dimensions, the TC-arrays also have analogues made by stacking $(d - 1)$-dimensional cross-polytopes. However, the construction is too involved to present here. A simple example is shown in FIGURE 7.

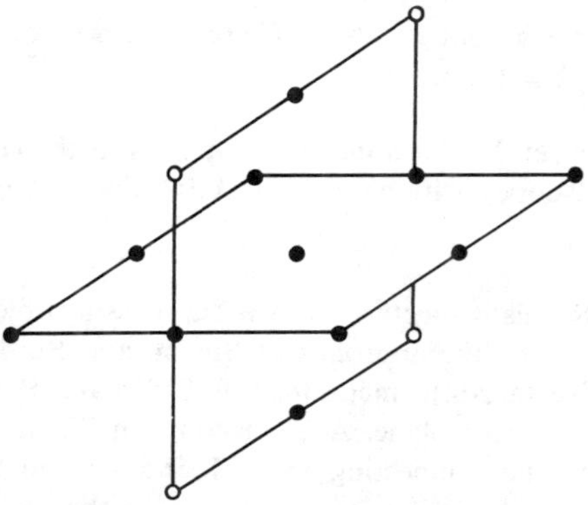

N = 15

FIGURE 7. A 3-critical configuration.

Analogous to the centered polygons, one might expect certain highly symmetric polytopes to yield d-critical configurations in higher dimensions. This is true to a limited extent. The centered cross-polytope is a trivial extension of the quincunx $P_0(4)$ and hence is d-critical for all d. The centered d-cube has $2^d + 1$ points and determines $(3^d - 1)/2$ directions. Hence it is d-critical only in dimensions 2, 3, and 4. One may also derive 3-critical sets from the icosahedron, the cuboctahedron, the hexagonal prism, and certain configurations associated with Kepler's stella octagula. However, the number of sporadic d-critical configurations seems to be quite limited for $d > 2$, and none are known for $d > 4$.

For higher dimensions, the range $d < n < 2d$ is also of interest.

Conjecture 6.5. For $n = d + k$ with $1 \leq k \leq d$, $k_d(n) = \binom{d}{2} + dk - \binom{k}{2}$.

The number in Conjecture 6.5 is the number of directions occurring in the "partial cross"

$$\pm e_1, \pm e_2, \ldots, \pm e_k, e_{k+1}, e_{k+2}, \ldots, e_d \tag{6.6}$$

of $n = d + k$ points. Indeed, among the d "positive" $+e_i$, there are $\binom{d}{2}$ directions. These include the directions among the k "negative" $-e_j$. Between the d "positives" and k "negatives" there are dk directions of the form $e_i + e_j$. But for $i < j \leq k$, the slope $e_i + e_j$ is counted twice, making a total of $\binom{k}{2}$ duplications. Thus, $\binom{d}{2} + dk - \binom{k}{2}$ is the number of directions determined by Configuration 6.6.

7. Slopes and Modular Lines in the Complex Plane

The bound $m \geq n - 1$ does not hold for configurations in the complex plane, as may be seen from the following toroidal variants of the exponential cross. For any positive integer r, the *trimodular array* $TM(r)$ consists of these $n = 2r + 1$ points:

$$(0, 0), (\omega^k, 0), (0, \omega^k), \qquad k = 1, \ldots, r, \tag{7.1}$$

where ω is a primitive rth root of unity. These determine only $r + 2 = (n + 3)/2$ slopes: 0, ∞, and $-\omega^k$, $k = 1, \ldots, r$.

Conjecture 7.2. Any set X of n noncollinear points in the complex plane determines at least $(n + 3)/2$ slopes, with equality iff X is affinely equivalent to $TM(r)$ for some r.

To lend credence to this conjecture, we shall discuss a projective variant of the Slope Problem suggested by the approach of Burton and Purdy [2] and Cordovil [6]. (A related projective variant is mentioned by Erdös and Purdy [9].) Let Y be a set of points in the projective plane. A connecting line L of Y is a *modular line* provided it meets every other connecting line of Y at a point in Y. (This terminology reflects the fact that such a line is a modular element in the geometric lattice of flats determined by Y.) Let us say Y is *trivial* if it lies on the union of two lines.

Any X in the affine plane may be extended to a projective configuration X^* by adding the points at infinity on the connecting lines of X. These new points determine a modular line in X^* which encodes the slope structure of X. Conversely, any nontrivial Y with a modular line L arises in essentially this way from $X = Y \backslash L$, except that L may have some "extraneous" points on it.

The Slope Problem may therefore be formulated projectively as the determination of the minimum cardinality of modular lines in nontrivial configurations Y. (Note that Y being nontrivial corresponds to X being noncollinear.) Now Ungar's THEOREM 2.3 says that any modular line in Y contains at least $(|Y|/2) - 1$ points of Y. Thus, modular lines are so large that only rarely can there be more than one of them. It is not hard, in fact, to classify all real configurations with two modular lines.

THEOREM 7.3. *A finite nontrivial configuration Y in the real projective plane has two or more modular lines iff $Y = X^*$, where X is a near-pencil or a bipencil.*

Now if Conjecture 7.2 is true, it would imply that a modular line in a nontrivial complex configuration Y must contain at least $(|Y|/3) + 1$ points. Thus, complex configurations should, with a few exceptions, contain at most three modular lines. It turns out that this can be proved directly, without reference to the Slope Problem (cf. [18]).

THEOREM 7.4. *If Y is a finite nontrivial configuration in the complex projective plane with three or more modular lines, then Y is projectively equivalent to $TM(r)^*$ for some r.*

Note that of the trimodular arrays, only $TM(1) = P(3)$ (the triangle) and $TM(2) = P_0(4)$ (the quincunx) lie in the real projective plane, and that the projective extensions of these have four and three modular lines, respectively.

If we drop the restriction of nontriviality, then we may have any number of modular lines, as the near-pencil illustrates. In fact, this characterizes the near-pencils.

THEOREM 7.5. *If a finite configuration Y in the complex projective plane has five or more modular lines, then Y is a near-pencil.*

8. Some Variants of the Slope Problem

There are two obstructions to a set of n points determining fully $\binom{n}{2}$ directions, namely, collinearity and parallelism. If we take these one at a time, the Slope Problem splits into two subproblems, each with a tidy solution. If we exclude parallelism, then the number of directions is the number of lines and we are back to the Line Problem whose solution may be phrased as follows:

THEOREM 8.1 (Erdös *et al.*). *If X has no parallel connecting lines, then $m \geq n$ with equality iff X is a near-pencil.*

Excluding collinearity leads to the following result from [19]:

THEOREM 8.2. *If X has no three points collinear, then $m \geq n$ with equality iff X is affinely equivalent to $P(n)$, the vertices of a regular n-gon.*

Having established the extremes for the possible number of slopes, it is natural to wonder what happens in-between. For the Line Problem, it is known (cf. Kelly and Moser [21]) that there are wide gaps at the low end of the spectrum, and the determination of the possible numbers of lines generated by n points appears to be very difficult (cf. Grünbaum [15, p. 13]). In this regard, the Slope Problem is much easier.

THEOREM 8.3. *Given integers n and m with $3 \leq n \leq m \leq \binom{n}{2}$, there is an X with n points and m slopes of connecting lines. Moreover,*

(a) *there is such an X with no three points collinear,*

(b) *there is such an X consisting of lattice points,*

(c) *if $m \geq 2n - 3$, there is such an X consisting of lattice points, no three of which are collinear.*

Proof. We first take the case $m \leq 2n - 4$. For (a), take n consecutive vertices of the regular m-gon $P(m)$. Since $n > m/2$, it is easy to see that these vertices determine all m slopes of $P(m)$.

For (b), write $m = 2s + \varepsilon$, where ε is 0 or 1, and consider the following $s + 2$ points:

$$(0, j), j = 1, \ldots, s, \qquad (1, 1), \quad \text{and} \quad (1, s + \varepsilon). \tag{8.4}$$

These determine m slopes. Any additional points of the form

$$(1, j), \qquad \text{where } 1 < j < s + \varepsilon, \tag{8.5}$$

when added to Configuration 8.4, will not change the number of slopes. Since $s + 2 \leq (m + 4)/2 \leq n \leq m = 2s + \varepsilon$, we find that $0 \leq n - (s + 2) \leq s + \varepsilon - 2$, so that there are enough points of the form (8.5) to add to Configuration 8.4 to make up the desired number n.

We shall settle (c) by exhibiting the required configuration in the form (a, a^2), $a \in A$, where A is a suitable set of integers. Since the slope from (a, a^2) to (b, b^2) is $a + b$, it suffices to find a set A of n positive integers having m distinct pairwise sums. We handle first the case $m \leq 3n - 7$. Let A_0 consist of integers $1, 2, \ldots, n - 1$. These determine $2n - 5$ sums, namely, $3, 4, \ldots, 2n - 3$. Let $k = m - (2n - 3)$. Then $0 \leq k \leq n - 4$, so $n - k - 2$ belongs to A_0. Now $n + k$ determines with $n - k - 2$, $\ldots, n - 1$ the $k + 2$ new sums $2n - 2, 2n - 1, \ldots, 2n + k - 1$. Thus, $A_0 \cup \{n + k\}$ determines $(2n - 5) + (k + 2) = m$ pairwise sums as desired.

For $m \geq 3n - 6$, we proceed by induction on n, starting from the trivial case $n = 3$. Letting $m' = m - (n - 1)$ we have $2(n - 1) - 3 \leq m' \leq \binom{n-1}{2}$, so by induction there is a set A_0 of $n - 1$ positive integers determining exactly m' pairwise sums. If $N = 2 \max A_0$, then N will determine all new sums with the $n - 1$ elements of A_0. Thus, $A = A_0 \cup \{N\}$ is the desired set. ∎

Question 8.6. What is the minimum number $M_0(n)$ of slopes which may be determined by n lattice points, no three of which are collinear?

As we shall now show, the bound in THEOREM 8.3(c) may be improved when n is even by taking our example on a hyperbola instead of a parabola.

THEOREM 8.7. *$M_0(n) \leq 2n - 4$ for even $n \geq 4$.*

Proof. Clearly, it suffices to exhibit our example in the rationals. To this end, consider the hyperbola $xy = 1$. The slope between two points on the same branch, say $(a, 1/a)$ and $(b, 1/b)$, is $-1/ab$. The slope between points on different branches, say $(a, 1/a)$ and $(-b, -1/b)$, is $1/ab$. In the latter case, we also have the possibility that $a = b$. If we choose our a's and b's to be integral powers of 2, the slopes then correspond to the possible pairwise sums of the exponents. Now for $n = 2s$, select these n points:

$$\pm(2^j, 2^{-j}), \qquad \text{where } j = 1, \ldots, s. \tag{8.8}$$

There are $2s - 3$ negative slopes (between points on the same branch) corresponding to the $2s - 3$ of distinct exponents and there are $2s - 1$ positive slopes (between branches) corresponding to the $2s - 1$ sums of not necessarily distinct exponents. Thus, Configuration 8.8 determines $4s - 4 = 2n - 4$ slopes as desired. ∎

By THEOREM 8.2, the only X satisfying $m = n$ with no three points collinear is an affinely regular n-gon. By the remarks preceding Conjecture 4.3, this is representable over the integers iff $\phi(n)/2 = 1$—that is, iff $n = 3$, 4, or 6. Thus, $M_0(n) > n$ for all other values of n. FIGURE 8 shows a rational octagon with only 10 slopes. Since removing a point cannot increase the number of slopes, we also have $M_0(5) = 5$ and $M_0(7) \leq 10$.

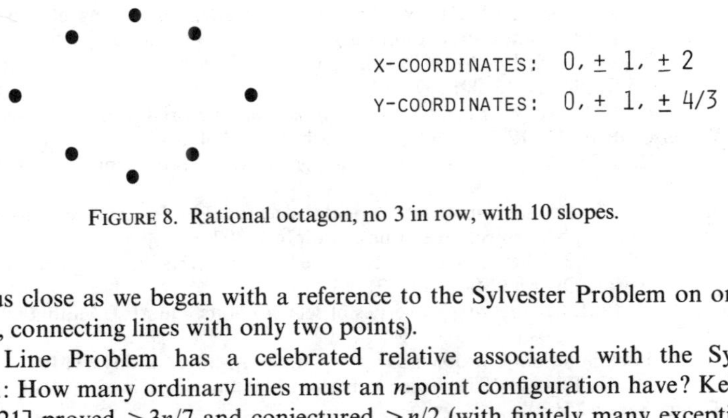

FIGURE 8. Rational octagon, no 3 in row, with 10 slopes.

Let us close as we began with a reference to the Sylvester Problem on ordinary lines (i.e., connecting lines with only two points).

The Line Problem has a celebrated relative associated with the Sylvester Problem: How many ordinary lines must an n-point configuration have? Kelly and Moser [21] proved $\geq 3n/7$ and conjectured $\geq n/2$ (with finitely many exceptions), a result recently claimed to have been proved by S. Hansen. It is also natural to ask: How many *slopes* of ordinary lines must there be? Dirac [7] showed that, under certain restrictions on X, there must be at least two. A referee has observed that this follows in general from Kelly and Moser's result. Namely, if all $3n/7$ ordinary lines in X were parallel, they would account for $6n/7$ points. Adding to X the point p at infinity common to these ordinary lines, we see in $X \cup p$ that any ordinary line must consist of p and one of the remaining $n/7$ points of X. This, however, yields too few ordinary lines for $X \cup p$.

Conjecture 8.9. Any noncollinear set X of n points determines at least $n/4$ slopes of ordinary lines.

Configurations with an odd number of points which are close to this conjectured minimum may be obtained from the regular arrangement $A_1(2k)$ consisting of the k edges and k lines of symmetry of the regular k-gon $P(k)$. One must delete an appropriate line, take the dual to obtain a point configuration, and then choose a suitable affine embedding. The only configuration presently known that exactly realizes the conjectured bound is obtained in a similar way from the Crowe-McKee configuration $A_4(13)$. (See Figure 2.9 of [15].) A slight modification of this arrangement involving an additional pseudoline leads to an allowable sequence with $n = 13$ and only 3 moves with ordinary lines. Thus Conjecture 8.9 fails for sequences.

It is also interesting to note that most slope-critical configurations are well above the conjectured minimum of ordinary slopes.

Acknowledgments

I wish to express my appreciation to Jacob Goodman for his suggestion to write this report, to Branko Grünbaum for his encouragement, to J. C. Fisher and Don Chakerian for helpful correspondence on regular polygons, and to Merwan S. Irani, without whom all would have been impossible.

References

1. Basterfield, J. G. & L. M. Kelly. 1968. A characterization of sets of n points which determine n hyperplanes. Proc. Cambridge Philos. Soc. **64:** 585–588.
2. Burton, G. R. & G. B. Purdy. 1979. The directions determined by n points in the plane. J. London Math. Soc. (2) **20:** 109–114.
3. de Bruijn, N. G. & P. Erdös. 1948. On a combinatorial problem. Nederl. Akad. Wetensch. Proc. **51:** 1277–1279 (Indag. Math. **10:** 421–423).
4. Coxeter, H. S. M. 1969. Affinely regular polygons. Abh. Math. Sem. Univ. Hamburg **34:** 38–58.
5. Cordovil, R. 1982. Sur les matroides orientes de rang 3 et les arrangements de pseudodroites dans le plan projectif reel. Eur. J. Comb. **3:** 307–318.
6. Cordovil, R. 1983. The directions determined by n points in the plane, a matroidal generalization. Discrete Math. **43:** 131–137.
7. Dirac, G. A. 1951. Collinearity properties of sets of points. Quart. J. Math. Oxford Ser. (2) **2:** 221–227.
8. Erdös, P. 1944. Solution to problem Nr. 4065. Amer. Math. Monthly **51:** 169.
9. Erdös, P. & G. Purdy. 1978. Some combinatorial problems in the plane. J. Combin. Theory Ser. A **25:** 205–210.
10. Folkman, J. & J. Lawrence. 1978. Oriented matroids. J. Combin. Theory Ser. B **25:** 199–236.
11. Goodman, J. E. & R. Pollack. 1980. On the combinatorial classification of nondegenerate configurations in the plane. J. Combin. Theory Ser. A **29:** 220–235.
12. Goodman, J. E. & R. Pollack. 1981. A combinatorial perspective on some problems in geometry. Congressus Numerantium **32:** 383–394. (*In* Proceedings, XII SE CGTC Conference.)
13. Goodman, J. E. & R. Pollack. Semispaces of configurations, cell complexes of arrangements. J. Combin. Theory Ser. A. In press.
14. Grünbaum, B. 1967. Convex Polytopes. John Wiley & Sons, Inc. London, England.
15. Grünbaum, B. 1972. Arrangements and Spreads. CBMS Regional Conference Series in Mathematics No. 10. American Mathematical Society. Providence, R.I.
16. Jamison, R. E. 1982. Planar Configurations Which Determine Few Slopes. Clemson Technical Report No. 397. (Geom. Dedicata. In press.)

17. JAMISON, R. E. 1983. Structure of Slope-Critical Configurations. Clemson Technical Report No. 427. (Geom. Dedicata. In press.)
18. JAMISON, R. E. 1984. Modular Lines in Complex Configurations. Clemson Technical Report No. 444.
19. JAMISON, R. E. Greedy sequences and critical configurations. In preparation.
20. JAMISON, R. E. & D. HILL. 1983. A catalogue of sporadic slope-critical configurations. Congressus Numerantium **40:** 101–125. (*In* Proceedings, XIV SE CGTC Conference.)
21. KELLY, L. M. & W. O. J. MOSER. 1958. On the number of ordinary lines determined by n points. Canad. J. Math. **10:** 210–219.
22. LEHMER, D. H. 1933. A note on trigonometric algebraic numbers. Amer. Math. Monthly **40:** 165–166.
23. LEHMER, D. N. 1900. Asymptotic evaluation of certain totient sums. Amer. J. Math. **22:** 293–335.
24. NIZETTE, N. 1977. Les polygones reguliers des plans affins arguesiens. Acad. Roy. Belg. Bull. Cl. Sci. (5) **63** (No. 11): 844–851.
25. ROBINSON, A. 1956. Complete Theories. North-Holland. Amsterdam, the Netherlands.
26. SCOTT, P. R. 1970. On the sets of directions determined by n-points. Amer. Math. Monthly **77:** 502–505.
27. SHANNON, R. W. 1974. Certain Extremal Problems in Arrangements of Hyperplanes. Doctoral Dissertation. University of Washington. Seattle, Wash.
28. UNGAR, P. 1982. $2N$ noncollinear points determine at least $2N$ directions. J. Combin. Theory Ser. A **33:** 343–347.

PROBLEMS ON EXTREMAL PROPERTIES OF A FINITE SET OF POINTS

W. O. J. Moser
Department of Mathematics
McGill University
Montreal, Quebec, Canada H3A 2K6

In 1977 I started a collection of problems in discrete geometry, distributing it and the subsequent revised versions to anyone who requested a copy. The most recent edition, "Research Problems in Discrete Geometry, 1981" (RPDG 1981) contains 68 problems; here I focus on problems from this collection that investigate extremal properties of a finite set of points. A revised version of each problem is given here to include whatever new information has become available. The designation number (#*n*) locates the problem in RPDG 1981.

Problem 1: (#1) Different Distances Determined by a Finite Set of Points

Let $\mathscr{P}$ be a finite set of points in the plane. We use the notation $(\mathscr{P}: n)$ to indicate that the cardinality of $\mathscr{P}$ is $|\mathscr{P}| = n$. Let $f(\mathscr{P}: n)$ denote the number of different distances determined by pairs of points in $(\mathscr{P}: n)$ and

$$f(n) = \min_{(\mathscr{P}: n)} f(\mathscr{P}: n).$$

Erdös [1] proved

$$\sqrt{n-1} - 1 < f(n) < cn/\sqrt{\log n} \tag{1}$$

(the upper bound is given by lattice points), L. Moser [3] improved the lower bound to

$$\frac{n^{2/3}}{2(9)^{1/3}} - 1 < f(n)$$

and recently Chung further improved this to

$$cn^{5/7} < f(n).$$

Erdös [2] conjectured: "It seems almost certain that $f(n) > n^{1-\varepsilon}$ for every $\varepsilon > 0$ if $n > n_0(\varepsilon)$ and in fact probably $f(n) > c_1 n\sqrt{\log n}$."

Let $d(p_i)$ be the number of distinct distances from p_i. Moser in fact proved that

$$\max_i d(p_i) \geq \frac{n^{2/3}}{2(9)^{1/3}} - 1.$$

Erdös showed that

$$\sum d(p_i) > n^{3/2}/2$$

and conjectured that

$$\sum d(p_i) > c_2 n^2/\sqrt{\log n}$$

which is stronger than (**1**).

For the corresponding problem in m dimensions Bannai and Bannai [63] recently proved

$$cn^{1/m} < f_m(n), \qquad c \text{ depends on } m.$$

Problem 2: (#2) More on Different Distances

Variations of Problem 1 arise when conditions are imposed on $\mathscr{P}$. Let $(\mathscr{P} : n, X)$ denote a set of n points satisfying condition X, e.g., X may be

- A: the points are vertices of a strictly convex polygon;
- B_k: no k of the points lie on a line;
- C: no three of the points are on a line and no four are on a circle;
- $D_{l,m}$: every l of the points determine at least m different distances.

Moser [3] proved $f(\mathscr{P} : n, A) \geq [(n+2)/3]$ and Altman [4, 5] proved $f(\mathscr{P} : n, A) \geq [n/2]$ (equality occurs when the points are vertices of a regular n-gon). Erdös [2] conjectured that any $(\mathscr{P} : n, A)$ has a point from which at least $[n/2]$ distinct distances are determined.

Szemerédi has shown that any $(\mathscr{P} : n, B_k)$ has a point from which $\varepsilon_k n$ different distances are determined (see Erdös [2] for the proof); his method can be used to show that for any $(\mathscr{P} : n, B_3)$

$$\max d(p_i) \geq [n/3],$$

and he has conjectured that for all $(\mathscr{P} : n, B_3)$,

$$f(\mathscr{P} : n, B_3) \geq [n/2].$$

Let $f(n : X)$ denote min $f(\mathscr{P} : n, X)$ taken over all $(\mathscr{P} : n, X)$. Erdös asked: Is it true that

$$\lim_{n\to\infty} f(n : C)/n = \infty ?$$

Hamburger and Ruzsa showed that $f(6 : D_{3,3}) \geq 6$, i.e., any six points, no three vertices of an isosceles triangle, determine at least six distinct distances. But bounds for $f(n : D_{3,2})$ (no equilateral triangle) or for $f(n : D_{4,5})$, etc., are not known.

Another variation is the following open problem of Erdös [2]: For each n, what is the largest number k such that from any n points in the plane one can select k of them with all mutual distances distinct?

If the points are restricted to lattice points, some results are known. Thus, Erdös asked: For each n, what is the largest number $k = k(n)$ of lattice points

$$(x_i, y_i), \qquad i = 1, 2, \ldots, k,\ 1 \leq x_i, y_i \leq n,\ x_i, y_i \text{ integers} \tag{1}$$

which can be chosen with all mutual distances distinct? Erdös and Guy [7] showed that

$$n^{2/3-\varepsilon} < k < cn/(\log n)^{1/4}$$

(for any $\varepsilon > 0$ and sufficiently large n). They also asked for configurations containing a minimum number of the lattice points in (**1**) determining distinct distances so that no point (with both coordinates in $\{1, 2, \ldots, n\}$) may be added without duplicating a distance. Can this be done with as few as $O(n^{1/2})$ points in two dimensions or with $O(n^{1/3})$ points in one dimension?

Purdy has raised the following problem. Let L_1 and L_2 be two intersecting lines in the plane, each containing n points, say $x_1, \ldots, x_n \in L_1, y_1, \ldots, y_n \in L_2$. Suppose the number of different distances $d(x_i, y_j)$ is $O(n)$. Does this imply that L_1 and L_2 are perpendicular? (In such a case, we can choose $x_k = (0, \sqrt{k})$ and $y_k = (\sqrt{k}, 0)$, for example.) What if we just assume the number is $O(n^{2-\varepsilon})$ for some $\varepsilon > 0$? It is not enough just to assume $o(n^2)$ as the example with L_1 the x-axis, L_2 the line $y = x$, and $x_k = (k, 0)$, $y_k = (k, k)$, $1 \leq k \leq n$, shows. In this case, the number of distances is only $O(n^2/\sqrt{\log n})$.

Problem 3: (#3) Repeated Distances

Again, with $\mathscr{P}$ denoting a finite set of points, let $g(\mathscr{P})$ be the number of unit distances determined by the pairs of points and

$$g(n) = \max_{|\mathscr{P}|=n} g(\mathscr{P}).$$

Erdös asked for bounds on $g(n)$. From (**1**) of Problem 1 it follows that $g(n) > cn\sqrt{\log n}$. Erdös [1] proved

$$n^{1+c/\log \log n} \leq g(n) \leq n^{3/2},$$

Szemerédi improved the upper bound to $o(n^{3/2})$, and Beck and Spencer [8] established

$$g(n) \leq n^{1.454\ldots}.$$

Given n points at the vertices of a strictly convex polygon, is it true that no distance can occur more than $5(n-1)/3$ times? (If $n \equiv 1 \pmod 3$ there is an n-gon in which some distance is determined $5(n-1)/3$ times.) This is a problem of Erdös and Moser [9].

L. Moser ([11]; see also Guy [10, p. 242]) conjectured that for any n points on a sphere the same distance can occur at most $2n - 2$ times.

For given $k \geq 3$ does there exist a convex n_k-gon so that every vertex has k other vertices equidistant from it? L. Danzer has an example that shows that the answer is yes when $k = 3$. If the answer is yes for some k, it would be interesting to find estimates for the smallest n_k.

Now let $p_i = (u_i, v_i)$ and

$$d(p_i, p_j) = \max(|u_i - u_j|, |v_i - v_j|),$$

i.e., the "distance" is not $[(u_i - u_j)^2 + (v_i - v_j)^2]^{1/2}$ but the maximum difference of the coordinates. Ulam asked: For how many pairs $\{i, j\}$ can $d(p_i, p_j) = 1$? Erdös proved that this maximum is $(n^2 + n)/4$ for $n > n_0$, $n \equiv 0 \pmod 4$.

Problem 4: Repeated Maximum Distance

Suppose that the finite set $\mathscr{P}$ of points in k dimensions has diameter 1 (i.e., the maximum distance between pairs of points of $\mathscr{P}$ is 1). Let $M_k(\mathscr{P})$ denote the number of pairs of points of $\mathscr{P}$ determining distance 1, and $M_k(n)$ the maximum of $M_k(\mathscr{P})$ taken over all $|\mathscr{P}| = n$. The problem is to find or estimate $M_k(n)$.

Of course, $M_1(n) = 1$.

Pannwitz (see Hopf and Pannwitz [14], Sutherland [16]) and Vázsonyi (see Erdös [1]) conjectured that $M_3(2n) = 2n - 2$ and this was proved independently by Grünbaum [12], Heppes [13], and Straszewicz [15]. Lenz observed that $M_4(n) \geq [n^2/4]$ and Erdös [2] proved

$$\lim_{n\to\infty} M_k(n)/n^2 = \frac{1}{2} - \frac{1}{2[k/2]}.$$

Problem 5: (#5) Repeated Minimum Distance

Suppose that the set $\mathscr{P}$ of n points in k dimensions determines the smallest distance 1. Let $m_k(\mathscr{P})$ denote the number of pairs of points in $\mathscr{P}$ determining distance 1, and $m_k(n)$ the maximum of $m_k(\mathscr{P})$ taken over all such $\mathscr{P}$.

Clearly, $m_1(n) = n - 1$, and $m_2(n) < 3n$ (because from every point there are at most six points at distance 1). Erdös [1] proved $m_2(n) \leq 3n - 6$. Reutter [18] asked if $m_2(n) = [3n - \sqrt{12n - 3}]$, and this was proved by Harborth [17].

Estimate $m_k(n)$ for $k \geq 3$.

Problem 6: (#6) On the Minimum Distance Determined by n Points in a Square (Circle, Rectangle, Cube)

Let $\mathscr{S}$ denote a finite set of points in the unit square, $f(\mathscr{S})$ the minimum distance between pairs of points of $\mathscr{S}$, and $f(n) = \max_{|\mathscr{S}|=n} f(\mathscr{S})$. L. Moser [25] asked for (exact) values of $f(n)$ for small n. The values of $f(n)$ are known for $n \leq 9$.

n	2	3	4	5	6	7	8	9
$f(n)$	$\sqrt{2}$	$\sqrt{6} - \sqrt{2}$	1	$\sqrt{2}/2$	$\sqrt{13}/6$	$2(2 - \sqrt{3})$	$(6 - \sqrt{2})/2$	1/2

The cases $n = 2, 3, 4, 5$ are easy (in fact, $n = 3$ and $n = 5$ appeared as Putnam problems); the solution for $n = 6$ was found by Graham, that for $n = 7$ by Schaer, that for $n = 8$ by Schaer and Meir [33], and that for $n = 9$ by Schaer [28].

For $n \geq 10$ exact values of $f(n)$ are not known.

Lower bounds (and the corresponding configurations) are given by Goldberg [21] for $10 \leq n \leq 27$ and for some larger values of n. His bound $f(10) \geq 5/12 = 0.4166\ldots$ was improved to $f(10) \geq (4 + \sqrt{2}) \cdot (\sqrt{8^{1/2} + 1} - \sqrt{2})/7 = 0.41954\ldots$ by Schaer [32] and independently by K. Schuter (written communication, 1977). The latter also has obtained the bound $f(13) \geq (\sqrt{3} - 1)/2 = 0.3660\ldots$ which improves Goldberg's $f(13) \geq \sqrt{2}/4 = 0.3535\ldots$.

The similar problem for a circle (instead of a square) has been investigated by Pirl [26]. Let $\mathscr{S}$ denote a finite set of points inside or on the unit circle, $g(\mathscr{S})$ the minimum distance determined by pairs of points of $\mathscr{S}$, and $g(n) = \max_{|\mathscr{S}|=n} g(\mathscr{S})$.

Pirl proved that

$$g(n) = 2 \sin \frac{\pi}{n} \qquad \text{for } 2 \le n \le 6,$$

$$g(n) = 2 \sin \frac{\pi}{n-1} \qquad \text{for } 7 \le n \le 9;$$

he also obtained $g(10)$ and has some results and conjectures for $11 \le n \le 17$. (See also Kravitz [24] and Goldberg [22].)

For the similar problem of packing circles on the surface of the sphere, see Goldberg [19, 20].

Ruda [27] proved that if n nonoverlapping congruent circles are in a rectangle of area A, then for $n \le 8$ the area covered by the circles is at most $\frac{1}{4}\pi A$. Ruda conjectured that equality holds for $n = 9$, 10 and perhaps $n = 12$; he showed that it fails for $n = 11$ and all $n \ge 14$.

The similar problem of n points in a unit cube has been investigated and solved for $n = 2, 3, \ldots, 9$ by Schaer [29–32]. Partial results for $n \ge 10$ are given by Goldberg [23].

Problem 7: (#7) Heilbronn's Triangle Problem

Let $\mathscr{S} = \{p_1, p_2, \ldots, p_n\}$ be a set of n points in a closed disk of unit area, $\Delta(\mathscr{S})$ the area of the smallest triangle $p_i p_j p_k$, and

$$\Delta(n) = \max_{|\mathscr{S}|=n} \Delta(\mathscr{S}).$$

The problem is to find bounds on $\Delta(n)$. Heilbronn (circa 1950) conjectured $\Delta(n) \ll 1/n^2$. ($X \ll Y$ denotes $X \le cY$ with c an absolute constant.) By selecting $n - 2$ nonoverlapping triangles $p_i p_j p_k$ we obtain $\Delta(n) \ll 1/n$.

The following have been proven:

$\Delta(n) \ll 1/n\sqrt{\log \log n}$, Roth [37];

$\Delta(n) \ll 1/n\sqrt{\log n}$, Schmidt [42];

$\Delta(n) \ll 1/n^{\mu}$, Roth [38], where $\mu = 2 - (2/\sqrt{5}) - \varepsilon = 1.105\ldots - \varepsilon$;

$\Delta(\mu) \ll 1/n^{\mu}$, Roth [39], where $\mu = (17 - \sqrt{65})/8 - \varepsilon = 1.117\ldots - \varepsilon$;

(see Roth [40] for simplified versions of the proofs; Roth [41] is a fine survey);

$\Delta(n) \ll 1/n^{\mu}$, Komlós *et al.* [35], where $\mu = 8/7 - \varepsilon = 1.142\ldots - \varepsilon$.

On the other side, Erdös (see Appendix in Roth [37]) proved

$$1/n^2 \ll \Delta(n),$$

and Komlós *et al.* [36] established

$$(\log n)/n^2 \ll \Delta(n),$$

thus disproving Heilbronn's conjecture.

For small n, the equivalent problem for n points in a square is investigated by Goldberg [34].

Problem 8: (#8) The Largest Convex Region Not Containing Any One of n Points in a Square

Let $\mathscr{S}$ be a finite set of points in the unit square, $f(\mathscr{S})$ the area of the largest convex region (in the square) not containing (in its interior) any point of $\mathscr{S}$, and

$$f(n) = \min_{|\mathscr{S}|=n} f(\mathscr{S}).$$

Find $f(n)$ for small n, and bounds or asymptotic values for large n.

Chung has proven

$$1/n \ll f(n) \ll \sqrt{\log n}/n^{3/4}$$

and has obtained $f(4) = 1/3$,

$$f(3) = \omega = 1/2 - a(1 - \sqrt{1 - 2a - a^2}) = 0.3784132\ldots,$$

where a is one root of a certain polynomial of degree 12.

Similarly, consider $f_r(n)$, the corresponding number in r dimensions. Komlós conjectured that for large n, $f_r(n) \sim cn^{-2/(r+1)}$, a result that would have applications in probability theory.

The similar problem for the disk may be investigated.

Problem 9: (#10) Directions Determined by n Points

Let $\mathscr{P}$ be a finite set of noncollinear points in r dimensions and $k_r(\mathscr{P})$ the number of different directions realized by the nonoriented lines determined by $\mathscr{P}$. Scott [45] asked for

$$k(n) = k_2(n) = \min_{|\mathscr{P}|=n} k_2(\mathscr{P});$$

he showed that

$$(1 + \sqrt{8n - 7})/2 \le k(n) \le 2[n/2]$$

and conjectured $k(n) = 2[n/2]$. Burton and Purdy [43] obtained $[n/2] \le k(n)$, Corrovil [44] proved $[n/2] + 2 \le k(n)$, and Ungar [46] settled the problem completely.

In three dimensions (Scott [45]),

$$\sqrt{6n - 6} \le k_3(n) \le 2n - 2.$$

In the plane, if $\mathscr{P}$ is "convex" (no point of $\mathscr{P}$ lies inside the convex hull of $\mathscr{P}$), then the least number of different directions is $k^*(n) = n$ (Scott [45]).

In three dimensions, if $\mathscr{P}$ is strictly convex, the least number $k_3^*(n)$ of different directions satisfies (Scott [45])

$$n + 2 \leq k_3^*(n) \leq 2n - 2.$$

A related open problem of Grünbaum asks for the least number of vectors that a set of n points determines.

Problem 10: (#62) Different Triangular Areas Determined by n Points

Let $\mathscr{P}$ be a finite set of noncollinear points in the plane, $g(\mathscr{P})$ the number of distinct positive triangular areas with vertices in $\mathscr{P}$, and

$$g(n) = \min_{|\mathscr{P}|=n} g(\mathscr{P}).$$

Erdös and Purdy [47] proved $g(n) \geq cn^{3/4}$; Burton and Purdy [43], using the bounds for $k(n)$ in Problem 9, proved $g(n) > 0.32n - 1$. Erdös *et al.* [48] showed $g(n) \leq [(n-1)/2]$. In fact, they showed that for the corresponding number in r dimensions (distinct volumes of nondegenerate simplices)

$$g_r(n) \leq [(n-1)/r],$$

and conjectured that equality holds for sufficiently large n.

Let T be the maximal area and t the minimal area of nondegenerate triangles with vertices in $\mathscr{P}, f(\mathscr{P}) = T/t$, and

$$f(n) = \inf_{|\mathscr{P}|=n} f(\mathscr{P}).$$

Erdös and Purdy asked (see Straus [49]) for bounds on $f(n)$. Erdös *et al.* [48] proved

$$f(n) = [(n-1)/2], \qquad n > 37,$$

and conjectured that the equality holds for $n > 5$.

For the corresponding number in r dimensions (inf ratio max and min volumes of nondegenerate simplices), they showed that

$$f_r(n) \leq [(n-1)/r]$$

[by taking equally spaced points on parallel lines through the vertices of an $(r-1)$-simplex] and conjectured that equality holds for sufficiently large n.

Problem 11: (#11) Large Subsets of Points Determining No Isosceles Triangle

Prove that from any set of n points in the plane one can always choose $n^{1-\varepsilon}$ of them which determine no isosceles triangle. Note the relation to Roth–Szemerédi [50, 51] results on arithmetic progressions in integers.

PROBLEM 12: (#18) VISIBILITY PROBLEMS

Let $\mathscr{S}$ be a finite set of noncollinear points in the plane. For points $p, q \in \mathscr{S}$ we say "p is visible from q" if there is no point of $\mathscr{S}$ in the open segment between p and q.

(a) Let $f(\mathscr{S}, p)$ denote the number of points in $\mathscr{S}$ visible from p, $f(\mathscr{S}) = \max_{p \in \mathscr{S}} f(\mathscr{S}, p)$, and

$$f(n) = \min_{|\mathscr{S}|=n} f(\mathscr{S}).$$

Dirac [53] proved $\sqrt{n} < f(n) \leq [n/2]$ and conjectured $f(n) = [n/2]$.

[**Editors' note:** E. Szemerédi and W. T. Trotter have recently succeeded in proving $f(n) \geq cn$. On the other hand, it has been known for some time that Dirac's conjecture fails (strictly speaking) for low values of n. Hence, what remains is to prove (or disprove) $f(n) \geq n/2 - c'$.]

Let $\mathscr{S}$ and $\mathscr{T}$ be disjoint sets of points. We say "$\mathscr{S}$ is visible from $\mathscr{T}$" if for every $q \in \mathscr{S}$ there is a $p \in \mathscr{T}$ such that q is visible from p.

(b) Let

$$h(\mathscr{S}) = \min_{\mathscr{T} \subseteq \mathscr{S}} \{|\mathscr{T}|: \mathscr{T} \subseteq \mathscr{S}, \mathscr{S} - \mathscr{T} \text{ is visible from } \mathscr{T}\}$$

and

$$h(n) = \max_{|\mathscr{S}|=n} h(\mathscr{S}).$$

L. Moser [11, Problem 12] conjectured $h(n) = O(\log n)$.

(c) Let $\mathscr{S}_n = \{(i, j): i, j = 1, 2, \ldots, n\}$ be the set of n^2 lattice points. L. Moser [11, Problem 47] asked for $h(\mathscr{S}_n)$, and conjectured $h(\mathscr{S}_n) = O(\log n)$. Abbott [52] proved

$$\frac{\log n}{2 \log \log n} < h(\mathscr{S}_n) < 4 \log n.$$

(d) Let

$$k(\mathscr{S}) = \min \{|\mathscr{T}|: \mathscr{T} \subseteq \mathscr{S}; \text{ for some } p \in \mathscr{S} - \mathscr{T},$$
$$\mathscr{S} - \mathscr{T} - \{p\} \text{ is visible from } \{p\}\}$$

and

$$k(n) = \max_{|\mathscr{S}|=n} k(\mathscr{S}).$$

M. Cohen suggested this problem [to find $k(n)$]. For the set $\mathscr{S}$ of $[n/2]$ points on a line and $n - [n/2]$ points on a parallel line,

$$k(\mathscr{S}) = [n/2] - 3$$

and hence $k(n) \geq [n/2] - 3$. Cohen conjectured that equality holds for all $n \geq 6$.

PROBLEM 13: (#29) THE SMALLEST NUMBER OF POINTS ALWAYS DETERMINING A CONVEX n-GON

E. Klein showed (see Erdös and Szekeres [55]) that from any five points (in the plane), no three collinear, it is always possible to select a (strictly) convex subset of four of the points. She suggested the following problem. Let $f(n)$, $n \geq 3$, be the smallest integer such that from any set of $f(n)$ points (in the plane), no three collinear, one can always choose a convex subset of $f(n)$ points. The Klein result is $f(4) = 5$. It is known (see Kalbfleisch *et al.* [58]) that $f(5) = 9$. Erdös and Szekeres [55] conjectured $f(n) = 2^{n-2} + 1$ and constructed an example (Erdös and Szekeres [56]) that showed $f(n) \geq 2^{n-2} + 1$.

For related results see Chvatal and Komlós [54] and Steele [59].

A related question asked by Erdös was: For $n \geq 3$ what is the smallest integer $g(n)$ such that from any set $\mathscr{S}$ of $g(n)$ points, no three collinear, one can always choose a convex subset of n of the points whose interior (i.e., the interior of the convex cover of the subset) contains no point of $\mathscr{S}$? $g(3) = 3$ is trivial and $g(4) = 5$ is easy. Harborth [57] showed $g(5) = 10$. It is not known whether $g(6)$ exists! However, $g(7)$ does not exist!! (Horton [64].)

PROBLEM 14: (#41) MAXIMUM NUMBER OF k-LINES DETERMINED BY n POINTS

Let $\mathscr{P}$ denote a finite set of points in the plane, $t_k(\mathscr{P})$ the number of lines each containing precisely k points of $\mathscr{P}$, and

$$t_k(n) = \max_{|\mathscr{P}|=n} t_k(\mathscr{P}).$$

$t_3(n)$ has been studied for more than 150 years. Burr *et al.* [60] proved

$$1 + [n(n-3)/6] \leq t_3(n) \leq \frac{1}{3}\left(\binom{n}{2} - \lceil 3n/7 \rceil\right)$$

($\lceil x \rceil$ denotes the least integer not less than x) and conjectured

$$t_3(n) = 1 + [n(n-3)/6], \qquad n \neq 7, 11, 16, 19.$$

Erdös and Croft proved

$$t_k(n) > c_k n^2, \qquad c_k \text{ an absolute constant.}$$

PROBLEM 15: (#44) ANGLES DETERMINED BY A FINITE SET OF POINTS

Let $\mathscr{S}$ be a set of n points in the plane, no three collinear and thus determining $N = n(n-1)(n-2)/2$ angles. For any $0 < \alpha < \pi$, let $f(n, \alpha, \mathscr{S})$, $g(n, \alpha, \mathscr{S})$ be the number of these angles $>\alpha$, $<\alpha$, respectively. Let

$$f(n, \alpha) = \min_{\mathscr{S}} f(n, \alpha, \mathscr{S}), \qquad g(n, \alpha) = \min_{\mathscr{S}} g(n, \alpha, \mathscr{S})$$

and

$$F(\alpha) = \lim_{n\to\infty} \frac{f(n, \alpha)}{N}, \qquad G(\alpha) = \lim_{n\to\infty} \frac{g(n, \alpha)}{N}.$$

These functions (and related ones) were studied by Conway *et al.* [61], who showed that: the limits exist; $F(\alpha)$ is increasing; $G(\alpha)$ is decreasing; $F(\alpha) = 1/3$, $0 < \alpha \leq \pi/3$; $G(\alpha) = 2/3$, $\alpha > \pi/2$; $1/9 < F(\pi/2) \leq 4/27$; F is discontinuous at $\pi/3$.

The problem is how to obtain more information about these functions. For example, are they discontinuous for other α or for infinitely many values of α? Find better bounds for $F(\pi/2)$, etc.

Problem 16: (#51) Minimum Spanning Trees

Given n points in a unit square, there is a polygon through them of length $\leq\sqrt{2}n^{1/2} + 7/4$. This is a result of K. Few. Improve the constant $\sqrt{2}$.

Graham commented: Few also studied minimum spanning trees and Steiner minimal trees for these sets. For Steiner minimal trees on n points in a unit square, he proved that the maximum length possible, say $s(n)$, satisfies $s(n) \leq \sqrt{n} + 7/4$. This was improved by Chung and Graham [62] to $s(n) \leq 0.995\sqrt{n} + O(1)$.

Problem 17: (#52) Longest Edge-Paths on a Convex Polyhedron with n Vertices

What is the largest $f(n)$ such that every convex polyhedron with n vertices has a simple path along edges passing through $f(n)$ vertices? Moon and Moser have shown that $f(n) < cn^{(\log 2)/\log 3}$. (This problem appears in L. Moser [11] and Guy [10].)

Problem 18: (#61) Packing Points in E^n

Given m points in E^n, the distance between every pair is at least 2. Let $r_1, r_2, \ldots, r_m$ be distances of these points from the origin. Blichfeldt showed that $\sum r_i^2 \geq 2(m-1)$. Can this inequality be strengthened in some way? The above inequality is best possible for $n = 2, 3, 4$. Find the corresponding best possible estimates for the next few values of n.

Problem 19: (#67) No Midpoints in the (Sub)set

Estimate the largest integer $k(n)$ such that from any set $\mathscr{S}$ of n points in the plane one can always choose a subset $\mathscr{A}$ of size $k(n)$ so that no point in $\mathscr{A}$ is the midpoint of two points in $\mathscr{A}$. Similar problems arise with "two points in $\mathscr{A}$" replaced by "two points in $\mathscr{S}$," or "two points in $\mathscr{S} - \mathscr{A}$."

REFERENCES

1. ERDÖS, P. 1946. On sets of distances of n points. Amer. Math. Monthly **53:** 248–250. MR **7:** 471. (This paper is also in: ERDÖS, P. 1973. The Art of Counting. MIT Press. Cambridge, Mass.)
2. ERDÖS, P. 1975. On some problems of elementary and combinatorial geometry. Ann. Mat. Pura Appl. Ser. IV **103:** 99–108. MR **54** #113.
3. MOSER, L. 1952. On different distances determined by n points. Amer. Math. Monthly **59:** 85–91. MR **13:** 768.
4. ALTMAN, E. 1963. On a problem of P. Erdös. Amer. Math. Monthly **70:** 148–157. MR **27** #4142.
5. ALTMAN, E. 1972. Some theorems on convex polygons. Canad. Math. Bull. **15:** 329–340. MR **47** #2495.
6. ERDÖS, P. 1957. Néhány geometriai problémáról. Mat. Lapok **8:** 86–92. MR **20** #6056.
7. ERDÖS, P. & R. GUY. 1970. Distinct distances between lattice points. Elem. Math. **25:** 121–123. MR **43** #7407.
8. BECK, J. & J. SPENCER. Unit distances. Preprint, 1982.
9. ERDÖS, P. & L. MOSER. 1959. Problem 11. Canad. Math. Bull. **2:** 43.
10. GUY, R. K. 1974. Problems. *In* The Geometry of Metric and Linear Spaces. Pp. 233–244. Lecture Notes in Mathematics No. 490. Springer-Verlag. West Berlin, FRG. MR **53** #6419.
11. MOSER, L. 1966. Poorly formulated problems of combinatorial geometry. Mimeographed.
12. GRÜNBAUM, B. 1956. A proof of Vázsonyi's conjecture. Bull. Res. Counc. Isr. Sect. A **6:** 77–78. MR **19:** 304.
13. HEPPES, A. 1956. Beweis einer Vermutung von A. Vázsonyi. Acta Math. Acad. Sci. Hungar. **7:** 463–466. MR **19:** 304.
14. HOPF, H. & E. PANNWITZ. 1934. Problem 167. Jahresber. Deutsch. Math. Verein. **43:** 114.
15. STRASZEWICZ, S. 1957. Sur un problème geometrique de P. Erdös. Bull. Acad. Polon. Sci. Cl. 3 **5:** 39–40. MR **19:** 304.
16. SUTHERLAND, J. W. 1935. Problem 167. Jahresber. Deutsch. Math. Verein. **45:** 33.
17. HARBORTH, H. 1974. Solution to Problem 664A. Elem. Math. **29:** 14–15.
18. REUTTER, O. 1972. Problem 664A. Elem. Math. **27:** 19.
19. GOLDBERG, M. 1967. Axially symmetric packing of equal circles on a sphere. Ann. Univ. Sci. Budapest, Eötvös Sect. Math. **10:** 37–48. MR **39** #7510.
20. GOLDBERG, M. 1969. Axially symmetric packing of equal circles on a sphere. Ann. Univ. Sci. Budapest, Eötvös Sect. Math. **12:** 137–142. MR **41** #9110.
21. GOLDBERG, M. 1970. The packing of equal circles in a square. Math. Mag. **43:** 24–30.
22. GOLDBERG, M. 1971. Packing of 14, 16, 17 and 20 circles in a circle. Math. Mag. **44:** 134–139.
23. GOLDBERG, M. 1971. On the densest packing of equal spheres in a cube. Math. Mag. **44:** 199–208. MR **45** #7614.
24. KRAVITZ, S. 1967. Packing cylinders into cylindrical containers. Math. Mag. **40:** 65–71.
25. MOSER, L. 1960. Problem 24. Canad. Math. Bull. **3:** 78.
26. PIRL, U. 1969. Der Mindesabstand von n in der Einheitskreisscheibe gelegenen Punkten. Math. Nachr. **40:** 111–124. MR **40** #6379.
27. RUDA, M. 1970. The packing of circles in rectangles. Magy. Tud. Akad. Mat. Fiz. Tud. Oszt. Közl. **19:** 73–87. (In Hungarian with an English summary.) MR **40** #6380.
28. SCHAER, J. 1965. The densest packing of 9 circles in a square. Canad. Math. Bull. **8:** 273–277. MR **31** #6164.
29. SCHAER, J. 1966. On the densest packing of spheres in a cube. Canad. Math. Bull. **9:** 265–270. MR **34** #683.
30. SCHAER, J. 1966. On the densest packing of five spheres in a cube. Canad. Math. Bull. **9:** 271–274. MR **34** #684.
31. SCHAER, J. 1966. The densest packing of six spheres in a cube. Canad. Math. Bull. **9:** 275–280. MR **34** #1929.
32. SCHAER, J. 1971. On the packing of ten equal circles in a square. Math. Mag. **44**: 139–140.
33. SCHAER, J. & A. MEIR. 1965. On a geometric extremum problem. Canad. Math. Bull. **8:** 21–27. MR **30** #5125.

34. GOLDBERG, M. 1972. Maximizing the smallest triangle made by points in a square. Math. Mag. **45:** 135–144. MR **45** #5875.
35. KOMLÓS, J., E. SZEMERÉDI & J. PINTZ. 1981. On Heilbronn's triangle problem. J. London Math. Soc. (2) **24:** 385–396.
36. KOMLÓS, J., E. SZEMERÉDI & J. PINTZ. 1982. A lower bound for Heilbronn's problem. J. London Math. Soc. (2) **25:** 13–24. ZM 483.52008.
37. ROTH, K. F. 1951. On a problem of Heilbronn. J. London Math. Soc. **26:** 198–204. MR **13:** 16.
38. ROTH, K. F. 1972. On a problem of Heilbronn, II. Proc. London Math. Soc. **25:** 193–212. MR **46** #3452.
39. ROTH, K. F. 1972. On a problem of Heilbronn, III. Proc. London Math. Soc. **25:** 543–549. MR **46** #3452.
40. ROTH, K. F. 1973. Estimation of the area of the smallest triangle obtained by selecting three out of n points in a disc of unit area. Amer. Math. Soc. Proc. Symp. Pure Math. **24:** 251–262.
41. ROTH, K. F. 1976. Developments in Heilbronn's triangle problem. Adv. in Math. **22:** 364–385.
42. SCHMIDT, W. M. 1971/1972. On a problem of Heilbronn. J. London Math. Soc. (2) **4:** 545–550. MR **45** #7612.
43. BURTON, G. R. & G. B. PURDY. 1979. The directions determined by n points in the plane. J. London Math. Soc. **20:** 109–114. MR 81d:52006.
44. CORDOVIL, R. The directions determined by n points in the plane. A matroidal generalization. Preprint, 1981.
45. SCOTT, P. R. 1970. On the sets of directions determined by n points. Amer. Math. Monthly **77:** 502–505. MR **41** #7538.
46. UNGAR, P. 1982. $2N$ noncollinear points determine at least $2N$ directions. J. Combin. Theory Ser. A **33:** 343–347.
47. ERDÖS, P. & G. B. PURDY. 1975. Some extremal problems in geometry IV. *In* Proceedings, 7th SE Conference on Combinatorics, Graph Theory and Computing, Baton Rouge. Pp. 307–322. Congress Numericum No. XIV. Utilitas Mathematica. Winnipeg, Manitoba, Canada. MR **43** #1045.
48. ERDÖS, P., G. B. PURDY & E. STRAUS. 1984. On a problem in combinatorial geometry. Discrete Math. In press.
49. STRAUS, E. G. 1978. Some extremal problems in combinatorial geometry. *In* Proceedings, International Conference on Combinatorial Mathematics, Canberra, 1977. Pp. 308–312. Lecture Notes in Mathematics No. 686. Springer-Verlag. West Berlin, FRG. ZM 407.05014.
50. SZEMERÉDI, E. 1974. On sets of integers containing no k elements in arithmetic progression. *In* Proceedings, ICM Vanc. **2:** 503–505. Canadian Mathematical Society. Ottawa, Ontario, Canada.
51. SZEMERÉDI, E. 1975. On sets of integers containing no k elements in arithmetic progression. Acta Arith. **27:** 199–245. MR **51** #5547.
52. ABBOTT, H. L. 1974. Some results in combinatorial geometry. Discrete Math. **9:** 199–204.
53. DIRAC, G. A. 1951. Collinearity properties of sets of points. Quart. J. Math. Oxford Ser. (2) **2:** 221–227. MR **13:** 270.
54. CHVATAL, V. & J. KOMLÓS. 1971. Some combinatorial theorems on monotonicity. Canad. Math. Bull. **14:** 151–157.
55. ERDÖS, P. & G. SZEKERES. 1935. A combinatorial problem in geometry. Compositio Math. **2:** 463–470. (This paper also appears in The Art of Counting; see Reference 1.)
56. ERDÖS, P. & G. SZEKERES. 1960/1961. On some extremum problems in elementary geometry. Ann. Univ. Sci. Budapest **3–4:** 53–62. MR **24 A** #3560D. (This paper also appears in The Art of Counting; see Reference 1.)
57. HARBORTH, H. 1978. Konvexe Fünfecke in ebenen Punktmengen. Elem. Math. **33:** 116–118. MR 80 a:52003.
58. KALBFLEISCH, J. D., J. G. KALBFLEISCH & R. G. STANTON. 1970. A combinatorial problem on convex n-gons. *In* Proceedings, Louisiana Conference on Combinatorics, Graph Theory and Computing, Baton Rouge. Pp. 180–188. MR **42** #8394.
59. STEELE, J. M. 1978. Existence of submatrices with all possible columns. J. Combin. Theory Ser. A **24:** 84–88. MR 80 f:05055.

60. BURR, S., B. GRÜNBAUM & N. J. A. SLOANE. 1974. The orchard problem. Geom. Dedicata **2:** 357–424.
61. CONWAY, J. H., H. T. CROFT, P. ERDÖS & M. J. T. GUY. 1979. On the distribution of values of angles by coplanar points. J. London Math. Soc. (2) **19:** 137–143. MR 80h:51021.
62. CHUNG, F. R. K. & R. GRAHAM. 1981. On Steiner trees for bounded points sets. Geom. Dedicata **11:** 353–361.
63. BANNAI, E. & E. BANNAI. 1981. An upper bound for the cardinality of an *s*-distance subset in real Euclidean space. Combinatorica **1:** 99–102.
64. HORTON, J. D. 1983. Sets with no empty convex 7-gons. Canad. Math. Bull. **26:** 482–484.

SOME COMBINATORIAL PROBLEMS IN THE PLANE

George B. Purdy

Mathematics Department
Texas A and M University
College Station, Texas 77843

1. Introduction

The subsets of a d-element set under set inclusion form a poset whose Hasse diagram consists of the edges of a d-dimensional cube. The projection of such a cube into three-dimensional space produces a zonohedron—see [5] for the definition of a zonohedron. Since there is a one-to-one correspondence between arrangements of lines in the plane and zonohedra (see [7]), it seems reasonable to look for a partial ordering on the regions generated by an arrangement of lines, since these correspond to the vertices of the zonohedron.

In this paper we define a partial ordering on the regions and discuss the size of its largest antichain under the assumption that not too many lines of the arrangement are parallel or concurrent (pass through the same point).

Edelman has also discussed this poset in the plane and higher dimensions [3] and the complicated relationship between the zonotopes and Hasse diagrams in some special cases.

2. Main Results

Given n lines in the Euclidean plane, they divide the plane up into a number of regions, some of which may be bounded, and $2n$ of which will be unbounded. Let us arbitrarily pick R_0 as one of the unbounded regions, and using it we will impose a partial order on the regions themselves in the following manner:

With each line L we associate a closed half space $H(L)$ which is determined uniquely if we say that L is its boundary and it does not contain R_0. This will define n half planes from the n different lines, and we define the rank $r(R)$ of a region R to be the number of these half planes that the region lies in. Thus, every region has a rank between zero and n.

The example of $n - 1$ lines passing through a point P and one line not passing through P provides (with appropriate choice of the rank zero region) a partial ordering that can be covered by three chains. The example of $n - k$ lines through P and $k + 1$ lines through Q (one line through both P and Q) gives a partial ordering that can be covered by $k + 2$ chains. We make the following conjecture:

Conjecture. Given n lines, with at most $n - k$ of the lines concurrent and at most $n - k$ parallel, the size of the largest antichain is at least $k + 2$.

Theorem 1. *If n lines are given and at most $n - k$ are concurrent or parallel, and if*

$$n \geq k(4k + 1) + 1, \qquad (*)$$

then the size S of the largest antichain satisfies

$$nS \geq (k + 1)(n - k).$$

Proof. Theorem 1 in Reference 8 asserts that given n lines in the real projective plane, with at most $n - k$ being concurrent, if (∗) holds, then at least $(k + 1)(n - k)$ regions are determined. When we remove the line at infinity, some of the $2n$ unbounded regions in the Euclidean plane may combine in pairs—the ones without parallel sides—but no new regions are created. Hence, under the hypothesis, there are at least $(k + 1)(n - k) - 1$ regions of positive rank, and S chains of length at most n covering them. The result follows. ▮

Although we are not able to prove the conjecture without the restriction (∗), we are able to prove the following weaker result:

THEOREM 2. *If at most $n - k$ lines are concurrent or parallel, then the size of the largest antichain in the ranked poset described above is at least ck, where c is a positive number independent of n and k.*

We will deduce THEOREM 2 from:

THEOREM 3. *Given n points in the plane, with at most $n - k$ on any line, if any two points are joined by a line, then there are at least $c'kn$ distinct lines.*

Remark. THEOREM 3 settles in the affirmative a conjecture of Erdös, stated in [4], of obvious independent interest. In our proof of THEOREM 3, we shall make use of a new result proved simultaneously by Szemerédi and Trotter [9, 10] and Beck [1]:

THEOREM 4 (Szemerédi, Trotter, and Beck). *Given n points in the plane, not all on a line, if any two points are joined by a line, then there is at least one point that is on at least $c''n$ different lines. Szemerédi and Trotter have*

$$c'' = 10^{-186}$$

and Beck has

$$c'' = 2^{-1000}.$$

THEOREM 4 settles in the affirmative the weak form of Dirac's conjecture stated, for example, in [6]. The original stronger form in Dirac's original paper [2] has $c'' = \frac{1}{2}$, and is false for some small values of n—see [6].

Proof of THEOREM 3. Let n points be given in the plane, and suppose that no more than $n - k$ lie on any line. By THEOREM 4, there is a point—call it X_n—that is on at least $c''n$ distinct lines, induced by the n points. Let $L_1, \ldots, L_t$ be all of the lines induced by the n points, and let the incidence matrix A be defined as follows:

$$a(i, j) = 1 \text{ if point } X_i \text{ lies on line } L_j \text{ and } a(i, j) = 0 \text{ otherwise.}$$

The removal of X_n from the system causes the matrix A to lose a whole row of ones, reducing the total number of ones in A by at least $c''n$, since there were at least that many ones in the nth row.

If we now apply THEOREM 4 to the set of $n - 1$ points, we see that there is a point—call it X_{n-1}—that lies on at least $c''(n - 1)$ lines, and whose removal causes the matrix A to lose at least that many ones. We keep removing points in this way, until all of the remaining points are collinear. Now suppose that we have removed r points in this way, so that at least

$$\begin{aligned} c''(n) + c''(n-1) + \cdots + c''(n-r+1) &= c''n(n+1)/2 - c''(n-r+1)(n-r)/2 \\ &= c''nr - c''r(r-1)/2 \end{aligned}$$

ones have been removed from the matrix, and suppose that the $n-r$ remaining points are all collinear. By the original hypothesis that at most $n-k$ points are collinear, we know that r must be at least k. The number of ones in the original matrix A is therefore at least $c''nk - c''k(k-1)/2 > c''kn/2$.

We now return to the original set of n points and let t_i be the number of lines incident with exactly i points for $i = 2, 3, \ldots$. (This is the dual of the t_i in the proof of THEOREM 2.) Then the number of ones in the matrix A can also be expressed as $2t_2 + 3t_3 + 4t_4 + \cdots$, and so this sum is at least $c''kn/2$. Now it is a well-known consequence of Euler's polyhedral formula (see, e.g., [6, p. 17]) that

$$t_2 = 3 + t_4 + 2t_5 + 3t_6 + \cdots + p_4 + 2p_5 + 3p_6 + \cdots,$$

where $p_4, p_5, \ldots$ are nonnegative numbers (that count the regions in the dual formulation). If we now add $2t_2 + 3t_3 + 3t_4 + 3t_5 + \cdots$ to both sides, we see that $3t = 3t_2 + 3t_3 + 3t_4 + \cdots > 2t_2 + 3t_3 + 4t_4 + 5t_5 + \cdots$, and so $3t > c''kn/2$, and the result follows. ∎

Proof of THEOREM 2. Suppose we are given n lines in the Euclidean plane, with at most $n-k$ being concurrent or parallel. Then we may also consider the lines to lie in the real projective plane, and, as in the proof of THEOREM 1, this can only cause us to underestimate the number of regions.

Another consequence of Euler's polyhedral formula applied to the projective plane is that

$$p_3 > p_5 + 2p_6 + 3p_7 + \cdots,$$

where p_i is the number of i-sided regions (in the projective plane)—see, e.g., [6, p. 32]. Adding $3p_3 + 4p_4 + 4p_5 + 4p_6 + \cdots$ to both sides yields

$$4F > 3p_3 + 4p_4 + 5p_5 + \cdots.$$

If t_i now denotes the number of intersection points of exactly i lines, so that the total number of intersection points is $t = t_2 + t_3 + t_4 + \cdots$, then by counting the number of line segments lying between intersection points in two different ways, we see that

$$4F > 3p_3 + 4p_4 + 5p_5 + \cdots = 2(2t_2 + 3t_3 + 4t_4 + \cdots) \geq 4t,$$

so that $F > t$. Hence, by the dual of THEOREM 3, the total number of regions in the projective setting F is at least $c''kn$, and the number of regions in the Euclidean setting is therefore at least that. Arguing as in the proof of THEOREM 1, we conclude that at least $c''k$ chains are needed to cover all of the elements, and the largest antichain has at least $c''k$ elements. ∎

3. SUMMARY

We define a partial ordering on the regions determined by n lines in the Euclidean plane and find a lower bound on the size of its largest antichain. We use as a tool a conjecture due to Erdös, which we prove using the weak version of Dirac's conjecture which has recently been proven by Beck, Szemerédi, and Trotter.

REFERENCES

1. BECK, J. On a problem of G. A. Dirac in combinatorial geometry. Unpublished manuscript.
2. DIRAC, G. A. 1951. Collinearity properties of sets of points. Quart. J. Math. **2:** 221–227.
3. EDELMAN, P. H. A partial order on the regions of R^n dissected by hyperplanes. Unpublished manuscript.
4. ELLIOTT, P. D. T. A. 1967. On the number of circles determined by n points. Acta Math. Acad. Sci. Hungar. **18:** 181–188.
5. GRÜNBAUM, B. 1967. Convex Polytopes. Interscience. London, England.
6. GRÜNBAUM, B. 1972. Arrangements and Spreads. American Mathematical Society. Providence, R.I.
7. KELLY, L. M. & W. O. J. MOSER. 1958. On the number of ordinary lines determined by n points. Canad. J. Math. **10:** 210–219.
8. PURDY, G. B. 1980. On the number of regions determined by n lines in the projective plane. Geom. Dedicata **9:** 107–109.
9. SZEMERÉDI, E. & W. T. TROTTER. A combinatorial distinction between the Euclidean and projective planes. Submitted for publication.
10. SZEMERÉDI, E. & W. T. TROTTER. Extremal problems in discrete geometry. Unpublished manuscript.

EXTREMAL ARRANGEMENTS OF HYPERPLANES[a]

Thomas Zaslavsky

Department of Mathematics
The Ohio State University
Columbus, Ohio 43210

Introduction

An *arrangement of hyperplanes* is a finite set of hyperplanes in real projective or Euclidean (affine) space, regarded as dissecting the space into *cells* (or *faces*) of various dimensions. The *flats* of the arrangement are the subspaces formed by intersecting the hyperplanes. I will discuss how by abstract methods of matroid theory one can find the extremal arrangements—those with the most or the fewest flats and cells given the number of hyperplanes and the dimension—and I will indicate how this abstract approach can and cannot solve some related geometrical questions.

Some Notation and Terminology

Real projective and affine spaces are $\mathbb{P}^d$ and $\mathbb{E}^d$. I consider the whole space a flat of any arrangement. The empty set is a flat (and a cell) only of arrangements that are projective and whose hyperplanes have intersection $\emptyset$. Flats with dimensions $d-3$ and $d-2$ I call *coplanes* and *colines*. Cells of dimensions 0 and d I call *vertices* and *regions*. For the numbers of k-dimensional flats, cells, and (in the Euclidean case) bounded cells, I write a_k, f_k, and b_k. For the set of all flats of an arrangement $\mathscr{H}$, ordered by reverse inclusion, I write $L(\mathscr{H})$.

To avoid getting false answers to minimization questions, it is necessary to exclude arrangements that belong in a lower dimension. A *d-arrangement* is an arrangement of hyperplanes in d-space that has no cross section by a proper subspace meeting all the cells except the vertices. For a projective arrangement that means the hyperplanes have empty intersection. For a Euclidean arrangement it means the arrangement has a vertex. As an example consider an arrangement $\mathscr{A}$ of $n \geq 3$ lines through a common point p. If $\mathscr{A}$ is in the projective plane, it is not a 2-arrangement, for any line not passing through p gives a faithful cross section. And $\mathscr{A}$ has $f_2(\mathscr{A}) = n$ regions, less than the minimum number $2(n-1)$ for a projective 2-arrangement. On the other hand, if $\mathscr{A}$ is in $\mathbb{E}^2$, it is a 2-arrangement; and it has $f_2(\mathscr{A}) = 2n$ regions, which is the minimum possible for a Euclidean 2-arrangement of n lines.

By adjoining the ideal hyperplane ∞, we turn any Euclidean arrangement $\mathscr{E}$ of n hyperplanes into an arrangement $\mathscr{E}_{\mathbb{P}}$ of $n+1$ projective hyperplanes. I call $\mathscr{E}_{\mathbb{P}}$ the *projectivization* of $\mathscr{E}$. Both $\mathscr{E}$ and $\mathscr{E}_{\mathbb{P}}$ have the same number of regions, although their other invariants a_k and f_k will differ.

One virtue of the abstract approach to arrangements of hyperplanes is that its results generalize to similar topological structures. Consider the following definitions

[a] Research partially supported by National Science Foundation Grant MCS-8003109 and SGPNR Grants.

of an arrangement of "topological hyperplanes." I will write $\cong$ for homeomorphism of topological spaces and of pairs of spaces. Let S denote a space homeomorphic to $\mathbb{P}^d$ or $\mathbb{E}^d$ and let h be a true hyperplane of $\mathbb{P}^d$ or $\mathbb{E}^d$ (with $h = \emptyset$ if $d = 0$). A *topological hyperplane* of S is a subset $x \subseteq S$ such that $(x, S) \cong (h, \mathbb{P}^d)$ or $(h, \mathbb{E}^d)$. A set $\mathscr{H}$ of topological hyperplanes induces a decomposition of S into *faces*, defined as the connected components of the equivalence classes of the relation on points of S given by $p \sim q$ if p and q lie in precisely the same members of $\mathscr{H}$.

A *projective topological arrangement* is a finite set $\mathscr{A}$ of topological hyperplanes of $\mathbb{P}^d$ satisfying (T1) and (T2′).

(T1) Every face of $\mathscr{A}$ is a topological cell.

(T2′) For each $\mathscr{A}' \subseteq \mathscr{A}$ and $h \in \mathscr{A} \backslash \mathscr{A}'$, setting $z' = \bigcap \mathscr{A}'$, we have either $h \supseteq z'$ or else $h \cap z'$ is a topological hyperplane of z'.

This definition is based on Lawrence's definition of arrangements of pseudo-hemispheres [15] and my study of topological dissections [32]. (See also Mandel's simplified definition [25].) It seems that the requirements can be weakened without admitting new examples (for instance, how necessary is (T1)?), but that is outside the scope of this paper.

Flats of $\mathscr{A}$, $L(\mathscr{A})$, a_k, and f_k are defined as for a straight projective arrangement. I call $\mathscr{A}$ a *d-arrangement* if $\bigcap \mathscr{A} = \emptyset$.

A *Euclidean topological arrangement* is harder to define in a way that is appropriate for our purposes. The weakest possible definition seems to be that it is a finite set $\mathscr{E}$ of topological hyperplanes of $\mathbb{E}^d$ satisfying (T1) and (T2). I will call this kind of arrangement *weak*.

(T2) For each $\mathscr{E}' \subseteq \mathscr{E}$ and $h \in \mathscr{E} \backslash \mathscr{E}'$, setting $z' = \bigcap \mathscr{E}'$, we have either $h \supseteq z'$ or $h \cap z' = \emptyset$ or else $h \cap z'$ is a topological hyperplane of z'.

Flats, $L(\mathscr{E})$, a_k, and f_k are defined as for straight arrangements. I call $\mathscr{E}$ a *d-arrangement* if it has a vertex.

We might forbid $h \cap z' = \emptyset$ (except when z' is a point) by requiring (T2′). I will call such an arrangement *parallelism-free*.

We might also impose the following requirement, which for a projective topological arrangement is redundant; I will call such an arrangement *strong*.

(T3) Whenever $h \not\supseteq z'$ and $h \cap z' \neq \emptyset$ in (T2), z' meets both of the open half spaces formed by h in $\mathbb{E}^d$.

The definitions of Euclidean topological arrangements are based on topological dissection theory [32] and are chosen to yield matroid-related enumeration formulas.

THE STOUTEST ARRANGEMENTS (PROLOGUE)

I will say an arrangement has *general intersection in dimension* $k + 1$ if every $d - k$ hyperplanes meet in a k-dimensional subspace; if this holds true for all $k \geq -1$, the arrangement has *general intersection*. Geometrically it is obvious that the arrangements with the most flats and cells are those having general intersection, and this has been the basis for many investigations of the maximum values of f_k and b_k going back to the nineteenth century [19]. The earliest complete evaluation I know of was carried out by Buck [7] (although he neglected to prove rigorously that the union of all bounded cells has Euler characteristic 1, an essential and not entirely trivial lemma). Rather than discuss any of the geometrical proofs, I will state the result and later explain how it reflects an abstract fact of matroid theory.

THEOREM 1. *The maximum numbers of k-dimensional flats and cells of an arrangement of n hyperplanes in d-space, for all $k = 0, 1, \ldots, d$, are those of an arrangement having general intersection:*

$$\max a_k = \binom{n}{d-k}$$

for the flats of a projective or Euclidean arrangement;

$$\max f_k = \binom{n}{d-k} \sum_{j=0}^{[k/2]} \binom{n-d+k}{k-2j}$$

for the cells of a projective arrangement;

$$\max f_k = \binom{n}{d-k} \sum_{j=0}^{k} \binom{n-d+k}{k-j},$$

$$\max b_k = \binom{n}{d-k} \binom{n-d+k-1}{k} \qquad \textit{if } n \geq d,$$

for the cells and bounded cells of a Euclidean arrangement. Provided $n > d - k$, the maxima of f_k and b_k are attained only by arrangements with general intersection and that of a_k only by arrangements with general intersection in dimension k.

THE SLIMMEST ARRANGEMENTS (PROLOGUE)

In 1967 Grünbaum [18] conjectured that the smallest possible number of k-cells of a projective d-arrangement of n hyperplanes is that achieved by the *near pencil*, consisting of $n - d + 1$ hyperplanes concurrent in a coline and $d - 1$ more hyperplanes in general position to make the whole thing a d-arrangement. Shannon [28] proved this conjecture, characterized all the minimal arrangements, and showed that the near pencil also minimizes the number of k-flats. His method is highly geometrical, based on the exploration in some detail of the structure of d-arrangements. Meanwhile, before I knew of Shannon's solution, I had realized that Grünbaum's conjecture for regions, the corresponding conjecture for flats, and descriptions of the minimal examples all followed at once from Dowling and Wilson's results [11] on the minimal values of the Whitney numbers of geometric lattices, by way of my thesis [30]. (Brylawski had come to the same realization in a dual form [5]. Cordovil [8] has given a short direct proof by geometric lattices of the minimum of f_d.) Thus, I looked for extensions of their abstract theorems that would settle not only Grünbaum's full conjecture but also the analogues for flats and cells of Euclidean arrangements. It is this approach that I want to describe.

First I will state what the slimmest arrangements are. For this I have to define an example. Take p skew colines $l_1, \ldots, l_p$ in $\mathbb{P}^d$ [where $0 \leq p \leq (d+1)/2$] and $n_i \geq 2$ hyperplanes on each l_i so that $\sum n_i = n - d - 1 + 2p$, and take $d + 1 - 2p$ more hyperplanes in general position with respect to $l_1 \cap \cdots \cap l_p$. These n hyperplanes form a d-arrangement I call a (projective) *near pencil-sum.* A *Euclidean near pencil-sum* is similar but with p colines in $\mathbb{E}^d$ (where $p \leq d/2$), $\sum n_i = n - d + 2p$, and $d - 2p$ general hyperplanes; if $p = 1$ (so $n_1 = n - d + 2$), we have a *Euclidean near pencil.*

THEOREM 2 (Shannon [28], Zaslavsky [33]). *The minimum numbers of flats and cells of a projective d-arrangement of n hyperplanes are those of the near pencil:*

$$\min a_k = \binom{d+1}{k+1} + (n-d-1)\binom{d-1}{k},$$

$$\min f_k = 2^k\binom{d+1}{k+1} + 2^{k-1}(n-d-1)\left\{\binom{d}{k} + \binom{d-1}{k}\right\},$$

for $k = 0, 1, \ldots, d$. A d-arrangement attains the minimum only if it is a near pencil, except that $a_d = 1$ and $a_{d-1} = n$ for all arrangements and a_0 and f_0 (which are equal) are minimal precisely for near pencil-sums.

THEOREM 3 (Zaslavsky [34]). *The minimum numbers of flats and cells of a Euclidean d-arrangement of n hyperplanes are those of the Euclidean near pencil:*

$$\min a_k = \binom{d}{k} + (n-d)\binom{d-2}{k-1},$$

$$\min f_k = 2^k\binom{d}{k} + 2^{k-1}(n-d)\left\{\binom{d-1}{k-1} + \binom{d-2}{k-1}\right\},$$

for $k = 0, 1, \ldots, d$. A d-arrangement attains the minimum only if it is a Euclidean near pencil, except for the following cases: $a_d = 1$ and $a_{d-1} = n$ for all arrangements; a_0 and $f_0 = 1$ attained by any d-arrangement whose hyperplanes meet in a point; a_1 and f_1 minimized by any Euclidean near pencil-sum; and

$$\min f_d = 2^{d-1}(n-d+2),$$

attained by any Euclidean d-arrangement whose projectivization is a near pencil.

THE LATTICE AND MATROID THEORY OF ARRANGEMENTS

The abstract approach to arrangements of hyperplanes is based on the fact that the set $L(\mathscr{A})$ of flats of a projective arrangement $\mathscr{A}$, when ordered by reverse inclusion, is a geometric lattice: a ranked, atomic lattice with certain additional properties which we need not go into [10, 29]. The 0 and 1 elements of $L(\mathscr{A})$ are $\mathbb{P}^d$ and $\bigcap \mathscr{A}$; the atoms are the members of $\mathscr{A}$; join in $L(\mathscr{A})$ is intersection of flats; the rank function is codimension. And $\mathscr{A}$ is a d-arrangement if, and only if, $L(\mathscr{A})$ has rank $d + 1$, the largest possible.

Let L be any (finite) geometric lattice and A its set of atoms. An element $x \in L$ can be regarded equivalently as a set of atoms (or *points*), specifically as $\{p \in A : p \leq x\}$; these sets of atoms are called *closed.* The set A with this closure [or any of several equivalent structures, such as the rank function $\text{rk}(S) = L\text{-rank of} \bigvee S$ for $S \subseteq A$] is a *simple matroid* [or (simple) combinatorial geometry]. For this I like to use the letter G, emphasizing that one has a set with a structure on it. Both the lattice and matroid viewpoints are useful in different situations, sometimes together; I refer to lattices and matroids more or less interchangeably.

A *basepointed geometric lattice* is a pair (L, b) where b is an atom of the geometric lattice L. The *ordinary part* of L, the set

$$L_b = \{x \in L : x \ngeq b\},$$

is a (lower) order ideal in L. A Euclidean arrangement $\mathscr{E}$ gives rise to a basepointed geometric lattice, namely $(L(\mathscr{E}_{\mathbb{P}}), \infty)$, and $L(\mathscr{E})$ as I defined it earlier is the ideal $L(\mathscr{E}_{\mathbb{P}})_\infty$.

If $x \in L$, the quotient of L by x is

$$L/x = \{y \in L : y \geq x\}.$$

The abstract equivalents of the face and flat numbers are *Whitney numbers.* I will define them for any *geometric semilattice* L_0, that is, a meet semilattice in which every interval is a geometric lattice. Those of the *second kind* are

$$W_j(L_0) = \#\{x \in L_0 : \mathrm{rk}(x) = j\} = W_{0j}(L_0),$$
$$W_{ij}(L_0) = \#\{(z, x) \in L_0 \times L_0 : \mathrm{rk}(z) = i, \mathrm{rk}(x) = j, z \leq x\},$$

and in particular

$$n(L_0) = W_1(L_0) = \text{the number of atoms of } L_0.$$

To define those of the first kind I will give an algorithm. The *Möbius function* of L is a function $\mu : L \times L \to \mathbb{Z}$. It satisfies $\mu(z, x) = 0$ if $x \ngeq z$, while $\mu(z, x)$ for $x \geq z$ is computed recursively upward from $x = z$ by the following rules:

(1) $\mu(z, z) = 1$. (Rule for $x = z$.)

(2) If $\mu(z, y)$ is known for all y of rank $<j$, then $\mu(z, x)$ is defined for x of rank j by

$$\mu(z, x) = -\sum_{z \leq y < x} \mu(z, y).$$

(Rule for $x > z$.)

Thus, for instance, $\mu(0, 0) = 1$ and $\mu(0, p) = -1$ for every atom p, but after that $\mu(0, x)$ becomes dependent on the structure of L. The Whitney numbers of the *first kind* are

$$w_j(L_0) = \sum \{\mu(0, x) : x \in L_0, \mathrm{rk}(x) = j\} = w_{0j}(L_0),$$
$$w_{ij}(L_0) = \sum \{\mu(z, x) : z, x \in L_0, z \leq x, \mathrm{rk}(z) = i, \mathrm{rk}(x) = j\}.$$

These are signed numbers, but Rota proved [27] that

$$|w_{ij}(L_0)| = (-1)^{j-i} w_{ij}(L_0) = \sum |\mu(z, x)|.$$

Some trivial evaluations: $w_{jj} = W_{jj} = W_j$, $|w_{01}| = W_1 = n(L_0)$. For a lattice of rank r, also $W_{jr} = W_j$.

Another important quantity is Crapo's invariant [9] of a geometric lattice L, which we define by [30]:

$$\beta(L) = (-1)^{r-1} \sum_{j=0}^{r-1} w_j(L_b)$$

for any atom b of L, where $r = \text{rk}(L)$. It is nonnegative and is independent of the choice of b. I will define also

$$\beta_i(L, b) = (-1)^{r-i-1} \sum_{j=i}^{r-1} w_{ij}(L_b)$$
$$= \sum \{\beta(L/x) : \text{rk}(x) = i, x \ngeq b\}.$$

The fundamental connection between arrangements and geometric lattices is this:

BASIC THEOREM [30]. *For an arrangement* $\mathscr{A}$ *in* $\mathbb{P}^d$ *and for* $\dim(\bigcap \mathscr{A}) < k \leq d$ *we have*

$$a_k(\mathscr{A}) = W_{d-k}(L(\mathscr{A})),$$
$$f_k(\mathscr{A}) = \tfrac{1}{2} \sum_{j=d-k}^{d+1} |w_{d-k,j}(L(\mathscr{A}))|$$
$$= \sum_{l=0}^{[k/2]} |w_{d-k,d-2l}(L(\mathscr{A}))|.$$

For an arrangement $\mathscr{E}$ *in* $\mathbb{E}^d$ *we have*

$$a_k(\mathscr{E}) = W_{d-k}(L(\mathscr{E})),$$
$$f_k(\mathscr{E}) = \sum_{j=d-k}^{d} |w_{d-k,j}(L(\mathscr{E}))|,$$

and, provided $\mathscr{E}$ *is a d-arrangement,*

$$b_k(\mathscr{E}) = (-1)^{k-1} \sum_{j=d-k}^{d} w_{d-k,j}(L(\mathscr{E})) = \beta_{d-k}(L(\mathscr{E}_{\mathbb{P}}), \infty),$$

$$b_d(\mathscr{E}) = \beta(L(\mathscr{E}_{\mathbb{P}})).$$

(The formulas for a_k are obvious; those for f_k are not difficult. That for b_k is decidedly nontrivial.)

For a projective topological arrangement $\mathscr{A}$, $L(\mathscr{A})$ is a geometric lattice. For a weak Euclidean topological arrangement $\mathscr{E}$, $L(\mathscr{E})$ is a geometric semilattice. We have therefore:

TOPOLOGICAL BASIC THEOREM [31, 32]. *The* BASIC THEOREM *holds good for topological arrangements* $\mathscr{A}$ *in* $\mathbb{P}^d$ *and weak topological arrangements* $\mathscr{E}$ *in* $\mathbb{E}^d$.

Let us call $\mathscr{E}$ *ordinary* if $L(\mathscr{E})$ equals the ordinary part of some basepointed geometric lattice. Such topological arrangements are important because they are subject to the matroidal bounds on Euclidean arrangements of straight hyperplanes. I regard it as a major unsolved problem to characterize ordinary arrangements. It is easy to see [by adding a basepoint b to $L(\mathscr{E})$ in general position] that *all parallelism-free topological arrangements are ordinary.* Also, $\mathscr{E}$ is ordinary if $d = 2$, or if the projectivization

$$\mathscr{E}_{\mathbb{P}} = \{\text{cl } H : H \in \mathscr{E}\} \cup \{\infty\},$$

where cl denotes topological closure in $\mathbb{P}^d$, is a projective topological arrangement (then $\mathscr{E}$ was necessarily strong).

Another interpretation of Whitney number sums comes from oriented matroid theory. I will state the main relevant theorem.

ORIENTATION THEOREM (Las Vergnas [22, 23]). *Let M be a loopless oriented matroid of rank r, L be its geometric lattice, and f_k be the number of acyclic reorientations of rank k contractions of M. Then*

$$f_k = \sum_{j=r-k}^{r} |w_{r-k,j}(L)|.$$

This result is actually quite geometrical, for oriented matroids are equivalent to topological arrangements in $\mathbb{P}^{r-1}$ and the acyclic reorientations of rank k contractions correspond to the $k-1$-cells [15]. Thus, the BASIC THEOREM for projective topological arrangements is equivalent to the ORIENTATION THEOREM.

Some more definitions are now necessary. A *copoint* of L is an element of rank $r-1$, where $\operatorname{rk}(L) = r$. The *direct sum* $L_1 \oplus L_2$, or $G_1 \oplus G_2$, is (i) the direct product of the lattices, and (ii) the matroid whose closed sets are the unions of closed sets in G_1 and G_2, where it is assumed that G_1 and G_2 have disjoint sets of atoms. If L is not a direct sum, it is called *irreducible.* The *deletion* $G \backslash p$ (p an atom) is obtained by removing p from every closed set. The *contraction* G/p is obtained by keeping as closed sets only those of G that contain p, and deleting p; but it is better regarded as having the lattice $L(G/p) = L(G)/p$. An *isthmus* is an atom p such that $G = (G\backslash p) \oplus p = (G/p) \oplus p$. A matroid is *free up to rank j* if every set of at most j points is closed. A geometric lattice is *modular* if its order dual is also geometric (this is a theorem [10], not the definition). Main examples of matroids (or geometric lattices) are:

(1) The *free* one, F^r of rank r, having r points and in which every set is closed.

(2) The *uniform* one, $U^r(n)$ of rank r on n points, which is the unique matroid of rank r free up to rank $r-1$. It has the most closed sets of any matroid of its rank on the same set.

(3) The *line* Λ_n, which is $U^2(n)$.

(4) A *plane* Π_n, which is an n-point simple matroid of rank 3 (not unique!).

THE STOUTEST ARRANGEMENTS AND GEOMETRIC LATTICES

Since a projective d-arrangement $\mathscr{A}$ has general intersection if and only if $L(\mathscr{A})$ is uniform, THEOREM 1 is an immediate corollary (via the BASIC THEOREM) of:

THEOREM 4. *For geometric lattices L of rank r on n points, the maximum magnitudes of the Whitney numbers are those of the uniform lattice $U^r(n)$; that is,*

$$\max W_{ij} = \max |w_{ij}| = \binom{n}{j}\binom{j}{i} \qquad \text{for } 0 \le i \le j < r;$$

$$\left.\begin{aligned} \max W_{ir} &= \binom{n}{i}, \\ \max |w_{ir}| &= \binom{n}{i}\binom{n-i-1}{r-i-1}, \end{aligned}\right\} \qquad \text{for } 0 \le i < r;$$

and $W_{rr} = w_{rr} = 1$. *Equality holds for* W_{ij} *if and only if* $i, j \in \{0, 1, r\}$, *or L is uniform, or* $j \le r - 2$ *and L is free up to rank j. Equality holds for* $\sum_j |w_{ij}|$ *with* $i < r$ *if and only if L is uniform.*

Moreover, if (L, b) is basepointed of rank r with $n + 1$ *points (including b), then*

$$\max W_{ij}(L_b) = \max |w_{ij}(L_b)| = \binom{n}{j}\binom{j}{i} \qquad \text{for } 0 \le i \le j < r,$$

$$\max \beta_i(L) = \binom{n}{i}\binom{n-i-1}{r-i-1} \qquad \text{for } 0 \le i < r,$$

with equality if L is uniform. Equality holds for W_{ij}, $|w_{ij}|$, *and* $\sum_j |w_{ij}|$ *under the same conditions as above, and for* β_i *if and only if L is uniform.*

Proof. The evaluations for uniform lattices are routine. For W_{ij} the theorem is obvious.

The theorem for $|w_j|$ and $\sum_j |w_j|$ is an aspect of the fundamental work of Dean Lucas on the effect of specializing the position of a matroid (technically, of a "weak map"). Lucas's Propositions 7.3 and 7.4 [24] have the results I want as a particular case. THEOREM 4 for $|w_j(L_b)|$ and $\sum_j |w_j(L_b)|$ is easily deduced.

The same propositions of Lucas's imply the maximality of $\beta(U^r(n + 1))$. To get the unique maximality one needs induction and a theorem of Crapo's [9] that if $\beta(G) > 0$, then

$$\beta(G) = \beta(G \backslash p) + \beta(G/p).$$

Suppose $\beta(G)$ is maximal; then so are $\beta(G\backslash p)$ and $\beta(G/p)$, and G/p must have n atoms. By induction on n, $G\backslash p$ and G/p must be uniform. It follows that p can only be added to $G\backslash p$ so as to make G a uniform matroid. (This is a common type of argument which I will use again.)

The (unique) maximality of $|w_{ij}(U^r(n))|$ and its sum on j and the basepointed analogues are easy consequences of the theorem for W_i and for $|w_{j-i}(L/x)|$, where $x \in L$ has rank i. ▮

I feel that in the purely geometrical approach to THEOREM 1 the cleverness lies in evaluating the invariants of arrangements with general intersection, while in the matroidal approach that is easy and (granting the BASIC THEOREM) the part that requires thought is to prove the (strict) decrease of the invariants as the position becomes more special. I also think that this is a nice way to treat bounded cell maxima, especially recalling the oversight in Buck's proof.

THE SLIMMEST ARRANGEMENTS AND GEOMETRIC LATTICES

Lucas's theory shows that the minimum values of the Whitney number magnitudes are attained by geometric lattices having maximally special position in the given rank. Unfortunately, there is no present hope of finding all such "tight" lattices or comparing their Whitney numbers. Instead we turn to the work of Dowling and Wilson [11], who by a complicated inductive argument determined the lattices that minimize W_j and $|w_j|$. (Those for W_{r-1} were found earlier by Greene [16].) This in light of the BASIC THEOREM is enough to prove THEOREM 2 for a_k and f_d. To settle the minima for all f_k one needs the minimal lattices for all $|w_{ij}|$. The results are summarized in THEOREM 5. THEOREM 6 is the extension to basepointed lattices.

THEOREM 5 [33]. *Let L be a geometric lattice of rank r with n points. For* $0 \le i \le j \le r$ *we have*

$$|w_{ij}(L)| \ge \binom{r}{i}\binom{r-i}{j-i} + (n-r)\left\{\binom{r-2}{i-1}\binom{r-i}{j-i} + \binom{r-2}{i}\binom{r-i-1}{j-i-1}\right\},$$

$$W_{ij}(L) \ge \binom{r}{i}\binom{r-i}{j-i} + (n-r)\left\{\binom{r-2}{i-1}\binom{r-i}{j-i} + \binom{r-2}{i}\binom{r-i-2}{j-i-2}\right\},$$

with equality when $L = \Lambda_{n-r+2} \oplus F^{r-2}$ *and also in precisely the following cases: For* W_j *when* $j = 0$, 1, *or* r, *or* $j = r - 1$ *and L is modular, or* $2 \le j \le r-2$ *and* $L = \Pi_{n-r+3} \oplus F^{r-3}$ *where* Π_{n-r+3} *is a modular plane. For* $|w_{ij}|$ *(considering only* $i < j, j \ge 2$*), only for* $|w_{r-1,r}|$ *when L is modular.*

(Björner [3] and Purdy [26] have shown that $\Lambda_{n-r+2} \oplus F^{r-2}$ also minimizes the number of independent sets of atoms of each rank k, uniquely so if $k \ge 2$.)

THEOREM 6 [34]. *Let* (L, b) *be a basepointed geometric lattice of rank* $d+1$ *with* $n+1$ *points including b. For* $0 \le i \le j \le d$ *we have*

$$|w_{ij}(L_b)| \ge \binom{d}{i}\binom{d-i}{j-i} + (n-d)\left\{\binom{d-2}{i-1}\binom{d-i}{j-i} + \binom{d-2}{i}\binom{d-i-1}{j-i-1}\right\},$$

$$W_{ij}(L_b) \ge \binom{d}{i}\binom{d-i}{j-i} + (n-d)\left\{\binom{d-2}{i-1}\binom{d-i}{j-i} + \binom{d-2}{i}\binom{d-i-2}{j-i-1}\right\},$$

with equality when $L = \Lambda_{n-d+2} \oplus F^{d-1}$ *and* $b \in F^{d-1}$ *and also in precisely the following cases: For* W_{ij} *(considering only* $i < j, j \ge 2$*) when b is an isthmus, and then only for* W_{1d}, *for* $W_{d-1,d}$ *when* $L = \Lambda_{n_1} \oplus \cdots \oplus \Lambda_{n_p} \oplus F^{d+1-2p}$, *and for* W_{id} *when* $2 \le i \le d-2$ *and* $L = \Pi_{n-d+3} \oplus F^{d-2}$ *where* Π_{n-d+3} *is a modular plane. For* $|w_{ij}|$ *(considering only* $i < j$, $j \ge 2$*), only for* $|w_j|$ *when* $L = \Lambda_{n-d+2} \oplus F^{d-1}$, *and for* $|w_{d-1,d}|$ *when* $L = \Lambda_{n_1} \oplus \cdots \oplus \Lambda_{n_p} \oplus F^{d+1-2p}$ *and* $b \in F^{d+1-2p}$.

The theorems describing the slimmest arrangements are immediate corollaries via the BASIC THEOREM. Moreover, they apply to projective and ordinary Euclidean topological arrangements.

A sketch of the proofs: The cases $i = 0$ and $0 < i = j - 1$ are crucial. The case $i = 0$ in THEOREM 5 is Dowling and Wilson's [11]; in THEOREM 6 it follows (by an argument) from their paper and two lemmas [12, 34] concerning points and lines. To prove the case $j = i + 1$ we group the $x \in L$ of rank $i + 1$ according to the size of $\{e \in E : e \le x\}$, where E is a fixed basis for L (with $b \in E$ in the basepointed case). Thence we deduce the structure of the sublattice $\{y \in L : y \le x\}$ when minimality holds, and consequently, by varying E, the structure of L. Then the general case is fairly routine.

These results suggest three problems.

Problem 1 (theoretical). The lower bounds for W_{jr} and $|w_{jr}|$ are identical respectively to those for W_{0j} and $|w_{0j}|$. This can be explained by the fact that the slimmest lattices are modular and hence their order duals are again geometric and slimmest with the same Whitney numbers. That is rather post hoc but does suggest that an argument of the following kind might give a new proof of THEOREM 5:

(1) $W_{r-1}(L) \geq W_1(L)$, with equality if and only if L is modular (Greene [16]).
(2) If W_j or $|w_j|$ is minimal, so is W_{r-1}.
(3) If W_{ij} or $|w_{ij}|$ is minimal, so are W_i and W_j.
(4) Among modular geometric lattices there are few possibilities.

Problem 2 (theoretical). The minima for L_b where $n(L) = n + 1$ and $\mathrm{rk}(L) = r + 1$ are the same as those for L where $n(L) = n$ and $\mathrm{rk}(L) = r$. The reason is that b is an isthmus in (almost all) the extreme cases. Is there a direct proof of this, and (hence?) a derivation of THEOREM 6 directly from THEOREM 5?

Problem 3: *Bounded Cells.* The general lower bound for $b_k(\mathscr{E})$, $k > 0$, is 0. If one imposes the natural requirement that $\mathscr{E}$ have at least one bounded region, determining the minimum of b_k becomes quite an interesting and difficult problem. Abstractly one is asking for the minimum of $\beta_i(L, b)$ over all *irreducible* basepointed geometric lattices of rank $d + 1$ with $n + 1$ points. The solution for $k = d$, that is the minimum of Crapo's invariant $\beta(L)$, found by Brylawski [6] without classifying the examples, is all that is known. We have [6, 34]:

$$\min b_d(\mathscr{E}) = \max(1, n - 2d + 1).$$

One would also like the minimum of $b_k(\mathscr{E})$ given that $\mathscr{E}$ has a bounded $d - 1$-cell. The abstract problem (as shown by Theorem E in [30]) is to minimize $\beta_i(L, b)$ given that either L is irreducible or $L = L_1 \oplus F^1$, where $b \in L_1$ and L_1 is irreducible. I expect the minima to be attained by $L_1 \oplus F^1$, where L_1 is a solution to the first part of this problem.

And what is the minimum of $b_k(\mathscr{E})$ given that $\mathscr{E}$ has a bounded e-cell where $e \leq d - 2$? I expect that the minimal arrangements correspond to basepointed lattices (L, b) where $L = U^{e+1}(e + 2) \oplus L_2$ and $b \in U^{e+1}(e + 2)$. For these we have $b_k = \binom{e+1}{k+1}$ (by THEOREM 1) since $b_k(\mathscr{E}) = b_k(\mathscr{E}_0)$, where $\mathscr{E}_0$ is the Euclidean e-arrangement corresponding to $(U^{e+1}(e + 2), b)$.

On the related problems of minimizing f_k and a_k given the existence of a bounded region, all that is known is the minimum of f_d, from Brylawski's work [6, 34].

NEARLY THE SLIMMEST

Ideally we would like to know not merely the extrema but the complete list of possible values of the invariants of an arrangement of hyperplanes. This problem seems hopelessly difficult but there are interesting partial results at both the high and low ends. Grünbaum, for instance, proved [20] for a projective arrangement of lines that $f_2 > 2n - 2$ (the minimum) implies $f_2 \geq 3n - 6$, and that $f_2 > 3n - 5$ implies $f_2 \geq 4n - 12$ for large enough n (a restriction reduced by Cordovil [8], using matroid theory, to $n \geq 9$, the best possible). These numbers are related to the maximum size of a concurrent set of lines: $n - 1$ for the minimum f_2, $n - 2$ for $f_2 \geq 3n - c$, etc. Recently Beck [1] has shown (not by matroid theory) that if at most $n - k$ lines concur, then for large n we have $a_0 > c_0(k - 1)n$ and consequently $f_2 > ckn$ (because $f_2 > a_0 + n$). (This result is also implicit in work of Szemerédi and

Trotter [35].) These planar results do not seem to generalize readily to higher dimensions. Nevertheless, Cordovil has made good progress by considering the effect of having $n - k$ hyperplanes of an arrangement concurrent in a point. Letting $k = 2$ he found the "second minimum" (the second smallest value) of $f_d(\mathscr{A})$. His method is entirely matroidal (barely mentioning real geometry): He found the second minimum of $\sum_j |w_j|$. I will use his proof technique to find the second minima of the $|w_j|$ themselves.

Let

$$\Pi_k^r(n) = \Pi_{n-r-l+4,\, l} \oplus F^{r-3} \qquad \text{when } l \geq 2,\ r \geq 3,\ n \geq l + r - 2,$$

where $\Pi_{m,l}$ is a plane formed of two lines of m and l points having one common point, and

$$\Pi_*^r(r+1) = U^3(4) \oplus F^{r-3} \qquad \text{when } r \geq 3,$$

where $U^3(4)$ is a plane of four points in general position. A copoint $x \in L$ is *modular* when every line determined by points outside x meets x in a point.

LEMMA. *Let L be a geometric lattice of rank r with n points and let $0 \leq j \leq r$. Suppose L has a copoint x of size $n - k$ points. Then*

$$|w_j(L)| \geq \binom{r-1}{j} + k\binom{r-1}{j-1}$$
$$+ (n - r - k + 1)\binom{r-2}{j-1} + k(n - r - k + 1)\binom{r-2}{j-2},$$

with equality precisely when $j < 2$ or x is a modular copoint of the form $\Lambda_{n-r-k+3} \oplus F^{r-3}$ [for instance, L may equal $\Pi_{k+1}^r(n)$].

Proof. Equality when $j < 2$ is trivial. To settle the case $j \geq 2$ we pull out two basic tools of matroid invariant theory, the deletion-contraction or "Tutte-Grothendieck" law and the law of direct sums [4]. For the w_j these reduce to the same formula: If p is a point in the simple matroid G, then

$$|w_j(G)| = |w_j(G \backslash p)| + |w_{j-1}(G/p)|. \tag{1}$$

(The two laws cover the cases where p is not, or is, an isthmus.)

We choose G so that $L(G)$ is as in the lemma. Let the points not in x be $p_1, p_2, \ldots, p_k$. Following Cordovil, by k applications of (1) we have

$$|w_j(G)| = |w_{j-1}(G/p_1)| + |w_{j-1}((G \backslash p_1)/p_2)| + \cdots$$
$$+ |w_{j-1}((G \backslash p_1 \cdots p_{k-1})/p_k)| + |w_j(G \backslash p_1 \cdots p_k)|.$$

This, by the Dowling–Wilson case $i = 0$ of THEOREM 5, is at least equal to the right side of the inequality of the lemma. Moreover, there is equality if and only if every $(G \backslash p_1 \cdots p_{i-1})/p_i$ had $n - k$ points, for $i = 1, 2, \ldots, k$; hence every line through p_i meets x, and $x = G \backslash p_1 \cdots p_k$ consists of a line plus $r - 3$ points. ∎

THEOREM 7. *Let L be a geometric lattice of rank r with n points such that $|w_j(L)|$ does not achieve the lower bound. Then $n > r \geq 3$, $j \geq 2$, and the minimum $|w_j(L)|$ is given by* (A) *and* (B):

(A) *If* $n = r + 1$, *we have*

$$|w_j(L)| \geq \binom{r}{j} + \binom{r-1}{j-1} + \binom{r-2}{j-2}$$

with equality precisely when $L = \Pi_*^r(r+1)$.

(B) *If* $n \geq r + 2$, *we have*

$$|w_j(L)| \geq \binom{r}{j} + \binom{r-1}{j-1} + (n-r-1)\left\{\binom{r-1}{j-1} + \binom{r-2}{j-2}\right\},$$

with equality precisely when L has a modular copoint of the form $\Lambda_{n-r+1} \oplus F^{r-3}$ *or* $\Lambda_3 \oplus F^{r-3}$ [*for instance, L may equal* $\Pi_3^r(n)$].

Proof of (A). The only lattices of rank r with $r + 1$ points are those of the form $U^k(k+1) \oplus F^{r-k}$, $k \geq 2$, whose Whitney numbers are easily found by means of **(1)** to be

$$|w_j| = \sum_{i=j-k+1}^{j} \binom{r}{i}.$$

This is an increasing function of k with second minimum at $k = 3$.

Proof of (B). The lower bound of the lemma is a quadratic function $f(k)$ with minimum (for feasible k) at $k = 1$ and $k = n - r + 1$ and second minimum at $k = 2$ and $k = n - r$. Since $2 < n - r + 1$, so $f(2) > f(1)$, and since $f(2)$ is attained [e.g., by $\Pi_3^r(n)$], the second minimum of $|w_j|$ is at most $f(2) = f(n-r)$. Suppose L has a copoint x with $n - k$ points where $2 \leq k \leq n - r$. If $2 < k < n - r$, then $|w_j| \geq f(k) > f(2)$, while if $k = 2$ or $n - r$ and x is not of the form in the theorem, then $|w_j| > f(2)$. Thus, we will have the theorem if we show that $|w_j| > f(2)$ under the assumption that every copoint of L has $n - 1$ or $r - 1$ points.

Such an L has the form $U^{r-l}(n-l) \oplus F^l$ for some l, $0 \leq l \leq r - 3$. (We rule out $l = r - 2$ because that gives minimal $|w_j|$.) By applying **(1)** to a point p in the uniform part of L, we have

$$|w_j(L)| = |w_j(U^{r-l}(n-l-1) \oplus F^l)| + |w_{j-1}(U^{r-l-1}(n-l-1) \oplus F^l)|$$

$$\geq \binom{r}{j} + (n-r-1)\binom{r-1}{j-1} + \binom{r-1}{j-1} + (n-r)\binom{r-2}{j-2}$$

by THEOREM 5. This exceeds $f(2)$ by the positive quantity $\binom{r-2}{j-2}$. ∎

Since all $|w_j(L)|$ have their second minima at the same lattices L, we deduce:

COROLLARY. *Let L be a geometric lattice of rank r with n points such that* $\gamma(L) = \sum_j |w_j(L)|$ *does not achieve the lower bound,* $2^{r-1}(n-r+2)$. *Then* $n > r \geq 3$ *and the minimum of* $\gamma(L)$ *is given by* (A) *and* (B).

(A) *If* $n = r + 1$, *then*

$$\gamma(L) \geq 2^{r-2}7,$$

with equality precisely when $L = \Pi_*^r(r+1)$.

(B) (Cordovil [8]) *If* $n \geq r + 2$, *then*

$$\gamma(L) \geq 2^{r-2}3(n - r + 1),$$

with equality precisely when L has a modular copoint of the form $\Lambda_{n-r+1} \oplus F^{r-3}$ *or* $\Lambda_3 \oplus F^{r-3}$.

THEOREM 8. *Let* $\mathscr{A}$ *be a projective d-arrangement of n hyperplanes for which* $f_d(\mathscr{A})$ *is not minimal. Then* $d \geq 2$, $n \geq d + 2$, *and we have*

$$f_d(\mathscr{A}) \geq 2^{d-2}7 \qquad \text{if } n = d + 2,$$

with equality precisely when $\mathscr{A}$ *consists of* 4 *hyperplanes concurrent in a coplane (and otherwise general) and* $d - 2$ *more hyperplanes in general position; and*

$$f_d(\mathscr{A}) \geq 2^{d-1}3(n - d) \qquad \text{if } n \geq d + 3,$$

with equality precisely when $\mathscr{A}$ *consists of* 3 *(or* $n - d$*) hyperplanes concurrent in a coline l,* $d - 2$ *more hyperplanes whose common intersection meets l at a single vertex v, and* $n - d - 1$ *(or* 2*) more hyperplanes not containing v, each of whose pairwise intersections lies in a hyperplane of* $\mathscr{A}$ *containing v.*

By similar arguments but with increasing difficulty one can determine the third and fourth minima of $|w_j|$ and f_d.

One would hope for similar results for the other cell numbers, and I suggest that the arrangements that work for f_d give the second minima in general. However, there are exceptions even in the plane. Kelly and Moser [21] showed (in dual form) that the second minimum of f_0 is $2n - 4$ for large n ($n \geq 10$ according to Elliott [13]) but that it is $2n - 5$ for $n = 7, 8, 9$. This suggests that the number of k-flats is less regular than the number of k-cells. There is a second difficulty: A purely matroidal method is intrinsically unable to solve the problem for flats of arrangements because the second minimum of W_j over all geometric lattices can be less than that over the lattices of real point sets. This is already true for W_2 of rank 3 lattices. Most likely the correct level of abstraction here is to oriented matroids.

STOUT

Erdös [14] showed (in the dual version) that for a projective arrangement of n lines, f_0 can take on any value from the maximum of $\binom{n}{2}$ down to $cn^{3/2}$, except for $\binom{n}{2} - 1$ and $\binom{n}{2} - 3$ which are easily seen to be impossible. And he deduced from a result of Kelly and Moser [21] that $n^{3/2}$ can be replaced by no asymptotically smaller function. His proof of the former part generalizes to Whitney numbers of geometric lattices and hence to all dimensions and all f_k (as Grünbaum noted for arrangements of lines [20]).

A *paving matroid* (whose lattice I will call a *paving lattice*) is free up to rank $r - 2$, where r is the total rank. An *essential* copoint is one of greater size than the minimum, $r - 1$. Remember that

$$\max |w_{ij}(L)| = \binom{n}{j}\binom{j}{i}.$$

THEOREM 9. *Given $j \geq 2$, $0 \leq i \leq j$, and $n > j$, consider all paving lattices L of rank $j+1$, having n points, whose essential copoints are pairwise disjoint. There are a constant c_{ij} and a finite set A_{ij} (void if $i = 0$) such that every number s in the range*

$$c_{ij} n^{j-1/j} \leq s \leq \binom{n}{j}\binom{j}{i},$$

with the exceptions $\binom{n}{j}\binom{j}{i} - a$ for $a \in A_{ij}$, is a value of $|w_{ij}(L)|$ for some such L.

I am interested in paving lattices because they apply readily to examples, as can be seen in THEOREM 10 and the discussion of separation in the last section.

Setting $i = 0$ and $i = j$, THEOREM 9 describes the upper range of values of $|w_j(L)|$ and $W_j(L)$. It can be adapted to lattices of rank $r > j + 1$, free up to rank $j - 1$. The set A_{ij} can in principle be calculated from number theory since it is the set of natural numbers that cannot be written as a nonnegative combination of the quantities

$$Q_m = \binom{m}{i}\binom{m-i-1}{j-i} \quad \text{if } i < j,$$

$$= \binom{m}{j} - 1 \quad \text{if } i = j,$$

for $m > j$. This A_{ij} is always finite, void if $i = 0$.

Proof. Erdös's idea [14] is to find some large copoints and some small ones that combine to reduce $|w_{ij}|$ by just the right amount. Given s, we want to attain the *deficiency*, compared to the maximum possible $|w_{ij}|$, of

$$t = \binom{n}{j}\binom{j}{i} - s.$$

A copoint of m points contributes exactly Q_m to the deficiency.

Let $J = 1 - 1/j$ and take q to be the smallest integer

$$q > 1 - \frac{\log j}{\log J}. \qquad \textbf{(a)}$$

Let $\alpha = 1 + \max A_{ij}$ (or 0 if $A_{ij} = \emptyset$) and Q_M be the last Q_m necessary to represent all positive integers not in A_{ij}. From now on I will assume $t > \alpha + Q_{j+1}$; making n large enough will take care of smaller t. Let $t_0 = t - \alpha$.

Given t_{k-1}, let m_k be the largest number such that $Q_{m_k} \leq t_{k-1}$, and set

$$t_k = t_{k-1} - Q_{m_k} < Q_{m_k+1} - Q_{m_k}.$$

Continue from $k = 1$ to q. This construction implies $Q_{m_{k+1}} < Q_{m_k+1} - Q_{m_k}$, from which we can deduce

$$\frac{m_{k+1} - j}{j} < \left(\frac{m_k - j}{j}\right)^J \qquad \textbf{(b)}$$

We take for essential copoints one of each size $m_1, m_2, \ldots, m_q$ and, to make up the remaining deficiency, n_m of sizes $m \le M$, where the n_m are the solutions of

$$\sum_{m=j+1}^{M} Q_m n_m = t_q + \alpha.$$

These latter copoints require n' points where

$$n' = \sum_{j+1}^{M} m n_m \le (t_q + \alpha) \max_{j<m\le M} \frac{m}{Q_m} < c_1 m_q^{j-1} + c_2. \qquad \textbf{(c)}$$

Now let us suppose $m_1 \le n - cn^J$. With **(a)**–**(c)** this yields

$$m_2 + \cdots + m_q \le (q-1)m_2 < (q-1)j^{1/j}n^J,$$

$$n' < c_3(n^{J^{q-1}})^{j-1} < c_3 n^J$$

for large n. Thus, the copoint of size m_1, those of sizes $m_2, \ldots, m_q$, and those of sizes $m \le M$ between them call for no more than n points, if c is suitably chosen and n is large.

If we had $m_1 > n - cn^J$, then

$$Q_{m_1} \ge \binom{j}{i}\frac{n^j}{j!} - \binom{j}{i}\frac{c}{(j-1)!}\, n^{j-1/j} + O(n^{j-2/j}).$$

On the other hand, we know

$$Q_{m_1} < t \le \binom{j}{i}\binom{n}{j} - c_{ij} n^{j-1/j} = \binom{j}{i}\frac{n^j}{j!} - c_{ij} n^{j-1/j} + O(n^{j-1}).$$

This is an inconsistency, proving $m_1 \le n - cn^J$, if we choose

$$c_{ij} > \binom{j}{i}\frac{c}{(j-1)!}.$$

Thus, there exists a c_{ij} that works for large n, and by increasing c_{ij} we can obviate the latter requirement. ∎

THEOREM 10. *Given $d \ge 2$, $0 \le k \le d$, and $n > d$. Every number s in the range*

$$c_{d-k,d}\, n^{d-1/d} \le s \le \max f_k(\mathscr{A}) = \binom{n}{d-k} \sum_{j=0}^{[k/2]} \binom{n-d+k}{k-2j},$$

except for $s = \max f_k(\mathscr{A}) - a$ where $a \in A_{d-k,d}$, is equal to $f_k(\mathscr{A})$ for some projective d-arrangement $\mathscr{A}$ of n hyperplanes in which, moreover, every d hyperplanes meet in a single point.

It may be that THEOREMS 9 and 10 as stated here are sharp. However, one can do much better by allowing essential copoints to overlap in an intricate pattern, as Beck has recently demonstrated [2]. He showed that a rank 4 paving lattice consisting of the circles determined by n points in the plane can have any value of $W_3 = w_{33}$ from $cn^{5/2}$ up to $\binom{n}{3}$, with finitely many necessary exceptions as in THEOREM 9, and that $cn^{5/2}$ is best possible for circles. It remains to extend Beck's result to the other Whitney numbers $|w_{i3}|$ in rank 4 and to higher ranks, where the problem of the best exponent of n is very much open.

What is certain is that one cannot simply transfer results on these questions from matroids to real point sets or (dually) arrangements of hyperplanes. As Erdös remarked [14], with planar matroids one can attain W_2 down to $n + cn^{3/4}$ at least, much less than his sharp limit of $cn^{3/2}$ for real planar point sets (dually, a_0 of arrangements of lines). Thus, results in real geometry will have to be explored geometrically, and matroids may be of little help.

Separation by Hyperplanes and Hyperspheres

The face numbers of an arrangement of hyperplanes have a straightforward dual interpretation that I hate to leave out. For a finite set $S \subseteq \mathbb{E}^d$ let $L(S)$ be the geometric lattice of all subspaces determined by S, including the empty subspace (dimension -1), ordered by inclusion. Let $t_k(S)$ be the number of ways to separate S into subsets by a hyperplane whose intersection with S is k-dimensional. Then

$$t_k(S) = \tfrac{1}{2} \sum_j |w_{k+1,j}(L(S))| \qquad \text{for } -1 \leq k < d$$

if S lies in no hyperplane. One immediately has for $t_k(S)$ the same conclusions about extrema and near extrema as for $f_{d+1-k}(\mathscr{A})$. [Brylawski [5] noted the extremality results for $t_{-1}(S)$.]

Separation by hyperspheres can be treated analogously. By a *hypersphere* I will mean a $d-1$-dimensional sphere in $\mathbb{E}^d$. Suppose $S \subseteq \mathbb{E}^d$ lies in no hypersphere or hyperplane. Let $L_0(S)$ be the set of all spheres and flats determined by S, including $\emptyset$, points, and point pairs (0-spheres). Then $L_0(S)$ is a geometric lattice, and a k-flat (except $\emptyset$ and points) or k-sphere has rank $k+2$. To see this, regard $\mathbb{E}^d$ as the stereographic projection of the Riemann sphere $S^d \subseteq \mathbb{E}^{d+1}$. If $S_1 \subseteq S^d$ is the set that projects to S, then $L_0(S) \cong L(S_1)$. Let $\sigma_k(S)$ be the number of ways to separate S by a hypersphere or hyperplane h such that $S \cap h$ lies in a sphere or flat of dimension k but no less (if $k > 0$), or is a point pair (if $k = 0$) or a point ($k = -1$) or void ($k = -2$). By stereographic projection of the Basic Theorem we get

$$\sigma_k(S) = \tfrac{1}{2} \sum_j |w_{k+2,j}(L_0(S))| \qquad \text{for } -2 \leq k < d.$$

One can draw the obvious conclusions about maxima (from Theorem 1, say, by duality and projection) and near maxima [from Theorem 10: values of σ_k down to $c_{k+2,d+1}n^{d+d/(d+1)}$ are attainable]. But at the other extreme there is a crucial new fact: $L_0(S)$ is free up to rank 2 and consequently cannot achieve the general minima. Thus we need another matroid theorem. Unfortunately I know only a limited result.

Theorem 11 (Björner [3]). *Let L be a geometric lattice of rank r with n points, free up to rank $l-1$, where $2 \leq l \leq r$. Then*

$$|w_j(L)| \geq \sum_{i=0}^{l-1} \binom{n-r-1+i}{i} \binom{r-i}{j-i} \qquad \textit{for } j = 0, 1, \ldots, r,$$

with equality for all $j < l$, and for $j \geq l$ if and only if $L = U^l(n-r+l) \oplus F^{r-l}$.

Proof. This is another place to apply (**1**). The cases $l = r$ and $n = r$ are trivial; that where $l = 2$ is Theorem 5. Thus we can assume $n > r > l > 2$ and carry out induction on n. Let $w_j(n, r, l) = \min |w_j|$.

Let $L = L(G)$ have minimal $|w_j|$ and let p be a nonisthmus. Since $L(G/p)$ is free up to rank $l - 2 \geq 1$, it has $n - 1$ atoms. Applying (**1**) twice,

$$\begin{aligned} w_j(n, r, l) &= |w_j(G\backslash p)| + |w_{j-1}(G/p)| \\ &\geq w_j(n-1, r, l) + w_{j-1}(n-1, r-1, l-1) \qquad (*) \\ &= |w_j(U^l(n-r+l) \oplus F^{r-l})| \end{aligned}$$

by induction. Consequently, $U^l(n - r + l) \oplus F^{r-l}$ is indeed minimal and equality holds in (*). This permits the calculation of $w_j(n, r, l)$ by induction.

The uniqueness of the minimum also follows since by induction

$$L(G\backslash p) = U^l(n - r + l - 1) \oplus F^{r-l}$$

and

$$L(G/p) = U^{l-1}(n - r + l - 1) \oplus F^{r-l}.$$

Adding p must leave F^{r-l} free, so p must be added to $U^l(n - r + l - 1)$ while leaving G free up to rank $l - 1$. This makes $L = U^l(n - r + l) \oplus F^{r-l}$. ▮

THEOREM 12. *For a set S of n points in* $\mathbb{E}^d$ *contained in no hyperplane or hypersphere,*

$$\sigma_{-2}(S) \geq 2^{d+1}\left[1 + \binom{n-d+1}{2}\right],$$

with equality if and only if all but d points of S lie on a circle.

It seems likely that the number of k-spheres and $(k - 1)$-flats determined by S is minimized by the same configuration, with small exceptions as an example of Segre shows (mentioned by Elliott [13]): for $n = 8$, $d = 2$, take the vertices of two concentric, parallel squares. The same should hold if only spheres and not flats are counted. Indeed, Elliott proved this [13] for $d = 2$ and $n > 393$. I would conjecture that for n points in $\mathbb{E}^d$, no l of them affinely dependent, the minimum number of k-flats occurs when all but $d - l + 1$ points lie in an $l - 1$-dimensional subspace, provided n is large enough.

CONCLUSION

Why have I emphasized, for instance, the Whitney numbers $|w_j|$, when it is their sum that counts the regions of an arrangement, and the sum is easier to study? One reason is that the matroidal approach to extrema naturally suggests consideration of Whitney numbers, especially since one already wants to know about the second kind, the W_j, for their geometrical significance. And since the doubly indexed numbers appear in the enumeration of cells and are, I believe, not much harder to treat individually than in sums, one might as well do them. But the best reasons are: the challenge of getting a better result, and the fact that, as Greene and I have shown, the numbers $|w_j|$ and $|w_{ij}|$ have, in themselves, interesting geometrical interpretations [17].

Notes and References

1. Beck, J. 1983. On the lattice property of the plane and some problems of Dirac, Motzkin and Erdös in combinatorial geometry. Combinatorica **3:** 281–297.
2. Beck, J. 1984. On the number of circles determined by *n* points. Studia Math. Acad. Hungar. In press.
3. Björner, A. 1980. Some matroid inequalities. Discrete Math. **31:** 101–103. MR 81h:05039. Zbl. 446.05013. This article announces results to be published in the chapter: Homology of matroids. *In* Combinatorial Geometries. N. White, Ed. Forthcoming.
4. Brylawski, T. 1972. A decomposition for combinatorial geometries. Trans. Amer. Math. Soc. **171:** 235–282. MR **46** #8869. Zbl. 224.05007, 248.05019.
5. Brylawski, T. 1976. A combinatorial perspective on the Radon convexity theorem. Geom. Dedicata **5:** 459–466. MR **55** #13340. Zbl. 361.52002.
6. Brylawski, T. 1977. Connected matroids with the smallest Whitney numbers. Discrete Math. **18:** 243–252. MR **58** #27566. Zbl. 373.05023.
7. Buck, R. C. 1943. Partition of space. Amer. Math. Monthly **50:** 541–544. MR **5:** 105.
8. Cordovil, R. 1980. Sur l'evaluation $t(M; 2, 0)$ du polynôme de Tutte d'un matroïde et une conjecture de B. Grünbaum relative aux arrangements de droites du plan. Eur. J. Combinatorics **1:** 317–322. MR 82j:05046. Zbl. 457.05019. In this article our $\gamma(L) = \sum_j |w_j(L)|$ is denoted $t(M; 2, 0)$.
9. Crapo, H. H. 1967. A higher invariant for matroids. J. Combin. Theory **2:** 406–417. MR **35** #6579. Zbl. **168:** 262.
10. Crapo, H. H. & G.-C. Rota. 1968. On the Foundations of Combinatorial Theory: Combinatorial Geometries. Preliminary edit. MIT Press. Cambridge, Mass. MR **45** #74. Zbl. **216:** 21.
11. Dowling, T. A. & R. M. Wilson. 1974. The slimmest geometric lattices. Trans. Amer. Math. Soc. **196:** 203–216. MR **49** #10579. Zbl. 289.05020.
12. Dowling, T. A. & R. M. Wilson. 1975. Whitney number inequalities for geometric lattices. Proc. Amer. Math. Soc. **47:** 504–512. MR **50** #6900. Zbl. 297.05010.
13. Elliott, P. D. T. A. 1967. On the number of circles determined by *n* points. Acta Math. Acad. Sci. Hungar. **18:** 181–188. MR **35** #4793. Zbl. **163:** 147.
14. Erdös, P. 1972. On a problem of Grünbaum. Canad. Math. Bull. **15:** 23–25. MR **47** #5709. Zbl. 233.05017.
15. Folkman, J. & J. Lawrence. 1978. Oriented matroids. J. Combin. Theory Ser. B **25:** 199–236. MR 81g:05045. Zbl. 325.05019, 382.05020.
16. Greene, C. 1970. A rank inequality for finite geometric lattices. J. Combin. Theory **9:** 357–364. MR **42** #1727. Zbl. 254.05016.
17. Greene, C. & T. Zaslavsky. 1983. On the interpretation of Whitney numbers through arrangements of hyperplanes, zonotopes, non-Radon partitions, and orientations of graphs. Trans. Amer. Math. Soc. **280:** 97–126.
18. Grünbaum, B. 1967. Convex Polytopes. Interscience. New York, N.Y. MR **37** #2085. Zbl. **163:** 166. See Chapter 18: Arrangements of hyperplanes.
19. Grünbaum, B. 1971. Arrangements of hyperplanes. *In* Proceedings of the Second Louisiana Conference on Combinatorics, Graph Theory and Computing (Baton Rouge, La., 1971). R. C. Mullin *et al.*, Eds.: 41–106. Louisiana State University. Baton Rouge, La. MR **47** #9428. Zbl. 289.52004.
20. Grünbaum, B. 1972. Arrangements and Spreads. CBMS Regional Conference Series in Mathematics, No. 10. American Mathematical Society. Providence, R.I. MR **46** #6148. Zbl. 249.50011.
21. Kelly, L. M. & W. O. J. Moser. 1958. On the number of ordinary lines determined by *n* points. Canad. J. Math. **10:** 210–219. MR **20** #3494. Zbl. **81:** 151.
22. Las Vergnas, M. 1975. Matroïdes orientables. C. R. Acad. Sci. Paris Ser. A **280:** 61–64. MR **51** #7910. Zbl. 304.05013. The Orientation Theorem for $k = d$ is stated here in dual form.
23. Las Vergnas, M. 1980. Convexity in oriented matroids. J. Combin. Theory Ser. B **29:** 231–243. MR 82f:05027. Zbl. 443.05026. The Orientation Theorem for $k = d$ is proved here; the case $k < d$ is an immediate corollary.

24. LUCAS, D. 1975. Weak maps of combinatorial geometries. Trans. Amer. Math. Soc. **206:** 247–279. MR **51** #7911. Zbl. 329.05015.
25. MANDEL, A. 1982. Topology of Oriented Matroids. Ph.D. Thesis. University of Waterloo. Waterloo, Ontario, Canada.
26. PURDY, G. 1982. The independent sets of rank *k* of a matroid. Discrete Math. **38:** 87–91. Zbl. 473.05018.
27. ROTA, G.-C. 1964. On the foundations of combinatorial theory. I. Theory of Möbius functions. Z. Wahrsch. Verw. Gebiete **2:** 340–368. MR **30** #4688. Zbl. **121:** 24.
28. SHANNON, R. W. 1976. A lower bound on the number of cells in arrangements of hyperplanes. J. Combin. Theory Ser. A **20:** 327–335. MR **53** #14321. Zbl. 336.50009.
29. WELSH, D. J. A. 1976. Matroid Theory. Academic Press, Inc. London, England. MR **55** #148. Zbl. 343.05002.
30. ZASLAVSKY, T. 1975. Facing up to arrangements: face-count formulas for partitions of space by hyperplanes. Mem. Amer. Math. Soc. **1**(1): No. 154. MR **50** #9603. Zbl. 296.50010.
31. ZASLAVSKY, T. 1975. Counting the faces of cut-up spaces. Bull. Amer. Math. Soc. **81:** 916–918. MR **53** #3901. Zbl. 317.05012. Results are proved in [32].
32. ZASLAVSKY, T. 1977. A combinatorial analysis of topological dissections. Adv. in Math. **25:** 267–285. MR **56** #5310. Zbl. 406.05004.
33. ZASLAVSKY, T. 1983. The slimmest arrangements of hyperplanes. I. Geometric lattices and projective arrangements. Geom. Dedicata **14:** 243–259.
34. ZASLAVSKY, T. 1981. The slimmest arrangements of hyperplanes. II. Basepointed geometric lattices and projective arrangements. Mathematika **28:** 169–190. Zbl. 483.51012.
35. SZEMERÉDI, E. & W. T. TROTTER, JR. 1983. Extremal problems in discrete geometry. Combinatorica **3:** 381–392.

THE SUM OF DISTANCES DETERMINED BY POINTS ON A SPHERE

G. D. Chakerian

Department of Mathematics
University of California
Davis, California 95616

M. A. Ghandehari

Department of Mathematics
University of San Francisco
San Francisco, California 94117

Some properties of the midpoint polyhedron of a set of r points on the unit sphere in a normed linear space are used to show that the sum of the distances determined by pairs of the points is at least $2r - 2$ if the convex hull contains the origin.

Let X be a real normed linear space, and for each finite subset $\{x_1, \ldots, x_r\} \subset X$ let $s = s(x_1, \ldots, x_r)$ denote the sum of all distances determined by pairs from $\{x_1, \ldots, x_r\}$. That is, let

$$s(x_1, \ldots, x_r) = \sum \| x_i - x_j \|, \tag{1}$$

where the sum is taken over all integers, i, j satisfying $1 \leqq i < j \leqq r$. Let $S = \{x : \| x \| = 1\}$ be the unit sphere of X. The following theorem concerning distances determined by points on S was conjectured in [1] by the first author and Klamkin.

THEOREM 1. *Let $x_1, \ldots, x_r$ be points on the unit sphere S of a normed linear space X, and suppose the origin o belongs to the convex hull of $\{x_1, \ldots, x_r\}$. Then*

$$s(x_1, \ldots, x_r) \geqq 2r - 2. \tag{2}$$

We note that the inequality is sharp, since equality holds in (2) if x_1 is any point of S and $x_2 = x_3 = \cdots = x_r = -x_1$.

The special case of THEOREM 1 where X is either a Euclidean space or is two dimensional was proved in [1], and the theorem was later proved in full generality by Wolfe [3] using the concept of metric dependence. In this paper we establish a geometric property of subsets of the unit sphere (see THEOREM 2) which in conjunction with a general inequality for sums of distances (see THEOREM 3) has THEOREM 1 as a direct consequence.

For $\{x_1, \ldots, x_r\} \subset X$, we define the *midpoint polyhedron* to be the convex hull of the set of midpoints of line segments determined by the set, that is, the convex hull of all points of the form $(x_i + x_j)/2$ with $1 \leqq i < j \leqq r$. The midpoint polyhedron of a set of points on the unit sphere has the following property.

THEOREM 2. *Suppose $x_1, \ldots, x_r$ are points on the unit sphere S of a normed linear space X, with o in the convex hull of $\{x_1, \ldots, x_r\}$. Then o belongs to the midpoint polyhedron of $\{x_1, \ldots, x_r\}$.*

Thus, for example, if a triangle T inscribed in a centrally symmetric convex curve C contains the center of C, then the triangle determined by the midpoints of the sides of T also contains the center of C.

The following theorem gives a bound on the sum of the distances from the points of a set to a point in its midpoint polyhedron.

THEOREM 3. *Let p belong to the midpoint polyhedron of $\{x_1, \ldots, x_r\} \subset X$. Then*

$$(2r - 2) \sum_{i=1}^{r} \| p - x_i \| \leqq rs(x_1, \ldots, x_r). \tag{3}$$

As in the case of THEOREM 1, this result is sharp, since equality holds in (3) if x_1 is any point of X and $x_2 = x_3 = \cdots = x_r = -x_1$, and $p = o$.

To illustrate THEOREM 3 in a simple case, suppose p belongs to the triangle determined by the midpoints of the sides of a given triangle T. Then the sum of the distances from p to the vertices of T is at most three-quarters of the perimeter of T.

We next show that THEOREM 1 is an immediate consequence of THEOREMS 2 and 3.

Proof of THEOREM 1. If $\{x_1, \ldots, x_r\} \subset S$ with the origin o in their convex hull, then THEOREM 2 implies that o belongs to the midpoint polyhedron of $\{x_1, \ldots, x_r\}$. But then, by THEOREM 3, we may apply (3) with $p = o$, which gives the required inequality (2), completing the proof. ▮

We proceed now to the proofs of THEOREMS 2 and 3.

Proof of THEOREM 2. Since the affine hull of $\{x_1, \ldots, x_r\}$ is finite dimensional, by Caratheodory's theorem (see, for example, [2]) there is no loss of generality in assuming that the origin o belongs to the relative interior of a nondegenerate simplex with vertices $\{p_0, \ldots, p_n\} \subset \{x_1, \ldots, x_r\}$. We may then restrict ourselves to the affine hull of $\{p_0, \ldots, p_n\}$ and assume without loss of generality that X is n-dimensional. We need only prove that o belongs to the midpoint polyhedron of $\{p_0, \ldots, p_n\}$ and we will be done.

Let M be the midpoint polyhedron of $\{p_0, \ldots, p_n\}$ and suppose $o \notin M$. Then there exists a hyperplane H passing through o with $H \cap M = \varnothing$ and such that none of $p_0, \ldots, p_n$ belong to H. Let H^+ and H^- be the open half spaces determined by H, and suppose $M \subset H^+$. Since o belongs to the convex hull of $\{p_0, \ldots, p_n\}$, some member of this set, say p_0, lies in H^-. But then no other p_i belongs to H^-, since otherwise the midpoint determined by p_0 and p_i would be in H^-. Thus we have $p_0 \in H^-$ and $\{p_1, \ldots, p_n\} \subset H^+$.

Since $\{(p_0 + p_1)/2, \ldots, (p_0 + p_n)/2\} \subset H^+$, the part of the convex hull of $\{p_0, \ldots, p_n\}$ contained in H^- is contained in the simplex with vertices p_0, $(p_0 + p_1)/2, \ldots, (p_0 + p_n)/2$, and in particular o is in this simplex. Thus,

$$o = \lambda_0 p_0 + \lambda_1 \left(\frac{p_0 + p_1}{2}\right) + \cdots + \lambda_n \left(\frac{p_0 + p_n}{2}\right), \tag{4}$$

with $\lambda_i \geqq 0$, $i = 0, 1, \ldots, n$, $\lambda_0 + \lambda_1 + \cdots + \lambda_n = 1$, and $\lambda_0 > 0$ since $o \notin H^+$. Rearrangement of (4) gives

$$-p_0 = \lambda_0 p_0 + \cdots + \lambda_n p_n, \tag{5}$$

with $\lambda_i \geqq 0$, $i = 0, 1, \ldots, n$, $\lambda_0 + \lambda_1 + \cdots + \lambda_n = 1$, and $\lambda_0 > 0$. This means that $-p_0$ belongs to the simplex with vertices $p_0, \ldots, p_n$ but not to the face with vertices $p_1, \ldots, p_n$; hence, $-p_0$ belongs to the relative interior of some face having p_0 as a vertex. It then follows that the entire line segment $[p_0, -p_0]$ belongs to the unit sphere S, implying that o belongs to S, a contradiction. Hence $o \in M$, as we wanted to prove. ▮

Proof of THEOREM 3. We begin by first proving (3) in the case where p is the midpoint of some segment determined by a pair from $\{x_1, \ldots, x_r\}$; suppose $p = (x_{r-1} + x_r)/2$. It will be convenient to denote $a_i = \| x_{r-1} - x_i \|$ and $b_i = \| x_r - x_i \|$, $i = 1, \ldots, r-2$. We note now that

$$s(x_1, \ldots, x_r) = s(x_1, \ldots, x_{r-2}) + \| x_r - x_{r-1} \| + \sum_{i=1}^{r-2} (a_i + b_i). \tag{6}$$

Observing that $s(x_1, \ldots, x_{r-2}) \geqq 0$ and multiplying both sides of (6) by r gives

$$rs(x_1, \ldots, x_r) \geqq r \| x_r - x_{r-1} \| + r \sum_{i=1}^{r-2} (a_i + b_i). \tag{7}$$

From the triangle inequality we have $a_i + b_i \geqq \| x_r - x_{r-1} \|$, $i = 1, \ldots, r = 2$, and summing this inequality over i gives

$$\sum_{i=1}^{r-2} (a_i + b_i) \geqq (r-2) \| x_r - x_{r-1} \|. \tag{8}$$

From (7) and (8) one easily deduces that

$$rs(x_1, \ldots, x_r) \geqq (2r-2) \| x_r - x_{r-1} \| + (r-1) \sum_{i=1}^{r-2} (a_i + b_i). \tag{9}$$

As another easy consequence of the triangle inequality we have

$$a_i + b_i \geqq 2 \| p - x_i \|, \qquad i = 1, \ldots, r-2. \tag{10}$$

Summing both sides of (10) over i and applying the result to (9) yields

$$rs(x_1, \ldots, x_r) \geqq (2r-2) \| x_r - x_{r-1} \| + (2r-2) \sum_{i=1}^{r-2} \| p - x_i \|. \tag{11}$$

But $\| x_r - x_{r-1} \| = \| p - x_r \| + \| p - x_{r-1} \|$, so (11) is in fact the same as (3), with $p = (x_r + x_{r-1})/2$.

We have shown that (3) holds when p is the midpoint of any segment determined by a pair from $\{x_1, \ldots, x_r\}$. Thus the function $f(p) = \sum_{i=1}^{r} \| p - x_i \|$, $p \in X$, satisfies

$$f(p) \leqq \left(\frac{r}{2r-2} \right) s(x_1, \ldots, x_r) \tag{12}$$

when $p = (x_i + x_j)/2$, $1 \leqq i < j \leqq r$. But f is convex, being the sum of convex functions. It follows readily then that $f(p)$ satisfies (12) when p is any convex combination of points of the form $(x_i + x_j)/2$, $1 \leqq i < j \leqq r$, in other words, when p belongs to the midpoint polyhedron of $\{x_1, \ldots, x_r\}$. This completes the proof. ∎

Acknowledgment

The authors would like to express their appreciation to Professor Richard Pfiefer for some helpful discussions concerning the results in this paper.

References

1. Chakerian, G. D. & M. S. Klamkin. 1973. Inequalities for sums of distances. Amer. Math. Monthly **80:** 1009–1017.
2. Eggleston, H. G. 1959. Convexity. Cambridge Tracts in Mathematics and Mathematical Physics No. 47. Cambridge University Press. New York, N.Y.
3. Wolfe, D. 1975. Metric dependence and a sum of distances. *In* The Geometry of Metric and Linear Spaces (Proceedings, Conference at Michigan State University, East Lansing, Michigan, 1974): 206–211. Lecture Notes in Mathematics No. 490. Springer-Verlag. West Berlin, FRG.

THE INTEGRAL-GEOMETRIC DENSITIES OF TANGENT FLATS TO CONVEX BODIES[a]

William J. Firey

Mathematics Department
Oregon State University
Corvallis, Oregon 97331

This paper describes a motion-invariant integral-geometric density for m-dimensional flats tangent to a sufficiently smooth convex body K in Euclidean n-space $\mathbb{E}^n$, where $1 \leq m < n$. It generalizes work of Borisov [2] and gives an explicit representation for Weil's measure of support flats described in [5], when K has sufficient smoothness.

1

To determine an oriented, m-dimensional plane L^m in $\mathbb{E}^n$ proceed as follows. First, choose an m-direction specified by an ordered set

$$u_{n-m+1}, \ldots, u_n \tag{1}$$

of m orthonormal vectors; next, enlarge this set to an orthonormal basis $u_1, \ldots, u_n$ of $\mathbb{E}^n$ by adjoining the vectors

$$u_1, \ldots, u_{n-m};$$

then, in the span Π of this last set of vectors, choose

$$x = \xi_1 u_1 + \cdots + \xi_{n-m} u_{n-m}; \tag{2}$$

finally, take L^m to be the set of all points of the form

$$x + \alpha_1 u_{n-m+1} + \cdots + \alpha_m u_n,$$

where $\alpha_1, \ldots, \alpha_m$ are arbitrary real numbers. L^m supports the convex, compact set K if L^m meets K but not the interior of K.

In what follows, write H for the support function of K given by

$$H(z) = \max_{y \in K} \langle y, z \rangle,$$

where

$$\langle y, z \rangle = y_1 z_1 + \cdots + y_n z_n$$

denotes the inner product of $\mathbb{E}^n$. Assume H is twice continuously differentiable away from the origin o. In fact, assume further: the Hessian matrix

$$\mathscr{H}(v) = [\partial^2 H(v)/\partial z_i \partial z_j], \qquad i, j = 1, 2, \ldots, n,$$

[a] This work was supported in part by U.S. National Science Foundation Grant MCS-8301706.

has maximal possible rank. In more detail, $\mathscr{H}(v)$ necessarily has zero as an eigenvalue corresponding to the normalized eigenvector v because H is homogeneous of degree one; then $\mathscr{H}(v)$ has maximal rank $n-1$. Consequently, the remaining eigenvalues

$$R_1(v), \ldots, R_{n-1}(v) \tag{3}$$

of $\mathscr{H}(v)$ are positive—H is convex—and are the positive principal radii of curvature of the bounding surface ∂K of K at the point

$$y(v) = \text{grad } H(v) = [\partial H(v)/\partial z_i]$$

of ∂K at which the support hyperplane to K has outer normal direction v. The eigenvectors corresponding to (3) make up the orthonormal set

$$v_1, \ldots, v_{n-1}$$

of principal directions for ∂K at $y(v)$ and are all tangential directions. These directions, together with v, form an orthonormal basis for $\mathbb{E}^n$. Details are in [1]. Under the smoothness assumptions, support m-dimensional flats to K are tangent to K; in what follows, I call them tangent m-flats. The assumptions also imply that each tangent m-flat to K meets K in a single point.

2

Let $K(\Pi)$ signify the orthogonal projection of K onto Π. In the notation of the preceding section, $\mathbb{L}^m$ meets K precisely when x is in $K(\Pi)$, and $\mathbb{L}^m$ is tangent to K precisely when x lies in the boundary $\partial K(\Pi)$ of $K(\Pi)$. In the latter case write

$$\bar{x} = \bar{\xi}_1 u_1 + \cdots + \bar{\xi}_N u_N \tag{4}$$

for x; here

$$N = n - m.$$

Thus the set of tangent m-flats to K, each of whose direction is specified by (1), is in 1–1 correspondence with the points of $\partial K(\Pi)$. Consequently, if we write $d\sigma$ and $d\mathbb{L}^m$ for the motion-invariant densities of points in $\partial K(\Pi)$ and of m-directions, respectively, then the outer product

$$d\sigma \wedge d\mathbb{L}^m \tag{5}$$

will be a motion-invariant density for tangent m-flats to ∂K. The density of m-directions is

$$d\mathbb{L}^m = \bigwedge \langle du_p, u_q \rangle,$$

where the exterior product of Pfaffians is taken over

$$N + 1 \le p \le n, \qquad 1 \le q \le N. \tag{6}$$

See Santaló [4, p. 202, (12.27)].

3

A motion-invariant density $d\sigma$ is the $(N-1)$-dimensional area element of $\partial K(\Pi)$ and in (5) it is to be calculated at the point $\bar{x}$ of $\partial K(\Pi)$ given by (4). Here the outer normal to $K(\Pi)$ in Π, and for that matter to K, has the form of the direction

$$v = \eta_1 u_1 + \cdots + \eta_N u_N,$$

where $\eta_1^2 + \cdots + \eta_N^2 = 1$.

For what follows it is convenient to introduce the support function $\bar{H}$ of $K(\Pi)$ defined over Π by, cf. (2),

$$H(x) = H(\xi_1 u_1 + \cdots + \xi_N u_N) = \bar{H}(\xi_1, \ldots, \xi_N).$$

We have, for $q, r = 1, 2, \ldots, N$:

$$\frac{\partial \bar{H}}{\partial \xi_q}(\xi_1, \ldots, \xi_n) = \langle u_q, \operatorname{grad} H(x) \rangle,$$

$$\frac{\partial^2 \bar{H}}{\partial \xi_r \partial \xi_q}(\xi_1, \ldots, \xi_n) = \langle u_q, \mathscr{H}(x) u_r \rangle. \tag{7}$$

Also,

$$\mathscr{H}(v) v_k = R_k(v) v_k, \qquad k = 1, 2, \ldots, n-1,$$

and

$$\mathscr{H}(v) v_k = 0.$$

Since

$$u_q = \sum_{k=1}^{n-1} \langle u_q, v_k \rangle v_k + \eta_q v,$$

the preceding equations and (7) yield

$$\langle u_q, \mathscr{H}(v) u_r \rangle = \sum_{k=1}^{n-1} R_k(v) \langle u_q, v_k \rangle \langle u_r, v_k \rangle. \tag{8}$$

Because $\bar{H}$ is the support function of $K(\Pi)$ relative to a Cartesian coordinate system in Π with basis vectors $u_1, \ldots, u_N$, the eigenvalues of

$$\mathscr{H}(\eta) = [\partial^2 \bar{H}(\eta_1, \ldots, \eta_n)/\partial \eta_r \, \partial \eta_q]$$

are $0, \rho_1, \ldots, \rho_{N-1}$, where the latter are the principal radii of curvature of $\partial K(\Pi)$ at the point $\bar{x}(v)$ at which the outer normal direction is v; the positivity of the radii in (3) guarantees the positivity of the radii $\rho_1(v), \ldots, \rho_{N-1}(v)$. In these terms the area element $d\sigma$ of $\partial K(\Pi)$ at $\bar{x}(v)$ is

$$\rho_1(v) \cdots \rho_{N-1}(v) d\omega(v) = \operatorname{tr}_{N-1} \mathscr{H}(\eta_1, \ldots, \eta_N) d\omega(v), \tag{9}$$

in which $d\omega(v)$ signifies the area element at v on the $(N-1)$-dimensional sphere in Π of outer normal directions to $K(\Pi)$ and $\operatorname{tr}_{N-1}\mathscr{H}$ signifies the trace of order $N-1$ of

$\mathscr{H}$, that is, the sum of its principal minors of order $N-1$. (**7**) and (**8**) show that the trace in (**9**) will be a sum of N determinants of the form

$$\det\left[\sum_{k=1}^{n-1} R_k(v)\langle u_p, v_k\rangle\langle u_q, v_k\rangle\right],$$

where, in each determinant, p, q range independently over an index set obtained by deleting one number from $\{1, 2, \ldots, N\}$.

Standard properties of determinants give

$$\rho_1 \cdots \rho_{N-1} = \sum_{*} R_{i_1} \cdots R_{i_{N-1}} \sum_{**}\left(\det\begin{bmatrix}\langle u_{j_1}, v_{i_1}\rangle & \cdots & \langle u_{j_1}, v_{i_{N-1}}\rangle \\ \cdots & \cdots & \cdots \\ \langle u_{j_{N-1}}, v_{i_1}\rangle & \cdots & \langle u_{j_{N-1}}, v_{i_{N-1}}\rangle\end{bmatrix}\right)^2, \tag{10}$$

in which the starred summation extends over $i_1, \ldots, i_{N-1}$ such that

$$1 \le i_1 < \cdots < i_{N-1} \le n-1, \tag{11}$$

and the double-starred summation extends over $j_1, \ldots, j_{N-1}$ such that

$$1 \le j_1 < \cdots < j_{N-1} \le N. \tag{12}$$

If K is the n-dimensional unit ball, then all principal radii in (**10**) equal one and so

$$\sum_{*}\sum_{**}\left(\det\begin{bmatrix}\langle u_{j_1}, v_{i_1}\rangle & \cdots & \langle u_{j_1}, v_{i_{N-1}}\rangle \\ \cdots & \cdots & \cdots \\ \langle u_{j_{N-1}}, v_{i_1}\rangle & \cdots & \langle u_{j_{N-1}}, v_{i_{N-1}}\rangle\end{bmatrix}\right)^2 = 1.$$

To summarize:

The density for oriented m-dimensional tangent flats to the convex body K, computed at the point $y(v)$ of ∂K at which the outward normal direction is v, and for the tangential m-direction specified by the ordered set

$$u_{N+1}, \ldots, u_n, \qquad N = n - m \ge 0, \tag{13}$$

of orthonormal vectors, each orthogonal to v, is given by the differential form

$$\sum_{*} R_{i_1} \cdots R_{i_{N-1}} \sum_{**}\left(\det\begin{bmatrix}\langle u_{j_1}, v_{i_1}\rangle & \cdots & \langle u_{j_1}, v_{i_{N-1}}\rangle \\ \cdots & \cdots & \cdots \\ \langle u_{j_{N-1}}, v_{i_1}\rangle & \cdots & \langle u_{j_{N-1}}, v_{i_{N-1}}\rangle\end{bmatrix}\right)^2 d\omega(v) \wedge \langle du_p, u_q\rangle. \tag{14}$$

In this expression $*$ denotes the index set described by (**11**), $**$ denotes the index set described by (**12**), the exterior product extends over the set in (**6**), and $u_1, \ldots, u_N$ extends (**13**) to an orthonormal basis. Also, $v_1, \ldots, v_{n-1}$ are principal directions at $y(v)$ corresponding to the principal radii $R_1(v), \ldots, R_{n-1}(v)$.

Observe that integration of (**14**) over all m-directions specified in (**1**) gives

$$C \sum_{*} R_i(v) \cdots R_{i_{N-1}}(v) d\omega(v)$$

for a suitable positive constant c. This accords with [3].

References

1. Bonnesen, T. & W. Fenchel. 1934. Theorie der konvexen Körper. Springer-Verlag. Berlin, Germany.
2. Borisov, A. V. 1981. Integral formulas for a convex hypersurface in E_n. C. R. Acad. Bulgare Sci. **34:** 173–175. (In Russian.)
3. Firey, W. 1972. An integral-geometric meaning for lower order area functions. Mathematika **19:** 205–212.
4. Santaló, L. A. 1976. Integral Geometry and Geometric Probability. Addison-Wesley Publishing Co. Reading, Mass.
5. Weil, W. 1981. Zufällige Berührung konvexer Körper durch q-dimensionale Ebenen. Resultate Math. **4** (No. 1): 84–101.

VOLUME AND CIRCUMRADIUS OF SIMPLICIAL POLYTOPES

Johann Linhart

Mathematisches Institut der Universität Salzburg
A-5020 Salzburg, Austria

By a d-polytope we mean a convex polytope in $\mathbb{R}^d$ containing interior points. It is called simplicial if all of its faces are simplices. The $(d-1)$-dimensional faces of a d-polytope are called facets.

For $d = 2$ or 3 the regular d-polytopes may be characterized by a number of extremal properties [4]. In higher dimensions only a few corresponding results have been obtained so far. Here we shall prove the following extremal property of regular simplicial d-polytopes:

THEOREM 1. *Each regular simplicial d-polytope has the greatest volume among all simplicial d-polytopes with the same number of facets and the same circumradius.*

Remark. The corresponding theorem for the inradius was proved by Fejes Tóth [3] in 1955 (cf. also [4]).

Proof of THEOREM 1. The validity of the theorem is well known for $d \leq 3$, as well as for simplices of arbitrary dimension [4]. For crosspolytopes in $\mathbb{R}^d$ it may be shown in the following manner: Let P be a simplicial d-polytope with 2^d facets and circumradius R. We may assume that the center of the circumsphere coincides with the origin O. For each facet F_i of P let Q_i be the convex hull of F_i and O. Denoting by V the d-dimensional volume, we have

$$V(P) = \sum_{i=1}^{2^d} V(Q_i).$$

If $A_1^{(i)}, \ldots, A_d^{(i)}$ are the vectors pointing from O to the vertices of Q_i,

$$V(Q_i) = \left| \frac{1}{d!} \det (A_1^{(i)}, \ldots, A_d^{(i)}) \right|.$$

Here the determinant reaches the maximal absolute value if $A_1^{(i)}, \ldots, A_d^{(i)}$ are pairwise orthogonal and have length R. Thus we have $V(Q_i) \leq (1/d!)R^d$ and therefore $V(P) \leq (2^d/d!)R^d$, with equality if and only if P is a regular d-crosspolytope. Besides the regular simplices and crosspolytopes there is only one regular simplicial d-polytope with $d \geq 4$, namely, the regular 600-cell in $\mathbb{R}^4$ [4]. In this case the assertion of THEOREM 1 is a consequence of THEOREM 2 below. ∎

Preliminaries to THEOREM 2. Let T be a regular spherical tetrahedron on the unit sphere $S^3 = \{x \in \mathbb{R}^4 : \|x\| = 1\}$ with volume τ. By $v(\tau)$ we denote the volume of the (Euclidean) convex hull of O and the vertices of T.

THEOREM 2. *Let V be the volume and R the circumradius of a simplicial 4-polytope P with n facets, $n \leq 8$ or $n \geq 16$. Then*

$$V/R^4 \leq n \cdot v\left(\frac{2\pi^2}{n}\right).$$

Here equality holds if and only if P is a regular polytope.

Remarks. It may of course be conjectured that this inequality holds for all $n \geq 5$. Possibly for a corresponding proof it would be necessary to investigate separately all combinatorial types of simplicial 4-polytopes with $9 \leq n \leq 15$. (There are 13 such types, cf. [5, 6].) One might even conjecture that the analogue of THEOREM 2 holds for arbitrary dimensions d and all $n \geq d + 1$. For $d = 2$ this may be seen very easily. For $d = 3$, L. Fejes Tóth, A. Florian, and H. Florian have proven a more general inequality for arbitrary (not necessarily simplicial) polytopes, where equality occurs exactly for the five regular 3-polytopes [4].

As to the inradius r, Fejes Tóth [3] has proven $V/r^4 \geq n \cdot \mathbf{v}(2\pi^2/n)$ for all $n \geq 5$ (and the analogous inequality for all $d > 4$). Here $\mathbf{v}(\tau)$ means the volume of the convex hull of O and a regular 3-simplex, which is touched by S^3 in its center and the radial projection of which onto S^3 has volume τ.

Proof of THEOREM 2 *for $n \leq 8$.* For $n = 5$ we have the above-mentioned extremal property of the regular simplex. If P is not a 4-simplex, then $n \geq 8$ [5]. For $n = 8$, P is necessarily a bipyramid over a 3-simplex T_0 [5]. The circumradius of T_0 is $\leq R$; thus, the volume of T_0 is at most equal to the volume of a regular tetrahedron with circumradius R, that is, $\leq (8/9\sqrt{3})R^3$. The heights of the two pyramidal parts of P with common basis T_0 together are $\leq 2R$, so we obtain

$$V \leq \frac{1}{4} \cdot 2R \cdot \frac{8}{9\sqrt{3}} R^3;$$

this means

$$V/R^4 \leq \frac{4}{9\sqrt{3}} = 0.2566\ldots < 8v\left(\frac{2\pi^2}{8}\right) = 0.2982\ldots.$$

(For the evaluation of $v(\tau)$ compare the proof of LEMMA 1.) ∎

For the proof of THEOREM 2 for $n \geq 16$:

LEMMA 1. *$v(\tau)$ is increasing and concave on the interval* $[0, \pi^2/8]$. *Furthermore, $v(\tau) \leq v(\pi^2/8)$ for all $\tau \in [0, \pi^2]$.*

Remark. $v(\tau)$ is defined on $[0, \pi^2]$, but not concave on the whole interval. There is a point of inflection near $0.34378\ldots\pi^2$.

Proof of LEMMA 1. Let T be a regular spherical tetrahedron on S^3 and 2ψ the (inner) angle at the edges of T. Then we have the following formula for the volume τ of T (cf. [1]):

$$\tau = 6 \int_{(1/2)\,\mathrm{arcsec}\,3}^{\psi} \mathrm{arcsec}\,(\sec 2x - 2)\,dx.$$

The angle ψ is connected with the circumradius ρ of T by $\tan \psi = \sqrt{\frac{1}{6}(3 + \tan^2 \rho)}$.

Setting $t := \tan \rho$, we may regard τ as a function of t and obtain after a short computation:

$$\frac{d\tau}{dt} = 12\sqrt{6}\,\frac{t}{(9 + t^2)\sqrt{3 + t^2}} \arctan \frac{t\sqrt{2}}{\sqrt{3 + t^2}}.$$

Now let T_1 be the (Euclidean) convex hull of the vertices of T, and Q the convex hull of O and T_1. T_1 is a regular Euclidean tetrahedron with circumradius $\sin \rho$ and hence has volume $(8/9\sqrt{3}) \sin^3 \rho$. Since the height of Q equals $\cos \rho$, we get for the volume v of Q:

$$v = \frac{2}{9\sqrt{3}} \sin^3 \rho \cos \rho = \frac{2}{9\sqrt{3}} \frac{t^3}{(1 + t^2)^2} \tag{1}$$

and

$$\frac{dv}{dt} = \frac{2}{9\sqrt{3}} \frac{t^2(3 - t^2)}{(1 + t^2)^3}. \tag{2}$$

Here we see that v reaches its maximum for $t = \sqrt{3}$. The corresponding value $\rho = \pi/3$ is the circumradius of the facets of the regular crosspolytope, where $\tau = \pi^2/8$. We conclude $v(\tau) \leq v(\pi^2/8)$ for all $\tau \in [0, \pi^2]$. By the way, this may also be inferred from the considerations in the proof of THEOREM 1.

Since τ is increasing as a function of t, we see by **(2)** that $v(\tau)$ is increasing for $0 \leq t \leq \sqrt{3}$, i.e., for $0 \leq \tau \leq \pi^2/8$.

Furthermore, we obtain

$$\frac{dv}{d\tau} = \frac{dv}{dt}\bigg/\frac{d\tau}{dt} = \frac{1}{162\sqrt{2}}\, \frac{t(3 - t^2)(9 + t^2)\sqrt{3 + t^2}}{(1 + t^2)^3}\bigg/ \arctan \frac{t\sqrt{2}}{\sqrt{3 + t^2}}. \tag{3}$$

To prove the concavity of $v(\tau)$ on $[0, \pi^2/8]$ it is sufficient to show that the function of t specified in **(3)** is decreasing on $[0, \sqrt{3}]$. By taking derivatives we see after some computations that the following inequality should be proved for $t \in [0, \sqrt{3}]$:

$$t(3 - t^2)(9 + t^2)\sqrt{2(3 + t^2)} \geq (81 - 405t^2 - 93t^4 + 9t^6) \arctan \frac{t\sqrt{2}}{\sqrt{3 + t^2}}.$$

If the factor in front of the arctan is negative, the inequality is trivial, since the left-hand side is ≥ 0 for $t \in [0, \sqrt{3}]$. So because of $\arctan x \leq x$ for $x \geq 0$ it is sufficient to prove the following inequality:

$$(3 - t^2)(9 + t^2)(3 + t^2) \geq 81 - 405t^2 - 93t^4 + 9t^6,$$

i.e.,

$$414 + 84t^2 - 10t^4 \geq 0.$$

Since the only positive root of this polynomial is $> \sqrt{3}$, the proof of LEMMA 1 is finished. ∎

Preliminaries to LEMMA 2. Let $U := \{x \in \mathbb{R}^2 : \|x\| \leq 1\}$ be the unit disk. By Z we denote the cylinder $U \times (-\pi, \pi]$. We use the following parametric representation of S^3:

$$\begin{aligned} \varphi \colon Z \to S^3 : x = (x_1, x_2, x_3) \\ \mapsto (x_1, x_2, \sqrt{1 - x_1^2 - x_2^2} \cos x_3, \sqrt{1 - x_1^2 - x_2^2} \sin x_3). \end{aligned}$$

φ leaves the volume invariant [2]. If M is an arbitrary subset of $\mathbb{R}^3$, let $M' := \varphi(M \cap Z)$.

Now let $T = ABCD$ be an arbitrary spherical tetrahedron on S^3, such that the lengths of the edges of T are $< \pi$. Replacing T by a congruent tetrahedron if necessary, we may assume

$$\begin{aligned} &A = (0, 0, \cos\theta, \sin\theta), \qquad B = (0, 0, \cos\theta, -\sin\theta), \\ &C = (0, c_2, \sqrt{1 - c_2^2} \cos\gamma, \sqrt{1 - c_2^2} \sin\gamma), \\ &D = (d_1, d_2, \sqrt{1 - d_1^2 - d_2^2} \cos\delta, \sqrt{1 - d_1^2 - d_2^2} \sin\delta) \end{aligned}$$

with $\theta \in (0, \pi)$, γ and $\delta \in (-\pi, \pi]$.

Let E be the plane $\{x \in \mathbb{R}^3 : x_3 = 0\}$. Then E' is the spherical plane of symmetry of AB. Now we consider two kinds of symmetrization of T with respect to E' and compare the results.

First, we apply Steiner's symmetrization to $\varphi^{-1}(T)$ with respect to E and denote the result by $\overline{\varphi^{-1}(T)}$. We then consider $\bar{T} := \varphi(\overline{\varphi^{-1}(T)})$ (cf. [2]). $\bar{T}$ has the same volume as T, but in general it is not a tetrahedron anymore (as will soon be apparent).

Second, we construct a tetrahedron $\tilde{T}$ which is symmetric with respect to E': Let p_E be the orthogonal projection from $\mathbb{R}^3$ onto E and $p'_E := \varphi \circ p_E \circ \varphi^{-1}$ the corresponding projection from S^3 onto E'. Now let $\tilde{C} := p'_E(C)$, $\tilde{D} := p'_E(D)$, and $\tilde{T}$ be the spherical tetrahedron $AB\tilde{C}\tilde{D}$.

The volume of T, respectively $\bar{T}$, is denoted by τ, that of $\tilde{T}$ by $\tilde{\tau}$.

LEMMA 2. *If T is not symmetric with respect to E', then $\tilde{\tau} < \tau$.*

Proof of LEMMA 2. Let G be a straight line in $\mathbb{R}^3$ parallel to the x_3-axis, which meets the interior of $\varphi^{-1}(T)$: $G = \{x \in \mathbb{R}^3 : x_1 = g_1, x_2 = g_2\}$, where (g_1, g_2, g_3) is an interior point of $\varphi^{-1}(T)$.

G together with the x_3-axis determines a plane F satisfying the equation $g_2 x_1 - g_1 x_2 = 0$. The corresponding spherical plane F' intersects T in a spherical triangle $\Delta = ABY$, where Y is a point on the edge $\overline{CD}$. Now we apply the two kinds of symmetrization explained above also to Δ. $\bar{\Delta}$ is the set obtained by Steiner's symmetrization, and $\tilde{\Delta}$ is the triangle $AB\tilde{Y}$ with $\tilde{Y} := p'_E(Y)$. G', Δ, $\bar{\Delta}$, and $\tilde{\Delta}$ are all lying in the plane F'. Denoting by l the usual measure of length, we have in view of the definition of Steiner's symmetrization

$$l(G \cap \varphi^{-1}(\bar{\Delta})) = l(G \cap \varphi^{-1}(\Delta)).$$

We will show

$$l(G \cap \varphi^{-1}(\tilde{\Delta})) < l(G \cap \varphi^{-1}(\Delta)), \tag{4}$$

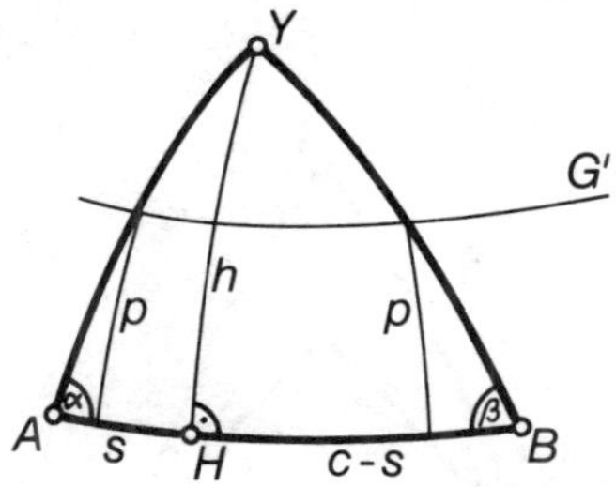

FIGURE 1

if Δ is not symmetric with respect to E' (i.e., Δ is not an isosceles triangle with basis AB). We denote the angles of Δ at A and B by α and β, respectively (FIGURE 1). The length of $\overline{AB}$ is denoted by c, the length of the altitude to AB by h $(0 < h \leq \pi/2)$, and the foot of h by H.

H is uniquely determined for $h < \pi/2$. Since Δ is not isosceles with basis AB, the case $h = \pi/2$ cannot occur. Let s be the length of the arc of a great circle with endpoints A and H lying in the plane AOB and containing relatively interior points of $\overline{AB}$. (For $H = A$ let $s := 0$.) So s may take on all values of the interval $[0, 2\pi]$. G' is an equidistant line to AB. By p we denote the distance from G' to AB ($p = \arcsin \sqrt{g_1^2 + g_2^2}$). Using some elementary spherical trigonometry we now obtain for $0 < p \leq h$:

$$l(G \cap \varphi^{-1}(\Delta)) = c - \arcsin(k \sin s) - \arcsin(k \sin(c - s))$$

with $k := \tan p/\tan h$.

It may happen that $p > h$ (and nevertheless G' meets the interior of Δ). But in this case $G' \cap \tilde{\Delta} = \varnothing$ and the assertion **(4)** becomes trivial. By the way, the case $p = h$ is also trivial, since then $G' \cap \tilde{\Delta}$ consists of one point only. So we may assume $0 < p < h$.

Setting $f(s) := \arcsin(k \sin s)$ we have $l(G \cap \varphi^{-1}(\tilde{\Delta})) = c - f(s) - f(c - s)$, and similarly $l(G \cap \varphi^{-1}(\tilde{\Delta})) = c - f(c/2) - f(c/2)$, since $\tilde{\Delta}$ has the same altitude h as Δ. So it is sufficient to show

$$g(s) := f(s) + f(c - s) < 2f\left(\frac{c}{2}\right)$$

for $s \neq c/2$.

$f'(s) = (k \cos s)/\sqrt{1 - k^2 \sin^2 s}$ is strictly decreasing on $[0, \pi]$ and strictly increasing on $[\pi, 2\pi]$. Furthermore, $f'(s) = f'(s + 2\pi) = f'(-s)$. Now let s_0 be a point where g reaches its maximum. Then $g'(s_0) = f'(s_0) - f'(c - s_0) = 0$. By the mentioned properties of f' it follows $s_0 = c - s_0 + 2m\pi$ or $-s_0 = c - s_0 + 2m\pi$ for an integer m. The second possibility cannot occur because $c \neq -2m\pi$. The first possibility yields $s_0 = c/2 + m\pi$. Because of the periodicity of f we have merely to consider $m = 0$ or 1.

$$g''\left(\frac{c}{2} + \pi\right) = f''\left(\frac{c}{2} + \pi\right) + f''\left(\frac{c}{2} - \pi\right) = 2f''\left(\frac{c}{2} + \pi\right) > 0,$$

since $c/2 + \pi \in (\pi, 2\pi)$. Consequently $s_0 = c/2$, which proves **(4)**.

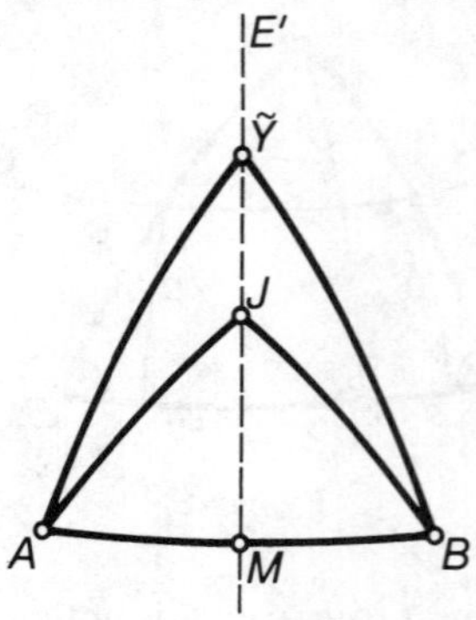

FIGURE 2

$\bar{\Delta} = F' \cap \bar{T}$, but in general $\tilde{\Delta} \neq F' \cap \tilde{T}$. We show

$$F' \cap \tilde{T} \subset \tilde{\Delta}. \tag{5}$$

These are two isosceles triangles in the plane F' with the same base $\overline{AB}$. The apex of $\tilde{\Delta}$ is $\tilde{Y}$. Let us denote the apex of $F' \cap \tilde{T}$ by J. $\tilde{Y}$ and J lie on the same side of the line $\overline{AB}$, so we have only to show that the distance from J to the midpoint M of $\overline{AB}$ does not exceed that from $\tilde{Y}$ to M (FIGURE 2).

$M = (0, 0, 1, 0)$, J, $\tilde{Y}$, $\tilde{C}$, and $\tilde{D}$ all lie in the plane E' (FIGURE 3). Since J lies on $\tilde{C}\tilde{D}$, we have only to prove that M and $\tilde{Y}$ do not lie in the same open half plane of E' bounded by $\tilde{C}\tilde{D}$. E' is the unit sphere in the hyperplane $\{X \in \mathbb{R}^4 : x_4 = 0\}$ which may be identified with $\mathbb{R}^3$ by simply omitting the fourth coordinate. Therefore the spherical straight line $\tilde{C}\tilde{D}$ is equal to $\{X \in E' : \det(\tilde{C}, \tilde{D}, X) = 0\}$. So we have to verify

$$\operatorname{sgn} \det(\tilde{C}, \tilde{D}, M) \neq \operatorname{sgn} \det(\tilde{C}, \tilde{D}, \tilde{Y}). \tag{6}$$

We have $\tilde{D} = (d_1, d_2, \sqrt{1 - d_1^2 - d_2^2})$ (using the identification mentioned above). $\tilde{C} = (0, c_2, \sqrt{1 - c_2^2})$, $M = (0, 0, 1)$, so we get

$$\det(\tilde{C}, \tilde{D}, M) = -c_2 d_1. \tag{7}$$

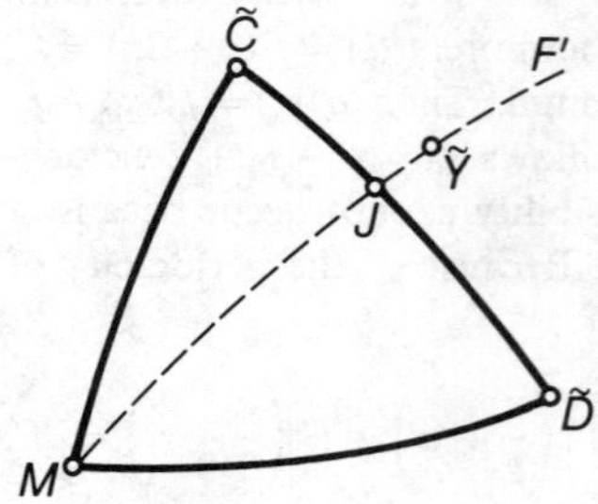

FIGURE 3

In an analogous notation $\tilde{Y} = (y_1, y_2, \sqrt{1 - y_1^2 - y_2^2})$. Y is lying on the spherical line segment $\overline{CD}$, so there is a $\lambda \in [0, 1]$ such that

$$Y = \frac{1}{a}(\lambda C + (1 - \lambda)D) \qquad \text{with } a = \|\lambda C + (1 - \lambda)D\|.$$

Thus we have

$$y_1 = \frac{1}{a}(1 - \lambda)d_1, \qquad y_2 = \frac{1}{a}(\lambda c_2 + (1 - \lambda)d_2).$$

From this we easily compute

$$\det(\tilde{C}, \tilde{D}, \tilde{Y}) = -c_2 d_1 \cdot \frac{1}{a}\Gamma,$$

where

$$\Gamma = \sqrt{a^2 - ((1 - \lambda)d_1)^2 - (\lambda c_2 + (1 - \lambda)d_2)^2} \\ -\lambda\sqrt{1 - c_2^2} - (1 - \lambda)\sqrt{1 - d_1^2 - d_2^2}. \tag{8}$$

In view of (**6**) and (**7**) we have to show $\Gamma \leq 0$. To this end we estimate a^2 from above.

$$a^2 = \lambda^2 + (1 - \lambda)^2 + 2\lambda(1 - \lambda)\langle C, D\rangle,$$

where $\langle\ ,\ \rangle$ denotes the usual inner product. Hence,

$$a^2 \leq 1 - 2\lambda(1 - \lambda)(1 - c_2 d_2 - \sqrt{(1 - c_2^2)(1 - d_1^2 - d_2^2)}).$$

Replacing a^2 in (**8**) by this upper bound, the resulting expression is identically zero, as may easily be checked. So $\Gamma \leq 0$ and (**5**) is proved. Hence, it follows $G' \cap \tilde{T} \subset G' \cap \tilde{\Delta}$. Together with (**4**) we obtain

$$l(G \cap \varphi^{-1}(\tilde{T})) < l(G \cap \varphi^{-1}(\Delta)) = l(G \cap \varphi^{-1}(T)) \tag{9}$$

if Δ is not symmetric relative to E'; and this must hold for all Δ with at most one exception, if T is not symmetric relative to E'. Now τ is equal to the volume of $\varphi^{-1}(T)$, so by Fubini's theorem

$$\tau = \int_{\mathfrak{G}} l(G \cap \varphi^{-1}(T))\, dG,$$

where $\mathfrak{G}$ is the set of all straight lines parallel to the x_3-axis and dG is the usual measure on $\mathfrak{G}$. Representing $\tilde{\tau}$ in a similar manner, we see that (**9**) implies the assertion of LEMMA 2. ▮

Preliminaries to LEMMA 3. Let Q be a 4-simplex with four vertices A, B, C, D on S^3 and one vertex coinciding with O. Let $T(Q)$ be the spherical tetrahedron $ABCD$ and $\tau(Q)$ its volume. For each number $v > 0$ we denote by $\mathfrak{S}_v$ the set of all 4-simplices of this kind having volume v. $\mathfrak{S}_v$ is a compact topological space and τ a continuous function on it, if we take the natural topology which is, e.g., induced by the Hausdorff metric [5]. Since τ is bounded from below, it attains its minimum on $\mathfrak{S}_v$.

LEMMA 3. *Let* $\hat{Q}$ *be an element of* $\mathfrak{S}_v$ *with the property*

$$\tau(\hat{Q}) = \min_{Q \in \mathfrak{S}_v} \tau(Q).$$

Then $T(\hat{Q})$ *is a regular spherical tetrahedron.*

Proof of LEMMA 3. Assume $\hat{T} := T(\hat{Q}) = ABCD$ is not regular. Then there is an edge of $\hat{T}$, say AB, such that $\hat{T}$ is not symmetric with respect to the plane of symmetry of this edge. Now let $\tilde{T} := AB\tilde{C}\tilde{D}$ in the notation of LEMMA 2, and $\tilde{Q}$ be the corresponding 4-simplex; thus, $\tilde{Q} = AB\tilde{C}\tilde{D}O$ and $\tilde{T} = T(\tilde{Q})$. $\tilde{Q}$ has the same volume as $\hat{Q}$, since the straight lines $C\tilde{C}$ and $D\tilde{D}$ are parallel to the plane ABO:

$$\tilde{C} - C = (0, 0, \sqrt{1 - c_2^2}(1 - \cos \gamma), -\sqrt{1 - c_2^2} \sin \gamma),$$
$$\tilde{D} - D = (0, 0, \sqrt{1 - d_1^2 - d_2^2}(1 - \cos \delta), -\sqrt{1 - d_1^2 - d_2^2} \sin \delta),$$

and the plane ABO is determined by the equation $x_1 = x_2 = 0$. Consequently $\tilde{Q} \in \mathfrak{S}_v$. But by LEMMA 2 we have $\tau(\tilde{Q}) < \tau(\hat{Q})$, contradicting the minimality of $\tau(\hat{Q})$. ∎

Proof of THEOREM 2. Without restriction of generality we may assume $R = 1$. Let $F_1, \ldots, F_n$ be the facets of P. We project the vertices of F_i radially onto the circumsphere S^3. The points so obtained together with O determine a 4-simplex Q_i, the volume of which we call v_i. Thus $Q_i \in \mathfrak{S}_{v_i}$ for $i \in \{1, \ldots, n\}$. Clearly, $V \leq \sum_{i=1}^n v_i$ and $\sum_{i=1}^n \tau(Q_i) = 2\pi^2$. Now let $\hat{\tau}_i := \min_{Q \in \mathfrak{S}_{v_i}} \tau(Q) \leq \tau(Q_i)$. By LEMMA 3 there is a 4-simplex $\hat{Q}_i \in \mathfrak{S}_{v_i}$ such that $\hat{\tau}_i = \tau(\hat{Q}_i)$, and $T(\hat{Q}_i)$ is a regular spherical tetrahedron. Using the notation introduced at the beginning, this means $v_i = v(\hat{\tau}_i)$. Now let

$$\bar{v}(\tau) := v(\tau) \qquad \text{for } \tau \in \left[0, \frac{\pi^2}{8}\right],$$
$$:= v\left(\frac{\pi^2}{8}\right) \qquad \text{for } \tau \in \left[\frac{\pi^2}{8}, \pi^2\right].$$

By LEMMA 1, $\bar{v}$ is concave and increasing. So we obtain

$$V \leq \sum_{i=1}^n v_i = \sum v(\hat{\tau}_i) \leq \sum \bar{v}(\tau_i) \leq n\bar{v}\left(\frac{1}{n} \sum \hat{\tau}_i\right)$$
$$\leq n\bar{v}\left(\frac{1}{n} \sum \tau(Q_i)\right) = n\bar{v}\left(\frac{2\pi^2}{n}\right).$$

For $n \geq 16$ we have $2\pi^2/n \in [0, \pi^2/8]$ and thus $\bar{v}(2\pi^2/n) = v(2\pi^2/n)$. This proves the inequality of THEOREM 2. The proof shows that equality can only occur if all facets are regular tetrahedra with vertices lying on the circumsphere. This is only the case for regular simplicial polytopes. ∎

REFERENCES

1. BÖHM, J. & E. HERTEL. 1981. Polyedergeometrie in n-dimensionalen Räumen konstanter Krümmung. Birkhäuser. Basel, Switzerland.
2. DINGHAS, A. 1961. Minkowskische Summen und Integrale. Mémorial des sciences Math., Fasc. CXLIX. Gauthier–Villars. Paris, France.

3. Fejes Tóth, L. 1955. Extremum properties of the regular polytopes. Acta Math. Acad. Sci. Hungar. **6**: 143–146.
4. Fejes Tóth, L. 1965. Reguläre Figuren. Akadémiai Kiadó. Budapest, Hungary.
5. Grünbaum, B. 1967. Convex Polytopes. Interscience. London, England.
6. Grünbaum, B. & V. P. Sreedharan. 1967. An enumeration of simplicial 4-polytopes with 8 vertices. J. Combin. Theory **2**: 437–465.

ON THE BLASCHKE–SANTALÓ INEQUALITY

Erwin Lutwak

Department of Mathematics
Polytechnic Institute of New York
Brooklyn, New York 11201

The Blaschke–Santaló inequality is one of the fundamental geometric inequalities. It has been a valuable tool in the fields of convexity (see, for example, [1, 9, 12, 15, 17, 19–26, 31–34, 40]), Minkowskian geometry (see, for example, [10, 18, 29]), functional analysis (see, for example, [6–8, 27, 28, 36–38]), stochastic geometry (see, for example, [43, 44]), and differential equations (see, for example, [11, 13, 16, 30, 41]).

The relation between the Blaschke–Santaló inequality and the affine isoperimetric inequality has been mentioned in several articles. Blaschke (for dimensions 2 and 3) and Santaló (for arbitrary dimensions) established the Blaschke–Santaló inequality by using the affine isoperimetric inequality in conjunction with the Minkowski (mixed volume) inequality. It has been noted by several investigators (see [15, 33, 34, 42]) that if one starts with the Blaschke–Santaló inequality one can derive the affine isoperimetric inequality as a consequence. It turns out, however, that the Blaschke–Santaló inequality will yield a more general inequality than the affine isoperimetric inequality.

Using a method similar to that of Blaschke and Santaló it can (and will presently) be shown that the Blaschke–Santaló inequality can be obtained quickly from a single geometric inequality—which we refer to as the general affine isoperimetric inequality. This general inequality is in a sense weaker than the Minkowski inequality. However, special cases of the general affine isoperimetric inequality include the affine isoperimetric inequality, Berwald's generalization of an inequality of Winternitz, and the Urysohn inequality.

This general affine isoperimetric inequality is a simple consequence of the affine isoperimetric inequality and the Minkowski inequality. However, what seems surprising is that a polar version of this general inequality, which can be obtained easily from the Hölder (integral) inequality, can be combined with the Blaschke–Santaló inequality to yield the general affine isoperimetric inequality. In fact, the polar version of the general affine isoperimetric inequality acts as a sort of intermediary between the Blaschke–Santaló inequality and the general affine isoperimetric inequality.

The setting for this paper will be Euclidean n-dimensional space, $\mathbb{R}^n$ ($n \geq 2$). We will use the letter K, possibly with superscripts, to denote a convex body (compact, convex set with nonempty interior) in $\mathbb{R}^n$. For a convex body K we will use $V(K)$, $S(K)$, and $B(K)$ to denote the volume, surface area, and mean width of K, respectively. We will use U to denote the unit ball and ω_n to denote its volume.

Associated with a convex body K is its support function h_K which is defined on $\mathbb{R}^n$ by

$$h_K(x) = \text{Max}\ \{x \cdot v : v \in K\},$$

where $x \cdot v$ denotes the usual inner product of x and v in $\mathbb{R}^n$. If K contains the origin in its interior, then also associated with K is its distance function d_K which is defined on $\mathbb{R}^n$ by

$$d_K(x) = \text{Min}\ \{\lambda > 0 : x \in \lambda K\}.$$

If K contains the origin in its interior, then the polar body of K (with respect to the unit sphere centered at the origin) is denoted by K^* and can be defined by

$$h_{K^*} = d_K.$$

For a convex body K and a point $x \in \text{int}\ K$, we write K^x for the polar body of K with x taken as the origin; i.e., $K^x = x + (-x + K)^*$. If $x \in \text{int}\ K$, and $y \in \text{int}\ K^x$, we write K^{xy} for $(K^x)^y$.

Two important points associated with a convex body K are its centroid, $c = \text{Cen}\ (K)$, and its Santaló point, $s = \text{San}\ (K)$. There are a number of characterizations of the Santaló point (see [3, 12, 32, 39]). The Santaló point of K is often defined as the unique point $s \in \text{int}\ K$, such that

$$V(K^s) = \text{Min}\ \{V(K^x) : x \in \text{int}\ K\}.$$

The Santaló point can also be defined as the unique point $s \in \text{int}\ K$ for which

$$\int_{S^{n-1}} u h_{-s+K}^{-(n+1)}(u)\ dS(u) = 0,$$

where $dS(u)$ denotes the area element on the unit sphere S^{n-1} at the point u. Another important characterization of the Santaló point of K is that it is the unique point $s \in \text{int}\ K$ for which $\text{Cen}\ (K^s) = s$.

From the last characterization it follows that for any convex body K we have

$$\text{San}\ (K^c) = \text{Cen}\ (K) \qquad \text{and} \qquad \text{Cen}\ (K^s) = \text{San}\ (K).$$

Hence,

$$K^{sc} = K,$$

where s is San (K) and c is Cen (K^s), and also

$$K^{cs} = K,$$

where c is Cen (K) and s is San (K^c). We note that if a convex body K is centrally symmetric, then its Santaló point, centroid, and center of symmetry all coincide.

The Blaschke–Santaló inequality is:

THEOREM 1. *If K is a convex body in $\mathbb{R}^n$, then*

$$V(K)V(K^s) \leq \omega_n^2,$$

with equality if and only if K is an ellipsoid.

The Blaschke–Santaló inequality is due to Blaschke [3] for $n \leq 3$ and Santaló [39] for $n \geq 2$ (see also the comments in [43, p. 552]). From the work of Blaschke [3] and Santaló [39] also follow the conditions for equality when K is sufficiently smooth. The conditions for equality for centrally symmetric K without smoothness assumptions were obtained by Saint Raymond [37]. The conditions for equality for arbitrary K are due to Petty [34].

As a consequence of the (previously mentioned) duality between the centroid and Santaló points, the Blaschke–Santaló inequality can also be written as:

THEOREM 1′. *If K is a convex body in* $\mathbb{R}^n$, *then*

$$V(K)V(K^c) \leq \omega_n^2,$$

with equality if and only if K is an ellipsoid.

A convex body A is said (see [5, p. 114]) to have positive continuous curvature function,

$$f_A : S^{n-1} \to (0, \infty),$$

provided that for each convex body K the mixed volume $V_1(A, K) = V(A, \ldots, A, K)$ has the integral representation

$$V_1(A, K) = \frac{1}{n} \int_{S^{n-1}} h_K(u) f_A(u)\, dS(u).$$

A convex body can have at most one curvature function (see [5, p. 115]). We will use the letter A to denote a convex body that has a positive continuous curvature function.

For a convex body A with positive continuous curvature function f_A, the affine surface area of A, $\Omega(A)$, can be defined by

$$\Omega(A) = \int_{S^{n-1}} f_A(u)^{n/(n+1)}\, dS(u).$$

The affine isoperimetric inequality states that, for a convex body A in $\mathbb{R}^n$ that has a positive continuous curvature function,

$$\Omega(A)^{n+1} \leq \omega_n^2 n^{n+1} V(A)^{n-1},$$

with equality if and only if A is an ellipsoid.

For a sufficiently smooth body A, the affine isoperimetric inequality is due to Blaschke [2] (for $n \leq 3$) and Santaló [39] (for $n \geq 2$). The version of the affine isoperimetric inequality given above (valid for any convex body A with positive continuous curvature function) is due to Petty [34].

The Minkowski (mixed volume) inequality [5, p. 91] states that for convex bodies K, $\bar{K}$ in $\mathbb{R}^n$

$$V(\bar{K})^{n-1} V(K) \leq V_1(\bar{K}, K)^n,$$

with equality if and only if K and $\bar{K}$ are homothetic.

The Minkowski inequality and the affine isoperimetric inequality can be trivially combined to yield the following general affine isoperimetric inequality:

THEOREM 2. *If A and K are convex bodies in* $\mathbb{R}^n$ *and A has a positive continuous curvature function, then*

$$\Omega(A)^{n+1} \leq \omega_n^2 n^{n+1} V_1^n(A, K)/V(K),$$

with equality if and only if A and K are homothetic ellipsoids.

Since $\Omega(A)^{n+1} \le \omega_n^2 n^{n+1} V(A)^{n-1}$, this general affine isoperimetric inequality makes a "weaker" statement than the Minkowski inequality. Using arguments similar to those of Blaschke [3] and Santaló [39], it can (and will) be shown that the Blaschke–Santaló inequality can be obtained directly from this general affine isoperimetric inequality by using the solution to the Minkowski problem [35]. However, the Hölder (integral) inequality can be used to establish a geometric inequality that, when combined with the Blaschke–Santaló inequality, will yield immediately this general affine isoperimetric inequality.

For the plane case ($n = 2$), Heil [15] obtained the Blaschke–Santaló inequality without using the affine isoperimetric inequality and derived from it the affine isoperimetric inequality (see also [42]). Petty [33] also showed that the affine isoperimetric inequality can be obtained as a consequence of the Blaschke–Santaló inequality.

By suitably choosing A and K in the general affine isoperimetric inequality, a number of important isoperimetric inequalities can be obtained. If we take $K = A$ and note that $V_1(A, A) = V(A)$, the general affine isoperimetric inequality reduces to the affine isoperimetric inequality. If we take $K = U$ and note that $nV_1(A, U) = S(A)$, then the general affine isoperimetric inequality reduces to

$$\Omega(A)^{n+1} \le n\omega_n S(A)^n,$$

with equality if and only if A is a ball. This is Berwald's generalization of an inequality of Winternitz (see [4, p. 206]). If we let $A = U$ in the general affine isoperimetric inequality and observe that $V_1(U, K) = \omega_n B(K)/2$, the Urysohn inequality is obtained:

$$V(K) \le 2^{-n}\omega_n B(K)^n,$$

with equality if and only if K is an n-ball. That the Urysohn inequality follows from the Blaschke–Santaló inequality can be seen (although possibly less directly) in [20, 21].

The Hölder integral inequality [14, p. 140] states that for positive real numbers p, q such that $p^{-1} + q^{-1} = 1$, and for positive continuous functions f, g on S^{n-1},

$$\int_{S^{n-1}} f(u)g(u)\, dS(u) \le \left[\int_{S^{n-1}} f^p(u)\, dS(u)\right]^{1/p} \left[\int_{S^{n-1}} g^q(u)\, dS(u)\right]^{1/q},$$

with equality if and only if $f^p(u)/g^q(u)$ is constant.

If for given A, K and $x \in \operatorname{int} K$, we take $p = (n+1)/n$, $q = n + 1$,

$$f = f_A^{n/(n+1)} h_{-x+K}^{n/(n+1)},$$

and

$$g = h_{-x+K}^{-n/(n+1)},$$

then by noting that

$$V_1(A, K) = V_1(A, -x + K) = \frac{1}{n}\int_{S^{n-1}} h_{-x+K}(u) f_A(u)\, dS(u)$$

and that

$$V(K^x) = \frac{1}{n}\int_{S^{n-1}} h_{-x+K}^{-n}(u)\, dS(u),$$

we obtain (from the Hölder inequality)

$$\Omega(A)^{n+1} \leq n^{n+1} V_1^n(A, K) V(K^x),$$

with equality if and only if $f_A(u) h_{-x+K}^{n+1}(u)$ is constant.

If we take x to be the point $s = \text{San}(K)$, then, by noting that $f_A(u) h_{-s+K}^{n+1}(u)$ being constant is equivalent to A having constant relative curvature with respect to $-s + K$ (see [33, p. 88]), we obtain the following polar version of the general affine isoperimetric inequality:

LEMMA. *If A and K are convex bodies in $\mathbb{R}^n$ and A has a positive continuous curvature function, then*

$$\Omega(A)^{n+1} \leq n^{n+1} V_1^n(A, K) V(K^s),$$

with equality if and only if A has constant relative curvature with respect to $-s + K$.

For a positive continuous function f on S^{n-1} that satisfies

$$\int_{S^{n-1}} u f(u)\, dS(u) = 0,$$

the solution of the n-dimensional Minkowski problem (see [5, 35]) guarantees the existence of a convex body (unique up to translation) with positive continuous curvature function f. For an arbitrary convex body K we always have

$$\int_{S^{n-1}} u h_{-s+K}^{-(n+1)}(u)\, dS(u) = 0.$$

Hence, it follows from the solution to the Minkowski problem that for a given convex body K there exists a convex body A with positive continuous curvature function

$$f_A = h_{-s+K}^{-(n+1)}.$$

Or, in other words, for a given convex body K, the solution to the Minkowski problem guarantees the existence of a convex body A (with a positive continuous curvature function) such that equality occurs in the polar version of the general affine isoperimetric inequality; i.e.,

$$\Omega(A)^{n+1} = n^{n+1} V_1(A, K)^n V(K^s).$$

If we combine this fact with the general affine isoperimetric inequality, we obtain immediately the Blaschke–Santaló inequality (with the conditions for equality). On the other hand, if we combine the polar version of the general affine isoperimetric inequality with the Blaschke–Santaló inequality, we immediately obtain the general affine isoperimetric inequality (with the conditions for equality). We observe that the equality conditions for the general affine isoperimetric inequality can be obtained from the equality conditions for the Blaschke–Santaló inequality (and the equality conditions for the polar version of the general affine isoperimetric inequality) by using the simple fact that if E is an ellipsoid and A is a convex body such that f_A is a constant multiple of $h_{-s+E}^{-(n+1)}$, then A and E must be homothetic ellipsoids. This follows from the observation that the curvature function f_E of an ellipsoid E is a constant multiple of $h_{-s+E}^{-(n+1)}$, while a pair of convex bodies with curvature functions that are constant multiples of each other must be homothetic (see [5, p. 115]).

References

1. Aleksandrov, A. D. 1967. On mean values of support functions. Dokl. Akad. Nauk SSSR **172**: 755–758. [English translation: Soviet Math. Dokl. **8** (1967): 149–153.]
2. Blaschke, W. 1916. Über Affine Geometrie I: Isoperimetrische Eigenschaften von Ellipse und Ellipsoid. Leipziger Ber. **68**: 217–239.
3. Blaschke, W. 1917. Über Affine Geometrie VII: Neue Extremeigenschaften von Ellipse und Ellipsoid. Leipziger Ber. **69**: 306–318.
4. Blaschke, W. 1923. Vorlesungen über Differentialgeometrie II. Springer-Verlag. Berlin, Germany.
5. Bonnesen, T. & W. Fenchel. 1934. Theorie der konvexen Körper. Springer-Verlag. Berlin, Germany.
6. Dilworth, S. J. 1984. On the dimension of almost Hilbertian subspaces of quotient spaces. J. London Math. Soc. In press.
7. Dilworth, S. J. 1984. The dimension of Euclidean subspaces of quasi-normed spaces. Math. Proc. Cambridge Philos. Soc. In press.
8. Dilworth, S. J. & S. Szarek. The cotype constant and an almost Euclidean decomposition for finite dimensional normed spaces. Preprint, 1984.
9. Firey, W. J. 1974. Shapes of worn stones. Mathematika **21**: 1–11.
10. Guggenheimer, H. The analytic geometry of the Minkowski plane. Unpublished manuscript, 1967.
11. Guggenheimer, H. 1971. Geometric theory of differential equations I: second order linear equations. SIAM J. Math. Anal. **2**: 233–241.
12. Guggenheimer, H. 1973. Polar reciprocal convex bodies. Israel J. Math. **14**: 309–316. [Correction: Israel J. Math. **29** (1978): 312.]
13. Guggenheimer, H. 1977. Applicable Geometry. Krieger Publishing Co. Huntington, N.Y.
14. Hardy, G. H., J. E. Littlewood & G. Pólya. 1934. Inequalities. Cambridge University Press. Cambridge, England.
15. Heil, E. 1967. Abschätzungen für einige affininvarianten konvexer Kurven. Monatsh. Math. **71**: 405–423.
16. Heil, E. 1975. Eigenvalue estimates for Hill's equation. J. Differential Equations **18**: 179–187.
17. Heil, E. 1976. Ungleichungen für die Quermassintegrale polarer Körper. Manuscripta Math. **19**: 143–149.
18. Holmes, R. D. & A. C. Thompson. 1979. *N*-dimensional area and content in Minkowski spaces. Pacific J. Math. **85**: 77–110.
19. Leichtweiss, K. 1983. Geometric convexity and differential geometry. *In* Convexity and Its Applications. P. M. Gruber & J. M. Wills, Eds.: 163–169. Birkhäuser. Basel, Switzerland.
20. Lutwak, E. 1975. Dual mixed volumes. Pacific J. Math. **58**: 531–538.
21. Lutwak, E. 1975. A general Bieberbach inequality. Math. Proc. Cambridge Philos. Soc. **78**: 493–496.
22. Lutwak, E. 1975. Width-integrals of convex bodies. Proc. Amer. Math. Soc. **53**: 435–439.
23. Lutwak, E. 1977. Mixed width-integrals of convex bodies. Israel J. Math. **28**: 249–253.
24. Lutwak, E. 1983. A width-diameter inequality for convex bodies. J. Math. Anal. Appl. **93**: 290–295.
25. Lutwak, E. 1984. A general isepiphanic inequality. Proc. Amer. Math. Soc. **90**: 415–421.
26. McKinney, J. R. 1974. On maximal simplices inscribed in a central convex set. Mathematika **21**: 38–44.
27. Milman, V. D. 1984. Geometrical inequalities and mixed volumes in local theory of Banach spaces. Asterisque. In press.
28. Milman, V. D. Almost euclidean quotient spaces of subsets of a finite dimensional normed space. Preprint, 1984.
29. Petty, C. M. 1955. On the geometry of the Minkowski plane. Riv. Mat. Univ. Parma **6**: 269–292.
30. Petty, C. M. & J. E. Barry. 1962. A geometrical approach to the second-order linear differential equation. Canad. J. Math. **14**: 349–358.

31. Petty, C. M. 1967. Projection bodies. *In* Proceedings, Colloquium on Convexity (Copenhagen, 1965). Pp. 234–241. Københavns Univ. Mat. Inst. Copenhagen, Denmark.
32. Petty, C. M. 1972. Isoperimetric problems. *In* Proceedings, Conference on Convexity and Combinatorial Geometry, University of Oklahoma, June 1971. Pp. 26–41.
33. Petty, C. M. 1974. Geominimal surface area. Geom. Dedicata **3:** 77–97.
34. Petty, C. M. Affine isoperimetric problems. Ann. N.Y. Acad. Sci. (this volume).
35. Pogorelov, A. V. 1978. The Minkowski Multidimensional Problem. John Wiley & Sons. New York, N.Y.
36. Rogalski, M. 1981. Sur le quotient d'un espace de dimension finie. Séminaire Choquet, 20e année (1980/81), Initiation à l'analyse, Comm. No. 3, 31 pp. Secrétariat Math., Paris, France.
37. Saint Raymond, J. 1981. Sur le volume des corps convexes symétriques. Séminaire Choquet, 20e année (1980/81), Initiation à l'analyse, Comm. No. 11, 25 pp. Secrétariat Math., Paris, France.
38. Saint Raymond, J. 1984. Le volume des ideaux d'operateurs classiques. Studia Math. In press.
39. Santaló, L. A. 1949. Un invariante afine para los cuerpos convexos del espacio de n dimensiones. Portugal. Math. **8:** 155–161.
40. Santaló, L. A. 1958. Un nuevo invariante afin para cuerpos convexos del plano y del espacio. Math. Notae **16:** 78–91.
41. Santaló, L. A. 1976. Integral Geometry and Geometric Probability. Addison-Wesley. Reading, Mass.
42. Schneider, R. 1968. Über die Finslerräume mit $S_{ijkl} = 0$. Arch. Math. **19:** 656–658.
43. Schneider, R. 1982. Random polytopes generated by anisotropic hyperplanes. Bull. London Math. Soc. **14:** 549–553.
44. Wieacker, A. Intersections of random hypersurfaces and visibility. Preprint.

AFFINE ISOPERIMETRIC PROBLEMS

C. M. Petty

Department of Mathematics
University of Missouri
Columbia, Missouri 65211

1. Introduction

The purpose of this paper is to remove regularity assumptions on the class of convex bodies for which equality is obtained in three closely related affine extremal problems. These three problems are described in Sections 2, 3, and 4. The need to remove regularity assumptions was recently pointed out by Schneider [15, p. 552].

Throughout this paper, a convex body is defined as a compact convex set with interior points in Euclidean n-dimensional space $\mathbb{E}^n$.

The oldest of these problems is called the affine isoperimetric problem and involves the affine surface area which is defined for a restricted class of convex bodies. Let $K \subset \mathbb{E}^n$ be a convex body whose surface ∂K is of class C^2 and has positive Gaussian curvature $G(u)$, where u is the outer normal to ∂K at the point of ∂K with Gaussian curvature $G(u)$. The affine surface area $A_a(K)$ of K is defined by

$$A_a(K) = \int_{\partial K} G(u)^{1/(n+1)}\, dS = \int_{\Omega} G(u)^{-n/(n+1)}\, d\omega, \tag{1.1}$$

where Ω is the unit sphere.

The solution of the affine isoperimetric problem is an inequality between the volume $|K|$ of K and its affine surface area $A_a(K)$. It has been proven that equality holds, subject to certain regularity assumptions, only for ellipsoids.

Blaschke ([1] or [3, Sections 26 and 73]) solved the affine isoperimetric problem for $n = 2, 3$. We may view Blaschke's proof as having three parts. The first part is an existence proof. The second part is a proof that the affine surface area does not decrease under Steiner symmetrization. The third part gives the condition under which equality is obtained in the second part. The latter is the condition that characterizes the ellipsoids. Santaló [14] observed that Blaschke's proof can be extended to the n-dimensional case and indicated how the inequality in the second part is obtained. This part of Blaschke's proof is also contained in Deicke [7] and, in particular, Deicke proves a regularity theorem which transfers regularity conditions on the surface of the convex body to regularity conditions on the surface of the body obtained by Steiner symmetrization. Here, we will only use a weak formulation of the second part.

Theorem 1.2. *Let $K \subset \mathbb{E}^n$ be a convex body whose surface ∂K is analytic and has positive Gaussian curvature. If $\bar{K}$ is obtained from K by Steiner symmetrization, then $A_a(\bar{K}) \geq A_a(K)$.*

The second problem (Section 3) is the Blaschke–Santaló inequality which involves the product of the volume of an arbitrary convex body K and the volume of its polar reciprocal with respect to a certain interior point of K. The proof for the n-dimensional case is due to Santaló [14] and rests on the solution to the affine isoperimetric problem. The third problem (Section 4) is similar to the affine isoperimetric problem but is formulated for arbitrary convex bodies. The proof in Petty [10] rests on the Blaschke–Santaló inequality. Unfortunately, regularity assumptions are not mentioned in [10] and [14], but this paper will justify the omission. Here, however, we will take a different path to prove these results.

2. The Affine Isoperimetric Problem

The problems described in Sections 3 and 4 apply to arbitrary convex bodies. In order for the affine isoperimetric problem to be equivalent to them, we need to extend the definition of affine surface area to a wider class of convex bodies.

A convex body $K \subset \mathbb{E}^n$ is said to possess a curvature function $f(u)$ provided $f(u)$ is a continuous nonnegative function defined on the unit sphere Ω such that for an arbitrary compact convex set L with supporting function $L(u)$ the mixed volume integral representation

$$V(L, K, \ldots, K) = \frac{1}{n} \int_{\Omega} L(u) f(u)\, d\omega \tag{2.1}$$

is valid. A convex body can have at most one curvature function. See Bonnesen and Fenchel [4, p. 115]. If a convex body $K \subset \mathbb{E}^n$ possesses a curvature function $f(u)$, we define the affine surface area $A_a(K)$ of K by

$$A_a(K) = \int_{\Omega} f(u)^{n/(n+1)}\, d\omega. \tag{2.2}$$

If the surface ∂K of K is of class C^2 and has positive Gaussian curvature, then the reciprocal Gaussian curvature as a function of the outer normal is a curvature function of K. Consequently, the definition **(2.2)** is an extension of **(1.1)**.

Let $|K|$ denote the volume of K and let π_n denote the volume of the unit ball in $\mathbb{E}^n$. We will prove the following solution to the affine isoperimetric problem.

THEOREM 2.3. *If $K \subset \mathbb{E}^n$ is a convex body which possesses a curvature function, then*

$$(n\pi_n)^2 (n|K|)^{n-1} \geq [A_a(K)]^{n+1}$$

with equality holding if and only if K is an ellipsoid.

3. The Blaschke–Santaló Inequality

This inequality is due to Blaschke [2] for dimensions $n = 2, 3$ and to Santaló for general n. Let $K \subset \mathbb{E}^n$ be a convex body and let p be an interior point of K. Let $I(K, p)$ be the volume of the polar reciprocal of K with respect to the unit sphere centered at p. Then there exists a unique interior point $s(K)$ of K, which we will call the Santaló point of K, such that $I(K, p)$ is minimal at $p = s(K)$. Let $I_m(K) = I(K, s(K))$. The product $|K| I_m(K)$ is an affine invariant of K. See Santaló [14].

Let $\sigma_p(u)$, $u \in \Omega$, be the supporting function of K with respect to the interior point p of K; i.e., $\sigma_p(u)$ is the distance from p to the supporting plane of K with outer normal u. A necessary and sufficient condition that $p = s(K)$ is that

$$\int_{\Omega} u\sigma_p^{-n-1}(u)\, d\omega = 0. \tag{3.1}$$

See Santaló [14]. The condition (**3.1**) may be interpreted in the following way. The point $p = s(K)$ if and only if p is the centroid of the polar reciprocal of K with respect to the unit sphere centered at p. The point $s(K)$ is an affine-invariant point of K and therefore it varies continuously with K. See Petty [10, p. 79].

We will prove the following result.

THEOREM 3.2. *If $K \subset \mathbb{E}^n$ is an arbitrary convex body, then*

$$|K|\, I_m(K) \leq \pi_n^2$$

with equality holding if and only if K is an ellipsoid.

E. Lutwak has informed the author about the work of J. Saint Raymond [13] where THEOREM 3.2 is established for centrally symmetric convex bodies.

4. Geominimal Surface Area

Let $\mathscr{T}_n$ be the set of all convex bodies $T \subset \mathbb{E}^n$ whose Santaló point $s(T)$ is at the origin and $I_m(T) = \pi_n$. In a given equivalence class of (directly) homothetic convex bodies, exactly one of them belongs to $\mathscr{T}_n$. If $K \subset \mathbb{E}^n$ is a convex body and $T \in \mathscr{T}_n$, the relative surface area $A(K, T)$ of K with respect to T is defined by

$$A(K, T) = \int_{\partial K} \sigma(u)\, dS = nV(T, K, \ldots, K), \tag{4.1}$$

where $\sigma(u)$ is the supporting function of T. The geominimal surface area $G(K)$ of K is defined by

$$G(K) = \inf\{A(K, T)\colon T \in \mathscr{T}_n\}. \tag{4.2}$$

Proofs of the following properties of the geominimal surface area are given in Petty [10].

(4.3) (a) *If $[K_i] \subset \mathbb{E}^n$ is a sequence of convex bodies and $K_i \to K$, where K is a convex body, then $G(K_i) \to G(K)$.* (b) *For a given convex body $K \subset \mathbb{E}^n$ there exists a unique $T = T(K)$ in $\mathscr{T}_n$ such that $G(K) = A(K, T(K))$.*

We will show that the following result is equivalent to THEOREMS 2.3 and 3.2.

THEOREM 4.4. *If $K \subset \mathbb{E}^n$ is an arbitrary convex body, then*

$$n^n \pi_n |K|^{n-1} \geq G^n(K)$$

with equality holding if and only if K is an ellipsoid.

5. The Minkowski Problem

We will summarize those results from the solution to the n-dimensional Minkowski problem which will be used here. In this paper, we use the term "Gaussian curvature" only for surfaces of class C^2, for which the various definitions of this term are equivalent.

First, we mention some results from the generalized solution. See Bonnesen and Fenchel [4, Section 13], Pogorelov [12, Chapter 2], and Cheng and Yau [6]. Let $f(u)$ be a positive continuous function on Ω which satisfies

$$\int_{\Omega} uf(u)\, d\omega = 0. \tag{5.1}$$

Then, there exists a convex body K, uniquely determined up to a translation, with the following properties. First, $f(u)$ is a curvature function of K and consequently the mixed volume representations (**2.1**) are valid. Second, there is a measure $S(K, \omega)$ on Ω, called the area function of K, defined for all Borel sets ω on Ω with this interpretation. Let $\Pi(u)$ be the supporting plane to K with outer normal u. Then $S(K, \omega)$ is the $(n-1)$-dimensional measure of

$$\bigcup_{u \in \omega} (\Pi(u) \cap K).$$

Moreover, $S(K, \omega)$ is absolutely continuous, i.e., $S(K, \omega) = 0$ if the area of ω is zero. The generalized solution says nothing about the regularity of the surface ∂K. However, it follows from the Gauss map that if ∂K is of class C^2, then $f(u)$ is the reciprocal Gaussian curvature at the point of ∂K with outer normal u.

The regular solution of the Minkowski problem transfers regularity conditions on $f(u)$ to regularity conditions on ∂K. From the regular solution, we will only use the following result.

(5.2) *If $f(u)$ is analytic, then ∂K is analytic.*

See Pogorelov [12, Chapter 3] and Cheng and Yau [6]. We mention that there is also a regularity theorem when $f(u)$ is of class C^k but for general n the known results are restricted to $k \geq 3$ and will not be of use to us here.

The Monge–Ampère equation is closely related to the Minkowski problem. We will use the following special result (see Pogorelov [12, Chapter 5, Theorem 5]).

(5.3) *If $F = F(x_1, \ldots, x_n)$ is a strictly convex function on an open convex domain $G \subset \mathbb{E}^n$ and F is of class C^2 and satisfies on G the equation*

$$\det\left(\frac{\partial^2 F}{\partial x_i\, \partial x_j}\right) = c > 0,$$

where c is a positive constant, then F is analytic on G.

6. THE CLASS $\mathscr{V}_n$

We study here certain classes of convex bodies which play an important role in affine extremal problems. The relation that two convex bodies are (directly) homothetic is an equivalence relation and determines equivalence classes, denoted by $\{K\}$. The similarity of conditions (**3.1**) and (**5.1**) suggests an injective mapping $\{K\} \to \{K^*\}$ defined in the following way. Let $\sigma(u)$ be the supporting function of a convex body K with respect to its Santaló point $s(K)$. Let $f(u) = c\sigma^{-n-1}(u)$, where c

is a positive constant. By (**3.1**) and (**5.1**), there exists a convex body K^* with curvature function $f(u)$ and this determines $\{K\} \rightarrow \{K^*\}$. The class of all convex bodies K^* obtained in this way from arbitrary convex bodies K will be denoted by $\mathscr{V}_n$. We define the subclass $\mathscr{F}_n \subset \mathscr{V}_n$ by the property that $K \in \mathscr{F}_n$ provided $\{K\} = \{K^*\}$; i.e., K and K^* are homothetic. The ellipsoids belong to $\mathscr{F}_n$.

We will need the following special case of Hölder's inequality for integrals. See [8, p. 140].

(6.1) *Let $g_1(u)$, $g_2(u)$ be continuous functions on the unit sphere Ω where $g_1(u)$ is positive and $g_2(u)$ is nonnegative but does not vanish identically. Then*

$$\left[\int_\Omega [g_1(u)]^{-n}\,d\omega\right]\left[\int_\Omega g_1(u)g_2(u)\,d\omega\right]^n \geq \left[\int_\Omega g_2^{n/(n+1)}(u)\,d\omega\right]^{n+1}$$

where the equality sign holds if and only if $g_2(u)g_1^{n+1}(u)$ is a positive constant.

In the following lemma, K and K^* may be any members of $\{K\}$ and $\{K^*\}$, respectively.

Lemma 6.2. *Let $K \subset \mathbb{E}^n$ be a convex body. Then*

$$|K^*|\,I_m(K^*) \geq |K|\,I_m(K), \qquad (\mathbf{6.3})$$

$$[A_a(K^*)]^{n+1} \geq n^{n+1}|K^*|^{n-1}(|K|\,I_m(K)), \qquad (\mathbf{6.4})$$

with equality in (**6.3**) *or* (**6.4**) *if and only if $K \in \mathscr{F}_n$.*

Proof. We may assume that both $s(K)$ and $s(K^*)$ are at the origin and $f(u) = \sigma^{-n-1}(u)$ is the curvature function of K^* where $\sigma(u)$ is the supporting function of K. Let $h(u)$ be the supporting function of K^*.

To prove (**6.3**), let $g_1(u) = h(u)$ and $g_2(u) = f(u)$. From (6.1), we obtain

$$I_m(K^*)\,|K^*|^n \geq [I_m(K)]^{n+1}$$

with equality if and only if $h(u)/\sigma(u)$ is a constant. From (**2.1**) and Minkowski's inequality for mixed volumes, we have

$$[V(K, K^*, \ldots, K^*)]^n = [I_m(K)]^n \geq |K|\,|K^*|^{n-1}$$

with equality if and only if K and K^* are homothetic. By combining these two inequalities, we obtain (**6.3**).

To prove (**6.4**), we combine the inequality $[I_m(K)]^n \geq |K|\,|K^*|^{n-1}$ given above with the equality $[A_a(K^*)]^{n+1} = n^{n+1}[I_m(K)]^{n+1}$ obtained from (**2.2**). ∎

Lemma 6.5. *If $K \subset \mathbb{E}^n$ is a convex body which possesses a curvature function, then*

$$n\pi_n G^n(K) \geq [A_a(K)]^{n+1} \qquad (\mathbf{6.6})$$

with equality if and only if $K \in \mathscr{V}_n$.

Proof. Let $f(u)$ be the curvature function of K. Since K is a convex body, $f(u)$ does not vanish identically. Let $T \in \mathscr{T}_n$ have supporting function $\sigma(u)$. Then, by (**4.1**) and (**2.1**), we have

$$A(K, T) = \int_\Omega \sigma(u) f(u)\,d\omega.$$

In (6.1), let $g_1(u) = \sigma(u)$, $g_2(u) = f(u)$. Then,

$$n\pi_n A^n(K, T) \geq [A_a(K)]^{n+1} \tag{6.7}$$

with equality if and only if $f(u)\sigma^{n+1}(u)$ is a positive constant; i.e., $K \in \{T^*\}$. The inequality (**6.6**) follows from the definition (**4.2**) since the right-hand side of (**6.7**) is independent of T. If $K \in \mathscr{V}_n$, then there exists $T \in \mathscr{T}_n$ such that $K \in \{T^*\}$. For this T, equality holds in (**6.7**) and consequently, equality holds in (**6.6**). Suppose equality holds in (**6.6**). By (4.3b), there exists $T(K) \in \mathscr{T}_n$ such that $G(K) = A(K, T(K))$. Then, equality holds in (**6.7**) with $T = T(K)$ and therefore $K \in \{T^*(K)\}$. This completes the proof. ▌

In Petty [10, p. 91], it is shown that a convex body that is not in $\mathscr{V}_n$ cannot be approximated by convex bodies in $\mathscr{V}_n$.

LEMMA 6.8. *If $K \in \mathscr{V}_n$, then K can be approximated by convex bodies in $\mathscr{V}_n$ whose surfaces are analytic and have positive Gaussian curvature.*

Proof. Let $f(u) = \sigma^{-n-1}(u)$ be the curvature function of K where $\sigma(u)$ is the supporting function of a convex body $\hat{K}$ with $s(\hat{K})$ at the origin. Let $\hat{K}_i \to \hat{K}$ where each convex body $\hat{K}_i$ has an analytic supporting function $\sigma_i(u)$ and $s(\hat{K}_i) = s(\hat{K})$. Then $f_i(u) \to f(u)$ uniformly where $f_i(u) = \sigma_i^{-n-1}(u)$. The convex body $K_i \in V_n$ is uniquely determined by the requirement that K_i have curvature function $f_i(u)$ and $s(K_i) = s(K)$. By (5.2), ∂K_i is analytic and $f_i(u)$ is its reciprocal Gaussian curvature. We will show that $K_i \to K$.

Let $p_i(u)$ be the area of the projection of K_i on a hyperplane normal to u. Since (**2.1**) is valid for K_i and with L a unit segment and since $f_i(u) \to f(u)$ uniformly, there exist positive constants k_1, k_2 such that $k_1 \leq p_i(u) \leq k_2$ for all i and all $u \in \Omega$. It follows since $s(K_i) = s(K)$, that the K_i are uniformly bounded and the volumes $|K_i|$ are bounded away from zero. See Pogorelov [12, pp. 30–32] for more details in a similar argument. By Blaschke's theorem, every subsequence of $[K_i]$ has a convergent subsequence. But any such convergent subsequence must converge to K. See Petty [10, p. 91]. Consequently, $K_i \to K$ and this completes the proof. ▌

Using the notation of (4.3b), we say that a convex body $K \subset \mathbb{E}^n$ is self-minimal provided K is homothetic to $T(K)$.

LEMMA 6.9. (a) *If $K \subset \mathbb{E}^n$ is a convex body, then*

$$n^n |K|^n I_m(K) \geq \pi_n G^n(K)$$

with equality if and only if K is self-minimal.

(b) *If $K \in \mathscr{V}_n$, then K is self-minimal if and only if $K \in \mathscr{F}_n$.*

Proof. (a) Let $s(K)$ be the origin and let $T = (\pi_n^{-1} I_m(K))^{1/n} K \in \mathscr{T}_n$. Then, by (**4.1**), $A^n(K, T) = n^n |T| |K|^{n-1} = n^n \pi_n^{-1} |K|^n I_m(K) \geq G^n(K)$ where, by (4.3b), the equality holds if and only if $T = T(K)$. (b) If $K \in \mathscr{V}_n$, then the equality holds in (**6.6**) and consequently equality holds in (**6.7**) with $T = T(K)$. Thus, $K \in \{T^*(K)\}$ and therefore K is self-minimal if and only if $K \in \mathscr{F}_n$. ▌

7. The Inequalities

We need to provide an existence proof for either Theorem 2.3, 3.2, or 4.4. We choose Theorem 3.2. Let $K \subset \mathbb{E}^n$ be a convex body with $s(K)$ at the origin and let K' be the polar reciprocal of K with respect to the unit sphere with center $s(K)$. Then the origin $s(K)$ is the centroid of K'. If E is the ellipsoid of smallest volume with center $s(K)$ which contains K', then K' contains the ellipsoid λE, where $\lambda = n^{-3/2}$. See Pogorelov [12, p. 91]. Now, $|K|I_m(K) = |K||K'|$ is an affine invariant of K. We may assume, therefore, that E is the unit ball B. See Petty [11, p. 236] for the transformation law for polar reciprocals. Consequently, $\lambda B \subset K' \subset B$ and $B \subset K \subset \lambda^{-1}B$. Let $C_n > 0$ be defined by $C_n^2 = \sup |K|I_m(K)$. The existence of an extremal body K for which $|K|I_m(K) = C_n^2$ follows from Blaschke's theorem and the fact that the centroid and the Santaló point vary continuously with the convex body.

If $|K|I_m(K) = C_n^2$, then $K \in \mathscr{F}_n$ by **(6.3)**. By Lemma 6.9, for an arbitrary convex body $K \subset \mathbb{E}^n$, we have

$$n^n|K|^{n-1}C_n^2 \geq \pi_n G^n(K) \tag{7.1}$$

with equality if and only if $|K|I_m(K) = C_n^2$.

Lemma 7.2. *If $K \subset \mathbb{E}^n$ is a convex body for which $|K|I_m(K) = C_n^2$ and $\bar{K}$ is obtained from K by Steiner symmetrization, then $|\bar{K}|I_m(\bar{K}) = C_n^2$.*

Proof. Since $K \in \mathscr{F}_n \subset \mathscr{V}_n$, there exists, by Lemma 6.8, a sequence $[K_i]$ of convex bodies in $\mathscr{V}_n$ whose surfaces are analytic and have positive Gaussian curvature with the property that $K_i \to K$. Let $\bar{K}_i$ be obtained from K_i as $\bar{K}$ is obtained from K. Then $\bar{K}_i \to \bar{K}$. See Leichtweiss [9, p. 223]. By Theorem 1.2 and Lemma 6.5, we have $n\pi_n G^n(\bar{K}_i) \geq A_a(\bar{K}_i)^{n+1} \geq A_a(K_i)^{n+1} = n\pi_n G^n(K_i)$. By (4.3a), $G(\bar{K}_i) \to G(\bar{K})$ and $G(K_i) \to G(K)$ and therefore $G(\bar{K}) \geq G(K)$. Since $|\bar{K}| = |K|$, equality is obtained in **(7.1)** for $\bar{K}$ and consequently, $|\bar{K}|I_m(\bar{K}) = C_n^2$. This completes the proof. ∎

We can now define explicitly the class $\mathscr{E}_n \subset \mathscr{F}_n$ of extremal bodies.

Definition 7.3. The convex body $K \subset \mathbb{E}^n$ belongs to $\mathscr{E}_n$ provided K has curvature function $f(u) = \pi_n^{-2}|K|^2\sigma^{-n-1}(u)$ where $\sigma(u)$ is the supporting function of K with respect to $s(K)$.

Lemma 7.4. *If $K \subset \mathbb{E}^n$ is a convex body, then $|K|I_m(K) \leq \pi_n^2$ with equality if and only if $K \in \mathscr{E}_n$.*

Proof. Let K have supporting function $\sigma(u)$ with respect to $s(K)$ and suppose $|K|I_m(K) = C_n^2$. Then $K \in \mathscr{F}_n$ and consequently K has curvature function $f(u) = c\sigma^{-n-1}(u)$ for some positive constant c. There exists a sequence $[K_i]$, with $K_1 = K$, of convex bodies obtained by successive Steiner symmetrizations which converge to a ball B. By Lemma 7.2, equality is obtained in **(7.1)** for each K_i and, by continuity, equality is obtained for the ball B. Hence $|B|I_m(B) = C_n^2 = \pi_n^2$. Now

$$|K| = \frac{1}{n}\int_\Omega \sigma(u)f(u)\,d\omega = cI_m(K),$$

and therefore $c = \pi_n^{-2}|K|^2$ and $K \in \mathscr{E}_n$. On the other hand, if $K \in \mathscr{E}_n$, then the calculation above shows that $|K|I_m(K) = \pi_n^2$. ∎

We will show that the following two lemmas are equivalent to LEMMA 7.4.

LEMMA 7.5. *If $K \subset \mathbb{E}^n$ is a convex body, then $n^n \pi_n |K|^{n-1} \geq G^n(K)$ with equality if and only if $K \in \mathscr{E}_n$.*

LEMMA 7.6. *If $K \subset \mathbb{E}^n$ is a convex body which possesses a curvature function, then $(n\pi_n)^2(n|K|)^{n-1} \geq A_a(K)^{n+1}$ with equality if and only if $K \in \mathscr{E}_n$.*

Proof. We will show that LEMMA 7.4 ⇒ LEMMA 7.5 ⇒ LEMMA 7.6 ⇒ LEMMA 7.4. LEMMA 7.4 and **(7.1)** with $C_n = \pi_n$ imply LEMMA 7.5. LEMMAS 7.5 and 6.5 imply LEMMA 7.6. Starting with **(6.4)**, LEMMA 7.6 with $K = K^*$ implies LEMMA 7.4. ▮

To prove THEOREMS 2.3, 3.2, and 4.4 we need to show that $\mathscr{E}_n$ consists only of ellipsoids. For this purpose, we restate LEMMA 7.2 together with the closure property.

(7.7) *If $K \in \mathscr{E}_n$ and $\bar{K}$ is obtained from K by Steiner symmetrization, then $\bar{K} \in \mathscr{E}_n$. Any convex body that can be approximated by members of $\mathscr{E}_n$ must also belong to $\mathscr{E}_n$.*

8. THE CLASS $\mathscr{F}_n$

In our proof that $\mathscr{E}_n$ consists only of ellipsoids, certain arguments apply as well to $\mathscr{F}_n$ and these are given in this section.

LEMMA 8.1. *The class $\mathscr{F}_2$ consists only of ellipses.*

Proof. Let $K \in \mathscr{F}_2$ with $s(K)$ at the origin. Let $u = (\cos\theta, \sin\theta)$ and let $\sigma(\theta)$ be the supporting function of K. Then, $f(\theta) = c\sigma^{-3}(\theta)$ is the curvature function of K for some positive constant c. Let

$$h(\theta) = C_1 \cos\theta + C_2 \sin\theta + \int_0^\theta \sin(\theta - t) f(t)\, dt.$$

Then $h(\theta)$ is of period 2π by **(5.1)**. Differentiating $h(\theta)$, we have $h''(\theta) + h(\theta) = f(\theta)$. This is a sufficient condition for $h(\theta)$ to be the supporting function of a planar convex body K_1 with $f(\theta)$ as the radius of curvature of its perimeter. By uniqueness, K_1 is a translate of K. We may assume, by the proper choice of C_1, C_2, that K_1 coincides with K and $\sigma(\theta)$ satisfies the differential equation

$$\sigma'' + \sigma = c\sigma^{-3}. \qquad \textbf{(8.2)}$$

Choose the reference axis, $\theta = 0$, through a point of ∂K at maximal distance from the origin. Then $\sigma'(0) = 0$, $\sigma''(0) \leq 0$. Let $a = \sigma(0)$ and $e^2 = |\sigma''(0)|a^{-1}$. From **(8.2)**, we have

$$(\sigma')^2 + c\sigma^{-2} + \sigma^2 = a^2(2 - e^2). \qquad \textbf{(8.3)}$$

From **(8.2)** and **(8.3)**, we have

$$(\sigma^2)'' + 4\sigma^2 = 2a^2(2 - e^2).$$

Consequently, $\sigma^2 = a^2(1 - e^2 \sin^2\theta)$ and σ is the supporting function of an ellipse with semimajor axis a and eccentricity e. ▮

LEMMA 8.4. *If $K \in \mathscr{F}_n$ and ∂K is of class C^2, then K is an ellipsoid.*

Proof. With $s(K)$ at the origin and supporting function $\sigma(u)$, the curvature function $f(u) = c\sigma^{-n-1}(u) = R_1 \ldots R_{n-1}$, where $R_1 \ldots R_{n-1}$ is the product of the principal radii of curvature of ∂K at the point y of ∂K with outer normal u. We extend $\sigma(u)$ to $\mathbb{E}^n$ by $H(x) = |x|\sigma(x/|x|)$, $x \neq 0$, and $H(0) = 0$. Then $H(x)$ is a positively homogeneous convex function on $\mathbb{E}^n$. If $\mathbb{E}'_n$ is $\mathbb{E}^n$ with the origin deleted, then $H(x)$ is of class C^2 on $\mathbb{E}'_n$ and the coordinates of y are given by $y_i = \partial H/\partial x_i$, where $u = x/|x|$. For computations, we write vectors as column vectors.

Let $H_{ij} = \partial^2 H/\partial x_i \, \partial x_j$ on $\mathbb{E}'_n$. Since H is positively homogeneous of degree one, we have

$$[H_{ij}]x = 0, \qquad |x| = 1. \tag{8.5}$$

Now $R_1 \ldots R_{n-1}$ is the sum of the principal $(n-1)$-rowed minors of $[H_{ij}]$, where H_{ij} is evaluated for the unit vector x. The rank $[H_{ij}] = n - 1$ and the rank adj $[H_{ij}] = 1$. From **(8.5)**, each column of adj $[H_{ij}]$ is a multiple of x. Since adj $[H_{ij}]$ is symmetric, adj $[H_{ij}] = [\lambda x_i x_j]$. The trace of adj $[H_{ij}]$ is $R_1 \ldots R_{n-1}$ and consequently $\lambda = R_1 \ldots R_{n-1}$.

Let $F = \frac{1}{2}H^2$. We will calculate det $[\partial^2 F/\partial x_i \, \partial x_j]$ which we may evaluate for unit vectors x since the elements of the matrix are positively homogeneous of degree zero. We obtain $\partial^2 F/\partial x_i \, \partial x_j = y_i y_j + HH_{ij}$ and therefore

$$\det\left[\frac{\partial^2 F}{\partial x_i \, \partial x_j}\right] = \sigma^{n-1}(x) \sum_{i=1}^{n} \det A_i, \qquad |x| = 1,$$

where the matrix A_i is the same as $[H_{ij}]$ except the ith column is $y_i y$. Now

$$\sum_{i=1}^{n} \det A_i = y^T \operatorname{adj}\,[H_{ij}]y = \lambda\left(\sum_{i=1}^{n} x_i y_i\right)^2 = R_1 \ldots R_{n-1}\,\sigma^2(x).$$

Consequently

$$\det\left[\frac{\partial^2 F}{\partial x_i \, \partial x_j}\right] = c > 0, \qquad x \in \mathbb{E}'_n. \tag{8.6}$$

Let x, y be any two distinct points in $\mathbb{E}^n$ and $0 < t < 1$. Since H is a convex function and $g(s) = (1/2)s^2$ is strictly convex, we obtain

$$F(tx + (1-t)y) \leq tF(x) + (1-t)F(y),$$

where the equality holds if and only if $H(sx + (1-s)y)$ is constant for $0 \leq s \leq 1$. But the latter is not possible since the surface of K is smooth. Thus, F is a strictly convex function on $\mathbb{E}^n$. Let G be an open convex domain in $\mathbb{E}'_n$. Then **(8.6)** and (5.3) imply that F is analytic on G. Consequently, F is analytic on $\mathbb{E}'_n$.

Brickell [5] proved that if H is a positive function on $\mathbb{E}'_n$ of class C^4 and H is positively homogeneous of degree one, then **(8.6)** implies that the matrix $[\partial^2 F/\partial x_i \, \partial x_j]$ is constant. It follows that F is a positive definite quadratic form and K is an ellipsoid with center at the origin. This completes the proof. ▮

We mention that the members of $\mathscr{F}_n$ that are sufficiently regular are those which are studied in affine differential geometry. The affine distance from $s(K)$ to a point of ∂K is constant, or equivalently, ∂K has constant mean affine curvature. Either property implies that K is a proper affine sphere (the affine normals to ∂K are concurrent). See Deicke [7].

LEMMA 8.7. *If $K \in \mathscr{F}_n$, $n \geq 3$, and K is body of rotation, then K is an ellipsoid.*

Proof. Let $s(K)$ be the origin and let L be the axis of rotation. Suppose ∂K contains a segment M. Let u be the outer normal to ∂K at the midpoint of M. Under the rotations about L which carry K onto itself, the segment M sweeps out a set of positive area but u sweeps out a set ω of zero area on the unit sphere Ω. This contradicts the absolute continuity of the area function $S(K, \omega)$ defined in Section 5. Hence, ∂K is a strictly convex surface. This implies that the supporting function $\sigma(u)$ is of class C^1 and consequently the curvature function $f(u)$ is of class C^1 since $f(u) = c\sigma^{-n-1}(u)$. The proof now follows from LEMMA 8.4 and the following result. ∎

LEMMA 8.8. *Let $K \subset \mathbb{E}^n$, $n \geq 3$, be a convex body of rotation. If K possesses a positive curvature function of class C^1, then ∂K is of class C^2.*

Proof. For computations, we again write vectors as column vectors. We parameterize the unit sphere Ω in $\mathbb{E}^n$ in the following way. First we parameterize the points $\xi = (\xi_1, \dots, \xi_{n-1}, 0)^T$ on the "great sphere" normal to the x_n-axis by

$$\begin{aligned}
\xi_1 &= \sin\theta_1 \sin\theta_2 \dots \sin\theta_{n-3} \cos\theta_{n-2},\\
\xi_2 &= \sin\theta_1 \sin\theta_2 \dots \sin\theta_{n-3} \sin\theta_{n-2},\\
\xi_3 &= \sin\theta_1 \sin\theta_2 \dots \sin\theta_{n-4} \cos\theta_{n-3},\\
\xi_4 &= \sin\theta_1 \sin\theta_2 \dots \sin\theta_{n-5} \cos\theta_{n-4},\\
&\vdots\\
\xi_{n-2} &= \sin\theta_1 \cos\theta_2,\\
\xi_{n-1} &= \cos\theta_1,
\end{aligned}$$

where $0 \leq \theta_{n-2} \leq 2\pi$, $0 \leq \theta_j \leq \pi$ for $j = 1, \dots, n-3$. Then, for $u \in \Omega$, we set

$$u = \begin{bmatrix} \sin\phi\xi_1 \\ \vdots \\ \sin\phi\xi_{n-1} \\ \cos\phi \end{bmatrix}, \tag{8.9}$$

where $0 \leq \phi \leq \pi$. Thus, we have

$$\det\left[u, \frac{\partial u}{\partial\phi}, \frac{\partial u}{\partial\theta_1}, \dots, \frac{\partial u}{\partial\theta_{n-2}}\right] = \sin^{n-2}\phi \sin^{n-3}\theta_1 \dots \sin\theta_{n-3}.$$

The surface area element $d\omega$ of Ω is given by

$$d\omega = \sin^{n-2}\phi \sin^{n-3}\theta_1 \dots \sin\theta_{n-3}\, d\phi\, d\theta_1 \dots d\theta_{n-2}.$$

Let the open region $G \subset \mathbb{R}^{n-1}$ be determined by $0 < \phi < \pi$,

$$0 < \theta_{n-2} < 2\pi, \qquad 0 < \theta_j < \pi \text{ for } j = 1, \dots, n-3.$$

The mapping (**8.9**) of G into Ω determines a patch of Ω which is analytic.

We may assume that the axis of K is parallel to the x_n-axis and therefore its curvature function $f(\phi)$ depends only on ϕ. By hypothesis, $f(\phi)$ is continuous for

$0 \le \phi \le \pi$ and has a continuous derivative for $0 < \phi < \pi$. Since $f(\phi)$ must satisfy **(5.1)**, we obtain

$$\int_0^\pi f(\phi) \sin^{n-2} \phi \cos \phi \, d\phi = 0. \tag{8.10}$$

Let $r(\phi) \ge 0$ be defined by

$$r^{n-1}(\phi) = (n-1) \int_0^\phi f(t) \sin^{n-2} t \cos t \, dt, \qquad 0 \le \phi \le \pi.$$

The function $r(\phi)$ is strictly increasing on $[0, \pi/2]$ and strictly decreasing on $[\pi/2, \pi]$ with $r(\pi) = 0$, by **(8.10)**. Consequently, $r(\phi)$ is of class C^2 in the open interval $(0, \pi)$. The inverse function

$$\phi\colon [0, r(\pi/2)] \to [0, \pi/2] \tag{8.11}$$

is continuous on $[0, r(\pi/2)]$ and is of class C^2 on $(0, r(\pi/2))$. Also, we find that

$$\lim_{\phi \to 0+} \frac{dr}{d\phi} = [f(0)]^{1/(n-1)}. \tag{8.12}$$

We define $g(\phi)$, $0 \le \phi \le \pi$, by $g(0) = g(\pi) = 0$ and

$$g(\phi) = \frac{f(\phi) \sin^{n-1} \phi}{r^{n-2}(\phi)}, \qquad 0 < \phi < \pi. \tag{8.13}$$

Then, we obtain

$$\cos \phi \, g(\phi) = \sin \phi \frac{dr(\phi)}{d\phi}, \qquad 0 < \phi < \pi, \tag{8.14}$$

$$\lim_{\phi \to 0+} \frac{r(\phi)}{\tan \phi} = \lim_{\phi \to 0+} \frac{g(\phi)}{\tan \phi} = [f(0)]^{1/(n-1)}. \tag{8.15}$$

Similar limits are obtained at the endpoint π. Consequently, $g(\phi)$ is continuous on $[0, \pi]$ and is of class C^1 on $(0, \pi)$.

Let

$$z(\phi) = \int_\phi^{\pi/2} g(t) \, dt, \qquad 0 \le \phi \le \pi.$$

It may be verified that if, in the (r, z) plane, we reflect the curve $r = r(\phi)$, $z = z(\phi)$, $0 \le \phi \le \pi$, across the z-axis, then we obtain a closed convex curve. This is the same curve that is obtained by the two-dimensional sections through the x_n-axis of the surface of rotation which we now define.

Let S be the closed convex surface of rotation, about the x_n-axis, whose position vector x is given by

$$x = \begin{bmatrix} r(\phi)\xi_1 \\ \vdots \\ r(\phi)\xi_{n-1} \\ z(\phi) \end{bmatrix} \tag{8.16}$$

where $0 \le \phi \le \pi$, $0 \le \theta_{n-2} \le 2\pi$, and $0 \le \theta_j \le \pi$ for $j = 1, \ldots, n-3$. Let $S_0 \subset S$ be the image of G determined by (**8.16**). Over G, the vector function $x = x(\phi, \theta_1, \ldots, \theta_{n-2})$ is of class C^2. We denote the inner product of vectors by $\langle x, y \rangle$. Using (**8.13**), (**8.14**), we find

$$\frac{\partial x}{\partial \phi} = \frac{g(\phi)}{\sin \phi} \frac{\partial u}{\partial \phi},$$

$$\frac{\partial x}{\partial \theta_i} = \frac{r(\phi)}{\sin \phi} \frac{\partial u}{\partial \theta_i}, \qquad i = 1, \ldots, n-2,$$

$$\left\langle \frac{\partial x}{\partial \phi}, u \right\rangle = \left\langle \frac{\partial x}{\partial \theta_1}, u \right\rangle = \cdots = \left\langle \frac{\partial x}{\partial \theta_{n-2}}, u \right\rangle = 0,$$

$$\det \left[u, \frac{\partial x}{\partial \phi}, \frac{\partial x}{\partial \theta_1}, \ldots, \frac{\partial x}{\partial \theta_{n-2}} \right] = f(\phi) \det \left[u, \frac{\partial u}{\partial \phi}, \frac{\partial u}{\partial \theta_1}, \ldots, \frac{\partial u}{\partial \theta_{n-2}} \right], \tag{8.17}$$

$$dS = f(\phi)\, d\omega,$$

where dS is the surface area element of S at the point $x \in S_0$. Since the determinant in (**8.17**) is positive over G, S_0 is of class C^2. Moreover, S_0 has reciprocal Gaussian curvature $f(\phi)$ at the point with outer normal u (which follows from the Gauss map). Since S is a surface of rotation, this is true for the region of S corresponding to $0 < \phi < \pi$. It remains to investigate the poles.

Let S^+ be the region of S above the coordinate hyperplane H normal to the x_n-axis. In H, we drop the nth coordinate and write $x = (x_1, \ldots, x_{n-1})^T$. Let $|x|$ be the norm in H. Let $U \subset H$ be the interior of the ball with center at the origin and radius $r(\pi/2)$. Let $x_n = x_n(x_1, \ldots, x_{n-1})$, $x \in U$, determine S^+. The overlap of S_0 and S^+ corresponds to the region $G_0 \subset G$ determined by $0 < \phi < \pi/2$. The transformation of coordinates, $x = r(\phi)\xi$, is of class C^2 over G_0 and the Jacobian is given by

$$\frac{\partial(x_1, \ldots, x_{n-1})}{\partial(\phi, \theta_1, \ldots, \theta_{n-2})} = f(\phi) \cos \phi \sin^{n-2} \phi \sin^{n-3} \theta_1 \ldots \sin \theta_{n-3}$$

which is positive on G_0.

To show that S^+ is of class C^2, we need to prove that

$$x_n = x_n(x_1, \ldots, x_{n-1}) = \int_\phi^{\pi/2} g(t)\, dt$$

is of class C^2 over U, where $\phi = \phi(|x|)$ is the inverse function of $r(\phi) = |x|$ defined by (**8.11**). From (**8.12**) and the continuity of g, we find $\partial x_n / \partial x_i = 0$ at the origin for $i = 1, \ldots, n-1$. From (**8.14**), we obtain, for $i = 1, \ldots, n-1$,

$$\frac{\partial x_n}{\partial x_i} = -\tan \phi \frac{\partial |x|}{\partial x_i}, \qquad 0 < |x| < r(\pi/2). \tag{8.18}$$

By (**8.18**) and (**8.15**), we find that, at the origin,

$$\left[\frac{\partial^2 x_n}{\partial x_i\, \partial x_j} \right] = -[f(0)]^{-1/(n-1)} I,$$

where I is the identity matrix of order $n - 1$. Using the identity

$$\left[|x|\frac{\partial^2|x|}{\partial x_i\,\partial x_j}\right]+\left[\frac{\partial|x|}{\partial x_i}\frac{\partial|x|}{\partial x_j}\right]=I,\qquad |x|\neq 0,$$

and (**8.14**), we obtain for $0 < |x| < r(\pi/2)$,

$$\left[\frac{\partial^2 x_n}{\partial x_i\,\partial x_j}\right]=-\frac{\tan\phi\sec^2\phi}{g(\phi)}I+\left(\frac{\tan\phi\sec^2\phi}{g(\phi)}-\frac{\tan\phi}{r(\phi)}\right)\left[|x|\frac{\partial^2|x|}{\partial x_i\,\partial x_j}\right]$$

From (**8.15**), we obtain

$$\lim_{|x|\to 0}\left[\frac{\partial^2 x_n}{\partial x_i\,\partial x_j}\right]=-[f(0)]^{-1/(n-1)}I,$$

and therefore S^+ is of class C^2. Similarly, the portion of S below H is of class C^2.

Since we have verified that the compatibility of the overlapping patches is of class C^2, S is of class C^2. In addition, the reciprocal Gaussian curvature is $f(\phi)$ at the point of S with outer normal u. By uniqueness, the convex body bounded by S is a translate of K and this completes the proof. ▮

9. The Class $\mathscr{E}_n$

In this section we complete the proofs of Theorems 2.3, 3.2, and 4.4 by showing that $\mathscr{E}_n$ consists only of ellipsoids.

Lemma 9.1. *If $K \in \mathscr{E}_n$ and K is centrally symmetric, then K is an ellipsoid.*

Proof. Let $C(K, u)$ denote the infinite cylinder $K + L$, where L is a line through the origin parallel to the direction u. Let $K(u)$ be the compact cylinder, obtained from $C(K, u)$, which is bounded by parallel supporting planes to K at the endpoints of the chord of K which is parallel to u and passes through the center z of K. Among all compact cylinders obtained from $C(K, u)$ and which contain K, $K(u)$ has minimal volume. In Petty [11, p. 237], it is shown that

$$\max_u |K(u)| \geq \frac{2\pi_{n-1}}{\pi_n}|K|, \tag{9.2}$$

with equality if and only if K is an ellipsoid. We will show that equality is obtained in (**9.2**).

Let H be the hyperplane through z and normal to u, where u is a (fixed) direction such that $|K(u)|$ is maximal. Let K_1 be obtained from K by Steiner symmetrization in H. Then K_1 has center z and $K_1 \in \mathscr{E}_n$ by (7.7). Moreover, $|K_1(u)| = |K(u)|$, $|K_1| = |K|$. Now, let K_2 be obtained from K_1 by Schwarz symmetrization with respect to the line through z parallel to u. Then K_2 has center z and $K_2 \in \mathscr{E}_n$ by (7.7) since the Schwarz symmetrization can be obtained as a limit of a sequence of Steiner symmetrizations. Moreover, $|K_2(u)| = |K_1(u)|$, $|K_2| = |K_1|$. But K_2 is a convex body of rotation and therefore K_2 is an ellipsoid by Lemma 8.7. Consequently, equality is obtained in (**9.2**) and this completes the proof. ▮

Let $K \subset \mathbb{E}^n$ be a convex body and let $S(H, K)$ denote the convex body obtained from K by Steiner symmetrization in the hyperplane H. Let $H_1, \ldots, H_n$ be the coordinate hyperplanes of a Cartesian coordinate system where H_i is normal to the x_i-axis. Let $K_i = S(H_i, K_{i-1})$, $i = 1, \ldots, n$, where $K = K_0$.

LEMMA 9.3. *Let $K \subset \mathbb{E}^n$ be a convex body. If for an arbitrary Cartesian coordinate system K_n is an ellipsoid, then K is an ellipsoid.*

Proof. Our proof is by induction on the dimension n. Let $n = 2$. Let $p(u)$ be the length of the projection of K on a line normal to the direction u and let $L(u)$ be the length of a maximal chord of K parallel to u. Let z be the origin and choose the x_i-axis parallel to u. By hypothesis, K_2 is an ellipse and must have axes of lengths $p(u)$ and $L(u)$. Consequently,

$$p(u)L(u) = 4\pi^{-1}|K|. \tag{9.4}$$

Let $T = \frac{1}{2}(K + (-K))$. Then T has center z and the chord of T through z in the direction u has length $L(u)$. Moreover,

$$|K| \leq |T|, \tag{9.5}$$

with equality if and only if K is a translate of T. Now, $p(u)$ is the supporting function of the projection body P of both K and T (remembering that $n = 2$). Hence, $|P| = 4|T|$. Let P_z be the polar reciprocal of P with respect to the center z of P. By (**9.4**), P_z differs from T only by a dilatation and a calculation gives $|P_z| = 4^{-1}\pi^2|K|^{-2}|T|$. Thus, $|P||P_z| = \pi^2|K|^{-2}|T|^2 \geq \pi^2$ by (**9.5**) with equality if and only if K is a translate of T. Now, LEMMAS 7.4 and 8.1 prove THEOREM 3.2 for $n = 2$ and it follows that K is an ellipse.

We now assume the validity of the lemma for dimension $n - 1$ and let $K \subset \mathbb{E}^n$. Let $H \subset \mathbb{E}^n$ be a hyperplane. We will show that the orthogonal projection Δ of K on H is an $(n - 1)$-dimensional ellipsoid. Let $z \in H$ and choose the x_1-axis through z normal to H and then complete the Cartesian coordinate system in any way so that z is the origin. Then $H = H_1$. By hypothesis, K_n is an ellipsoid and therefore $H \cap K_n$ is an $(n - 1)$-dimensional ellipsoid. By the induction assumption, Δ is an $(n - 1)$-dimensional ellipsoid. Since H was arbitrary, K is an ellipsoid. ▌

LEMMA 9.6. *Every member of $\mathscr{E}_n$ is an ellipsoid.*

Proof. Let $K \in \mathscr{E}_n$. For any Cartesian coordinate system, K_n is centrally symmetric and, by (7.7), $K_n \in \mathscr{E}_n$. By LEMMA 9.1, K_n is an ellipsoid. Consequently, by LEMMA 9.3, K is an ellipsoid. ▌

REFERENCES

1. BLASCHKE, W. 1916. Über affine Geometrie I: Isoperimetrische Eigenschaften von Ellipse und Ellipsoid. Ber. Verh. Sächs. Akad. Wiss. Leipzig Math. Phys. Kl. **68:** 217–239.
2. BLASCHKE, W. 1917. Über affine Geometrie VII: Neue Extremeigenschaften von Ellipse und Ellipsoid. Ber. Verh. Sächs. Akad. Wiss. Leipzig Math. Phys. Kl. **69:** 306–318.
3. BLASCHKE, W. 1923. Vorlesungen Über Differentialgeometrie II. Springer-Verlag. Berlin, Germany.
4. BONNESEN, T. & W. FENCHEL. 1934. Theorie der konvexen Körper. Ergebnisse der Mathematik, Vol. 3, No. 1. Springer-Verlag. Berlin, Germany.
5. BRICKELL, F. 1965. A new proof of Deicke's theorem on homogeneous functions. Proc. Amer. Math. Soc. **16:** 190–191.

6. Cheng, S. Y. & S. T. Yau. 1976. On the regularity of the solution of the n-dimensional Minkowski problem. Comm. Pure Appl. Math. **29:** 495–516.
7. Deicke, A. 1953. Über die Finsler-Räume mit $A_i = 0$. Arch. Math. **4:** 45–51.
8. Hardy, G. H., J. E. Littlewood & G. Pólya. 1934. Inequalities. Cambridge University Press. Cambridge, England.
9. Leichtweiss, K. 1980. Konvexe Mengen. Springer-Verlag. West Berlin, FRG.
10. Petty, C. M. 1974. Geominimal surface area. Geom. Dedicata **3:** 77–97.
11. Petty, C. M. 1967. Projection bodies. *In* Proceedings, Colloquium on Convexity (Copenhagen, 1965). Pp. 234–241. Kobenhavns Univ. Mat. Inst. Copenhagen, Denmark.
12. Pogorelov, A. V. 1978. The Minkowski Multidimensional Problem. V. H. Winston & Sons. Washington, D.C.
13. Saint Raymond, J. 1981. Sur le volume des corps convexes symétriques. Séminaire Choquet, 20e année (1980/81), Initiation à l'analyse, Comm. No. 11, 25 pp. Secrétariat Math., Paris, France.
14. Santaló, L. A. 1949. Un invariante afin para los cuerpos convexos del espacio de n dimensiones. Portugal. Math. **8:** 155–161.
15. Schneider, R. 1982. Random polytopes generated by anisotropic hyperplanes. Bull. London Math. Soc. **14:** 549–553.

ON SOME INVARIANTS UNDER SIMILITUDES FOR CONVEX BODIES

L. A. Santaló

Department of Mathematics
Ciudad Universitaria
Buenos Aires, Argentina

INTRODUCTION

In a paper on stereology, Franklin [2] introduced a functional for convex bodies in R^3 which is invariant under similitudes, i.e., it depends on the shape of the body and not on its size. We give a set of similitude invariants **(1.4)** of the type introduced by Franklin for convex bodies in R^n and prove the inequalities **(3.10)** for them.

1. THE INVARIANTS $I_i(K)$

Let K be a compact convex set with nonempty interior in the Euclidean space R^n. Let L_{n-1} denote a hyperplane [$(n-1)$-dimensional flat] in R^n and let dL_{n-1} be the density for measuring sets of hyperplanes in the sense of the integral geometry [3, p. 227; 5, p. 204], i.e.,

$$dL_{n-1} = dx\, d\Omega_{n-1}, \tag{1.1}$$

where x denotes the distance from the origin to L_{n-1} and $d\Omega_{n-1}$ is the volume element of the unit $(n-1)$-dimensional sphere corresponding to the direction orthogonal to L_{n-1}.

Let W'_i $(i = 0, 1, \ldots, n-1)$ denote the quermassintegrale of the intersection $K \cap L_{n-1}$ considered as a $(n-1)$-dimensional convex body in L_{n-1} (see [3, p. 209; 4, p. 178; 5, p. 217]). We have, for instance,

$$\begin{aligned} W'_0 &= \text{volume of } K \cap L_{n-1}, \\ W'_1 &= (1/n-1) \text{ surface area of } \partial K \cap L_{n-1}, \\ W'_{n-1} &= \omega_{n-1} = \text{volume } (n-1)\text{-dimensional unit ball.} \end{aligned} \tag{1.2}$$

Recall that

$$\omega_i = 2\pi^{i/2}(i\Gamma(i/2))^{-1}. \tag{1.3}$$

It is well known that W'_i is a functional homogene of degree $(n-1-i)$ in the sense that $W'_i(\lambda K) = \lambda^{n-1-i} W'_i(K)$ (see [4, p. 180]). Therefore, the integrals

$$I_i(K) = \int (W'_i(K \cap L_{n-1}))^{-1/(n-1-i)}\, dL_{n-1} \tag{1.4}$$

for $i = 0, 1, \ldots, n-2$, are invariant under similitudes. All the integrals in this paper are to be taken over the set of flats that meet K.

EXAMPLES. 1. For convex sets in the Euclidean plane, $n = 2$, we have the invariant

$$I_0(K) = \int \sigma^{-1}\, dL_1, \tag{1.5}$$

where dL_1 denotes the density for lines in the plane [5, p. 27] and σ denotes the length of the chord $K \cap L_1$.

2. For convex bodies in R^3 we have the similitude invariants

$$I_0(K) = \int A^{-1/2}\, dL_2, \qquad I_1(K) = 2 \int U^{-1}\, dL_2, \tag{1.6}$$

where dL_2 is the density for planes in R^3 [5, p. 210] and A and U denote respectively the area and the perimeter of the intersection $K \cap L_2$.

The functional $I_0(K)$ is the Franklin invariant introduced in [2].

2. THE CASE OF BALLS

Let K be a ball of radius R. The quermassintegrale W_i' of the $(n - 1)$-dimensional ball cut from K by the hyperplane L_{n-1} at distance x from the center of the ball, are [5, p. 224],

$$W_i' = \omega_{n-1}(R^2 - x^2)^{(n-1-i)/2} \tag{2.1}$$

and using (1.1) we have

$$\begin{aligned} I_i(\text{ball}) &= \omega_{n-1}^{-1/(n-1-i)} \int (R^2 - x^2)^{-1/2}\, dL_{n-1} \\ &= \omega_{n-1}^{-1/(n-1-i)} n\omega_n \int_0^R (R^2 - x^2)^{-1/2}\, dx \\ &= (\pi/2) n\omega_n \omega_{n-1}^{-1/(n-1-i)}. \end{aligned} \tag{2.2}$$

EXAMPLES. 1. For the circles of R^2 ($n = 2, i = 0$) we have

$$I_0(\text{circles}) = \pi^2/2. \tag{2.3}$$

2. For the balls in R^3 we have

$$I_0(\text{ball}) = 2\pi^{3/2}, \qquad I_1(\text{ball}) = 2\pi. \tag{2.4}$$

3. INEQUALITIES

Except for the simple case of balls, the invariants $I_i(K)$ cannot be easily computed exactly. Franklin [2] has given approximate values for ellipsoids in R^3 and noted that $I_0(K) = \infty$ for all polyhedra. He has also conjectured that for convex bodies in R^3 the least value of $I_0(K)$ corresponds to the balls, that is,

$$I_0(K) \geq I_0(\text{ball}) = 2\pi^{3/2} = 11.1344\ldots. \tag{3.1}$$

Probably this is true for all invariants $I_i(K)$ in R^n, that is,

$$I_i(K) \geq (\pi/2)n\omega_n\,\omega_{n-1}^{-1/(n-1-i)} \tag{3.2}$$

with equality if and only if K is a ball.

We have not been able to prove the inequalities (**3.2**). However, we will prove other less sharp inequalities.

The quermassintegrale for convex bodies in R^{n-1} (intersection $K \cap L_{n-1}$) satisfy the following inequalities of Minkowski [3, p. 278; 4, p. 270]:

$$W'_i \leq \omega_{n-1}^{i-n+2}(W'_{n-2})^{n-1-i} \tag{3.3}$$

where the equality holds only for balls. Thus, we have

$$I_i(K) \geq \omega_{n-1}^{(n-i-2)/(n-i-1)} \int (W'_{n-2})^{-1}\,dL_{n-1}. \tag{3.4}$$

Applying the Schwarz inequality [1, p. 21] we have

$$\left[\int dL_{n-1}\right]^2 < \left[\int W'_{n-2}\,dL_{n-1}\right]\left[\int (W'_{n-2})^{-1}\,dL_{n-1}\right], \tag{3.5}$$

where the integrals are always to be taken over the set of hyperplanes that intersect K. The inequality is strict since W'_{n-2} is not a constant.

Taking into account that [3, p. 232; 5, p. 248],

$$\int W'_{n-2}(K \cap L_{n-1})\,dL_{n-1} = (\pi/4)n\omega_{n-1}W_{n-2}(K), \tag{3.6}$$

and that [3, p. 240; 5, p. 233],

$$\int dL_{n-1} = nW_{n-1}(K), \tag{3.7}$$

we get from (**3.4**) and (**3.5**)

$$I_i(K) > (4/\pi)n\omega_{n-1}^{-1/(n-i-1)}(W_{n-1}^2/W_{n-2}). \tag{3.8}$$

Now we can apply the inequality of Fenchel and Aleksandrov [3, p. 282; 4, p. 282],

$$W_{n-1}^2 \geq W_{n-2}\,W_n = \omega_n\,W_{n-2}, \tag{3.9}$$

and we get

$$I_i(K) > (4n/\pi)\omega_n\,\omega_{n-1}^{-1/(n-1-i)} \qquad (i = 0, 1, \ldots, n-2). \tag{3.10}$$

This is our final result. Comparing the right side of the conjectured inequalities (**3.2**) with the right side of (**3.10**) we note that they differ by the factor $\pi^2/8$, which is independent of i and n.

EXAMPLES. 1. $n = 2$. For the invariant (**1.5**) we have the inequality

$$I_0(K) = \int \sigma^{-1}\,dL_1 > 4. \tag{3.11}$$

2. $n = 3$. For the invariants (**1.6**) we have

$$I_0(K) = \int A^{-1/2}\, dL_2 > 16\pi^{-1/2},$$

$$I_1(K) = 2\int U^{-1}\, dL_2 > 16/\pi. \qquad \textbf{(3.12)}$$

References

1. Beckenbach, E. F. & R. Bellman. 1965. Inequalities. Springer-Verlag. West Berlin, FRG.
2. Franklin, J. N. 1977. Some stereological principles in morphometric cytology. SIAM J. Appl. Math. **33:** 267–278.
3. Hadwiger, H. 1957. Vorlesungen über Inhalt, Oberfläche und Isoperimetrie. Springer-Verlag. West Berlin, FRG.
4. Leichtweiss, K. 1980. Konvexe Mengen. VEB Deutscher Verlag der Wissenschaften. East Berlin, GDR.
5. Santaló, L. A. 1976. Integral Geometry and Geometric Probability, Encyclopedia of Mathematics and Its Applications. Addison–Wesley Publishing Company. Reading, Mass.

ON THE ALEKSANDROV–FENCHEL INEQUALITY

Rolf Schneider

Mathematisches Institut der Albert-Ludwigs-Universität
D-7800 Freiburg im Breisgau, Federal Republic of Germany

1. Introduction

For convex bodies $K_1, \ldots, K_n$ in n-dimensional Euclidean space, the Aleksandrov–Fenchel inequality says

$$V(K_1, K_2, K_3, \ldots, K_n)^2 \geq V(K_1, K_1, K_3, \ldots, K_n)V(K_2, K_2, K_3, \ldots, K_n), \quad \textbf{(1.1)}$$

where V denotes the mixed volume. Aleksandrov's [1, 2] proofs are reproduced in the books of Busemann [6] and Leichtweiss [15]. This inequality, special cases of which go back to Minkowski, has numerous applications to extremal and other problems in the geometry of convex bodies. In recent years new interest in this inequality has arisen from different sides. Surprising connections with algebraic geometry have been discovered, which have led to new proofs of (**1.1**) by Teissier [21] and Hovanski [11] (the proof of Fedotov [8], reproduced by Burago and Zalgaller in [5], seems to be incomplete). Also, some unusual applications of (**1.1**) or the closely related inequality for mixed discriminants were found, of which we mention Egorychev's and Falikman's proofs of the van der Waerden conjecture on permanents (see Lagarias [13] for references and further discussion) and Stanley's [20] results of a combinatorial nature.

Despite fresh interest in the Aleksandrov–Fenchel inequality and the new approaches to it, there is one major problem connected with it which has remained open for decades, namely, the characterization of the equality case. Without very restrictive assumptions on the bodies, it is not known when equality occurs in (**1.1**). None of the known proofs yields general information in this respect, since (**1.1**) is first proved for special convex bodies (very smooth bodies in Aleksandrov's second proof, and special polytopes in the other cases) and then generally by approximation, which blurs the cases of equality. Not even a plausible conjecture concerning equality in (**1.1**) has been reported in the literature. It is the purpose of the present paper to recall the known results in that direction, to formulate a conjecture on the general case of equality, in part due to Loritz, and to collect some partial results in favor of it. Maybe a published conjecture will stimulate further study of this question.

2. Preliminaries

By $\mathbb{R}^n$ we denote n-dimensional Euclidean vector space with scalar product $\langle \cdot, \cdot \rangle$, by B its unit ball with center at the origin 0, and by $\Omega = \partial B$ its unit sphere (∂ stands for the boundary). The volume of B is denoted by κ_n, and the usual Lebesgue measure on Ω by ω. $\mathscr{K}^n$ is the space of convex bodies (nonempty, compact, convex subsets) in $\mathbb{R}^n$. For mixed volumes and for some of the results used in the following, one may consult Bonnesen and Fenchel [4] and Leichtweiss [15]. $S(K_1,$

$\ldots, K_{n-1}; \cdot)$ is the mixed area function of $K_1, \ldots, K_{n-1}$; this is a positive Borel measure on Ω and it satisfies

$$nV(K, K_2, \ldots, K_n) = \int_\Omega h(K, u)\, dS(K_2, \ldots, K_n; u) \tag{2.1}$$

for all $K \in \mathscr{R}^n$, where $h(K, \cdot)$ is the support function of K.

For $K \in \mathscr{R}^n$ and $u \in \Omega$, let K^u be the image of K under orthogonal projection onto the hyperplane through 0 orthogonal to u. The $(n-1)$-dimensional mixed volume of $n-1$ bodies $K_1^u, \ldots, K_{n-1}^u$ in such a hyperplane is denoted by $v(K_1^u, \ldots, K_{n-1}^u)$. Since $v(K_1^u, \ldots, K_{n-1}^u) = nV(K_1, \ldots, K_{n-1}, U)$, where U is a segment of length one parallel to u (Bonnesen and Fenchel [4, p. 45]), (**2.1**) gives

$$v(K_1^u, \ldots, K_{n-1}^u) = \frac{1}{2} \int_\Omega |\langle u, v\rangle|\, dS(K_1, \ldots, K_{n-1}; v). \tag{2.2}$$

Integration of (**2.2**) over all u together with an application of Fubini's theorem results in the generalized Kubota formula

$$\int_\Omega v(K_1^u, \ldots, K_{n-1}^u)\, d\omega(u) = n\kappa_{n-1} V(K_1, \ldots, K_{n-1}, B). \tag{2.3}$$

If $\mathscr{C} = (K_{r+1}, \ldots, K_n)$ is an $(n-r)$-tuple of convex bodies, we will often use the abbreviation

$$V(K_1, \ldots, K_r, \mathscr{C}) := V(K_1, \ldots, K_r, K_{r+1}, \ldots, K_n),$$

and similarly for mixed area functions and in lower dimensions. We will also write $\mathscr{C}^u := (K_{r+1}^u, \ldots, K_n^u)$.

Now let $\mathscr{C}$ be a fixed $(n-2)$-tuple of convex bodies. We are interested in the convex bodies K, L for which equality holds in the inequality

$$V(K, L, \mathscr{C})^2 \geq V(K, K, \mathscr{C}) V(L, L, \mathscr{C}). \tag{2.4}$$

We may exclude the case where one of the mixed volumes vanishes, since in that case the equality is easily discussed by means of a well-known criterion (Bonnesen and Fenchel [4, p. 41]). The following lemma collects useful information on the equality case in (**2.4**).

LEMMA 2.5. *Suppose that the $(n-2)$-tuple $\mathscr{C}$ and the convex bodies K, L, M satisfy* dim $M = n$ *and*

$$V(K, K, \mathscr{C}) > 0, \qquad V(L, L, \mathscr{C}) > 0. \tag{2.6}$$

Then

$$\frac{V(K, K, \mathscr{C})}{V(K, M, \mathscr{C})^2} - 2\frac{V(K, L, \mathscr{C})}{V(K, M, \mathscr{C})V(L, M, \mathscr{C})} + \frac{V(L, L, \mathscr{C})}{V(L, M, \mathscr{C})^2} \leq 0. \tag{2.7}$$

The following assumptions are equivalent:

(a) *equality in* (**2.4**),

(b) *equality in* (**2.7**),

(c) $V(K, K, \mathscr{C})/V(L, L, \mathscr{C}) = V(K, M, \mathscr{C})^2/V(L, M, \mathscr{C})^2$,

(d) *the measures* $S(K, \mathscr{C}; \cdot)$ *and* $S(L, \mathscr{C}; \cdot)$ *are proportional.*

Parts of this lemma go back to Favard [7] and Fenchel [9]; for the general case see Fenchel and Jessen [10, p. 25]; see also Leichtweiss [14; 15, Section 24]. A different proof of the implication (a) ⇒ (d) is due to Aleksandrov [1].

3. Conjectures

Suppose now that equality holds in (**2.4**), where (**2.6**) is satisfied. Then Lemma 2.5 shows that, after a suitable dilatation of K or L,

$$S(K, \mathscr{C}; \cdot) = S(L, \mathscr{C}; \cdot). \tag{3.1}$$

Therefore, one has to find out what this equality implies for the convex bodies K and L. The first conjecture below, as well as the lemma following it (for the special case $L = B$), is due to Loritz [16]. Here supp μ denotes the support of the measure μ.

Conjecture 3.2. Equality (**3.1**) holds if, and under the assumption (**2.6**) only if, after a suitable translation of K or L,

$$h(K, u) = h(L, u) \qquad \text{for each } u \in \operatorname{supp} S(B, \mathscr{C}; \cdot). \tag{3.3}$$

According to Aleksandrov [2], Conjecture 3.2 is true if $\mathscr{C}$ consists only of bodies with analytic support functions and positive radii of curvature.

A convex body is called *regular* if at each of its boundary points it has only one supporting hyperplane.

Lemma 3.4. *If* $L \in \mathscr{R}^n$ *is regular and strictly convex, then*

$$\operatorname{supp} S(K, \mathscr{C}; \cdot) \subset \operatorname{supp} S(L, \mathscr{C}; \cdot)$$

for all $K \in \mathscr{R}^n$.

Proof. Suppose there exists $u_0 \in \operatorname{supp} S(K, \mathscr{C}; \cdot)\setminus\operatorname{supp} S(L, \mathscr{C}; \cdot)$. Then there is an open neighborhood α of u_0 in Ω such that $S(L, \mathscr{C}; \alpha) = 0$. Let

$$\tilde{L} := \{x \in L : \langle x, u_0\rangle \le h(L, u_0) - \varepsilon\},$$

where $\varepsilon > 0$ is so small that every exterior unit normal vector to L at a point of $\operatorname{cl}(\partial L\setminus\tilde{L})$ belongs to α. Such an ε exists since L is regular. For $u \in \Omega\setminus\alpha$, the unique boundary point of L at which u is attained as an exterior normal vector, is also a boundary point of $\tilde{L}$. We deduce that $S(L, \mathscr{C}; \beta) = S(\tilde{L}, \mathscr{C}; \beta)$ for every Borel set $\beta \subset \Omega\setminus\alpha$. This yields

$$\begin{aligned} 0 &\le V(B, L, \mathscr{C}) - V(B, \tilde{L}, \mathscr{C}) \\ &= \int_\Omega h(B, u)\, dS(L, \mathscr{C}; u) - \int_\Omega h(B, u)\, dS(\tilde{L}, \mathscr{C}; u) \\ &= S(L, \mathscr{C}; \alpha) - S(\tilde{L}, \mathscr{C}; \alpha) \end{aligned}$$

and hence $S(\tilde{L}, \mathscr{C}; \alpha) = 0$. Thus we have $S(L, \mathscr{C}; \cdot) = S(\tilde{L}, \mathscr{C}; \cdot)$ and therefore $V(L, K, \mathscr{C}) = V(\tilde{L}, K, \mathscr{C})$ for any K by (**2.1**). On the other hand, we can choose an open neighborhood β of u_0 such that $h(L, u) > h(\tilde{L}, u)$ for $u \in \beta$. Since $u_0 \in \operatorname{supp} S(K, \mathscr{C}; \cdot)$, we have $S(K, \mathscr{C}; \beta) > 0$, which implies

$$\begin{aligned} nV(L, K, \mathscr{C}) &= \int_\Omega h(L, u)\, dS(K, \mathscr{C}; u) \\ &> \int_\Omega h(\tilde{L}, u)\, dS(K, \mathscr{C}; u) = nV(\tilde{L}, K, \mathscr{C}), \end{aligned}$$

a contradiction. This proves Lemma 3.4. ∎

Lemma 3.4 shows, in particular, that in (**3.3**) the unit ball B can be replaced by any other regular and strictly convex body, without changing the condition. It also shows that one direction of Conjecture 3.2 is true, namely, that (**3.3**) implies (**3.1**): Suppose that (**3.3**) holds. Then for any convex body M we have $\operatorname{supp} S(M, \mathscr{C}; \cdot) \subset \operatorname{supp} S(B, \mathscr{C}; \cdot)$ and thus

$$\begin{aligned} &\int_\Omega h(M, u)\, d[S(K, \mathscr{C}; u) - S(L, \mathscr{C}; u)] \\ &\quad = nV(M, K, \mathscr{C}) - nV(M, L, \mathscr{C}) \\ &\quad = \int_\Omega [h(K, u) - h(L, u)]\, dS(M, \mathscr{C}; u) = 0. \end{aligned}$$

Since M was arbitrary, (**3.1**) follows.

In order to interpret (**3.3**) geometrically, one would next require a geometric description of the support of the measure $S(B, \mathscr{C}; \cdot)$. For the formulation of a corresponding conjecture we need some more notation.

For $K \in \mathscr{K}^n$ and $u \in \Omega$, we consider a boundary point $x \in \partial K$ where u occurs as an exterior unit normal vector. Let N be the cone (with apex 0) of normal vectors to K at x. The vector u is contained in a unique relatively open face of the convex cone N (see, e.g., Theorem 18.2 of Rockafellar [17]). This maximal relatively open convex cone of normal vectors does not depend upon the choice of the point x (if this point is not unique); call it $N(K, u)$. For $K_1, \ldots, K_{n-1} \in \mathscr{K}^n$, we now say that the vector u is $(K_1, \ldots, K_{n-1})$-*extreme* if there exist $(n-1)$-dimensional linear subspaces $H_1, \ldots, H_{n-1} \subset \mathbb{R}^n$ with $N(K_i, u) \subset H_i$ for $i = 1, \ldots, n-1$ and $\dim H_1 \cap \ldots \cap H_{n-1} = 1$. This notion generalizes that of a p-extreme normal vector: The vector $u \in \Omega$ is called a p-*extreme* normal vector of the convex body K if $\dim N(K, u) \leq p + 1$. Thus, u is a p-extreme normal vector of K if and only if u is

$$(\underbrace{K, \ldots, K}_{n-1-p}, \underbrace{B, \ldots, B}_{p})\text{-extreme.}$$

Conjecture 3.5. $\operatorname{supp} S(K_1, \ldots, K_{n-1}; \cdot)$ is the closure of the set of $(K_1, \ldots, K_{n-1})$-extreme unit vectors.

This is true at least in the following cases:

(a) if $K_1, \ldots, K_{n-1}$ are polytopes,

(b) if $K_1 = \cdots = K_p$ and the bodies $K_{p+1}, \ldots, K_{n-1}$ are regular and strictly convex, for some $p \in \{0, \ldots, n-1\}$.

Suppose first that $K_1, \ldots, K_{n-1}$ are polytopes. Writing

$$F(K, u) := \{x \in \mathbb{R}^n : \langle x, u \rangle = h(K, u)\}$$

for the face of K with exterior unit normal vector u, we have

$$S(K_1, \ldots, K_{n-1}; \{u\}) = v(F(K_1, u), \ldots, F(K_{n-1}, u)),$$

and $S(K_1, \ldots, K_{n-1}; \cdot)$ is concentrated on the finite set of those u for which this is positive. Hence, $u \in \text{supp}\, S(K_1, \ldots, K_{n-1}; \cdot)$ if and only if segments $S_i \subset F(K_i, u)$, $i = 1, \ldots, n-1$, can be found with linearly independent directions (Bonnesen and Fenchel [4, p. 41]). Now $F(K_i, u)$ contains a (nondegenerate) segment S_i if and only if $N(K_i, u)$ is contained in the $(n-1)$-dimensional linear subspace orthogonal to S_i. The truth of Conjecture 3.5 for polytopes is now easy to see.

Now let $K \in \mathscr{R}^n$ and $p \in \{0, \ldots, n-1\}$ be given. Then

$$\text{supp}\, S(\underbrace{K, \ldots, K}_{p}, \underbrace{B, \ldots, B}_{n-1-p}; \cdot)$$

is the closure of the set of $(n-p-1)$-extreme normal vectors of K, as shown by Schneider [18]. By the remark made above, this is also the closure of the set of

$$(\underbrace{K, \ldots, K}_{p}, \underbrace{B, \ldots, B}_{n-1-p})\text{-extreme}$$

vectors. If here the last $n-1-p$ arguments are replaced by any regular and strictly convex bodies, neither the set of $(K, \ldots, K, \cdot, \ldots, \cdot)$-extreme vectors nor, by LEMMA 3.4, the support of $S(K, \ldots, K, \cdot, \ldots, \cdot; \cdot)$ changes. Hence, Conjecture 3.5 is also true in case (b).

Putting LEMMA 2.5 and Conjectures 3.2 and 3.5 together, we now end up with the following conjecture. Here a supporting hyperplane of a convex body is called *$(B, \mathscr{C})$-extreme* if its exterior unit normal vector is $(B, \mathscr{C})$-extreme.

Conjecture 3.6. If $V(K, K, \mathscr{C}) > 0$ and $V(L, L, \mathscr{C}) > 0$, then equality in **(2.4)** holds if and only if suitable homothets of K and L have the same $(B, \mathscr{C})$-extreme supporting hyperplanes.

Under each of the following assumptions, it is known that equality in **(2.4)** [**(2.6)** being satisfied] implies that K and L are homothetic:

(a) $\mathscr{C}$ consists of balls,

(b) $\mathscr{C}$ consists of strongly combinatorially isomorphic simple polytopes, and K, L are polytopes having the same system of normal vectors to the facets as the polytopes of $\mathscr{C}$,

(c) $\mathscr{C}$ consists of convex bodies with twice continuously differentiable support functions and positive radii of curvature, and K and L have twice continuously differentiable support functions.

For (a), see Bonnesen and Fenchel [4, p. 93]; (b) and (c) are due to Aleksandrov [1, 2] (compare Busemann [6]). In cases (a) and (c), the set of $(B, \mathscr{C})$-extreme vectors coincides with Ω, while in case (b) it is the set of exterior unit normal vectors to the facets of the polytopes in $\mathscr{C}$. Thus, Conjecture 3.6 is in agreement with these classical cases.

For the case $\mathscr{C} = (K, \ldots, K)$, Bol [3] (see also Knothe [12] for a special case with a different proof) proved that equality in **(2.4)** implies that K is homothetic to a $(n-2)$-tangential body of L, which means that for a suitable homothet L' of L, each 1-extreme supporting hyperplane of K supports L'. Since the 1-extreme vectors of K are precisely the $(B, \mathscr{C})$-extreme vectors, Conjecture 3.6 holds also in this case.

4. Special Results

We shall now prove a few new results which confirm Conjecture 3.6 in additional special cases. Our first theorem implies a generalization of cases (a) and (c) above. The convex body $\tilde{M}$ is called a *summand* of the convex body M if $M = \tilde{M} + M'$ for a suitable convex body M'.

Theorem 4.1. *Let $\mathscr{C} = (K_3, \ldots, K_n)$ and suppose that K_i has an n-dimensional summand $\tilde{K}_i$ $(i = 3, \ldots, n)$. If* **(2.6)** *is satisfied and equality holds in* **(2.4)**, *then also*

$$V(K, L, \tilde{\mathscr{C}})^2 = V(K, K, \tilde{\mathscr{C}})V(L, L, \tilde{\mathscr{C}})$$

with $\tilde{\mathscr{C}} = (\tilde{K}_3, \ldots, \tilde{K}_n)$.

Proof. If the assumptions are satisfied, then by Lemma 2.5 we may assume, after some dilatation of K or L, that

$$S(K, \mathscr{C}; \cdot) = S(L, \mathscr{C}; \cdot). \tag{4.2}$$

Write $\mathscr{C}' = (K_4, \ldots, K_n)$ (for $n = 2$ there is nothing to prove, and for $n = 3$ we modify the proof by just omitting $\mathscr{C}'$). Let Q be a convex body of dimension n. By **(2.7)** (replace K_3 by Q and M by K_3), we have

$$\frac{V(K, K, Q, \mathscr{C}')}{V(K, Q, \mathscr{C})^2} - 2\frac{V(K, L, Q, \mathscr{C}')}{V(K, Q, \mathscr{C})V(L, Q, \mathscr{C})} + \frac{V(L, L, Q, \mathscr{C}')}{V(L, Q, \mathscr{C})^2} \leq 0. \tag{4.3}$$

Since

$$V(K, Q, \mathscr{C}) = V(L, Q, \mathscr{C}) \tag{4.4}$$

by **(2.1)** and **(4.2)**, we get

$$V(K, K, Q, \mathscr{C}') - 2V(K, L, Q, \mathscr{C}') + V(L, L, Q, \mathscr{C}') \leq 0. \tag{4.5}$$

By approximation, this inequality holds also if Q is a lower-dimensional convex body.

Now suppose that $K_3 = \tilde{K}_3 + K'_3$ with convex bodies $\tilde{K}_3$, K'_3. Writing down **(4.5)** for $Q = \tilde{K}_3$ and for $Q = K'_3$ and adding the two inequalities, we get

$$V(K, K, K_3, \mathscr{C}') - 2V(K, L, K_3, \mathscr{C}') + V(L, L, K_3, \mathscr{C}') \leq 0.$$

By **(4.2)** the equality sign must hold here; hence, in particular,

$$V(K, K, \tilde{K}_3, \mathscr{C}') - 2V(K, L, \tilde{K}_3, \mathscr{C}') + V(L, L, \tilde{K}_3, \mathscr{C}') = 0.$$

Together with **(4.4)** for $Q = \tilde{K}_3$ this shows that **(2.7)**, with K_3 replaced by $\tilde{K}_3$ and M by K_3, holds with equality; hence,

$$V(K, L, \tilde{K}_3, \mathscr{C}')^2 = V(K, K, \tilde{K}_3, \mathscr{C}')V(L, L, \tilde{K}_3, \mathscr{C}')$$

by Lemma 2.5, provided that $\dim \tilde{K}_3 = n$. Thus we have replaced K_3 by an n-dimensional summand and still have equality in **(2.4)**. Repeating the argument with $K_4, \ldots, K_n$, we arrive at the conclusion of Theorem 4.1. ▌

Corollary 4.6. *Suppose that $\mathscr{C}$ consists only of outer parallel bodies (i.e., K_i has some ball as a summand, $i = 3, \ldots, n$). If* **(2.6)** *is satisfied and equality holds in* **(2.4)**, *then K and L are homothetic.*

This follows from THEOREM 4.1 and result (a) mentioned after Conjecture 3.6. Thus, the corollary is a common generalization of the classical cases (a) and (c) after Conjecture 3.6, since it is well known that every convex body with a twice continuously differentiable support function and with positive radii of curvature has some ball as a summand. Since under the assumption of COROLLARY 4.6 every unit vector is $(B, K_3, \ldots, K_n)$-extreme, COROLLARY 4.6 is in agreement with Conjecture 3.6.

Another specialization of Conjecture 3.6 which should be tested is the case where each component of $\mathscr{C}$ is equal to either K or L. It was conjectured that equality in

$$V(\underbrace{K, \ldots, K}_{n-i}, \underbrace{L, \ldots, L}_{i})^2 \geq V(\underbrace{K, \ldots, K}_{n-i+1}, \underbrace{L, \ldots, L}_{i-1})V(\underbrace{K, \ldots, K}_{n-i-1}, \underbrace{L, \ldots, L}_{i+1}) \quad \mathbf{(4.7)}$$

$(i = 1, \ldots, n-1)$ holds only if K is homothetic to an $(n-i-1)$-tangential body of L or L is homothetic to an $(i-1)$-tangential body of K (see Bonnesen and Fenchel [4, p. 92]). Recall that, for $i \in \{0, \ldots, n-1\}$, a convex body $K \in \mathscr{R}^n$ is called an *$(n-i-1)$-tangential body* of the convex body L if every i-extreme supporting hyperplane of K also supports L (for equivalent definitions, see Schneider [19]). For $i = 1$ and $i = n-1$ this was settled by Bol [3], but he conjectured [3, p. 56] that for $i \neq 1, n-1$ these are not the only cases of equality. With the further specialization $L = B$, (**4.7**) is the inequality

$$W_i(K)^2 \geq W_{i-1}(K)W_{i+1}(K) \quad \mathbf{(4.8)}$$

$(i = 1, \ldots, n-1)$ for the quermassintegrals of K. For these we have the following result.

THEOREM 4.9. *If $K \in \mathscr{R}^n$ is centrally symmetric and $i \in \{1, \ldots, n-1\}$, then equality in* (**4.8**) *holds if and only if either* dim $K < n-i$ *or K is an $(n-i-1)$-tangential body of a ball.*

It is very probable that the assumption of central symmetry, which is made for a technical reason, can be omitted. In that case our result would show that Conjecture 3.6 is true in the special case

$$\mathscr{C} = (\underbrace{K, \ldots, K}_{n-i-1}, \underbrace{B, \ldots, B}_{i-1}).$$

The proof generalizes a method of Favard [7] (compare also Leichtweiss [14, 15]). The first part is stated as a separate lemma since it extends to a more general situation.

LEMMA 4.10. *Let $\mathscr{C}$ be an $(n-3)$-tuple of convex bodies in $\mathbb{R}^n$ $(n \geq 3)$; let K, $L \in \mathscr{R}^n$ satisfy*

$$V(K, K, B, \mathscr{C}) > 0, \qquad V(L, L, B, \mathscr{C}) > 0 \quad \mathbf{(4.11)}$$

and

$$S(K, B, \mathscr{C}; \cdot) = S(L, B, \mathscr{C}; \cdot). \quad \mathbf{(4.12)}$$

Then

$$v(K^u, L^u, \mathscr{C}^u)^2 = v(K^u, K^u, \mathscr{C}^u)v(L^u, L^u, \mathscr{C}^u) \quad \mathbf{(4.13)}$$

for each $u \in \Omega$.

Proof. In the following, E_u denotes the $(n-1)$-dimensional linear subspace orthogonal to $u \in \Omega$. Let $u \in \Omega$. Inequality (**2.7**), applied in E_u, shows that

$$\frac{v(K^u, K^u, \mathscr{C}^u)}{v(K^u, B^u, \mathscr{C}^u)^2} - 2\frac{v(K^u, L^u, \mathscr{C}^u)}{v(K^u, B^u, \mathscr{C}^u)v(L^u, B^u, \mathscr{C}^u)} + \frac{v(L^u, L^u, \mathscr{C}^u)}{v(L^u, B^u, \mathscr{C}^u)^2} \leq 0,$$

provided that $v(K^u, K^u, \mathscr{C}^u) > 0$ and $v(L^u, L^u, \mathscr{C}^u) > 0$. By (**2.3**) and (**4.11**), this holds for ω-almost all u. From (**2.2**) and (**4.12**) we get

$$v(K^u, B^u, \mathscr{C}^u) = v(L^u, B^u, \mathscr{C}^u); \tag{4.14}$$

hence,

$$v(K^u, K^u, \mathscr{C}^u) - 2v(K^u, L^u, \mathscr{C}^u) + v(L^u, L^u, \mathscr{C}^u) \leq 0 \tag{4.15}$$

for almost all u and then by continuity for all $u \in \Omega$. In view of (**2.3**), integration yields

$$V(K, K, B, \mathscr{C}) - 2V(K, L, B, \mathscr{C}) + V(L, L, B, \mathscr{C}) \leq 0.$$

By (**2.1**) and (**4.12**), here the equality sign is valid; hence, equality holds in (**4.15**) for all $u \in \Omega$. Together with (**4.14**) this implies (**4.13**) by Lemma 2.5. ∎

Proof of Theorem 4.9. Since $W_i(K) > 0$ if and only if $\dim K \geq n - i$, we may assume that $\dim K \geq n - i + 1$. That equality in (**4.8**) holds for $(n-i-1)$-tangential bodies of balls, is known (Schneider [19, Theorem (3.8)]). We prove by induction with respect to n that equality in (**4.8**) and the assumption $\dim K \geq n - i + 1$ imply that K is an $(n-i-1)$-tangential body of a ball ($i \in \{1, \ldots, n-1\}$). For $n = 2$ this is well known. Assume that $n \geq 3$; the assertion has been proven in dimensions less than n, and equality holds in (**4.8**) for some $i \in \{1, \ldots, n-1\}$, where $\dim K \geq n - i + 1$. For $i = 1$, K is an $(n-2)$-tangential body of a ball by the result of Bol mentioned above; hence, we may assume that $i \geq 2$. According to Lemma 2.5, after a suitable dilatation of K we may assume that

$$S(\underbrace{K, \ldots, K}_{n-i}, \underbrace{B, \ldots, B}_{i-1}; \cdot) = S(\underbrace{K, \ldots, K}_{n-i-1}, \underbrace{B, \ldots, B}_{i}; \cdot). \tag{4.16}$$

By Lemma 4.10 this implies

$$v_{n-i}(K^u)^2 = v_{n-i+1}(K^u)v_{n-i-1}(K^u)$$

for $u \in \Omega$, where we have written

$$v_k(K^u) = v(\underbrace{K^u, \ldots, K^u}_{k}, \underbrace{B^u, \ldots, B^u}_{n-k-1}).$$

By the induction hypothesis, K^u is an $(n-i-1)$-tangential body of a homothet of B^u. If this homothet has radius r_u, then $r_u^{-1}K^u$ is an $(n-i-1)$-tangential body of a unit ball; hence (Schneider [19, Theorem (3.8)]),

$$v_{n-i}(r_u^{-1}K^u) = v_{n-i-1}(r_u^{-1}K^u).$$

Since (**2.2**) and (**4.16**) imply

$$v_{n-i}(K^u) = v_{n-i-1}(K^u),$$

we deduce that $r_u = 1$.

Let B_r be the maximal ball (r its radius) in K with center at the symmetry center of K. K and B_r have at least one common supporting hyperplane H, and by symmetry a second one parallel to H, say H'. Choose a unit vector u parallel to H. The intersection of H with E_u is a common supporting hyperplane (in E_u) of B_r^u and K^u and hence an extreme supporting hyperplane of K^u. The same holds true for the intersection of H' and E_u. Since both hyperplanes must support some ball (in E_u) of radius 1, it follows that $r = 1$. Hence, we may assume that $B_r = B$.

Now let $u \in \Omega$ be arbitrary. The projection K^u is an $(n - i - 1)$-tangential body of a unit ball, but since $B^u \subset K^u$, this is necessarily the ball B^u (observe that a convex body is the intersection of its supporting half spaces bounded by extreme supporting hyperplanes).

Finally, let H be an i-extreme supporting hyperplane of K and u its exterior unit normal vector. Choose a vector v orthogonal to u and, if $\dim N(K, u) \geq 2$, in the linear hull of $N(K, u)$. The intersection of H with E_v is then an $(i - 1)$-extreme supporting hyperplane of K^v (relative to E_v). Since K^v is an $(n - i - 1)$-tangential body of B^v, that hyperplane supports B^v; hence, H supports B. Thus, K is an $(n - i - 1)$-tangential body of B. ▮

References

1. Aleksandrov, A. D. 1937. Zur Theorie der gemischten Volumina von konvexen Körpern. II. Neue Ungleichungen zwischen den gemischten Volumina und ihre Anwendungen. Mat. Sb. N.S. **2:** 1205–1238. (In Russian.)
2. Aleksandrov, A. D. 1938. Zur Theorie der gemischten Volumina von konvexen Körpern. IV. Gemischte Diskriminanten und gemischte Volumina. Mat. Sb. N.S. **3:** 227–251. (In Russian.)
3. Bol, G. 1943. Beweis einer Vermutung von H. Minkowski. Abh. Math. Sem. Univ. Hamburg **15:** 37–56.
4. Bonnesen, T. & W. Fenchel. 1934. Theorie der konvexen Körper. Springer-Verlag. Berlin, Germany.
5. Burago, Ju. D. & V. A. Zalgaller. 1980. Geometric Inequalities. Nauka/Otdel. Leningrad, USSR. (In Russian.)
6. Busemann, H. 1958. Convex Surfaces. Interscience Publishers. New York, N.Y.
7. Favard, J. 1933. Sur les corps convexes. J. Math. Pures Appl. (9) **12:** 219–282.
8. Fedotov, V. P. 1979. A new method of proving inequalities between mixed volumes, and a generalization of the Aleksandrov-Fenchel-Shephard inequalities. Dokl. Akad. Nauk SSSR **245:** 31–41. (English translation: Soviet Math. Dokl. **20** (1979): 268–271.)
9. Fenchel, W. 1936. Généralisation du théorème de Brunn et Minkowski concernant les corps convexes. C.R. Acad. Sci. Paris **203:** 764–766.
10. Fenchel, W. & B. Jessen. 1938. Mengenfunktionen und konvexe Körper. Danske Vid. Selskab. Mat. Fys. Medd. **16**(3), 31 pp.
11. Hovanski, A. G. Newton polyhedra and Aleksandrov-Fenchel inequalities. To be published. (Abstract: Usp. Akad. Nauk **34**(4) (1979): 160–161.)
12. Knothe, H. 1949. Über eine Vermutung H. Minkowskis. Math. Nachr. **2:** 380–385.
13. Lagarias, J. C. 1982. The van der Waerden conjecture: two Soviet solutions. Notices Amer. Math. Soc. **29:** 130–133.
14. Leichtweiss, K. 1981. Zum Beweis eines Eindeutigkeitssatzes von A. D. Aleksandrov. *In* E. B. Christoffel: The Influence of His Work on Mathematics and the Physical Sciences, International Symposium, Aachen, 1979. P. L. Butzer & F. Fehér, Eds. Birkhäuser. Basel, Switzerland.
15. Leichtweiss, K. 1980. Konvexe Mengen. VEB Deutscher Verlag der Wissenschaften. East Berlin, GDR/Springer-Verlag. West Berlin, FRG.
16. Loritz, A. 1979. Über den Gleichheitsfall in Ungleichungen zwischen gemischten Volumina konvexer Körper. Diplomarbeit. Universität Freiburg. Freiburg, FRG.

17. Rockafellar, R. T. 1970. Convex Analysis. Princeton University Press. Princeton, N.J.
18. Schneider, R. 1975. Kinematische Berührmaße für konvexe Körper und Integralrelationen für Oberflächenmaße. Math. Ann. **218:** 253–267.
19. Schneider, R. 1978. Über Tangentialkörper der Kugel. Manuscripta Math. **23:** 269–278.
20. Stanley, R. P. 1981. Two combinatorial applications of the Aleksandrov-Fenchel inequalities. J. Combin. Theory Ser. A **31:** 56–65.
21. Teissier, B. 1979. From the Hodge index theorem to the isoperimetric inequalities. C.R. Acad. Sci. Paris **288:** 287–289.

Part III. Qualitative Convexity

KRASNOSEL'SKII-TYPE THEOREMS

Marilyn Breen

Department of Mathematics
University of Oklahoma
Norman, Oklahoma 73019

1. Introduction

Let S be a subset of R^d. Set S is said to be starshaped if and only if there is some point p in S such that p sees each point of S via a line segment in S, and the set of all such points p is called the convex kernel of S, denoted ker S. A well-known theorem of Krasnosel'skii [11] states that if S is a nonempty compact set in R^d, then S is starshaped if and only if every $d + 1$ points of S see a common point via S. The combinatorial flavor of this result is very appealing, and it has inspired some interesting questions: Can analogues of the Krasnosel'skii theorem be generated to reveal the dimension of ker S and the size of ker S? Can the nature of ker S be determined from the behavior of appropriately selected subsets of S? And can a combinatorial characterization theorem be provided for noncompact starshaped sets? The purpose of this article is to examine some recent work related to these questions and their answers.

The following definitions and terminology will be used throughout the paper. Let $S \subseteq R^d$. Point s in the closure of S, cl S, is said to be a *point of local convexity* of S if and only if there is some neighborhood N of s such that $N \cap S$ is convex. If no such neighborhood exists, then $s \in$ cl S is said to be a *point of local nonconvexity* (lnc point) of S. Lnc S denotes the set of lnc points of S. For points x and y in S, we say x *sees* y via S (x is *visible* from y via S) if and only if the corresponding segment $[x, y]$ lies in S. Similarly, for x in cl S and y in S, x is *clearly visible* from y via S if and only if there is some neighborhood N of x such that y sees each point of $N \cap S$ via S. As noted previously, set S is *starshaped* if and only if there is some point p in S such that p sees each point of S via S, and the set of all such points p is the convex kernel of S, ker S. The symbol dim ker S will denote the dimension of ker S.

2. The Dimension and Size of ker S

We begin with some results from [7] concerning the dimension of the kernel of a starshaped set. The following lemma will be helpful.

Lemma 1. *For each k and d, $1 \le k \le d$, let $f(d, d) = d + 1$ and $f(d, k) = 2d$ if $1 \le k \le d - 1$. Let $\mathscr{B}$ be a uniformly bounded collection of compact convex sets in R^d. Then for a k with $1 \le k \le d$, $\dim \bigcap \{B: B \text{ in } \mathscr{B}\} \ge k$ if and only if for some $\varepsilon > 0$, every $f(d, k)$ members of $\mathscr{B}$ contain a common k-dimensional ε-neighborhood.*

With this lemma, it is easy to establish the following result.

Theorem 1. *For each k and d, $1 \le k \le d$, let $f(d, d) = d + 1$ and $f(d, k) = 2d$ if $1 \le k \le d - 1$. Let S be a nonempty compact set in some linear topological space L.*

Then for a k with $1 \leq k \leq d$, dim ker $S \geq k$ *if and only if for some* $\varepsilon > 0$ *and some d-dimensional flat F in L, every* $f(d, k)$ *points of S see via S a common k-dimensional* ε*-neighborhood in F. If* $k = d$ *or* $k = 1$, *the result is best possible.*

Perhaps several observations should be made here. First note that the Krasnosel'skii number $d + 1$ is not good enough to yield our result, even in the planar case. Furthermore, it is easy to find examples that show that the uniform lower bound of ε is required in the lemma and in the theorem. And finally, examples from [9] illustrate that the results fail unless set S is both closed and bounded.

Note also that the results above provide no information concerning the size of ker S, and it is this kind of quantitative analogue of Krasnosel'skii's theorem that we examine now. We would like to study those compact sets in R^d whose kernels contain a k-dimensional ε-neighborhood, $0 \leq k \leq d$, and if $k = 0$ or $k = d$, characterizations will be easy consequences of Krasnosel'skii's theorem. (See [7, 15].) However, for $1 \leq k \leq d - 1$, the problem is more troublesome. One approach, appearing in [3, 4], involves the points of local nonconvexity of our set, and we have the following theorem.

THEOREM 2. *Let S be a nonempty compact set in* R^2 *having n lnc points. The kernel of S contains an interval of radius* $\varepsilon > 0$ *if and only if every* $f(n) = \max\{4, 2n\}$ *points of S see via S a common interval of radius* ε. *The number* $f(n)$ *is best possible for every* $n \geq 1$.

Since the number $f(n)$ above is best, in this setting no finite Krasnosel'skii theorem is possible independent of the number of lnc points of S. Fortunately, with suitable restrictions on the arrangement of these lnc points, THEOREM 2 may be generalized to R^d. Moreover, the procedure succeeds not only in showing that ker S contains a k-dimensional ε-neighborhood but also in showing that ker S contains a member of $\mathscr{A}$, where $\mathscr{A}$ is any family of compact subsets of S which converge in the sense of the Hausdorff metric. While a statement of the general result appears below, the reader is referred to [3] for necessary definitions and details.

THEOREM 3. *Let* $\mathscr{A}$ *denote a family of compact sets which has the Bolzano–Weierstrass property relative to the Hausdorff metric. Let S be a d-dimensional compact set whose set Q of lnc points is a union of n disjoint closed convex sets* $C_1, \ldots, C_n$, *each essential. If every 2n or fewer points of S see via S a member of* $\mathscr{A}$, *then* ker S *contains a member of* $\mathscr{A}$.

3. The Concept of Clear Visibility

In this section, we turn our attention to a second question: Can information concerning the set ker S be determined from the behavior of appropriately selected subsets of S? It is well known that versions of Krasnosel'skii's theorem may be obtained by replacing points of S with boundary points of S or with regular points of S [15]. Moreover, in [6, 13], other Krasnosel'skii-type theorems are obtained using the set of $(d - 2)$-extreme points of set S. In Section 2 above, we observed a relationship between the kernel of S and the set Q of lnc points of S. Although this result is dependent upon the cardinality of Q, under more favorable circumstances we might hope that Q could be used to produce a Krasnosel'skii-type theorem, independent of this cardinality. The notion of clearly visible, a concept which appears in work by Stavrakas [14] and by Falconer [10], is a valuable tool in this study, as the next theorem illustrates.

THEOREM 4. *Let S be a nonempty compact connected set in R^2. Then S is starshaped if and only if every 3 lnc points of S are clearly visible from a common point of S. The number 3 is best possible.*

Proof. The details of the argument appear in [2] and will not be presented here. However, we should remark that the approach resembles a familiar proof of Krasnosel'skii's theorem [15]. Defining

$$A_q = \{x\colon q \text{ is clearly visible from } x \text{ via } S\}$$

and

$$C_q = \bigcap \{H\colon H \text{ a closed half space with } A_q \subseteq H \text{ and } q \in \text{bdry } H\},$$

the main step in the proof is to show that $\ker S = \bigcap \{C_q \cap \text{conv } S\colon q \in \text{lnc } S\}$. The remainder of the argument follows easily from Helly's theorem in R^2. ▮

Similarly, the concept of clear visibility may be used to generate analogues to our previous results concerning the dimension of ker *S*. Recall that THEOREM 1 provided a Krasnosel'skii-type theorem for dim ker *S*, yet it required a uniform lower bound ε on the radii of the *k*-dimensional neighborhoods. Using clear visibility, this cumbersome bound may be eliminated, and we have the following results from [5]. Observe that LEMMA 2 is a variation of our earlier LEMMA 1.

LEMMA 2. *For each k and d, $1 \leq k \leq d$, let $f(d, d) = d + 1$ and $f(d, k) = 2d$ if $1 \leq k \leq d - 1$. Let $\mathscr{B}$ be a uniformly bounded collection of compact convex sets in R^d which is closed with respect to the Hausdorff metric. Then for a k with $1 \leq k \leq d$, $\dim \bigcap \{B\colon B \text{ in } \mathscr{B}\} \geq k$ if and only if every $f(d, k)$ members of $\mathscr{B}$ contain a common k-dimensional set.*

THEOREM 5. *For each k and d, $1 \leq k \leq d$, let $f(d, d) = d + 1$ and $f(d, k) = 2d$ if $1 \leq k \leq d - 1$. Let S be a nonempty compact set in R^d. Then for a k with $1 \leq k \leq d$, dim ker $S \geq k$ if and only if every $f(d, k)$ points of* bdry *S are clearly visible from each point of a common k-dimensional subset of S. If $k = 1$ or $k = d$, the result is best possible.*

Finally, note that THEOREMS 4 and 5 suggest a new method of generating analogues to existing theorems on starshaped sets in the following way: Suppose a familiar theorem states that the kernel of a compact starshaped set *S* has a particular configuration if and only if every *j*-member subset of *S* is visible from a set having another configuration. The analogue may be obtained by replacing *S* with lnc *S* (or bdry *S*) and visible with clearly visible.

4. THE CASE FOR NONCOMPACT SETS

In general, Krasnosel'skii-type theorems rely on properties of compact sets, and these results do not have direct analogues for sets that are not both closed and bounded. (See [9] for illustrative examples.) For this reason, it is somewhat surprising to learn that analogues may be obtained for bounded sets, again by using the concept of clear visibility. THEOREM 6 below is a generalization of THEOREM 4 to open bounded sets, while THEOREM 7 provides a characterization for bounded starshaped sets in the plane. The details appear in [1, 8].

THEOREM 6. *Let S be a nonempty bounded connected set in R^2, and assume that every 3 or fewer lnc points of S are clearly visible from a common point of S. Then for some point p in S, the set $A \equiv \{s: s \text{ in } S \text{ and } [p, s] \not\subseteq S\}$ is nowhere dense in S. Moreover, if S is open, then S is starshaped.*

THEOREM 7. *Let S be a nonempty bounded set in R^2. Then S is starshaped if and only if every 3 or fewer boundary points of S are clearly visible from a common point of S. The number 3 is best possible.*

5. CONCLUSION

We close with a few comments concerning some unsolved problems. One of the most enticing open questions is the following: For S compact in R^d, does the behavior of the lnc points of S determine whether S is starshaped? That is, does THEOREM 4 have a d-dimensional analogue? If the answer is affirmative, then d-dimensional versions of THEOREMS 6 and 7 should hold as well, and a Krasnosel'skii-type characterization for bounded starshaped sets in R^d could be established.

Another very interesting problem involves the existence of Krasnosel'skii-type theorems for sets that are neither closed nor bounded. While finite versions of Krasnosel'skii's theorem are not suitable for such sets, analogues may hold for sets that satisfy the weaker requirement of being finitely starlike, a concept introduced by Peterson [12]: A set S in R^d is finitely starlike if and only if every finite subset of S sees a common point via S. Peterson has conjectured that a finite Krasnosel'skii theorem exists to characterize these sets, and the problem appears to be a challenging one.

REFERENCES

1. BREEN, M. 1985. A characterization theorem for bounded starshaped sets in the plane. Proc. Amer. Math. Soc. In press.
2. BREEN, M. 1982. A Krasnosel'skii-type theorem for points of local nonconvexity. Proc. Amer. Math. Soc. **85:** 261–266.
3. BREEN, M. 1982. A quantitative Krasnosel'skii theorem in R^d. Geom. Dedicata **12:** 219–226.
4. BREEN, M. 1980. A quantitative version of Krasnosel'skii's theorem in R^2. Pacific J. Math. **91:** 31–37.
5. BREEN, M. 1982. Clear visibility and the dimension of kernels of starshaped sets. Proc. Amer. Math. Soc. **85:** 414–418.
6. BREEN, M. 1980. $(d-2)$-extreme points and a Helly-type theorem for starshaped sets. Canad. J. Math. **32:** 703–713.
7. BREEN, M. 1981. K-dimensional intersections of convex sets and convex kernels. Discrete Math. **36:** 233–237.
8. BREEN, M. 1982. Points of local nonconvexity and sets which are almost starshaped. Geom. Dedicata **13:** 201–213.
9. BREEN, M. 1979. The dimension of the kernel of a planar set. Pacific J. Math. **82:** 15–21.

10. Falconer, K. J. 1977. The dimension of the kernel of a compact starshaped set. Bull. London Math. Soc. **9:** 313–316.
11. Krasnosel'skii, M. A. 1946. Sur un critère pour qu'un domain soit étoilé. Mat. Sb. **61:** 309–310.
12. Peterson, B. B. 1982. Is there a Krasnosel'skii theorem for finitely starlike sets? *In* Proceedings, Conference on Convexity and Related Combinatorial Geometry. Marcel Dekker, Inc. New York, N.Y.
13. Stavrakas, N. 1984. Krasnosel'skii theorems for non-separating compact sets. Canad. Math. Bull. In press.
14. Stavrakas, N. 1972. The dimension of the convex kernel and points of local nonconvexity. Proc. Amer. Math. Soc. **34:** 222–224.
15. Valentine, F. A. 1964. Convex Sets. McGraw–Hill Book Co., Inc. New York, N.Y.

CHOQUET SIMPLICES IN FINITE DIMENSIONS

René Fourneau

Institut Supérieur Industriel Liégeois
B-4000 Liège, Belgium

Institut de Mathématique
Université de Liège
B-4000 Liège, Belgium

1. Introduction

Since 1956, many authors have given various descriptions of larger and larger classes of Choquet simplices, but as yet no characterization of the Choquet simplices of $\mathbb{R}^d$ is known.

The identification of the compact d-Choquet simplices has been proven by Rogers and Shephard [6]; the description of bounded d-Choquet simplices as bounded intersections of open or closed half spaces was given in 1970 by Simons [7]. In 1975, we described the closed d-Choquet simplices [3] and in 1976, we compiled a list of all open and all line-free d-Choquet simplices [4], answering questions raised by Goullet de Rugy [5].

The d-Choquet simplices with lines are much more complicated to describe. We give here many properties of the Choquet simplices which are used to provide the complete list of them in dimensions 2 and 3. These properties can be used to give shorter proofs of the classifications of Simons and Fourneau (Theorems 4.1 and 4.2, respectively). It is possible, in fact, to use them to describe all the d-Choquet simplices, but this is a very long task which is beyond the scope of this paper.

Many other problems are connected with the subject we are dealing with (see [8, 9] for instance) and can profit from the methods developed here.

2. Notation and Prerequisites

For the notation and terminology, we shall refer to [1, 2, 4] and we shall restrict our attention to $\mathbb{R}^d$ $(d > 0)$.

A *Choquet simplex* of $\mathbb{R}^d$ is a nonempty convex subset such that $S \cap (\lambda S + x)$ is either empty or equal to some $\mu S + y$ $(\mu \geq 0,\ y \in \mathbb{R}^d)$, for every $\lambda \geq 0$ and $x \in \mathbb{R}^d$. One can of course replace the assumption on λ by $\lambda > 0$.

The reader should note that in [3] a weaker requirement was used in the definition of Choquet simplices but the validity of the results of [3] is not affected by the present definition.

We recall that a d-convex set C is a *generalized* d-simplex if it can be written as

$$\bullet \quad C = {}^c\{x_1, \ldots, x_k\} + \sum_{i=k+1}^{d+1} [0\colon x_i).$$

We shall have to use very often the structure of the open Choquet simplices we obtained in [4, 3.7] so we shall state this theorem again for easy reference and we shall give in addition a description of the open Choquet simplices in terms of linear forms.

THEOREM 2.1. *The open Choquet simplices of* $\mathbb{R}^d$ *containing a flat of dimension* k *are the direct sum of a subspace of dimension* k *and the relative interior of a generalized* $(d-k)$*-simplex.*

Therefore, a d-convex set is an open Choquet simplex including a flat of dimension k $(k \leq d)$*, if and only if it is of one of the following types (up to a translation):*

$$\text{(a)}\quad S = \left\{ x \in \mathbb{R}^d : f_i(x) > 0,\ i = 1, \ldots, d-k,\ \sum_{i=1}^{d-k} f_i(x) < 1 \right\},$$

where the linear forms f_i $(i = 1, \ldots, d-k)$ *are linearly independent;*

$$\text{(b)}\quad S = \{x \in \mathbb{R}^d : f_i(x) > 0,\ i = 1, \ldots, d-k\},$$

where the f_i $(i = 1, \ldots, d-k)$ *are linearly independent;*

$$\text{(c)}\quad S = \left\{ x \in \mathbb{R}^d : f_i(x) > 0,\ i = 1, \ldots, l-1, l, \ldots, d-k,\ \sum_{i=1}^{l-1} f_i(x) < 1 \right\},$$

where the f_i $(i = 1, \ldots, d-k)$ *are linearly independent.*

The proof of the second part is trivial, using 4.1 of [4] and the decomposition $\mathbb{R}^d = \mathbb{R}^{d-k} \oplus \mathbb{R}^k$.

(2.2) It is useful, in order to avoid lengthy calculations in the sequel, to give explicitly the form of $S \cap (\lambda S + a)$ for an open Choquet simplex S of the type described above.

We shall denote by $r \vee s$ $(r, s \in \mathbb{R}^d)$ the supremum of the projections of r and s on a supplementary subspace L of the lineality space $\Gamma(S)$ of S for the order generated by the cone

$$\{x \in L : f_{i|L}(x) \geq 0,\ i = 1, \ldots, d-k\}.$$

Due to the various choices of L, the point $r \vee s$ is not uniquely defined; in fact, it is defined up to a translation within $\Gamma(S)$.

On the other hand, we write $a \vee b$, $a \wedge b$ for the greatest and the smallest of two real numbers a, b in $\mathbb{R}$.

PROPOSITION. *For any* $a \in \mathbb{R}^d$ *and* $\lambda > 0$*, the set* $S \cap (\lambda S + a)$ *is, for the sets described under* (a), (b), *and* (c) *in* THEOREM 2.1 *respectively,*

$$\text{(a)}\quad \varnothing \text{ if } \sum_{i=1}^{d-k} f_i(a \vee 0) \geq 1 \wedge \left[\lambda + \sum_{i=1}^{d-k} f_i(a) \right],$$

$$\left[\left(1 \wedge \left[\lambda + \sum_{i=1}^{d-k} f_i(a) \right] \right) - \sum_{i=1}^{d-k} f_i(a \vee 0) \right] S + (a \vee 0) \text{ otherwise;}$$

(b) $S + a \vee 0$;

(c) $\varnothing$ *if* $\sum_{i=1}^{l-1} f_i(a \vee 0) \geq 1 \wedge \left[\lambda + \sum_{i=1}^{l-1} f_i(a)\right]$,

$$\left[\left(1 \wedge \left[\lambda + \sum_{i=1}^{l-1} f_i(a)\right]\right) - \sum_{i=1}^{l-1} f_i(a \vee 0)\right] S + (a \vee 0) \text{ otherwise.}$$

We shall only give the proof for case (c). First,

$$\lambda S + a = \left\{x \in \mathbb{R}^d : f_i(x) > f_i(a),\ i = 1, \ldots, d - k;\ \sum_{i=1}^{l-1} f_i(x) < \lambda + \sum_{i=1}^{l-1} f_i(a)\right\};$$

hence, the set $S \cap (\lambda S + a)$ can be written

$$\left\{x \in \mathbb{R}^d : f_i(x) > f_i(a) \vee 0,\ i = 1, \ldots, d - k;\ \sum_{i=1}^{l-1} f_i(x) < 1 \wedge \left[\lambda + \sum_{i=1}^{l-1} f_i(a)\right]\right\}$$

$$= \left\{x \in \mathbb{R}^d : f_i(x) > f_i(a \vee 0),\ i = 1, \ldots, d - k;\ \sum_{i=1}^{l-1} f_i(x) < 1 \wedge \left[\lambda + \sum_{i=1}^{l-1} f_i(a)\right]\right\}$$

$$= \left\{x \in \mathbb{R}^d : f_i[x - (a \vee 0)] > 0,\ i = 1, \ldots, d - k;\ \sum_{i=1}^{l-1} f_i[x - (a \vee 0)] < 1 \wedge \left[\lambda + \sum_{i=1}^{l-1} f_i(a)\right] - \sum_{i=1}^{l-1} f_i(a \vee 0)\right\}$$

which gives at once the result stated above.

3. Properties of d-Choquet Simplices without Lineality Restriction

Proposition 3.1. *If S is a d-Choquet simplex and if*

$$S \cap (\alpha S + x) = \beta S + y \qquad (\alpha, \beta > 0),$$

one can write

$$S \cap (\alpha S + x) = \gamma S + z \qquad \text{or} \qquad S^0 \cap (\alpha S^0 + x) = \gamma S^0 + z$$

only if $z \in y + \Gamma(S)$.

Since

$$[S \cap (\alpha S + x)]^0 = S^0 \cap (\alpha S^0 + x) = \beta S^0 + y = \gamma S^0 + z,$$

it suffices to prove the property for an open Choquet simplex and this is plain from Theorem 2.1.

PROPOSITION 3.2. *Every nonvoid face F of a Choquet simplex is itself a Choquet simplex:*

$$F \cap (\alpha F + x) = \varnothing \quad \text{or} \quad F \cap (\alpha F + x) = \beta F + y,$$

for every β and y such that $S \cap (\alpha S + x) = \beta S + y$.

Let S be a Choquet simplex and F be a nonvoid face of S.

The case $F = S$ is trivial.

Let H be a support hyperplane of S such that $F = H \cap S$. We may assume that $0 \in F$. Whatever $x \in {}^sF$ and $\alpha > 0$ are chosen,

$$\begin{aligned} F \cap (\alpha F + x) &= (H \cap S) \cap [\alpha(H \cap S) + x] = (H \cap S) \cap [(H \cap \alpha S) + x] \\ &= H \cap S \cap [(H + x) \cap (\alpha S + x)] = H \cap S \cap (\alpha S + x). \end{aligned}$$

Since $S \cap (\alpha S + x)$ is either empty or equal to some $\beta S + y$ ($\beta \geq 0$, $y \in \mathbb{R}^d$),

$$F \cap (\alpha F + x) = \begin{cases} \varnothing \\ \text{or} \\ H \cap (\beta S + y). \end{cases}$$

Let us look at the second case. Let f be a linear form with kernel H such that $S \subset \{u: f(u) \geq 0\}$. If $f(y) > 0$, f is positive on $\beta S + y$; hence, $(\beta S + y) \cap H = \varnothing$. If $f(y) = 0$, $y \in H$ and

$$H \cap (\beta S + y) = (H \cap \beta S) + y = \beta F + y.$$

Moreover, it is impossible that $f(y) < 0$ since

$$y \in \beta S + y = S \cap (\alpha S + x) \subset \{u: f(u) \geq 0\}.$$

Therefore,

$$F \cap (\alpha F + x) = \begin{cases} \varnothing \\ \text{or} \\ \beta F + y. \end{cases}$$

Finally, it suffices to note that, if $z \notin {}^sF$,

$$F \cap (\alpha F + z) = \varnothing.$$

PROPOSITION 3.3. *Assume that the Choquet simplex S has been translated in such a way that S^0 is of one of the types described in* THEOREM 2.1. *If a face F of the closure $\bar{S}$ includes a proper face F' of S, such that ${}^lF \ni 0$,*

$$\bar{S} \cap (x + F') \subset {}^iF \quad \textit{and} \quad S \cap (x + F') \neq \varnothing \quad \textit{and} \quad \dim [(x + F') \cap S] = \dim F',$$

for every $x \in {}^iF$; hence, ${}^iF \cap S \neq \varnothing$. In fact,

$$S \cap (x + F') = \alpha F' + y,$$

where $\alpha > 0$ and $y \in x + \Gamma(S)$.

Let $x \in {}^iF$: $S \cap (S + x)$ has nonvoid interior since

$$S^0 \cap (S^0 + x) \neq \varnothing,$$

as shown by the following lemma.

LEMMA. *For any convex set A with nonvoid interior such that a nonsingleton proper face F of $\bar{A}$ contains* 0 *and a relative interior point x,*

$$A^0 \cap (A^0 + x) \neq \emptyset.$$

Indeed, $(0: x) \cap F$ includes a segment $[0: y]$ strictly including $[0: x]$. If $a \in A^0$, $[a: y[\subset A^0$ and $[a: 0[\subset A^0$; hence, $[a + x: x[\subset A^0 + x$ and z defined by

$$\{z\} = [a: y[\cap [a + x: x[$$

belongs to $A^0 \cap (A^0 + x)$.

Therefore,

$$S \cap (S + x) = \alpha S + y,$$

where $\alpha > 0$ and $y \in \mathbb{R}^d$.

Let H' be a support hyperplane of S such that $F' = H' \cap S$: $H' + x$ supports $S + x$. Hence, $H' + x$ is a contact hyperplane for $S \cap (S + x)$, since $x \in \bar{S} \cap (S + x) \cap (H' + x) = \bar{S} \cap (S + x) \cap (H' + x)$.

Now

$$(H' + y) \cap (\alpha S + y) = \alpha(S \cap H') + y = \alpha F' + y.$$

This shows that $H' + y$ supports $S \cap (S + x)$. The (contact and support, respectively) hyperplanes $H' + x$ and $H' + y$ are parallel and include $S \cap (S + x)$ on the same side, so $H' + x = H' + y$.

From there on,

$$\alpha F' + y = (H' + y) \cap (\alpha S + y) = (H' + x) \cap S \cap (S + x) = S \cap (F' + x).$$

Using (2.2), $S^0 \cap (S^0 + x) = \beta S^0 + (x \vee 0)$, where $\beta > 0$. Since $x \in S \subset \{u: f_i(u) \geq 0, i = 1, \ldots, d - k\}$, we have $x \geq 0$; hence, $x \vee 0 = x$. Therefore,

$$S^0 \cap (S^0 + x) = \beta S^0 + x$$

and PROPOSITION 3.1 gives $y \in x + \Gamma(S)$ when compared with $S \cap (S + x) = \alpha S + y$.

Let us show that $\bar{S} \cap (F' + x) \subset {}^iF$. From here on, we shall use the notation of THEOREM 2.1 and (2.2), changing the indexing if necessary.

Assume that

$$F = \bar{S} \cap \{u \in \mathbb{R}^d: f_1(u) = \cdots = f_m(u) = 0\} \qquad (m \leq d - k),$$

so that the minimal face $[F']$ of $\bar{S}$ including F' is

$$[F'] = \bar{S} \cap \{u \in \mathbb{R}^d: f_1(u) = \cdots = f_m(u) = \cdots = f_n(u) = 0\}$$

$(n \leq d - k)$.

But, if $x \in {}^iF$

$$f_i(x) = 0, \qquad i \in \{1, \ldots, m\},$$

$$f_i(x) > 0, \qquad i \in \{m + 1, \ldots, d - k\},$$

$$\sum_{i=1}^{d-k} f_i(x) < 1 \text{ or } \sum_{i=1}^{l-1} f_i(x) < 1 \text{ (if necessary)}.$$

Therefore,

$$x + [F'] = \{u: f_1(u) = \cdots = f_m(u) = 0, f_{m+1}(u) = f_{m+1}(x), \ldots, f_n(u) = f_n(x)\} \cap (\bar{S} + x)$$

so that

$$\bar{S} \cap (x + [F']) \subset {}^iF.$$

Now,

$$\alpha F' + y = S \cap (F' + x) \subset S \cap {}^iF;$$

hence, $S \cap {}^iF \neq \emptyset$.

The equality concerning dim $[S \cap (F' + x)]$ is trivial for $\alpha > 0$.

COROLLARY. *If a face F of $\bar{S}$ includes a proper face F' of the Choquet simplex S,*

$$\dim (F \cap S) = \dim ({}^iF \cap S) = \dim F.$$

Assume that S^0 is of one of the types described in THEOREM 2.1, using a translation if necessary.

Let $r = \dim F - \dim F'$. Choose $x_1, \ldots, x_r \in {}^iF \cap L$ linearly independent, where L is a convenient supplementary subspace of ${}^sF'$ in ${}^sF$:

$$(x_i + F') \cap S = \alpha_i F' + y_i \qquad (\alpha_i > 0,\ y_i \in \mathbb{R}^d),$$

for $i \in \{1, \ldots, r\}$. The sets $\alpha_i F' + y_i$ include respectively (dim F') − balls $B_1, \ldots, B_r$, parallel to ${}^sF'$, and are included in ${}^iF \cap S$, so

$$c(B_1 \cup \cdots \cup B_r) \subset {}^iF \cap S.$$

In the last inclusion, the set on the left-hand side is (r + dim F')-dimensional, so

$$\dim ({}^iF \cap S) = \dim F,$$

hence the conclusion.

PROPOSITION 3.4. *If S is a nonopen d-Choquet simplex, it has a $(d - 1)$-face: each proper face is included in such a face.*

Any $F \in \mathscr{F}(S)$ is included in a maximal face $\tilde{F}$ of $\bar{S}$. PROPOSITION 3.3 and translation invariance assert that ${}^i\tilde{F} \cap S \neq \emptyset$ and the corollary of PROPOSITION 3.3 gives $\dim (\tilde{F} \cap S) = d - 1$. Hence, $\tilde{F} \cap S$ is the required face of S.

LEMMA 3.5. *If a face of the Choquet simplex S is reduced to a point, it is an extreme point of $\bar{S}$ and any face of $\bar{S}$ containing it is included, except for its margin, in S. This implies that S is line-free.*

Let $\{e\}$ be a face of S. The corollary of PROPOSITION 3.3 gives $\dim (\{e\} \cap S) = \dim \{e\}$; hence, $\dim \{e\}$ must be 0 and e is an extreme point of $\bar{S}$. Obviously, S is line-free.

We can assume that $e = 0$ and that S has been translated in such a way that S^0 is of one of the types described in THEOREM 2.1. Let F' be a face of S containing 0. For any $x \in {}^iF'$, PROPOSITION 3.3 gives

$$\{x\} \cap S = (x + \{0\}) \cap S \neq \emptyset;$$

hence, $x \in S$ and ${}^iF' \subset S$.

PROPOSITION 3.6. *Any two proper faces of an unbounded Choquet simplex have a common point.*

Let F_1, F_2 be two proper faces of S, $x_1 \in {}^iF_1$ and $x_2 \in {}^iF_2$. Consider

$$S \cap (S + x_2 - x_1).$$

Since x_2 belongs to both S and $S + x_2 - x_1$, one can find $\alpha \geq 0$ and $x \in \mathbb{R}^d$ such that

$$S \cap (S + x_2 - x_1) = \alpha S + x.$$

According to THEOREM 2.1, the point x_1 belongs to the closure of an open set K, included in S, which is the intersection of an open ball with center x_1 with m open half spaces [$m \leq d - k + 1$, where $k = \dim \Gamma(S)$] whose bounding hyperplanes contain x and are independent. Therefore

$$K + x_2 - x_1 \subset S + x_2 - x_1$$

and

$$S^0 \cap (K + x_2 - x_1) \subset \alpha S + x.$$

The faces F_1 and F_2 have a common point or the set on the left-hand side is not empty; hence, $\alpha > 0$. Let us first give the proof for the case in which S can be translated in such a way that S^0 is of type 2.1.a; assume that the indexing is such that

$$K = B(x_1) \cap \{x : f_i(x) > 0,\ i = 1, \ldots, m\}.$$

One has

$$\begin{aligned} S^0 \cap (K + x_2 - x_1) &= S^0 \cap B(x_2) \cap \{x : f_i(x) > f_i(x_2),\ i = 1, \ldots, m\} \\ &= B(x_2) \cap \{x : f_i(x) > f_i(x_2),\ i = 1, \ldots, m;\ f_i(x) > 0,\ i = 1, \ldots, d - k; \\ &\qquad\qquad \textstyle\sum_{i=1}^{d-k} f_i(x) < 1\} \end{aligned}$$

and the second set in this last intersection is nonvoid (and open) if

$$\sum_{i=1}^{d-k} f_i(x) < 1;$$

in this case, indeed, a segment $]x_2 : a[$ is included in this set and $B(x_2) \cap\,]x_2 : a[\, \neq \varnothing$, hence the conclusion; if $\sum_{i=1}^{d-k} f_i(x_2) = 1$,

$$S \cap (S + x_2 - x_1) \subset \bar{S} \cap (\bar{S} + x_2 - x_1) \subset \left\{x : \sum_{i=1}^{d-k} f_i(x) = 1\right\},$$

and, since this last set is a hyperplane, $\alpha = 0$ and

$$S \cap (S + x_2 - x_1) = \{x_2\}.$$

Therefore, $F_2 = \{x_2\}$ and, using LEMMA 3.5, x_2 is an extreme point of $\bar{S}$, S is line-free, and it remains to use [4, 4.6] to see that the faces of S have a common point except when S is the closed d-simplex.

In case S^0 is (up to a translation) of type 2.1.b, the same proof shows that $S^0 \cap (K + x_2 - x_1) \neq \emptyset$ without restriction on x_2.

For case 2.1.c, $S^0 \cap (K + x_2 - x_1) \neq \emptyset$ for any x_2 but those for which

$$\sum_{i=1}^{l-1} f_i(x_2) = 1$$

and one can dispose of this as previously.

Let us now proceed with the proof when $\alpha > 0$. The preceding paragraph shows that, in these cases, $0 \in {}^lF_1 \cap {}^lF_2$.

Let g_i $(i = 1, 2)$ be linear forms on $\mathbb{R}^d$ such that

$$S \subset \{u \in \mathbb{R}^d: g_i(u) \geq 0\} \qquad (i = 1, 2),$$

and

$$F_i = H_i \cap S \qquad (i = 1, 2),$$

where

$$H_i = g_i^{-1}(\{0\}) \qquad (i = 1, 2).$$

In fact, we shall choose for the g_i's $(i = 1, 2)$ positive linear combinations of the f_j's.

As

$$S + x_2 - x_1 \subset \{u: g_1(u) \geq 0\} + x_2 - x_1 = \{y: g_1(y) \geq g_1(x_2)\} \ni x_2,$$

the hyperplane $H_1' = \{y: g_1(y) = g_1(x_2)\}$ supports $\alpha S + x$ in x_2; one can prove in the same way that H_2 supports $\alpha S + x$ in x_2.

Obviously,

$$x_2 \in H_1' \cap H_2 \cap (\alpha S + x).$$

Therefore

$$\frac{x_2 - x}{\alpha} \in S,$$

and, since

$$S^0 \cap (S^0 + x_2 - x_1) = \alpha S^0 + x,$$

$x = (x_2 - x_1) \vee 0$ [with the notation of (2.2)] and

$$g_i(x) = [g_i(x_2) - g_i(x_1)] \vee 0$$

or

$$g_1(x) = g_1(x_2) \qquad \text{and} \qquad g_2(x) = 0;$$

hence,

$$g_i\left(\frac{x_2 - x}{\alpha}\right) = \frac{1}{\alpha}\,[g_i(x_2) - g_i(x)] = 0$$

and one gets

$$\frac{x_2 - x}{\alpha} \in H_1 \cap H_2 \cap S = F_1 \cap F_2.$$

Proposition 3.7. *The nonvoid faces of an unbounded Choquet simplex S have a point in common*: $\bigcap[\mathscr{F}(S)\backslash\{\varnothing\}] \neq \varnothing$.

Let F_1, F_2 be two faces of S: $F_1 \cap F_2 \neq \varnothing$ and $F_1 \cap F_2$ is a face of S (see II.3.1 in [2] for instance). Hence, from step to step, one can show that $\bigcap[\mathscr{F}(S)\backslash\{\varnothing\}] \neq \varnothing$.

(3.8) Let us say that a linear form f is *bounded above on* a set A *at the level* α if $\sup f(A) = \alpha$. If f is bounded above on A at some level, we say that f *bounds* A.

Lemma. *Let* F_1, F_2 *be maximal faces of a Choquet simplex S which contain* 0 *and let* g_1, g_2, *respectively, be linear forms on* $\mathbb{R}^d$ *which determine* F_1, F_2 *in S. Let* f'_1 *be a linear form on* ${}^sF_1$ *linearly independent of* $g_{2|{}^sF_1}$ *and bounded above on* F_1 *at the level* 0. *If a linear form* f'_2 *on* ${}^sF_2$, *linearly independent of* $g_{1|{}^sF_2}$, *is bounded above on* F_2 *at the level* α, *then either* $\alpha = 0$ *or the dimension of* $V = f_1'^{-1}(\{0\}) \cap f_2'^{-1}(\{0\})$ *is less than* $d - 3$.

Assume $\dim V \geq d - 3$. Proposition 3.4 gives $\dim F_1 = \dim F_2 = d - 1$; hence, $\dim f_1'^{-1}(\{0\}) = \dim f_2'^{-1}(\{0\}) = d - 2$ and $\dim V = d - 3$.

Therefore,

$$\begin{aligned}\dim\,[f_1'^{-1}(\{0\}) + f_2'^{-1}(\{0\})] &= \dim f_1'^{-1}(\{0\}) + \dim f_2'^{-1}(\{0\}) + \dim V \\ &= d - 1,\end{aligned}$$

and one can find a nonzero linear form f on $\mathbb{R}^d$ such that

$$f^{-1}(\{0\}) = f_1'^{-1}(\{0\}) + f_2'^{-1}(\{0\}).$$

The restriction of f to ${}^sF_1$ (respectively, ${}^sF_2$) is a multiple of f'_1 (respectively, f'_2):

$$f_{|{}^sF_1} = \alpha_1 f'_1 \qquad \text{and} \qquad f_{|{}^sF_2} = \alpha_2 f'_2,$$

$\alpha_1, \alpha_2 \neq 0$.

One can choose f in such a way that $\alpha_2 > 0$.

Assume that α_1 is positive. If $x \in {}^i[\bar{F}_2 \cap f^{-1}(\{\alpha\alpha_2\})]$, the lemma of Proposition 3.3 shows that

$$S^0 \cap (S^0 + x) \neq \varnothing;$$

hence,

$$S \cap (S + x) = \beta S + y \qquad (\beta > 0,\ y \in \mathbb{R}^d).$$

Since

$$\beta S + y \subset S,$$

$$(\beta S + y) \cap {}^sF_2 \subset S \cap {}^sF_2 = F_2,$$

and f is bounded above on $(\beta S + y) \cap {}^sF_2$ at the level $\alpha\alpha_2$ [note that $x \in \bar{S} \cap (\bar{S} + x) \cap f^{-1}(\{\alpha\alpha_2\})$]. But

$$F_2 \supset (\beta S + y) \cap {}^sF_2 = \beta F_2 + y,$$

since $x \in {}^sF_2$, so $y \in {}^sF_2$, $f(y) \leq \alpha\alpha_2$, and $y \in x + \Gamma(S)$.

On the other hand, one can find s such that $[s\colon 0[\,\subset F_1$ and then, using the description of S^0, $s' \in [s\colon 0[$ such that

$$[s' + x\colon x[\,\subset S;$$

therefore,

$$[s' + x\colon x[\,\subset S \cap (S + x) \cap g_1^{-1}(\{x\})$$
$$= (\beta S + y) \cap g_1^{-1}(\{y\}) = \beta \bar{F}_1 + y,$$

for $y \in x + \Gamma(S)$ implies $g_1(x) = g_1(y)$, and so $x \in \beta \bar{F}_1 + y$. We can then find $s \in \bar{F}_1$ such that

$$x = \beta s + y,$$

so

$$\alpha\alpha_2 = f(x) = \beta f(s) + f(y) \leq f(y) < \alpha\alpha_2;$$

hence,

$$f(y) = \alpha\alpha_2.$$

But,

$$\sup f(\beta F_2 + y) = \beta\alpha\alpha_2 + f(y) = (1 + \beta)\alpha\alpha_2$$

and, since $\beta F_2 + y \subset F_2$,

$$\sup f(\beta F_2 + y) \leq \alpha\alpha_2;$$

hence,

$$(1 + \beta)\alpha\alpha_2 < \alpha\alpha_2.$$

This implies that $\alpha \leq 0$. Since $0 \in F_2$, $\alpha \geq 0$, so $\alpha = 0$.

Now, assume that $\alpha_1 < 0$. If $x \in {}^i[\bar{F}_1 \cap [F_2] \cap f^{-1}(\{\gamma\})]$, where $[F_2]$ is the smallest face of S including F_2 and where $\gamma \leq -\alpha\alpha_2$,

$$S^0 \cap (S^0 + x) \neq \varnothing$$

(lemma of Proposition 3.3); hence,

$$S \cap (S + x) = \delta S + z, \qquad \delta > 0,\ z \in \mathbb{R}^d.$$

We can write, using the same argument as above,

$$F_1 \supset (\delta S + z) \cap {}^sF_1 = \beta F_1 + z \qquad \text{and} \qquad F_2 \supset (\delta S + z) \cap {}^sF_2 = \beta F_2 + z$$

so $z \in F_1 \cap F_2$ and $0 \leq f(z) \leq \alpha\alpha_2$.

Moreover, $x = \beta s + z$, where $s \in F_1 \cap [F_2]$, so

$$-\alpha\alpha_2 \geq f(x) = \beta f(s) + f(z) \geq f(z) \geq 0;$$

hence, $\alpha \leq 0$.

Since $\alpha\alpha_2 \geq 0$, we must admit that $\alpha = 0$.

PROPOSITION 3.9. *Let S be a d-Choquet simplex: One can find* x_1, $i \in \{1, \ldots, d+1\}$, *affinely independent such that*

$$\bar{S} = {}^c\{x_i\colon i = 1, \ldots, 1\} + \sum_{i=1+1}^{k+1} [0\colon x_i) + \sum_{i=k+2}^{d+1} (-x_i\colon x_i).$$

The closure of every proper face F of S can be written as

$$\bar{F} = {}^c\{y_i\colon i \in A'\} + \sum_{i \in B'} [0\colon y_i) + \sum_{i \in C'} (-y_i\colon y_i),$$

where the y_i *are affinely independent and such that, if*

$$[F] = {}^c\{x_i\colon i \in A\} + \sum_{i \in B} [0\colon x_i) + \sum_{i=k+2}^{d+1} (-x_i\colon x_i)$$

is the minimal face of $\bar{S}$ *including F* ($A \subset \{1, \ldots, l\}$, $B \subset \{l+1, \ldots, k+1\}$), $A' = A$, $p(y_i) = p(x_i)$ *for each* $i \in A$, *where p is the canonical projection onto* $L = {}^l\{x_i\colon i = 1, \ldots, k+1\}$ *in the decomposition* $L \oplus \Gamma(S) = \mathbb{R}^d$, $\sum_{i \in B'} [0\colon y_i) \subset \sum_{i \in B} [0\colon x_i)$, *and* $\sum_{i \in C'} (-y_i\colon y_i) \subset \Gamma([F])$.

Assume that S has been translated in such a way that $\dot{S}^0$ is of one of the types described in THEOREM 2.1.

It is known [4, 3.7 p. 78] that

$$\bar{S} = {}^c\{x_1, \ldots, x_l\} + \sum_{i=l+1}^{k+1} [0\colon x_i) + \sum_{i=k+2}^{d+1} (-x_i\colon x_i)$$

where the x_i's are affinely independent.

Let $F \ni 0$ be a proper face of S: The minimal face $[F]$ of $\bar{S}$ including S can be written

$$[F] = {}^c\{x_i\colon i \in A\} + \sum_{i \in B} [0\colon x_i) + \sum_{i=k+2}^{d+1} (-x_i\colon x_i),$$

where $A \subset \{1, \ldots, l\}$ and $B \subset \{l+1, \ldots, k+1\}$, one of these inclusions being strict.

Whatever $x \in {}^i[F]$ is chosen, $\alpha > 0$ and $y \in x + \Gamma(S)$ can be found such that (PROPOSITION 3.3):

$$F \cap (F + x) = S \cap (F + x) = \alpha F + y.$$

Since $0 \in F$, $y \in F$, hence

$$y = \sum_{i \in A} \theta'_i x_i + \sum_{i \in B} \theta'_i x_i + \sum_{i=k+2}^{d+1} \alpha'_i x_i,$$

where $\sum_{i \in A} \theta'_i = 1$ and $\theta'_i \geq 0$ ($i \in A \cup B$). But,

$$x = \sum_{i \in A} \theta_i x_i + \sum_{i \in B} \theta_i x_i + \sum_{i=k+2}^{d+1} \alpha_i x_i,$$

where $\sum_{i \in A} \theta_i = 1$ and $\theta_i > 0$ ($i \in A \cup B$); hence

$$y - x = \sum_{i \in A} (\theta'_i - \theta_i) x_i + \sum_{i \in B} (\theta'_i - \theta_i) + \sum_{i=k+2}^{d+1} (\alpha'_i - \alpha_i) x_i \in \Gamma(S)$$

and this requires that

$$\theta_i' = \theta_i, \qquad i \in A \cup B,$$

since $\Gamma(S)$ and $L = {}^l\{x_i\colon i = 1, \ldots, k + 1\}$ are supplementary subspaces. Then the canonical projections $p(x)$ and $p(y)$ of x and y on L in the decomposition $\mathbb{R}^d = L \oplus \Gamma(S)$ coincide and

$$[F] \cap L \supset p(F) \supset {}^i[F] \cap L.$$

But we know that F is a Choquet simplex, so

$$\bar{F} = {}^c\{y_i\colon i \in A'\} + \sum_{i \in B'} [0\colon y_i) + \sum_{i \in C'} (-y_i\colon y_i)$$

where the y_i are affinely independent.

Since $\bar{F} \subset [F]$, $\Gamma(\bar{F}) \subset \Gamma([F])$, hence $y_i \in \Gamma([F])$, $(i \in C')$.

The projections on L of the y_i $(i \in A')$ are obviously the x_i $(i \in A)$ for, since

$$p(\bar{F}) \supset p(F) \supset {}^i[F] \cap L,$$

we have

$$p(\bar{F}) = [F] \cap L,$$

$p(\bar{F})$ being closed as a polyhedron.

Moreover, $C_{\bar{F}} \subset C_{[F]}$; hence $y_i \in C_{[F]}$ $(i \in B' \cup C')$.

PROPOSITION 3.10. *If F is a maximal face of a d-Choquet simplex S such that $\bar{S}$ has at least two maximal faces $[F]$ and F', the asymptotic cone $C_{\bar{F}}$ of $\bar{F}$ has all its edges in F', except perhaps one.*

Assume that $0 \in {}^i([F] \cap F')$.

Let f' be a linear form determining F': f' bounds F at the level 0.

Let $x \in {}^i([F] \cap F')$. The lemma of PROPOSITION 3.3 ensures that

$$S^0 \cap (S^0 + x) \neq \varnothing;$$

hence

$$S \cap (S + x) = \alpha S + y, \qquad \alpha > 0,\ y \in \mathbb{R}^d.$$

Using (2.2) and PROPOSITION 3.1, one sees that $y \in (x \vee 0) + \Gamma(S)$, with the notation of (2.2). Therefore, since f' is one of the linear forms determining the positive cone related to

$$f'(y) = f'(x \vee 0) = f'(x) \vee f'(0) = 0,$$

thence

$$\sup_{z \in \alpha F} f'(z) = 0.$$

Now, $F \cap (F + x) = \alpha F + y$; hence, $\alpha F + y \subset \bar{F} \cap (\bar{F} + x)$. If the g_i's $(i \in I)$ are linearly independent linear forms on ${}^sF$ which determine $\bar{F}$ (we choose g_i in such a way that g_i is nonnegative on $\bar{F}$):

$$\alpha F + y \subset \{z \in {}^sF\colon g_i(z) \geq g_i(x)\} \cap \{z \in {}^sF\colon g_i(z) \geq 0\}$$
$$= \{z \in {}^sF\colon g_i(z) \geq g_i(x) \vee 0\}.$$

One can find $\alpha_i > 0$ $(i \in I' \subset I)$ such that

$$f'_{|^sF} = \sum_{i \in I'} \alpha_i g_i,$$

(see [2, IX.3.6, p. 246]). But such an equality leads to

$$0 = \sup_{z \in \alpha F} f'(z) \geq \sum_{i \in I'} \alpha_i [g_i(x) \vee 0];$$

hence

$$g_i(x) \leq 0 \qquad \forall\, i \in I'.$$

Therefore, since the last inequality holds for any $x \in {}^i([F] \cap F')$,

$$g_i([F] \cap F') = \left\{ \begin{matrix} [a, 0] \\ \text{or} \\ ]-\infty, a[\end{matrix} \right\} \text{with } a \leq 0, \qquad \forall\, i \in I',$$

for $0 \in [F] \cap F'$. But, since $0 \in {}^i([F] \cap F')$, the only possibility is

$$g_i([F] \cap F') = \{0\}.$$

Since $\dim([F] \cap F') = d - 2$, we have

$$g_i = \beta_i f'_{|^sF} \qquad \forall\, i \in I';$$

this implies that $\# I' = 1$ (β_i is obviously positive).

Therefore, $f'_{|^sF}$ is one of the g_i's and all the edges of the cone $C_{\bar{F}}$, except perhaps one, are in F'.

PROPOSITION 3.11. *Let S be a d-Choquet simplex, let F and F' be maximal faces of S and $a \in F \cap F'$. If F is bounded above by a linear form g'_1 on ${}^lF$, linearly independent of the restriction to ${}^lF$ of the linear form which determines ${}^lF'$, and if a linear form g'_2 on ${}^lF'$, linearly independent of the restriction to ${}^lF'$ of the linear form which determines ${}^lF$, bounds F' above, then the dimension of $V = [g_1'^{-1}(\{0\}) + a] \cap [g_2'^{-1}(\{0\}) + a]$ is less than $d - 3$.*

Assume that $a = 0$. Let $x \in {}^i[F]$ be such that $g'_1(x) > 0$. Using the lemma of PROPOSITION 3.3, we have $S^0 \cap (S^0 + x) \neq \varnothing$; hence,

$$S \cap (S + x) = \alpha S + y, \qquad \alpha > 0.$$

Assume that g'_2 bounds F' above and $\dim V = d - 2$. Denote by g a linear form such that

$$g^{-1}(\{0\}) = {}^s[g_1'^{-1}(\{0\}) \cup g_2'^{-1}(\{0\})],$$

and

$$g_{|^sF} = \alpha_1 g'_1, \qquad \alpha_1 > 0.$$

Since $\alpha S + y \subset S$,

$$(\alpha S + y) \cap {}^sF = \alpha F + y \subset F;$$

hence, g is bounded above on $(\alpha S + y) \cap {}^sF$ at the level $g(y)$ less than 0. Then, using (3.8), g must be bounded above on $\alpha F' + y$ at the level $g(y) \leq 0$. But

$$\overline{\alpha F' + y} = \overline{(F' + x) \cap S} \ni x;$$

hence,

$$\sup g(\alpha F' + y) \geq g(x) > 0.$$

(3.12) A face F of a Choquet simplex S will be called *regular* if $\bar{F}$ coincides with the minimal face of $\bar{S}$ containing F.

PROPOSITION. *Let F_1, F_2 be two regular maximal faces of a Choquet simplex S. The part of the intersection of the minimal faces of $\bar{S}$ including F_1 and F_2 respectively, which is not contained in another maximal face of $\bar{S}$, is included in $F_1 \cap F_2$.*

Assume that $0 \in F_1 \cap F_2$ (PROPOSITION 3.6); let $a_1 \in {}^iF_1 \backslash ([F_1] \cap [F_2])$ and $a_2 \in {}^iF_2 \backslash ([F_1] \cap [F_2])$: The set

$$S \cap (S + a_2 - a_1)$$

is not reduced to a point. Indeed, for $\varepsilon > 0$ sufficiently small, an open $(d-1)$-half-ball B_1, of radius ε, centered at a_1, is included in F_1 and an open d-half-ball B_2, of radius ε, centered at a_2, is included in S; hence,

$$(B_1 + a_2 - a_1) \cap B_2 = B_1 + a_2 - a_1 \subset S \cap (S + a_2 - a_1);$$

therefore,

$$S \cap (S + a_2 - a_1) = \lambda S + u, \qquad \lambda > 0.$$

Moreover,

$$\begin{aligned} F_2 \cap (F_1 + a_2 - a_1) &= H_2 \cap S \cap [(H_1 \cap S) + a_2 - a_1] \\ &= H_2 \cap S \cap (H_1 + a_2 - a_1) \cap (S + a_2 - a_1) \\ &= H_2 \cap (H_1 + a_2 - a_1) \cap (\lambda S + u), \end{aligned}$$

where $H_i = {}^sF_i$ $(i = 1, 2)$, is a face of $\lambda S + u$, for it contains a_2. But this set is just a translate of $[F_1] \cap [F_2]$ limited by the maximal faces of $\bar{S}$ and $\bar{S} + a_2 - a_1$. The transformation $x \to (1/\lambda)(x - u)$ gives the result.

4. THE CHARACTERIZATION THEOREMS

Here are some characterization theorems for finite-dimensional Choquet simplices. The first three were proved elsewhere [2, 4, 7] by other methods, the last is new.

THEOREM 4.1. *The bounded d-Choquet simplices are the bounded intersections of $d + 1$ open or closed half spaces.*

The necessity is obvious from PROPOSITIONS 3.3, 3.6, and 3.9. The sufficiency is easily verified by a direct calculation.

THEOREM 4.2. *The line-free unbounded d-Choquet simplices are the line-free intersections of d open or closed half spaces or the unbounded sets of the type (up to a translation):*

$$S = \left\{ x \in \mathbb{R}^d : f_i(x) \geq 0,\ i = 1, \ldots, k;\ f_i(x) > 0,\ i = k+1, \ldots, d;\ \sum_{i \in I} f_i(x) < 1 \right\}$$

where the f_i are linearly independent and $I \subset \{1, \ldots, d\}$ (and, possibly, $I = \varnothing$).

The necessity is obvious since it is impossible to find a linear form on the affine hull of a maximal face of S linearly independent of the restrictions of the forms which define $\bar{S}$, and the faces of S must have a point in common (PROPOSITION 3.7).

The sufficiency is easily verified by a direct calculation (see, for instance, the end of the proof of 4.6 in [4]).

THEOREM 4.3. *The 2-Choquet simplices with lines are:*

(a) *a closed half plane, an open half plane, or an open half plane with a (pointed or not) ray of its bounding line;*

(b) *an open strip, an open strip with one of its bounding lines, or an open strip with a (pointed or not) ray of one of its bounding lines;*

(c) *the whole plane.*

To prove the necessity, let us note that a proper nonopen Choquet simplex with lines has one and only one 1-face and that this face is a 1-Choquet simplex (PROPOSITION 3.2). Moreover, PROPOSITION 3.2 shows that any two translates of this face within its affine hull must intersect; therefore, it cannot be bounded.

The sufficiency is obvious.

THEOREM 4.4. *The 3-Choquet simplices with lines are:*

(a) *an open half space or an open half space with, in its bounding plane, any conical 2-Choquet simplex;*

(b) *an open strip or an open strip with, in one of its bounding planes, any conical 2-Choquet simplex;*

(c) *an intersection of two open or closed nonparallel half spaces;*

(d) *a set of type* c *with a conical 2-Choquet simplex, part of a maximal face F of the closure, which is bounded by the other maximal face F', the bounding ray being included if and only if F' is a part of the simplex;*

(e) *the intersection of a set of type* c *or* d *with an open half space, including the edge of the dihedron, the bounding plane of which is parallel to this edge;*

(f) *the whole space.*

Let us first prove that a 3-Choquet simplex S with lines is of one of the types described. We shall assume that S is not open (see THEOREM 2.1 for this case).

Let us assume that $\bar{S}$ has only one maximal face: S^0 is therefore an open half space. The face F of S must be a 2-Choquet simplex which meets any of its translates within its affine hull (PROPOSITION 3.2); hence, F must be conical (see THEOREMS 4.1, 4.2, and 4.3).

Now assume that $\bar{S}$ has two parallel maximal faces: S^0 is an open strip. Using PROPOSITION 3.6, we see that one of the corresponding faces of S must be empty. The same argument as above shows that S is of type b.

Let us now turn to the case where $\bar{S}$ has two nonparallel maximal faces: S^0 is an open dihedron. If S has only one maximal face F, F must be a 2-Choquet simplex which meets any of its translates within its affine hull (PROPOSITION 3.2); hence, F is a cone. The bounding line or one of the bounding rays of F must be included in the other maximal face F' of $\bar{S}$ (PROPOSITION 3.10), but no part of it belongs to S, otherwise S would have two maximal faces (PROPOSITION 3.3). This proves that S is of type c or d.

If S has two maximal faces F and F', the same argument as in the preceding paragraph applies but the bounding line of F, or the bounding ray of F included in

$[F']$, is partly included in $F \cap F'$. In fact, if F and F' are both regular, $F \cap F' = [F] \cap [F']$ (3.12). PROPOSITION 3.11 asserts that a 3-Choquet simplex has at most one singular face. Assume that F is singular: $F \cap F'$ is nonvoid, hence F must contain a point of its bounding ray included in $[F']$; since F is a 2-Choquet simplex, we know that this ray must be included in F.

Finally, assume that $\bar{S}$ has three maximal faces. Since the lineality of S is positive, these faces must be parallel to a given straight line: $\bar{S}$ is a prism. The faces of S must have a common point; therefore, S has at most two maximal faces. The same argument as in the case where $\bar{S}$ is a dihedron can be used to describe S: It is of type e.

The proof that all the sets described are 3-Choquet simplices is left to the reader: It suffices to use a standard form of them in terms of linear forms.

Remark 4.5. In higher dimensions, the possibility of having many singular faces exists. The description of the d-Choquet simplices with lines is therefore intricate. In order to keep this paper to a reasonable length, we shall not go into this here.

REFERENCES

1. BAIR, J. & R. FOURNEAU. 1975. Etude géométrique des espaces vectoriels—une introduction. Lecture Notes in Mathematics No. 489. Springer-Verlag. West Berlin, FRG.
2. BAIR, J. & R. FOURNEAU. 1980. Etude géométrique des espaces vectoriels II: polyèdres et polytopes convexes. Lecture Notes in Mathematics No. 802. Springer-Verlag. West Berlin, FRG.
3. FOURNEAU, R. 1977. On the geometry of unbounded Choquet simplices. J. Geom. **9:** 143–147.
4. FOURNEAU, R. 1977. Nonclosed simplices and quasi-simplices. Mathematika **24:** 71–85.
5. GOULLET DE RUGY, A. 1968. Géométrie des simplexes. Centre de Documentation Universitaire. Paris, France.
6. ROGERS, C. A. & G. C. SHEPHARD. 1957. The difference body of a convex body. Arch. Math. **8:** 220–233.
7. SIMONS, S. 1970. Noncompact simplices. Trans. Amer. Math. Soc. **149:** 155–161.
8. WILKER, J. B. 1975. Sets homothetic to their intersection with a translate. Canad. Math. Bull. **18:** 739–748.
9. WILKER, J. B. 1978. Generalized symmetry. Bull. Austral. Math. Soc. **18:** 373–393.

RESULTS OF BAIRE CATEGORY TYPE IN CONVEXITY

Peter M. Gruber

Institut für Analysis
Technische Mathematik und Versicherungsmathematik
Technische Universität Wien
A-1040 Vienna, Austria

1

In the following, there is first a short survey of Baire category results in convexity (Section 2), then two new results are presented: one on minimal ellipsoids (Section 3), and one on lattice packings of convex bodies (Section 4).

A *convex body* in d-dimensional Euclidean space $\mathbb{E}^d$ is a compact convex subset of $\mathbb{E}^d$ with nonempty interior. Denote by $\mathscr{C}$ the space of all convex bodies in $\mathbb{E}^d$ endowed with its "natural" topology. This topology is induced by, e.g., the Hausdorff metric δ^H or the symmetric difference metric δ^S, defined by

$$\delta^H(C, D) = \max\left\{\sup_{x\in C}\inf_{y\in D}|x-y|,\ \sup_{y\in D}\inf_{x\in C}|x-y|\right\}$$

for $C, D \in \mathscr{C}$, and

$$\delta^S(C, D) = v(C\Delta D)$$

for $C, D \in \mathscr{C}$, respectively. Here $|\ |$ and v denote the Euclidean norm and the Lebesgue measure in $\mathbb{E}^d$. Blaschke's selection theorem implies that $\mathscr{C}$ is locally compact. Hence, we may apply to $\mathscr{C}$ a version of the Baire category theorem which says that in a locally compact or complete metric space, meager subsets may be considered "small," where a set is called *meager* or of *first Baire category* if it can be represented as a countable union of nowhere dense sets. We say that some property holds for *most* elements of a locally compact or complete metric space if it holds for all elements, except those from a meager set. See [12].

The concept of a meager set is a topological tool to distinguish between "large" and "small" sets. It would be desirable to construct measures on $\mathscr{C}$ serving the same purpose and which, at the same time, are easy to handle and are geometrically meaningful. A conjecture of the author's that some Hausdorff measure might do was readily disproved by Schneider (letter to J. M. Wills, 1977). A more general negative result of this type was communicated to the author by Bandt (letter to the author, 1982).

2

The results of Baire category type in convexity will be collected under several headings.

Differentiability

Historically speaking, the first category result in convexity seems to be due to Klee [17]. It implies that most convex bodies are strictly convex and smooth, that is (their boundaries are surfaces) of class $\mathscr{C}^1$. Independent proofs of this result are due to Choquet, Schneider (both communicated by Schneider in a letter to the author, 1976), Gruber [6], and Howe [13]. Gruber [6] proved that most convex bodies are not of class $\mathscr{C}^2$, at least at one boundary point. This was generalized in two different directions. On the one hand, it was shown by Klima and Netuka [19], using Hölder conditions, that most convex bodies are not of class $\mathscr{C}^{1+\alpha}$ for any $\alpha > 0$ at "many" boundary points. On the other hand, Schneider [25] and Zamfirescu [29, 30, 36] proved that most convex bodies have many boundary points with very unexpected curvature properties. Among the multitude of Zamfirescu's results, the following are particularly appealing. They will be stated without giving precise definitions. In [29] Zamfirescu showed that for most convex bodies for each boundary point p and each tangent direction t, the lower radius of curvature (of the boundary at p in direction t) is 0 or the upper radius of curvature is ∞. Thus by a result of Aleksandrov [1] for most convex bodies, for almost all boundary points (in the sense of $(d-1)$-dimensional surface measure) and all tangent directions, the upper and lower radii of curvature are both equal to ∞. In [30] it is shown that for most convex bodies at most boundary points and for all tangent directions, the lower radius of curvature is 0 and the upper radius of curvature is ∞. Note that the last two results do not contradict each other, since the boundary of any convex body may be partitioned into two disjoint sets, one of them meager and the other of measure zero. It seems to be difficult to explicitly construct convex bodies with nasty differentiability properties such as those just mentioned. An application of curvature results to the problem of concurrent normals of convex bodies is due to Heil [11].

Approximation of Convex Bodies by Polytopes

Let $\mathscr{P}_n$, $n = d + 1, d + 2, \ldots$, denote the set of all polytopes of $\mathscr{C}$ having at most n vertices. For $C \in \mathscr{C}$ and $n = d + 1, d + 2, \ldots$ there exist polytopes P_n, $Q_n \in \mathscr{P}_n$ such that

$$\delta^H(C, P_n) = \inf\{\delta^H(C, P): P \in \mathscr{P}_n\} = \delta^H(C, \mathscr{P}_n),$$
$$\delta^S(C, Q_n) = \inf\{\delta^S(C, Q): Q \in \mathscr{P}_n\} = \delta^S(C, \mathscr{P}_n).$$

P_n and Q_n are called best approximating polytopes of C with respect to δ^H and δ^S. Kenderov [16] and Gruber and Kenderov [10] proved that for $d = 2$, most convex bodies have unique best approximating polygons with respect to δ^H and δ^S and each $n = 3, 4, \ldots$. A refinement of this result in the case of the Hausdorff metric was given by Zhivkov [38]. Although by the results in [10, 16] most convex bodies in the plane have unique best approximating polygons with respect to δ^H and δ^S in $\mathscr{P}_n$, it seems to be difficult to give explicit descriptions of such bodies. Presumably the semicircle which is an example for the case of the metric δ^H is also an example for the case of the metric δ^S. Schneider and Wieacker [26] and Gruber and Kenderov [10] showed that for most convex bodies C, the asymptotic behavior of $\delta^H(C, \mathscr{P}_n)$ and $\delta^S(C, \mathscr{P}_n)$ as $n \to \infty$ is very irregular. An elaboration of the idea underlying this result and applications to various other branches of mathematics are contained in [8]. For a "construction" of a convex body C for which the asymptotic behavior of $\delta^H(C, \mathscr{P}_n)$ and $\delta^S(C, \mathscr{P}_n)$ is very irregular as $n \to \infty$, see [10].

Geodesics

A geodesic segment on the boundary of a convex body is a continuous curve connecting two points and having minimal length among all such curves. It is well known that a geodesic segment never passes through a vertex of a convex body. Taking into account that most convex bodies are smooth, the following results of Zamfirescu [35] are rather unexpected. For most convex bodies through most boundary points there passes no geodesic segment. For most convex bodies, for each boundary point p and for most tangent directions t there exists no geodesic segment with endpoint p having direction t at p. The latter result has been proven so far only for $d = 3$. For examples of convex bodies exhibiting properties of this type see Aleksandrov [2].

Convex Mirrors

A smooth planar convex body C is called a convex mirror. We suppose that a light ray enters and traverses the body without hindrance but is reflected at the interior side of the boundary in the usual way and then traverses the body again and leaves it without changing direction. A result of Zamfirescu [34] implies that, given a point p in the plane, most convex mirrors have the property that from p one can see infinitely many mirror images of most other points in the plane. In the proof of this result, differentiability properties of the type described above are employed. It is rather surprising that such mirrors exist at all. Examples are provided by smooth convex bodies having a dense set of boundary points with lower radius of curvature 0 and upper radius of curvature ∞. Recently, Zamfirescu (letter to the author, 1983) succeeded in extending his result to dimensions $d \geq 2$.

Miscellanea

From the Baire category results not related to the aforementioned ones, we cite the following. Perles [24] proved a category result in the context of so-called critical exponents of two-dimensional normed spaces. Zamfirescu [32] proved that most planar convex bodies touch the boundary of their minimal circumcircle at three points precisely. Schneider (letter to the author, 1976) considered the concept of reducible convex bodies (in the sense of Shephard [27]) from a category point of view. He proved that most convex bodies are, in fact, irreducible. Zamfirescu [33] showed that for most convex bodies C in the plane most points of C are contained in at least countably many diameters of C. Here a diameter of C is a chord of C at the endpoints of which are parallel supporting lines of C. For other results see [31, 37].

3

For a nonplanar compact subset C of $\mathbb{E}^d$, there exists a unique ellipsoid $E(C)$ containing C and having minimal volume. It is called the *minimal* or *Löwner ellipsoid* of C.

The first published proofs for the existence and the uniqueness seem to be due to Behrend [3] ($d = 2$) and Danzer *et al.* [5] ($d \geq 2$). The following interesting Helly-type theorem for minimal ellipsoids was given by Behrend [3] ($d = 2$) and John [14] ($d \geq 2$): For any convex body C, there exist points $p_1, \ldots, p_N$, $N \leq d(d+3)/2$, belonging to C and the boundary of $E(C)$, such that $E(C) = E(\{p_1, \ldots, p_N\})$.

The minimal ellipsoid is an important tool for characterizations of Euclidean norms and ellipsoids (see, e.g., [5, 9]). Other applications range from functional analysis (see [15]) to affine differential geometry (see [21]), statistics (see [20]), and, more recently, to the optimization algorithm of Shor and Khachyan, which is polynomial in time (see [18]).

I will now discuss the first result mentioned in Section 1.

Most convex bodies meet the boundary of their minimal ellipsoid in exactly $d(d+3)/2$ points. **(1)**

An analogous result holds for the case of centrally symmetric bodies with $d(d+3)/2$ replaced by $d(d+1)$. The result **(1)** continues to hold if the term "convex bodies" is replaced by "nonplanar compact sets."

The group of affinities leaving invariant a convex body also leaves invariant its minimal ellipsoid. Hence, the intersection of the convex body with the boundary of its minimal ellipsoid is also invariant under this group and thus **(1)** yields that for most convex bodies the group of affinities is finite. (Actually, it can be shown that it is trivial for most convex bodies.) An *optimal design* in statistics is a measure, the support of which consists of the intersection of a nonplanar compact set with the boundary of the ellipsoid with center at the origin which contains C and has minimal volume. Thus, a result related to **(1)** but for compact sets instead of convex bodies yields that in most cases the support of an optimal design consists of precisely $d(d+1)/2$ points.

The proof of **(1)** is somewhat lengthy. The main idea is to transfer the problem from $\mathbb{E}^d$ to $\mathbb{E}^{d(d+3)/2}$: To any ellipsoid of the form

$$\left\{x\colon \sum_{i,k=1}^{d} a_{ik} x_i x_k + \sum_{i=1}^{d} a_i x_i \leq 1\right\}, \tag{2}$$

we let correspond the point

$$A = (a_{11}, a_{12}, \ldots, a_{1d}, a_{22}, \ldots, a_{dd}, a_1, \ldots, a_d)^{\mathrm{tr}} \in \mathbb{E}^{d(d+3)/2}.$$

To the set of ellipsoids of given volume, there corresponds a hypersurface in $\mathbb{E}^{d(d+3)/2}$, a generalization of the discriminant surface, a well-known concept from the geometric theory of quadratic forms. If a point $p = (p_1, \ldots, p_d)^{\mathrm{tr}} \in \mathbb{E}^d$ belongs to an ellipsoid of the form **(2)**, then

$$\sum_{i,k=1}^{d} a_{ik} p_i p_k + \sum_{i=1}^{d} a_i p_i \leq 1 \qquad \text{or} \qquad A \cdot V(p) \leq 1,$$

where

$$V(p) = (p_1^2, 2p_1p_2, \ldots, 2p_1p_d, p_2^2, \ldots, p_d^2, p_1, \ldots, p_d)^{\mathrm{tr}} \in \mathbb{E}^{d(d+3)/2}$$

and $\cdot$ denotes the ordinary inner product in $\mathbb{E}^{d(d+3)/2}$. Hence, the set of all ellipsoids of the form (**2**) that contain p essentially corresponds to the half space

$$\{A\colon A \cdot V(p) \leqq 1\}$$

in $\mathbb{E}^{d(d+3)/2}$. By extensive use of this correspondence, one can show the existence of a dense set of convex bodies that meet the boundaries of their minimal ellipsoids in precisely $d(d + 3)/2$ points. From this and the theorem of Behrend and John [3, 14], one ultimately obtains (**1**).

4

A *lattice* L in $\mathbb{E}^d$ is the set of all linear combinations with integer coefficients of d fixed linearly independent vectors of $\mathbb{E}^d$. These vectors are said to form a *basis* of the lattice L, and the absolute value of their determinant is called the *determinant* $d(L)$ of L. Given a convex body C and a lattice L, the system of translates $C + l\colon l \in L$ is called a *lattice packing* of C by L if any two distinct translates of C have disjoint interior. Its *density* is then defined as $v(C)/d(L)$, that is, the proportion of space covered by the bodies of the packing. A translate of C different from C but intersecting C is called a *neighbor* of C.

Results of Minkowski yield that in any lattice packing of a convex body C, the body C has at most $3^d - 1$ neighbors, where $3^d - 1$ may be replaced by $2^{d+1} - 2$ if C is strictly convex. For each convex body C, there exists a (not necessarily unique) lattice packing having maximal density. A result of Swinnerton-Dyer implies that in any lattice packing of a convex body C having maximal density, C has at least $d(d + 1)$ neighbors. See [4, 7, 22, 23, 28].

The convex bodies which in any lattice packing of maximal density have at most $2d^2$ *neighbors form a dense open subset of* $\mathscr{C}$. (**3**)

Here the method of proof consists again of transferring the problem from $\mathbb{E}^d$ to $\mathbb{E}^{d^2}$. A lattice with basis $b_1, \ldots, b_d$ may be interpreted as a point of $\mathbb{E}^{d^2}$. The set of lattices of given determinant then forms a hypersurface in $\mathbb{E}^{d^2}$ and the set of lattices which contain a point of a given hyperplane in $\mathbb{E}^d$ corresponds—cum grano salis—to a hyperplane in E^{d^2}. Using this, it is possible to construct a dense set of convex bodies that have the property that in any lattice packing of maximal density they have at most $2d^2$ neighbors. Having shown this, the proof of (**3**) is a matter of routine, using compactness arguments.

Acknowledgments

For a discussion on (**1**), which inspired me to investigate lattice packings and thus led to (**3**), I am obliged to Professor Hlawka. For their advice, I would like to thank Professor Schnitzer and the referee.

References

1. Aleksandrov, A. D. 1939. Almost everywhere existence of the second differential of a convex function and some properties of convex surfaces connected with it. Učen. Zap. Leningr. Gos. Univ. Ser. Mat. Nauk **6:** 3–35.
2. Aleksandrov, A. D. 1948. Die innere Geometrie der konvexen Flächen. Gos. Izdat. Tehn.-Teor. Lit. Moscow-Leningrad, USSR. [Translation: Akademie-Verlag. Berlin, GDR.]
3. Behrend, F. 1938. Über die kleinste umbeschriebene und die grösste einbeschriebene Ellipse eines konvexen Bereichs. Math. Z. **115:** 379–411.
4. Cassels, J. W. S. 1959/1971. An Introduction to the Geometry of Numbers. Springer-Verlag. West Berlin, FRG.
5. Danzer, L., D. Laugwitz & H. Lenz. 1957. Über das Löwnersche Ellipsoid und sein Analogon unter den einem Eikörper einbeschriebenen Ellipsoiden. Arch. Math. **8:** 214–219.
6. Gruber, P. M. 1977. Die meisten konvexen Körper sind glatt, aber nicht zu glatt. Math. Ann. **229:** 259–266.
7. Gruber, P. M. 1979. Geometry of numbers. *In* Contributions to Geometry (Siegen, 1978). J. Tölke & J. M. Wills, Eds.: 186–225. Birkhäuser. Basel, Switzerland.
8. Gruber, P. M. In most cases approximation is irregular. Rend. Sem. Mat. Torino. In press.
9. Gruber, P. M. & J. Höbinger. 1976. Kennzeichnungen von Ellipsoiden mit Anwendungen. Jahrb. Überblicke Math. **1976:** 9–29. Bibl. Inst. Mannheim, FRG.
10. Gruber, P. M. & P. Kenderov. 1982. Approximation of convex bodies by polytopes. Rend. Circ. Mat. Palermo (2) **31:** 195–225.
11. Heil, E. Concurrent normals and critical points under weak smoothness assumptions. Ann. N.Y. Acad. Sci. (this volume).
12. Holmes, R. G. 1975. Geometrical Functional Analysis and Its Applications. Springer-Verlag. New York, N.Y.
13. Howe, R. 1982. Most convex functions are smooth. J. Math. Econom. **9:** 37–39.
14. John, F. 1948. Extremum problems with inequalities as subsidiary conditions. *In* Studies and Essays, Courant Anniversary Volume. K. Friedrichs, O. Neugebauer & J. J. Stoker, Eds.: 187–204. John Wiley & Sons, Inc. New York, N.Y.
15. Kadets, M. I. 1975. The geometry of normed spaces. Itogi Nauki Tekh. Mat. Analiz. **13:** 99–128. [Translation: J. Soviet Math. **7** (1977): 953–973.]
16. Kenderov, P. 1980. Approximation of plane convex compacta by polygons. C.R. Acad. Bulgare Sci. **33:** 889–891.
17. Klee, V. 1959. Some new results on smoothness and rotundity in normed linear spaces. Math. Ann. **139:** 51–63.
18. Klee, V. Invited Lecture. Jahrestagung Deutsch. Math. Verein. Dortmund, September 1980.
19. Klima, V. & I. Netuka. 1981. Smoothness of a typical convex function. Czechoslovak Math. J. **31**(106): 569–572.
20. Kraft, O. 1981. Dual optimization problems in stochastics. Jahresber. Deutsch. Math. Verein. **83:** 97–105.
21. Laugwitz, D. 1965. Differentialgeometrie in Vektorräumen. Vieweg. Braunschweig, FRG.
22. Lekkerkerker, G. C. 1969. Geometry of Numbers. North-Holland. Amsterdam, the Netherlands.
23. Minkowski, H. 1907/1961. Diophantische Approximationen. Physica-Verlag. Würzburg, FRG.
24. Perles, M. A. 1967. Critical exponents of convex sets. *In* Proceedings, Coll. Convexity (Copenhagen, 1965). W. Fenchel, Ed.: 221–228. Mat. Inst. Copenhagen, Denmark.
25. Schneider, R. 1979. On the curvature of convex bodies. Math. Ann. **240:** 177–181.
26. Schneider, R. & J. A. Wieacker. 1981. Approximation of convex bodies by polytopes. Bull. London Math. Soc. **13:** 149–156.
27. Shephard, G. C. 1963. Decomposable convex polyhedra. Mathematika **10:** 89–95.
28. Swinnerton-Dyer, P. 1953. Extremal lattices of convex bodies. Proc. Cambridge Philos. Soc. **49:** 161–162.

29. Zamfirescu, T. 1980. The curvature of most convex surfaces vanishes almost everywhere. Math. Z. **174:** 135–139.
30. Zamfirescu, T. 1980. Nonexistence of curvature in most points of most convex surfaces. Math. Ann. **252:** 217–219.
31. Zamfirescu, T. 1981. Intersections of tangent convex curves. J. Austral. Math. Soc. Ser. A **31:** 456–458.
32. Zamfirescu, T. 1980. Inscribed and circumscribed circles to convex curves. Proc. Amer. Math. Soc. **80:** 455–457.
33. Zamfirescu, T. Intersecting diameters in convex bodies. Submitted for publication.
34. Zamfirescu, T. 1982. Most convex mirrors are magic. Topology **21:** 65–69.
35. Zamfirescu, T. 1982. Many endpoints and few interior points of geodesics. Invent. Math. **69:** 253–257.
36. Zamfirescu, T. Curvature properties of typical convex surfaces. Submitted for publication.
37. Zamfirescu, T. Points on infinitely many normals. Submitted for publication.
38. Zhivkov, V. 1982. Plane polygonal approximation of bounded convex sets. C.R. Acad. Bulgare Sci. **35:** 1631–1634.

CONCURRENT NORMALS AND CRITICAL POINTS UNDER WEAK SMOOTHNESS ASSUMPTIONS

E. Heil

Fachbereich Mathematik
Technische Hochschule Darmstadt
61 Darmstadt, Federal Republic of Germany

It is shown that in any convex body K there is a point which is met by six normals. Essential tools are Morse's relations for the number of critical points. In Section 1 we assume for simplicity that the boundary of K is of class C^2. In Section 3 we show that convexity itself provides enough smoothness. Hereby an earlier paper [7] is corrected where it was erroneously claimed that the proof given there was valid for arbitrary convex bodies, but in fact only worked for bodies of class C^2. The proof given here is also simpler since it does not need the theory of degree of mappings. In Section 2 we discuss what further information is given by the degree. In Section 4 we list some open problems.

1. Normals of Smooth Convex Bodies

Let K be a convex body in Euclidean space E^n ($n \geq 3$). In this section we assume for simplicity that its boundary ∂K is of class C^2, i.e., twice continuously differentiable. A normal at a point $x \in \partial K$ is a straight line through x orthogonal to the tangent hyperplane at x. If it is also a normal at another point $x' \in \partial K$ it is counted twice (as in the normal bundle). A normal at $x \in \partial K$ meets $p \in \operatorname{int} K$ iff x is a critical point of

$$f_p\colon \partial K \to \mathbb{R}, \qquad f_p(x) = \|x - p\|^2, \tag{1.1}$$

i.e.,

$$\frac{\partial f_p}{\partial \xi_i}(\xi) = 0, \qquad i = 1, \ldots, m = n - 1.$$

Here $\xi = (\xi_1, \ldots, \xi_m)$ locally represents $x \in \partial K$. f_p is of class C^2, since ∂K has this property. Therefore, if f_p is nondegenerate, i.e., if

$$\det\left(\frac{\partial f_p^2}{\partial \xi_i\, \partial \xi_j}\right) \neq 0 \tag{1.2}$$

at all critical points, Morse's relations hold.

We repeat some details: If (**1.2**) holds at some critical point (which may be assumed to be 0), then *Morse's lemma* states that there is a diffeomorphic coordinate transformation in a neighborhood of 0 such that

$$f_p(\xi(\eta)) = f_p(0) - \eta_1^2 \cdots - \eta_i^2 + \eta_{i+1}^2 + \cdots + \eta_m^2. \tag{1.3}$$

i is called the *index* of the critical point. For a minimum $i = 0$, for a maximum $i = m$, otherwise we have saddle points. (Morse's lemma has been proven for C^2-functions by Kuiper [11] and Ostrowski [14]. This proof is much more difficult than that for C^3-functions.) We now quote Morse's relations involving the Betti numbers β_i of the compact manifold where the nondegenerate function is defined though we need only a special case. Let μ_i be the number of critical points of index i. Then

$$\begin{aligned} \mu_0 &\geq \beta_0, \\ \mu_1 - \mu_0 &\geq \beta_1 - \beta_0, \\ &\vdots \\ \mu_{m-1} - \mu_{m-2} + - \cdots &\geq \beta_{m-1} - \beta_{m-2} + - \cdots, \\ \mu_m - \mu_{m-1} + - \cdots &= \beta_m - \beta_{m-1} + - \cdots = \chi, \end{aligned}$$

where χ is the Euler characteristic. For ∂K, which is essentially S^m, we have $\beta_0 = \beta_m = 1$, $\beta_i = 0$ else, and

$$\chi = \begin{cases} 0 & \text{for } m \text{ odd}, \\ 2 & \text{for } m \text{ even}. \end{cases}$$

We need only the following two relations:

$$\mu_1 - \mu_0 \geq -1, \tag{1.4}$$

$$\mu_m - \mu_{m-1} + - \cdots + (-1)^m \mu_0 = \chi. \tag{1.5}$$

The following lemma relates the total number of critical points to the number of extrema.

LEMMA 1. *For any nondegenerate function $f: S^m \to \mathbb{R}$ of class C^2 ($m \geq 2$),*

$$\sum_{i=0}^{m} \mu_i \geq 2(\mu_0 + \mu_m - 1).$$

Proof. Let $m \geq 3$ be odd. (The case $m \geq 2$ even is similar.) From **(1.5)** we have

$$\begin{aligned} \mu_m + \mu_{m-2} + \cdots + \mu_1 &= \mu_{m-1} + \mu_{m-3} + \cdots + \mu_0, \\ \mu_m \qquad\qquad + \mu_1 &\leq \mu_{m-1} + \mu_{m-3} + \cdots + \mu_0, \end{aligned}$$

$$2(\mu_m + \mu_1) \leq \sum_{j=0}^{m} \mu_j.$$

Together with **(1.4)** this gives the assertion. ∎

Remark. For m-dimensional compact manifolds (instead of S^m), we have by the same reasoning

$$\sum_{i=0}^{m} \mu_i \geq 2(\mu_m + \mu_0 - 1) - \chi \qquad (m \text{ odd}),$$

$$\sum_{i=0}^{m} \mu_i \geq 2(\mu_m + \mu_0) - \chi \qquad (m \text{ even}).$$

It is remarkable that for S^m both inequalities can be combined into one, which moreover does not involve χ.

We now come to the question of how to assure the nondegeneracy of f_p. The first step is:

LEMMA 2. *f_p is nondegenerate iff p is not on the focal surface $\mathscr{F}$. ($\mathscr{F}$ is explained in the proof.)*

Proof. In classical surface theory this corresponds to the relation between second fundamental form and Weingarten map. But with respect to the weakening of smoothness conditions, in Section 3 we shall sketch the proof without relying much on the classical theory. Let orthonormal coordinates be chosen such that $x = 0$ is a critical point of f_p and the nth basis vector e_n has the direction of the inner normal vector n at 0. A neighborhood of 0 in ∂K may be represented by

$$x(\xi) = (\xi, \zeta(\xi)), \qquad \xi = (\xi_1, \ldots, \xi_m), \qquad m = n - 1. \tag{1.6}$$

The mapping Φ of the normal bundle into E^n is locally represented by

$$\phi(\xi, \lambda) = x(\xi) + \lambda n(\xi). \tag{1.7}$$

Let

$$p = \phi(0, \lambda_p) = x(0) + \lambda_p n(0).$$

The focal surface $\mathscr{F}$ may be defined as the set of those image points at which the functional determinant Δ vanishes. We write

$$x_i = \frac{\partial x}{\partial \xi_i}, \text{ etc.}$$

Then

$$x_i(0) = e_i, \qquad x_{ij}(0) = (0, \ldots, 0, \zeta_{ij}(0)),$$

and differentiating

$$nx_i = 0, \qquad nn = 1$$

we derive

$$n_i(0) = -\sum_{j=1}^{m} \zeta_{ij}(0)e_j. \tag{1.8}$$

By definition,

$$\Delta(\xi, \lambda) = \det\left(\phi_i, \frac{\partial \phi}{\partial \lambda}\right) = \det\,(x_i + \lambda n_i, n),$$

where ϕ_i stands for the columns $\phi_1, \ldots, \phi_m$ and so on. From **(1.8)** we have for p

$$\Delta(0, \lambda_p) = \det\,(e_i - \sum_j \lambda_p \zeta_{ij}(0)e_j, e_n)$$

$$= \det\left(\sum_j (\delta_{ij} - \lambda_p \zeta_{ij}(0))e_j, e_n\right).$$

Since $e_1, \ldots, e_n$ form a basis, we have $\Delta(0, \lambda_p) = 0$ iff

$$\det\,(\delta_{ij} - \lambda_p \zeta_{ij}(0)) = 0. \tag{1.9}$$

This is thus equivalent to $p \in \mathscr{F}$. (Incidentally, this shows that λ_p is one of the principal curvature radii iff $p \in \mathscr{F}$. $\mathscr{F}$ therefore is also called a centrosurface.)

Now we have to show that f_p is degenerate iff $p \in \mathscr{F}$. We consider the critical point $x = 0$ and show that **(1.2)** and **(1.9)** are equivalent. From the choice of the representation **(1.6)** we have

$$\zeta(\xi) = \frac{1}{2} \sum_{i,j} \zeta_{ij}(0)\xi_i \xi_j + \|\xi\|^2 h_1(\xi), \tag{1.10}$$

where $h_1(\xi) \to 0$ for $\xi \to 0$. Inserting **(1.10)** into **(1.1)**,

$$f_p(x(\xi)) = \|x(\xi) - p\|^2 = (x(\xi) - p)(x(\xi) - p),$$

where as before $x(\xi) = (\xi, \zeta(\xi))$, $p = (0, \lambda_p)$, we obtain

$$f_p(x(\xi)) = \lambda_p^2 + \sum_{i,j} (\delta_{ij} - \lambda_p \zeta_{ij}(0))\xi_i \xi_j + \|\xi\|^2 h_2(\xi). \tag{1.11}$$

Comparing this with **(1.9)** we see that the assertion of LEMMA 2 is true. ∎

If in the following applications of LEMMA 1 p happens to lie on $\mathscr{F}$ so that f_p is degenerate, then we have to go to a point q close to p such that f_q is nondegenerate. The following lemma assures that this is always possible.

LEMMA 3. *If f_p is degenerate, then there is a point $q \in \operatorname{int} K$ arbitrarily close to p such that f_q is nondegenerate and has at least as many extrema as f_q.*

Proof. The mapping Φ of the normal bundle into E^n, represented by **(1.7)**, is of class C^1 since n is of class C^1. Therefore, according to Sard's lemma (see, e.g., Sternberg [16, p. 47]) the set of its critical values, i.e., the focal surface $\mathscr{F}$, has measure 0 and, thus, does not contain inner points. Then, arbitrarily close to p, there are points $q \in \operatorname{int} K$ with $q \notin \mathscr{F}$. If q is close enough, it may be shown by elementary geometry that to every extremum of f_p there corresponds an extremum of f_q (compare Wegner [19, p. 784]). By LEMMA 2, f_q is nondegenerate. This proves LEMMA 3. ∎

We are now able to give applications of LEMMA 1.

THEOREM 1. *Let K be a convex body ($\partial K \in C^2$) in E^n ($n \geq 3$). Then there is a point $q \in \operatorname{int} K$ such that at least 6 points of ∂K send their normals through q.*

Proof. We consider the *minimal spherical shell.* It consists of two concentric spheres, one inscribed, the other circumscribed to ∂K, such that the difference of radii is minimal (Bonnesen and Fenchel [3, p. 54]). Kritikos [9, p. 585] has shown that ∂K has at least two points in common with each of the spheres. (His proof for $n = 3$ works also for $n \geq 3$.) The ellipsoid in E^n shows that one cannot expect more. We choose p to be the center of the minimal spherical shell. Then f_p has at least two minima and two maxima. If $p \in \mathscr{F}$ we apply LEMMA 3. Thus, there is $q \in \operatorname{int} K$ such that f_q is nondegenerate and $\mu_0 \geq 2$, $\mu_m \geq 2$. LEMMA 1 then gives

$$\sum_{i=0}^{m} \mu_i \geq 2(2 + 2 - 1) = 6.$$

This proves THEOREM 1. ∎

In some sense it is the regular case that the inscribed, respectively circumscribed, sphere has exactly $n+1$ points of contact with ∂K. (Compare Zamfirescu [21] and the remarks in Section 3 concerning THEOREM 2′.)

THEOREM 2. *If the inscribed or circumscribed sphere of a convex body K in E^n ($\partial K \in C^2$) has $n+1$ points of contact with ∂K, then there is a point $q \in \operatorname{int} K$ such that at least $2n+2$ points of ∂K send their normals through p.*

Proof. For the center of one of the spheres $\mu_0 \geq 1$, $\mu_m = n+1$. Applying LEMMA 3, if necessary, LEMMA 1 gives

$$\sum_{i=0}^{m} \mu_i \geq 2(1+n+1-1) = 2n+2. \quad \blacksquare$$

THEOREM 3. *If $K \subset E^n$ ($\partial K \in C^2$) has constant width, then there is a point $q \in \operatorname{int} K$ such that at least 10 points of ∂K send their normals through q.*

Proof. Inscribed and circumscribed spheres are concentric and each has more than two points of contact with ∂K. Thus for their center p $\mu_0 \geq 3$, $\mu_m \geq 3$. (It follows from a theorem of Vrećica [18, footnote on p. 289], that the number 3 is possible.) Similar to the above reasoning,

$$\sum_{i=0}^{m} \mu_i \geq 2(3+3-1) = 10. \quad \blacksquare$$

2. THE SET OF NON-FOCAL POINTS

For the ellipse in E^2 there pass four normals through any point inside the evolute (an astroid) and two outside. The focal surface $\mathscr{F}$ of the ellipsoid in E^3 was studied by Caley and others. It divides E^3 into regions with two, four, or six normals through each point; see [7] for an illustration and more details. The situation is similar in the general case ($\partial K \in C^2$) as we shall see. The mapping Φ, locally represented by (**1.7**), is of class C^1. For $p \notin \mathscr{F}$ we have by definition $\Delta \neq 0$ for all pre-images of p. By the implicit function theorem, Φ is a local homeomorphism at each pre-image. From this and the compactness of ∂K it follows that any $p \notin \mathscr{F}$ has only a finite number of pre-images. Furthermore, it is easily shown that this number is constant on each connected component of the complement of $\mathscr{F}$.

More can be said using the theory of degree of mappings. Some details can be found in [7]. The normal bundle is orientable and the degree of Φ is even, namely, $1-(-1)^n$. Therefore the number of pre-images is even, too. Thus we have:

THEOREM 4. *The number of normals passing through a point $p \notin \mathscr{F}$ is even and is constant on the open sets which are the connected components of the complement of $\mathscr{F}$ ($\partial K \in C^2$).*

For sets of constant width the number of normals through a non-focal point is congruent to 2 modulo 4. For details see [7].

3. The Non-smooth Case

Now we show that the results of Section 1 are still true if the smoothness condition ($\partial K \in C^2$) is dropped. (This was also claimed in [7], but the proof given there is valid only for $\partial K \in C^2$.) For an arbitrary convex body $K \subset E^n$ ($n \geq 3$), a normal is a straight line orthogonal to a supporting hyperplane at one of its points of support. Any parallel body $K_\varepsilon = K + \varepsilon B$ ($\varepsilon > 0$, B unit ball) has the same set of normals as K. The supporting hyperplanes are unique at every point of ∂K; i.e., they are tangent hyperplanes. Therefore, we may assume $\partial K \in C^1$ without loss of generality (see Bonnesen and Fenchel [3, p. 26]).

It is of course trivial that infinitely many normals pass through a point with more than one supporting hyperplane. But as in Theorems 1, 2, and 3 we will assure the existence of a point $q \in \operatorname{int} K$ which is met by a certain number of normals. We will still obtain inner points of K even if we have to pass over to a K_ε. The radii of the minimal spherical shell, etc., of K have just to be enlarged by ε.

Henceforth we assume that the local representations of **(1.6)**, $\zeta(\xi)$, $\xi = (\xi_1, \ldots, \xi_m)$, are of class C^1. Alexandrov [1] showed that in fact they are almost everywhere twice differentiable in the following sense (see also Bangert [2, p. 319]):

$$\frac{\partial \zeta}{\partial \xi_1} \text{ is (totally) differentiable a.e.}$$

and at such a point (for simplicity assumed to be 0)

$$\zeta(\xi) = \zeta(0) + \sum_{i=1}^{m} \frac{\partial \zeta}{\partial \xi_i}(0)\xi_i + \frac{1}{2} \sum_{i,j=1}^{m} \frac{\partial^2 \zeta}{\partial \xi_i \, \partial \xi_j}(0)\xi_i \xi_j + R(\xi)$$

with

$$R(\xi)/\|\xi\|^2 \to 0 \qquad \text{for } \xi \to 0. \tag{1.12}$$

Then the $\partial R/\partial \xi_i$ vanish at 0, are differentiable at 0, and it follows that

$$\frac{\partial R}{\partial \xi_i}\Big/ \|\xi\| \to 0 \qquad \text{for } \xi \to 0. \tag{1.13}$$

We define a generalized focal set in the following manner. Let $E \subset \partial K$ be the exceptional set where ∂K is not twice differentiable and let Φ denote again the map of the normal bundle $\partial K \times \mathbb{R}$ into E^n. The set of all points that are covered by normals starting at points of E is $\Phi(E \times \mathbb{R})$. If Φ is not locally Lipschitzian, then we take a parallel body and easily show that it is. Since $E \times \mathbb{R}$ is a null set in $\partial K \times \mathbb{R}$, so is $\Phi(E \times \mathbb{R})$. Now consider those points of $(\partial K \backslash E) \times \mathbb{R}$ where the functional determinant Δ of Φ vanishes. We call this set $(\Delta = 0)$. There is a generalization of Sard's theorem which says that $\Phi(\Delta = 0)$ has measure 0; see Varberg [17, Theorem 2], also Bangert [2, p. 320]. We define the *generalized focal set* to be

$$\mathscr{F}_g = \Phi(E \times \mathbb{R}) \cup \Phi(\Delta = 0).$$

As explained it is a null set. This definition is not only suitable for our purposes, but it is also geometrically meaningful to include $\Phi(E \times \mathbb{R})$ in $\mathscr{F}_g$. Consider, for instance, a parallel body of a cube. The regions with a constant number of normals through each point are bounded by the planes that contain the faces of the cube, i.e., in this case $\mathscr{F}_g = \Phi(E \times \mathbb{R})$. (The relations between different concepts of focal sets in

E^2 are discussed in [5]. Hentschke [8] gives a topological definition of a focal surface in E^3.)

The differentiability properties of ∂K carry over to the functions of **(1.1)**, f_p. Then, as we shall see, Morse's relations and thereby LEMMA 1 are still valid. There is an approach to Morse's relations for continuous functions with isolated critical points by Kuiper [12]. Since he does not assume that critical points are (in some sense) nondegenerate, the μ_i count critical points with multiplicities. For instance, the monkey saddle counts with $\mu_1 = 2$. This is, of course, not in the sense we wish it to be since $\sum \mu_i$ shall give the total number of different critical points, i.e., points sending their normals through p. (There also are approaches for continuous functions by Morse himself assuming a topologically defined nondegeneracy. But it seems to be more convenient to work with Kuiper's version.)

We now call f_p nondegenerate if, at all critical points, f_p is twice differentiable (as specified above) and **(1.2)** holds.

LEMMA 2′. *f_p is nondegenerate iff $p \notin \mathscr{F}_g$.*

We have given the proof of LEMMA 2 in such a manner that it is easily checked that LEMMA 2′ is true. One has to remember the definition of $\mathscr{F}_g$ and that f_p has the same differentiability properties as ∂K.

LEMMA 3′ is literally the same as LEMMA 3, except for the reduced differentiability assumptions.

We have already explained that F_g is a null set. LEMMA 2′ and the proof of LEMMA 3 establish LEMMA 3′.

Before we can generalize THEOREMS 1, 2, and 3 we have to show that nondegeneracy will imply $\mu_1 \leq 1$ for any critical point. The usual procedure, i.e., reducing the given function f locally to a quadratic function **(1.3)** by means of a coordinate transformation, probably does not work. Seifert and Threlfall [15, p. 35] continuously deform the sublevels $\{\xi; f(\xi) < f(0)\}$ to those of the quadratic terms. This proof for C^2-functions really needs only **(1.12)** and **(1.13)** and not the existence of second derivatives in a whole neighborhood.

We have thus shown that all steps in the proofs of THEOREMS 1, 2, and 3 can be carried through for arbitrary convex bodies and that the resulting number $\sum \mu_i$ is the number of different boundary points sending their normals through q.

THEOREMS 1′ *and* 3′ *are literally the same as* THEOREMS 1 *and* 3, except that there is no smoothness condition.

Zamfirescu [21] has shown that "most" convex bodies have exactly $n + 1$ points of contact with the circumscribed sphere. *Most* means that the set of bodies without this property is a set of first Baire category in the space of all convex bodies endowed with the Hausdorff metric. Since most convex bodies are not C^2-smooth (Gruber [6]) it was meaningless to use Zamfirescu's result in Section 1, but here we may state:

THEOREM 2′. *For most convex bodies K there is a point $q \in \operatorname{int} K$ such that at least $2n + 2$ points of ∂K send their normals through q.*

4. OPEN QUESTIONS

If K is symmetric with respect to p, then one may consider the length l of all chords of K through p as a function defined on the projective space P^m. The number of critical points of l is at least equal to the Lusternik–Schnirelmann category of P^m, which is $m + 1 = n$. Thus for symmetric K we may replace the number 6 in THEOREM 1 by $2n$. More gene ally, Lusternik and Schnirelmann [13, p. 49] show that a C^2-embedded sphere S^m possesses n double normals, and Kuiper [10] shows that the same is true for general convex bodies.

Conjecture 1. For a convex body in E^n $(n \geq 2)$ there is a point $q \in \text{int } K$ such that at least $2n$ boundary points send their normals through q.

This is proved for $n = 2$, 3 and, as remarked above, for centrally symmetric convex bodies.

Conjecture 2. For a body of constant width in E^n $(n \geq 2)$ the number $2n$ in Conjecture 1 may be replaced by $4n - 2$.

This is true for $n = 2$, 3.

For most convex bodies in the sense of Baire category, the number $2n + 2$ established in THEOREM 2′ may be much too small. For Zamfirescu [22, p. 68] has shown that for most convex curves in E^2 most points lie on infinitely many normals. This might as well be the case for higher dimensions because for most convex bodies most boundary points have lower curvature 0 and upper curvature ∞ for every direction (Zamfirescu [20]).

[**Note added in proof:** Meanwhile Zamfirescu has shown that, for most convex surfaces, most points lie on infinitely many normals. (ZAMFIRESCU, T. 1984. Points on infinitely many normals to convex surfaces. *Journal für die Reine und Angewandte Mathematik* **350:** 183–187.)]

Question 1. Is it possible to work in the space of all smooth (C^2) convex bodies endowed with a metric including second derivatives and to show that most bodies in this space have exactly $n + 1$ points of contact with the circumscribed sphere?

The best possible number for most convex bodies (in this sense) could then be $2n + 2$ (as in THEOREM 2) or another finite number.

The same may be asked for bodies of constant width. If it is true that for most (or most smooth) bodies K of constant width the circumscribed (and inscribed) sphere has $n + 1$ points of contact with ∂K, then there are at least $4n + 2$ boundary points sending their normals through a point q.

Question 2. Is Conjecture 1 true for C^2-immersions of the sphere S^m?

This is shown by Deo and Klamkin [4] for $n = 2$ $(m = 1)$ and there is a sketch of a proof for $n = 3$ by Hentschke [8]. In both cases q is found near the diameter of K. Our proof of THEOREM 1 does not work since the theorem of Kritikos is probably not true for nonconvex closed surfaces.

Question 3. What can be said if the normals are replaced by the affine normals?

It has been shown in [5, p. 380], that in E^2 there exists a point through which pass at least six affine normals of the given oval.

ACKNOWLEDGMENTS

I have to thank several colleagues, especially S. Hentschke, W. Kühnel, R. Schneider, B. Wegner, and T. Zamfirescu.

References

1. Alexandrov, A. D. 1939. Über das fast überall existierende zweite Differential einer konvexen Funktion und gewisse damit zusammenhängende Eigenschaften konvexer Flächen. Učen. Zap. Leningr. Gos. Univ. Ser. Mat. Nauk **6:** 3–35. (In Russian.) (German translation: Universitätsbibliothek Hannover, Ostsprachenabteilung-Übersetzungsstelle 800/4630.)
2. Bangert, V. 1979. Analytische Eigenschaften konvexer Funktionen auf Riemannschen Mannigfaltigkeiten. J. Reine Angew. Math. **307/308:** 309–324.
3. Bonnesen, T. & W. Fenchel. 1934/1974. Theorie der konvexen Körper. Springer-Verlag. Berlin.
4. Deo, N. & M. S. Klamkin. 1970. Existence of four concurrent normals to a smooth closed curve. Amer. Math. Monthly **77:** 1083–1084.
5. Douglas, V., E. Heil & C. Lübbert. 1976. Geradenfamilien, Enveloppen und Evoluten. J. Reine Angew. Math. **283/284:** 370–383.
6. Gruber, P. 1977. Die meisten konvexen Körper sind glatt, aber nicht zu glatt. Math. Ann. **229:** 259–266.
7. Heil, E. 1979. Existenz eines 6-Normalenpunktes in einem konvexen Körper. Arch. Math. **32:** 412–416. [Erratum: Arch. Math. **33** (1979): 496.]
8. Hentschke, S. Topologische Geradenfamilien. Preprint 138, Fachbereich Math. TH Darmstadt, 1974.
9. Kritikos, N. 1927. Über konvexe Flächen und einschließende Kugeln. Math. Ann. **96:** 582–586.
10. Kuiper, N. H. 1964. Double normals of convex bodies. Israel J. Math. **2:** 71–80.
11. Kuiper, N. H. 1972. C^1-equivalence of functions near isolated critical points. *In* Proceedings, Symposium on Infinite Dimensional Topology. R. D. Anderson, Ed. Ann. of Math. Studies. Princeton University Press. Princeton, N.J.
12. Kuiper, N. H. 1971. Morse relations for curvature and tightness. *In* Proceedings, Liverpool Singularities Symposium II. C.T.C. Wall, Ed.: 77–89. Lecture Notes in Mathematics No. 209. Springer-Verlag. West Berlin, FRG.
13. Lusternik, L. & L. Schnirelmann. 1934. Méthodes topologiques dans les problèmes variationnels. Hermann. Paris, France.
14. Ostrowski, A. 1968. On the Morse-Kuiper theorem. Aequationes Math. **1:** 66–76.
15. Seifert, H. & W. Threlfall. 1938. Variationsrechnung im Großen. (Reprinted 1971 by Chelsea, New York, N.Y.)
16. Sternberg, S. 1964. Lectures on Differential Geometry. Prentice-Hall, Inc. Englewood Cliffs, N.J.
17. Varberg, D. E. 1966. On differentiable transformations in R^n. Amer. Math. Monthly **73,** Slaught Memorial Papers: 111–114.
18. Vrećica, S. 1981. A note on sets of constant width. Publ. Inst. Math. (Beograd) (N.S.) **29:** 289–291.
19. Wegner, B. 1973. Existence of four concurrent normals to a smooth closed hypersurface of E^n. Amer. Math. Monthly **80:** 782–785.
20. Zamfirescu, T. 1980. Nonexistence of curvature in most points of most convex surfaces. Math. Ann. **252:** 217–219.
21. Zamfirescu, T. 1980. Inscribed and circumscribed circles to convex curves. Proc. Amer. Math. Soc. **80:** 455–457.
22. Zamfirescu, T. 1982. Most convex mirrors are magic. Topology **21:** 65–69.

GENERALIZATIONS OF TIETZE'S THEOREM ON LOCAL CONVEXITY FOR SETS IN R^d

David C. Kay

Department of Mathematics
The University of North Carolina at Asheville
Asheville, North Carolina 28804

Introduction

The author generalizes earlier results for closed sets having finitely many points of local nonconvexity. Suppose S is a closed set in R^d having full dimension d in a neighborhood of every point, and that its (infinite) set of points of local nonconvexity Q is decomposable into a finite union of convex subsets $Q_1, \ldots, Q_n$ where for each i, $\dim Q_i \leq d-2$, Q_i and Q_j are a positive distance apart for $i \neq j$, and there exists a hyperplane which locally separates S into convex components and does not contain or separate Q_j for $j \neq i$. Then if $S \backslash Q$ has k components and Q contains no subset of dimension $\leq d-3$ which locally separates S, S is the union of $n+k$ or fewer convex sets. This leads to the result that if S has the properties (1) S is locally starshaped, (2) $S \backslash Q$ has k components, (3) Q is the union of n convex subsets no one of which contains the other, and (4) Q contains no subsets of dimension $\leq d-3$ which locally separate S, then S is the union of $k+pn-1$ or fewer convex sets, where

$$p = \binom{n}{0} + \binom{n}{1} + \cdots + \binom{n}{d}.$$

An interesting side result (generalizing a result for finite Q) is: If $S \backslash Q$ is connected and Q is a countable union of arbitrary sets each having dimension $\leq k \leq d-3$ for all i, then either S is convex or $\dim S \leq k+2$.

1. Preliminary Discussion

Let S be any set in a linear topological space $\mathscr{L}$ and Q its set of points of local nonconvexity (lnc set); that is, for each $q \in Q$ and any neighborhood U of x, $S \cap U$ is not convex. Tietze [5] proved for R^d that if S is locally convex (that is, $Q = \varnothing$), closed, and connected, then S is convex, a result later extended to sets in $\mathscr{L}$ (see [6, pp. 48–51]). It was proved by the author in [2] that if S is closed in $\mathscr{L}$, $S \backslash Q$ is connected, and Q has $1 \leq n < \infty$ points, then S is planar and is the union of $n+1$ convex sets. We now extend this result to the case when Q is infinite and $S \subset R^d$.

Other related results have been obtained by Valentine, Stavrakas, and Breen. In [7] Valentine proved that (1) if S is closed and connected in R^d and $Q = \bigcup_{i=1}^n Q_i$, where each Q_i is a singleton, then S is an L_{n+1} set (a set having the property that each two of its points may be joined by a polygonal arc in the set having $n+1$ or fewer sides), and (2) if $Q = \bigcup_{i=1}^n Q_i$, where each Q_i is convex, then S is an L_{2n+1} set. In [4] Stavrakas proved essentially that if $S \backslash Q$ is connected and $Q = \bigcup_{i=1}^n Q_i$, where each Q_i is a closed convex set, then S is an L_{n+1} set. In [1] Breen proved that if $Q = \bigcup_{i=1}^{n} Q_i$, where each Q_i is convex and essential (defined below), then S is the union of 2^n or fewer convex sets. All these results are best possible under their respective hypotheses.

Unless stated otherwise, we shall assume throughout that S is a closed set in R^d of dimension d having a (closed) lnc set Q that is decomposable into a union $Q_1 \cup Q_2 \cup \cdots$ of connected lnc subsets of dimension $\leq d-1$, not necessarily pairwise disjoint, but with $Q_i \not\subset Q_j$ for all $i \neq j$. If $\mathcal{Q} = \{Q_i \mid i \geq 1\}$, then dim $\mathcal{Q}$ denotes the maximum dimension of Q_i for $i \geq 1$. For the most part, Valentine's notation in [6] will be adhered to as in previous papers, where cl A, int A, and relint A denote, respectively, the closure of A, the set of interior points of A, and the relative interior of A as a subset of the smallest flat containing A. The terms conv A, aff A, and dim A stand for the convex hull of A, the affine hull of A (smallest flat containing A), and the dimension of $A \equiv$ dimension of the smallest flat containing A. The local dimension of S at a point $x \in S$ is the number inf (dim $U \cap S$) for all neighborhoods U of x.

2. Examples

If S is the solid torus in R^3, then Q consists of all boundary points lying on the inside rim; all other points are points of local convexity. Here, any decomposition $\mathcal{Q}$ of Q into closed connected subsets of dimension ≤ 2 is necessarily infinite (even uncountable) and S is not the union of a finite number of convex sets, although $S \backslash Q$ is connected. If S consists of a sequence of 2-simplexes $T_1, T_2, \ldots$ converging to a single 2-simplex T, with T joining all the T_i along an edge of each, then S is connected, $Q = T$ and $\mathcal{Q} = \{Q\}$ with dim $Q = 2$. Again, S is not the union of a finite number of convex sets. In this example, $S \backslash Q$ is disconnected.

To obtain an example in R^3 where both $|\mathcal{Q}| < \infty$ and $S \backslash Q$ is connected, a little more effort is required. Begin with an infinite-sided polygon and interior, inscribed in a circle (FIGURE 1), with consecutive pairs of sides congruent ($AB = BC$, $CD = DE, \ldots$). Let this generate two congruent but oppositely oriented cylindrical solids S_1 and S_2 with parapets added, which form $Q = AB \cup BC \cup CD \cup \cdots$. Then $S = S_1 \cup S_2$ where S_1 is turned upside down and placed on top of S_2 so that the protruding parapets interlock. Again S is not the union of finitely many convex sets but $S \backslash Q$ is connected and we can take $\mathcal{Q} = \{Q\}$ with dim $Q = 2$.

Positive results can be obtained in R^3 if we assume essentially that $S \backslash Q$ is connected and dim $\mathcal{Q} = 1$ (or in R^d that dim $\mathcal{Q} = d - 2$). Interestingly enough, if dim $\mathcal{Q} \leq d - 3$ is demanded in R^d, then S, if full d-dimensional, must be convex. Unfortunately, our results do not cover even such a simple case as the union of two solid, right circular cylinders of unit radii whose axes meet at right angles. Here, Q consists of a double figure eight wrapped around each of the cylinders, but, like the torus, dim $Q = 3$ and Q has no finite or countable decomposition $\mathcal{Q}$ with dim $\mathcal{Q} = 2$. The only result we have for dim $\mathcal{Q} = d - 1$ in R^d, is a rather crude one; much work remains to be done for this case, and the case dim $\mathcal{Q} = d$ appears to be hopeless at the present time. (That is, it is not known what general hypotheses concerning Q and its decomposition into some finite number of connected subsets permit the conclusion that S is a finite union of convex sets.)

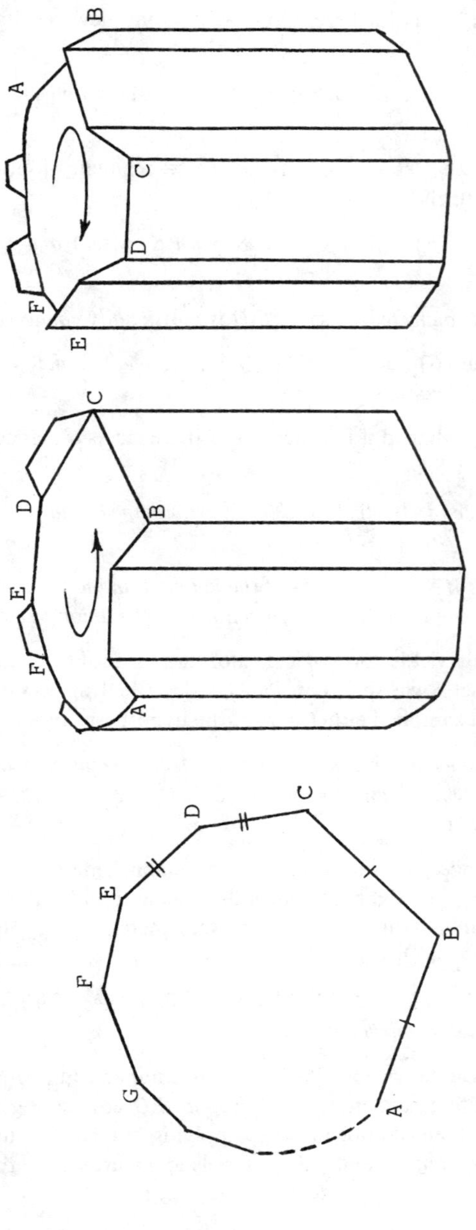

FIGURE 1. Example for dim $\mathscr{Q} = d - 1$.

3. Preliminary Lemmas

We list for convenience some basic results which have appeared in previous papers and which will be useful here, along with some new results. The first was already mentioned above.

Lemma 1 (Tietze [5]). *If a closed set in R^d is connected and locally convex, it is convex.*

After Valentine, we let $xy = \text{conv}(x \cup y) \equiv$ segment joining x and y, $xyz = \text{conv}(x \cup y \cup z)$, and so on.

Lemma 2 (Valentine [6]). *If $xy \cup yz \subset S$ (with S closed) and Q does not meet $xyz \backslash xz$, then $xyz \subset S$.*

Lemma 3. *Any open connected subset of $S \backslash Q$ is polygonally connected.*

Lemma 4 (Valentine [6] and Kay [2]). *If S is connected and $Q \neq \varnothing$, then $S = \bigcup_{q \in Q} S_q$, where $S_q = \{x \mid xq \subset S\}$.*

A set S is locally starshaped if for each $x \in S$ there exists a neighborhood U of x such that $U \cap S = U \cap S_x$.

Lemma 5 (Stavrakas [4]). *If S is locally starshaped and Q is convex, then* $\dim Q \leq d - 2$.

Lemma 6 (Kay [2]). *If S is the closure of the union of an increasing sequence of sets which are each the union of n convex sets, then S is also the union of n convex sets.*

The following result yields topological information about the lnc set Q for the most general type of set S we shall consider (as described in Section 1). A circular arc provides a counterexample if either part of the hypothesis about Q is omitted.

Lemma 7. *If S has local dimension d and Q has a countable decomposition of connected subsets $Q_1, Q_2, \ldots$ and for each i,* $\dim Q_i \leq d - 1$ *(with $Q_i \not\subset Q_j$ for all $i \neq j$), then $S \backslash Q$ is dense in S.*

Proof. For the purpose of proof it suffices to demonstrate that $S \neq Q$. But since each point of S possesses a neighborhood of dimension d and $\dim Q_i \leq d - 1$, each Q_i is nowhere dense in S. Since S is a complete metric space, Baire's Category Theorem implies $S \neq \bigcup_{i=1}^{\infty} Q_i$. ∎

Corollary 1. *If S is connected and Q has a finite decomposition $\mathcal{Q}$ described above, then S is polygonally connected.*

Proof. One may assume without loss that S is compact. Let $x, y \in S$. By applying Lemma 7 one can find sequences $\{x_n\}$, $\{y_n\}$ in $S \backslash Q$ converging to x, y, respectively. The proof is then completed by applying Lemma 3 to the adjacent components of $S \backslash Q$, joining x_n and y_n by a polygonal arc in S, then using compactness. ∎

Lemma 8. *If S satisfies the hypothesis of* Lemma 7 *and $S \backslash Q$ is connected, then S has local dimension d at all points.*

Proof. Observe that the local dimension is constant along any polygonal arc in $S\backslash Q$. Suppose the local dimension of S at some point $x \in S\backslash Q$ is $<d$. Then some relative neighborhood of x in S lies in a hyperplane, from which it follows by our opening remark that $S\backslash Q$, and therefore S, lies in that same hyperplane, a contradiction. Hence, $S\backslash Q$ has local dimension d at all points, and LEMMA 7 completes the proof. ∎

LEMMA 9 (Kay [2]). *If $S\backslash Q$ is connected and Q is countable and nonempty, then S is planar.*

Proof. The proof in [2] may be extended to the infinite case by observing that if $xy \cup yz \cup zw$ is a three-dimensional polygonal arc in $S\backslash Q$, for some sequence $\{z_i\}$ on z_w converging to z, $yz_i \in S\backslash Q$ and xyz_i is free of points of Q; hence, by Valentine's lemma and the fact that S is closed, $xz \subset S$. ∎

4. RESULTS FOR dim $\mathcal{Q} \le d - 3$

It seems surprising that under the hypothesis $S\backslash Q$ connected, a restriction on the dimension of Q forces one on all of S (unless S is convex and $Q = \varnothing$). Breen observed in [1] under similar hypotheses that if dim $S = d$, then dim $Q_i = d - 2$ for each i. For convenience, let the connected component of A which contains the connected subset $B \subset A$ be denoted comp (A, B).

THEOREM 1. *Suppose S is any closed nonconvex set in R^d, with Q its set of points of local nonconvexity, that $S\backslash Q$ is connected, and the lnc set Q may be expressed as a countable union $\bigcup_{i=1}^{\infty} Q_i$ of closed, connected subsets such that* dim $Q_i \le k \le d - 3$ *for each i. Then* dim $S \le k + 2$.

Proof. The proof is by induction on k, with $k = 0$ being covered by LEMMA 9. Assume that dim $Q_i \le k$ and dim $S \ge k + 3$ to gain a contradiction. Thus, the local degree of S is at least $k + 3$ at all points and there exist two points x and y in $S\backslash Q$ such that some polygonal arc P in $S\backslash Q$ joining x and y has dimension $\ge k + 3$. Assume that P has consecutive vertices $x = x_0, x_1, \ldots, x_n = y$ and that its order n is least among all such $(k + 3)$-dimensional arcs joining x and y. Thus for all $i \le j + 2$, $x_i x_j \not\subset S$. Let x_m for $1 \le m < n$ be the first vertex such that dim $\{x_m \cdots x_n\} = k + 2$ and x_l the last vertex such that dim $(x_m \cup \cdots \cup x_l) = k + 1$. Wolog, $m = 1$ and $l + 1 = n$. Set $F = \text{aff}\,(x_1 \cup \cdots \cup x_n)$. By considering $(k + 2)$-dimensional neighborhoods in $F \cap (S\backslash Q)$ of each of the vertices of P it is clear that one may perturb P so as to retain all of the aforementioned properties of P. Thus, as we perturb $x_2, \ldots, x_{n-1}$ in $(k + 2)$-dimensional neighborhoods, there is generated an uncountable number of $(k + 1)$-flats passing through $x_1, x_2, \ldots, x_{n-1}$, whose intersection is just x_1. Hence, there must be one that does not contain any set Q_i. It may then be assumed that P has the following properties (FIGURE 2): $x \notin F = \text{aff}\,(x_1 \cup \cdots \cup x_n)$, $y \notin F' = \text{aff}\,(x_1 \cup \cdots \cup x_{n-1})$, dim $F' = k + 1$, $Q_i \not\subset F'$ for all i and $x_1 x_3 \not\subset S$. Again, by perturbing x_1 we may further assume $x_1 \notin \text{aff}\,(x_2 \cup \cdots \cup x_{n-1})$.

Choose $w \in F\backslash F'$ so close to x_1 that $xw \cup wx_2 \subset S\backslash Q$ and $wx_3 \not\subset S$. For each $u \in (xw)$, consider the $(k + 2)$-flat $F_u = \text{aff}\,(u \cup F')$. A simple application of the

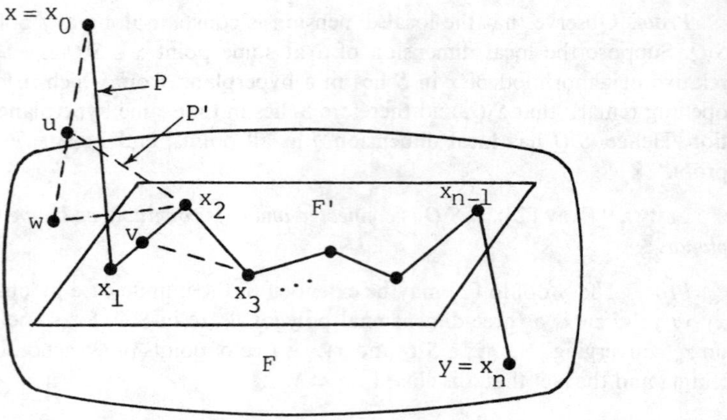

FIGURE 2. Figure for proof of THEOREM 1.

exchange lemma shows that $F_u \cap F = F'$. It follows, therefore, that for any two choices of u on (xw) the corresponding flats F_u are distinct and meet in F'. Since there are uncountably many such F_u's, it follows again that some F_u does not contain Q_i for all i, and we can take u so close to w that $ux_2 \subset S\backslash Q$ and $ux_3 \not\subset S$. By a simple dimension argument, $\dim (Q_i \cap F_u) \leq k - 1$. Take $v \in (x_1 x_2)$ so close to x_2 that $vk_3 \subset S\backslash Q$. Therefore, we obtain a polygonal arc $P' = ux_2 \cup x_2 v \cup vx_3 \cup \cdots \cup x_{n-2} x_{n-1}$ in $(S\backslash Q) \cap F_u$ of dimension $k + 2$.

Define $S' = \text{cl } S''$ where $S'' = \text{comp } [(S\backslash Q) \cap F_u, P']$, having lnc set Q'. Clearly, $Q' \subset Q \subset F_u$ and Q' is a countable union $\bigcup_{j=1}^{\infty} Q'_j$ of closed connected subsets such that for each j, $Q'_j \subset Q_i \subset F_u$ for some i. Also, $\text{comp } [(S\backslash Q) \cap F_u, P']\backslash Q$ is connected, so $S'\backslash Q$ and $S'\backslash Q'$ are connected, and S' is not convex since $ux_3 \not\subset S$. But $\dim Q'_j \leq k - 1$ for all j, producing the contradiction $\dim P' \leq \dim S' \leq (k - 1) + 2 = k + 1$ by the induction hypothesis. ▮

5. RESULTS FOR $\dim \mathcal{Q} = d - 2$

The first step is to show that if $\mathcal{Q}$ has but one member (that is, Q itself is connected with dimension $d - 2$) and $S\backslash Q$ is connected, then S is the union of two convex sets. This proceeds in substantially the same manner as in the two-dimensional case in [2], except that the line L in that proof is replaced by a hyperplane and T becomes the intersection of two half spaces. The details are rather transparent, so we merely outline the steps.

Since $\dim Q = d - 2$, we let $L = \text{aff } Q$. With $x, y \in \text{int } S\backslash L$ such that $xy \not\subset S$, determine the hyperplanes $H_x = \text{aff } (x \cup L)$ and $H_y = \text{aff } (y \cup L)$, with H_x^+ and H_y^+ the open half spaces containing y and x, respectively, and put $T = H_x^+ \cap H_y^+$. The sets $S_x \cap T$ and $S_y \cap T$ are easily shown to be nonempty, disjoint convex subsets of S by Valentine's lemma. Thus, there exists a hyperplane H of separation between $S_x \cap T$ and $S_y \cap T$, which must therefore contain Q, and we let H^+ be the side of H containing x and H^- that containing y. If $z \in S \cap H^+ \cap T$ and $z \notin S_x$, then $S_z \cap T$ is convex and disjoint from both $S_x \cap T$ and $S_y \cap T$, with H_1 the hyperplane of separation between $S_z \cap T$ and $S_x \cap T$, and H_2 that between $S_z \cap T$ and $S_y \cap T$. A polygonal arc in $S \backslash Q$ joining z and x must cut H_x or H_y (say H_x) at some point x' such that $zx' \subset S$, which implies that H_1 cuts L in a $(d-3)$-dimensional set and $Q \subset H_1 \cap H_x \cap H_y \subset H_1 \cap L$ or $\dim Q \leq d-3$, a contradiction. Hence $z \in S_x$. Finally, let $z \in H^+ \cap H_x^-$ and consider a polygonal arc P of least order in $S \backslash Q$ joining z and x. Then P cannot have a vertex in T or in H^- or else that vertex could be eliminated by Valentine's lemma contradicting the assumption of minimal order for P. Hence, $P = xz \subset S$ and $z \in S_x$ so that $H^+ \cap S$ is convex. Similarly, $H^- \cap S$ is convex and S is the union of the two convex sets $\mathrm{cl}\,(H^+ \cap S)$ and $\mathrm{cl}\,(H^- \cap S)$. Using Valentine's lemma, it also easily follows that Q is convex. This proves:

LEMMA 10. *If $S \backslash Q$ is connected and* $\dim Q = d-2$, *then Q is convex and S is the union of two convex sets.*

It is natural to conjecture at this point that if $Q = \bigcup_{i=1}^n Q_i$ and each Q_i is closed and connected with dimension $d-2$, then S is the union of $n+1$ convex sets. That this is false is shown by the three-dimensional example in FIGURE 3, where $S = \bigcup_{i=1}^4 C_i$ with C_i convex and $Q = \bigcup_{i=1}^2 Q_i$. The other example in FIGURE 4 is due to Breen [1] and is self-explanatory. Thus, it will be necessary to assume two more properties about Q. We use the following terminology: $A \subset S$ *separates* $B \subset S$ if $B \backslash A$ is disconnected; $A \subset S$ *locally separates* $B \subset S$ if A separates some neighborhood $U(B) \equiv U \cap S$ of B relative to S (U open in R^d), and if some neighborhood $U(A)$ of A in S is separated by A, then A is called a *local cut-set* of S. The first property we need is:

(1) For each i there exists a hyperplane H_i containing Q_i which locally separates S into convex components and does not contain or separate Q_j for $j \neq i$.

The preceding property rules out both examples in FIGURES 3 and 4, but still allows a set of the type illustrated in FIGURE 5. This set consists of two adjacent pointed stakes cut from a dowel with a wedge carved out from the center of each, forming an lnc set Q_1 along the axis; these two stakes are beveled at the top and along the four slanted sides, the part of the dowel that was cut off is replaced along the flat top as indicated by the arrows, forming two more lnc sets Q_2, Q_3. The resulting set is the union of five convex sets, not $n+1=4$ (in this case, $n=3$). Thus, we assume:

(2) Q contains no local cut-set of S of dimension $\leq d-3$.

As in [2] we distinguish between types of lnc sets Q_i in $\mathcal{Q}$.

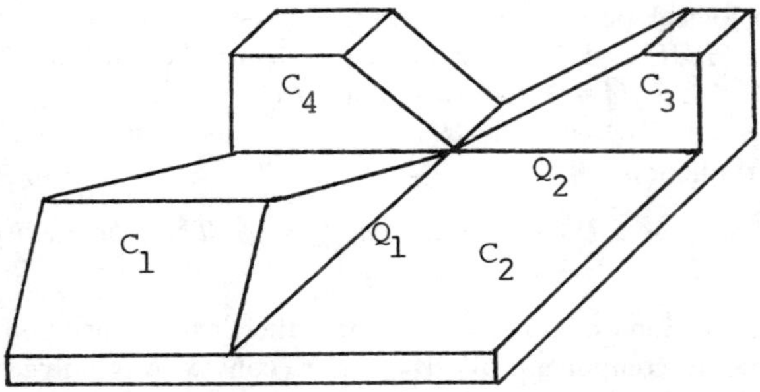

FIGURE 3. Example showing necessity of hypothesis (1).

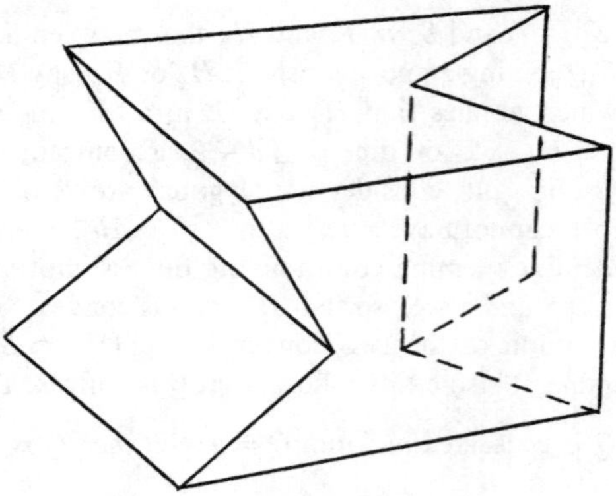

FIGURE 4. Breen's example.

DEFINITION 1. If for all sufficiently small relatively convex neighborhoods $U(Q_i)$ of Q_i in S (obtained by intersecting S with a convex neighborhood of Q_i) the set $U(Q_i)\backslash Q_i$ is connected, then Q_i is called a *simple* lnc set of $\mathcal{Q}$. If $U(Q_i)\backslash Q_i$ has at least one nonconvex component, Q_i is said to be *essential*, and if for some relatively convex neighborhood $U(Q_i)$ of Q_i in S all the components of $U(Q_i)\backslash Q_i$ are convex, Q_i is called *inessential.*

LEMMA 11. *Suppose $S\backslash Q$ is connected, that property* (2) *above is satisfied, and $Q_1 \in \mathcal{Q}$ is any nonsimple lnc set of dimension $d - 2$ such that some neighborhood V of Q_1 is devoid of points of Q_i for $i \neq 1$. Then for all sufficiently small relatively convex neighborhoods $U(Q_1)$ of Q_1 in S there is at most one nonconvex component of $U(Q_1)\backslash Q_1$.*

Proof. Suppose $U(Q_1)\backslash Q_1$ has two nonconvex components, W and W'. As in the notation of the proof outlined for LEMMA 10, let H be the hyperplane separating W into convex sets, C_1 and C_2, with H_x, H_y and T as in that proof, and H' the corresponding hyperplane separating W' into convex sets C'_1 and C'_2. Recall that H_x and H_y are the boundaries of T and meet in the (d-2)-dimensional flat L that contains Q_1^*, the lnc set of W. If W' lies wholly inside cl T, then it is easily verified by Valentine's lemma that W' is convex, contrary to hypothesis. Therefore, there exists a point $u \in W'$ which lies in H_x^- or H_y^-. Let us assume that $u \in H_x^- \cap C'_1\backslash Q_1$, and choose v any point in $H_x^- \cap C_1\backslash H$. Now, the disjoint convex sets $C_1\backslash Q_1$ and $C'_1\backslash Q_1$ possess a separating hyperplane H^* with u and v on opposite sides. By examining the logical possibilities of the relative positions of u, v and points of $H_x^- \cap C_1\backslash Q_1$ if H, H_x and H^* all contain L, a contradiction follows and we conclude that $H^* \cap L$ is ($d - 3$)-flat such that $U(Q_1)\backslash(H^* \cap L \cap Q_1)$ has the components $C_1\backslash Q_1$ and $C_1\backslash Q_1$ making $H^* \cap L \cap Q_1$ a local cut-set of S of dimension $\leq d - 3$, a contradiction. ∎

LEMMA 12. *Suppose $S\backslash Q$ is connected and $Q \neq \varnothing$. Then there exists at least one essential lnc set Q_i in $\mathcal{Q}$.*

Proof. (By induction on n.) If $n = 1$, suppose there exists a neighborhood $U(Q)$ of Q in S such that all components W_1, W_2, ... are convex. Any polygonal arc in $S\backslash Q$ joining W_1 and W_2 can be reduced by LEMMA 2 to a segment; hence, $W_1 = W_2$, a

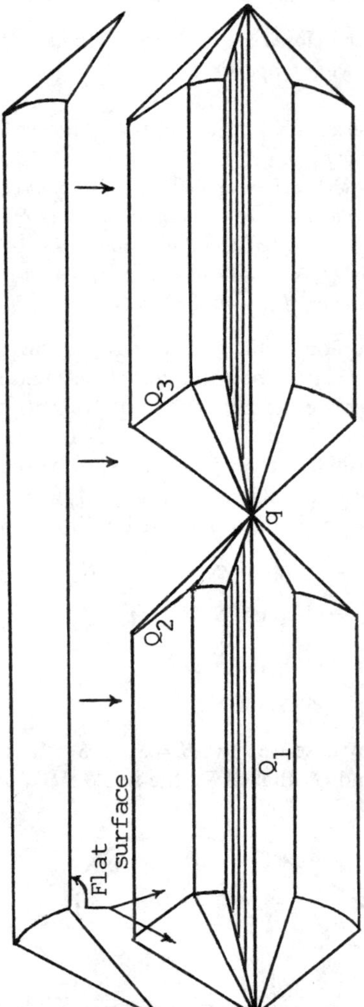

FIGURE 5. Example showing necessity of hypothesis (2).

contradiction. Hence, assume true for $k < n$, and consider Q_1. By the exact same construction as used in the proof of the theorem below, except that H is merely some hyperplane containing Q_1, one can define a set S_2 having the same properties as S but with a decomposition $\mathcal{Q}'$ having one less member. Hence, by the induction hypothesis, S_2, and therefore S, has an essential lnc set. ▌

THEOREM 2. *Let S be any closed set in R^d with Q its set of local nonconvexity, and suppose S satisfies the following properties:*

(a) *$S \backslash Q$ is connected.*

(b) *$Q = \bigcup_{i=1}^{n} Q_i$ where each Q_i is connected, of dimension $d - 2$, and lying at a positive distance from Q_j for $j \neq i$.*

(c) *For each i there exists a hyperplane H_i containing Q_i which locally separates S into convex components and does not contain or separate Q_j for $j \neq i$.*

(d) *Q contains no local cut-set of S of dimension $\leq d - 3$.*

Then each essential lnc set Q_i is convex and S is the union of $m + 1$ or fewer convex sets, where $m \leq n$ is the number of essential lnc sets.

Proof. We use induction on m, the case for $m = 1$ being LEMMA 10. Let Q_1 be an essential lnc set and let $H \equiv H_1$ be the hyperplane guaranteed in hypothesis (c), where W is the only nonconvex component of $U(Q_1) \backslash Q_1$ (LEMMA 11). By LEMMA 10, $W = K_1 \cup K_2$, where K_1, K_2 are convex, and it is clear that $K_1 \subset \text{cl } H^+$ and $K_2 \subset \text{cl } H^-$. For each positive integer j, define M_j to be the closed, parallel strip of width $1/j$ lying in cl H^+. For all sufficiently large j, M_j misses some fixed, preselected point $p \in \text{int } K_1$. For all such j define (illustrated in FIGURE 6 for $d = 2$):

$$M_j' = \text{comp } (S \cap M_j \backslash Q, \text{ int } K_1 \cap M_j),$$

$$S_{1j} = \text{cl comp } (S \backslash M_j' \backslash Q, p),$$

$$S_1 = \text{cl } (\textstyle\bigcup_{j=1}^{\infty} S_{1j}),$$

$$S_2 = \text{cl } (S \backslash S_1)$$

(S_2 may be empty). In any case we have $S = S_1 \cup S_2$. It is routine to show that if Q'_{1j} is the lnc set of S_{1j} and Q'_2 that of S_2, the sets $S_{1j} \backslash Q'_{1j}$ and $S_2 \backslash Q'_2$ are connected

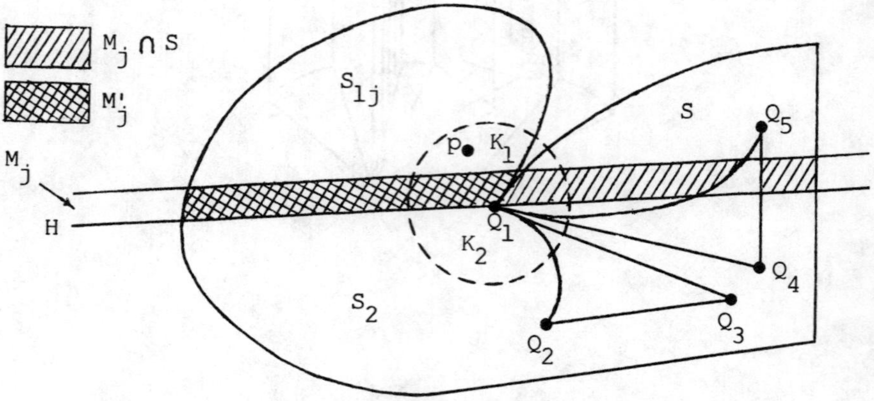

FIGURE 6. Figure for proof of THEOREM 2.

[one uses the principle that if $A\backslash B$ is connected, then (cl $A)\backslash B$ is connected]. It is clear that Q'_{1j} and Q'_2 are contained in $Q_1 \cup \cdots \cup Q_n$, and since Q'_{1j} and Q'_2 are disjoint, and because no Q_k meeting Q_1 is separated by H, the Q_k's may be relabeled so that $Q'_{1j} = (Q_1 \cup \cdots \cup Q_r) \cap S_{1j}$ and $Q'_2 = (Q_{r+1} \cup \cdots \cup Q_n) \cap S_2$. Therefore, it follows that the original hypothesis for S is satisfied by S_{1j} and S_2, for all sufficiently large j. But Q_1 is no longer an essential lnc set for either S_{1j} or S_2; hence, say S_{1j} and S_2 possess, respectively, $m_1 < m$ and $m_2 < m$ essential lnc sets, with $m_1 + m_2 < m - 1$. By the induction hypothesis, S_{1j} (therefore, S_1 in view of LEMMA 6) and S_2 are, respectively, the union of $m_1 + 1$ and $m_2 + 1$ convex sets, and S is the union of $m_1 + m_2 + 2 < m + 1$ convex sets. ∎

It is easy to see that if $S\backslash Q$ is not connected but has k components we can apply the above theorem to each component to yield the bound $(m_1 + 1) + \cdots + (m_k + 1) = m + k$. This leads to the result:

COROLLARY 2. *Let S satisfy the same hypothesis as* THEOREM 2 *except for the connectedness of $S\backslash Q$. Then if $S\backslash Q$ has k components and S has local degree d at all points, S is the union of $k + n$ convex sets.*

By using a different argument we can prove a result which does not require hypothesis (a) or (c) of THEOREM 2. First, assume that $S\backslash Q$ is connected. Let $Q = \bigcup_{i=1}^n Q_i$ with $Q_1, \ldots, Q_m$ essential, $m \le n$, and dim $Q_i = d - 2$ for each i. For $i \le m$ let H_i be the hyperplane of separation of the two convex subsets K_{i1}, K_{i2} of W_i—the only nonconvex component of $U(Q_i)\backslash Q_i$, where $U(Q_i)$ is a sufficiently small neighborhood of Q_i in S. If $B_1, \ldots, B_p$ denote the open regions that are formed in R^d after removing the hyperplanes $H_1, \ldots, H_m$, where, as is well known,

$$p \le \binom{m}{0} + \binom{m}{1} + \cdots + \binom{m}{d},$$

consider $B_i \cap S$ for each i, $1 \le i \le p$. Then the components of $B_i \cap S$ are associated with some Q_j, so there cannot be more than m such components. Hence, taking the closure of these components, S is the union of all these closures, which consists of not more than m convex sets for each of the p regions, or mp convex sets altogether. Now if $S\backslash Q$ is not connected and has k components $S_1, \ldots, S_k$, we can take S_1 as the first one having lnc points, if any, and thus, if S_i has m_i essential lnc sets, by the above argument S is the union of not more than $k - 1 + \sum_{i=1}^k m_i p_i$ convex sets, where

$$p_i = \binom{m_i}{0} + \binom{m_i}{1} + \cdots + \binom{m_i}{d} \le p.$$

(Hence, $\sum_{i=1}^k m_i p_i \le p \sum_{i=1}^k m_i \le mp$.)

COROLLARY 3. *Let S have local degree d at all points and suppose Q contains no cut-set of S of dimension $d - 3$. Further, suppose $Q = \bigcup_{i=1}^n Q_i$ where Q_i is connected and* dim $Q_i = d - 2$, *with $S\backslash Q$ having k components. Then S is the union of $k + pn - 1$ or fewer convex sets, where*

$$p = \binom{n}{0} + \binom{n}{1} + \cdots + \binom{n}{d}.$$

We mention one last result which takes advantage of LEMMA 5. It allows us to substitute local starshapedness for the hypothesis dim $\mathcal{Q} = d - 2$ since, by LEMMA 5, requiring Q_i to be convex implies dim $Q_i = d - 2$.

COROLLARY 4. *Let $S \subset R^d$ be locally starshaped, and suppose Q contains no cut-set of S of dimension $d - 3$, with $Q = \bigcup_{i=1}^{n} Q_i$ where Q_i is convex and* dim $Q_i = d - 2$, *and with k components in $S \backslash Q$. Then S is the union of $k + pn - 1$ or fewer convex sets, where p is defined as in* COROLLARY 3.

It is interesting to compare the bound of COROLLARY 4 with that obtained by Breen [1] in the case $k = 1$ (as required in [1]). The bound obtained by Breen was 2^n, where it was assumed that all Q_i are essential. If $n \leq d$, then

$$p = \binom{n}{0} + \binom{n}{1} + \cdots + \binom{n}{n} = 2^n$$

and our bound becomes $n \cdot 2^n$, which is larger than the bound in [1]. But if $n > d$, then our bound reduces to a polynomial $p_d(n)$ in n and $p_d(n) < 2^n$ for all n sufficiently large. For $d = 3$ one can easily deduce the formula

$$p_3(n) = \tfrac{1}{6}(n^4 + 5n^2 + 6n),$$

and for $n \geq 12$, $p_3(n) < 2^n$ since

$$p_3(12) = 3588 < 4096 = 2^{12}.$$

6. A RESULT FOR dim $\mathcal{Q} = d - 1$

Only the following weak result has been obtained for the case dim $\mathcal{Q} = d - 1$.

THEOREM 3. *Let S be any closed set in R^d having local degree d at all points, with Q its set of points of local nonconvexity, and suppose $Q = \bigcup_{i=1}^{n} Q_i$ where each Q_i is connected, of dimension $d - 1$, and for each $i \neq j$,* conv $Q_i \cap$ conv $Q_j \cap$ relint $Q_i = \varnothing$. *For each i let H_i be the unique hyperplane* aff Q_i. *Then if $S \backslash \bigcup_{i \geq 1}$* comp $(S \cap H_i, Q_i)$ *has k components, S is the union of k convex sets.*

Proof. Merely make the trivial observation that the interiors of the k components of $S \backslash \bigcup_{i \geq 1}$ comp $(S \cap H_i, Q_i)$ lie in the open convex polyhedra defined by $R^d \backslash \bigcup_{i \geq 1} H_i$ and, as such, their individual closures are free of points of local convexity and must be convex by Tietze's theorem. ∎

REFERENCES

1. BREEN, M. 1975. Points of local nonconvexity and finite unions of convex sets. Canad. J. Math. **27**(No. 2): 376–383.
2. KAY, D. C. 1971. On sets having finitely many points of local nonconvexity and property P_m. Israel J. Math. **10**(No. 2): 196–209.
3. STAVRAKAS, N. M. 1973. A generalization of Tietze's theorem on convex sets in R^3. Proc. Amer. Math. Soc. **40**: 565–567.
4. STAVRAKAS, N. M. 1975. L_n sets and the closures of open connected sets. Canad. J. Math. **27**: 1–5.

5. TIETZE, H. 1928. Über Konvexheit im kleinen und im grossen und über gewisse den Punkten einer Menge zugeordnete Dimensionszahlen. Math. Z. **28:** 697–707.
6. VALENTINE, F. A. 1964. Convex Sets. McGraw–Hill Book Co., Inc. New York, N.Y.
7. VALENTINE, F. A. 1965. Local convexity and L_n sets. Proc. Amer. Math. Soc. **16:** 1305–1310.

PART IV. POLYHEDRAL GEOMETRY

COLORING POLYHEDRAL MANIFOLDS[a]

D. W. Barnette

Department of Mathematics
University of California
Davis, California 95616

1. INTRODUCTION

It is well known that the four color conjecture for maps on the sphere is equivalent to the four color conjecture for convex three-dimensional polyhedra (see Section 4). It is also well known that every map on the torus can be colored with seven or fewer colors, and that there are toroidal maps requiring seven colors [3]. One might guess that there exist toroidal polytopes that require seven colors. Surprisingly, this is not the case. We shall prove that every toroidal polytope can be colored with six or fewer colors.

2. DEFINITIONS

If a graph is embedded in a surface, the connected components of the complement of the graph are called its *faces*. If every face is a cell, we shall call the graph a *map*. The *dual* of a graph in the torus is constructed the same way as the dual of a graph in the plane. We place a vertex in each face and join vertices when the corresponding faces meet on an edge. The vertices are joined by one edge for each edge that the two faces have in common. Generally, the edges of the dual are drawn so that they cross the corresponding edges of the original graph.

A *toroidal polytope* is a topological torus consisting of convex polygons such that:

(i) The intersection of two polygons is either an edge of both, a vertex of both, or empty.

(ii) No two polygons that meet lie on the same plane.

The polygons will be called the *faces* of the polytope.

A *polyhedral immersion* of a torus is the continuous image in E^3 of a toroidal map such that:

(i) The image of each n-sided country of the map is a convex n-gon with the vertices of the country taken one-to-one onto the vertices of the n-gon.

(ii) No two n-gons in the immersion that meet on an edge are coplanar.

By a *vertex*, *edge*, or *face* of the immersion we shall mean the images of a vertex edge or country of the map. The *valence* of a vertex of the immersion is the valence of its pre-image in the map.

A toroidal polytope is *simple* provided each vertex has valence 3. A polyhedral immersion is *simple* provided it is the image of a map on the torus all of whose vertices are 3-valent.

[a] Research supported by National Science Foundation Grant MCS76-07466.

Since the vertices and edges of a toroidal polytope form a graph on the torus, we can speak of the dual graph of a toroidal polytope. Since two polygons can meet only on a vertex or an edge or not at all in a polytope, we see that the dual graph has no multiple edges or loops.

We shall need an inequality that follows from Euler's equation for graphs embedded in the torus. For any graph embedded in the torus, let V be the number of vertices, E the number of edges, and F the number of faces of the graph.

LEMMA 1. *For any graph in the torus we have*

$$V - E + F \geq 0,$$

with equality if each face is a cell.

(The reader is referred to [1] where a good treatment of Euler's equation for orientable surfaces is given.) We shall call this *Euler's inequality*. In the case where we have equality it is known as *Euler's equation* for the torus.

3. The Main Result

LEMMA 2. *If v_i is the number of i-valent vertices in a graph without loops and multiple edges on the torus, then*

$$\sum (6 - i)v_i = 0$$

if and only if the graph is a triangulation of the torus. Otherwise, the sum on the left is greater than 0.

Proof. Since the graph is without loops and multiple edges, every face has at least three edges, and we have

$$3F \leq 2E$$

with equality if and only if every face is a triangle. We also have Euler's inequality, $V - E + F \geq 0$, with equality if each face is a cell.

The sum $\sum (6 - i)v_i$ equals $6V - 2E$. Combining our two inequalities, we get $6V - 2E \geq 0$, with equality if and only if every face is a triangle. From this the desired result follows. ▮

LEMMA 3. *There are no simple polyhedral immersions of the torus.*

Proof. We add the sum of the two-dimensional angles of the faces of such an immersion in two different ways and get a contradiction. Since there are exactly three faces meeting at each vertex, the sum of the angles is less than 2π at each vertex. Thus the sum of the angles is less than $2\pi V$. The sum of the angles of an n-sided face is $\pi(n - 2)$. Summing these quantities over all faces gives us

$$\sum \pi(n - 2) = 2\pi E - 2\pi F.$$

We conclude that $2\pi V > 2\pi E - 2\pi F$, which contradicts Euler's equation for the torus. ▮

THEOREM. *Toroidal polytopes are 6-colorable.*

Proof. Let T be a toroidal polytope and let G be its dual graph. We shall show that we can remove vertices from G, one at a time, such that at each step the vertex to be removed will have valence at most 5 in the graph that we have at that step.

Once this has been shown, the result is immediate because we can remove vertices one at a time until at most six remain, color them, and then return them one at a time assigning a color to each as it is returned. Since the returned vertex is at most 5-valent, there is always a color available for it.

We can choose a vertex of valence at most 5 in G because, by LEMMA 3, T is not simple; thus, G is not a triangulation of the torus; thus, $\sum (6 - i)v_i > 0$, for G, and so such a vertex exists in G. What we must show is that as we proceed, we will always be able to choose such a vertex at later steps.

By LEMMA 2, we know that $\sum (6 - i)v_i > 0$, provided there is a face with at least four edges. Suppose that at some step we have produced a proper subgraph H of G such that H has no faces with at least four edges. In this case, H is a graph whose faces are all triangles. If H is planar, then it is well known that there will be a vertex of valence at most 5 (see, for example, [1, Section 4.1]). If the graph is not planar, then H is a triangulation of the torus which is a proper subgraph of G.

We shall show that this is impossible.

We do this by showing that it implies the existence of an immersion of a simple toroidal polytope. Suppose that u, v, and w are three vertices of G that determine a face of H, but do not determine a face of G. Corresponding to these three vertices will be three faces U, V, and W, of the polytope T. These three faces will form a region that is a topological annulus. This annulus encloses a topological cell C consisting of faces of T that correspond to the vertices of G enclosed by the circuit determined by u, v, and w. We now modify the polytope. The faces U, V, and W meet pairwise on three edges. Each of these three edges has one vertex in C. We take a plane through these three vertices. This plane may cut some of the faces U, V, and W into two pieces. In each case where a face is cut into two pieces we throw away the piece that has an edge in common with C. We also throw away all of C, and fill in the "hole" with the convex hull of the three vertices.

We carry out this procedure for each triangle in G that is a face of H but not of G. When we are done we will have faces corresponding to each of the vertices of H and no other faces. The graph H will be the dual graph of the immersion of the toroidal polytope that we have constructed. Since H is a triangulation, we have created a simple polyhedral immersion of the torus, a contradiction. ▮

4. REMARKS

If we substitute "2-manifold" for "torus" in the definition of toroidal polytope we get the definition of a *polyhedral 2-manifold*. The author knows of no polyhedral manifolds of any genus requiring more than four colors. Recently, McMullen *et al.* [4] developed some new techniques of construction that produce polyhedral 2-manifolds all of whose faces are hexagons or all octagons. Perhaps among these is an example requiring five or six colors. The few that the author has tried can be colored with four colors. The author conjectures that all toroidal polytopes are 4-colorable.

In view of the results in [4], there exist polyhedral 2-manifolds whose average face size is at least 8. Do there exist polyhedral 2-manifolds with arbitrarily large average face sizes?

LEMMA 2 is a special case of a theorem of Grünbaum [2, Chapter 11, exercise 7] that every simple cell complex is the boundary complex of a convex polytope.

To prove the equivalence of the four color problem for maps in the plane and for convex polytopes, one first reduces the four color conjecture to the case of 3-connected graphs (a simple argument will do this), then one uses the theorem of Steinitz [5] which states that the planar 3-connected graphs are the graphs of the convex three-dimensional polytopes.

REFERENCES

1. BEHZAD, M., G. CHARTRAND & L. LESNIAK-FOSTER. 1979. Graphs and Digraphs. Prindle, Weber & Schmidt. Boston, Mass.
2. GRÜNBAUM, B. 1967. Convex Polytopes. John Wiley & Sons, Inc. New York, N.Y.
3. HEAWOOD, P. J. 1880. Map-colour theorem. Quart. J. Math. Oxford Ser. **24:** 332–338.
4. MCMULLEN, P., C. SCHULZ & J. M. WILLS. Equivelar polyhedral manifolds in E^3. To appear.
5. STEINITZ, E. & H. RADEMACHER. 1934. Vorlesungen über die Theorie der Polyeder. Springer-Verlag. Berlin, Germany.

CONVEX BODIES AND ALGEBRAIC GEOMETRY

Günter Ewald
Mathematisches Institut
Ruhr Universität Bochum
D4630 Bochum, Federal Republic of Germany

During the last decade a new area of research has developed relating two subjects which until now had very little in common: convexity and algebraic geometry. An initial success was Stanley's solution of the so-called upper bound conjecture for combinatorial spheres ([16]; for polytopes solved by McMullen, published in [14]): Among all combinatorial spheres with v vertices (faces are convex polytopes), the convex hulls of v points on a moment curve $\{(t, t^2, \ldots, t^n) | t \in \mathbb{R}\}$ (called cyclic polytopes) possess maximal numbers of faces in all dimensions. This result has been pursued further by Kind and Kleinschmidt [10]. Another result is the solution of McMullen's conjecture about characterizing those vectors $f(P) = (f_0(P), \ldots, f_{d-1}(P))$ for which $f_j(P)$ is the number of j-faces of a polytope P. The necessity of McMullen's condition has been shown by Stanley [17] and the sufficiency by Billera and Lee [2].

A foundation for Stanley's, Billera's, and Lee's work was laid by Hochster [8]. This paper by Hochster was also one of the starting points for a development which is to be outlined in what follows. We emphasize two main achievements: First, a characterization of Milnor's number of critical points of complex algebraic functions by numbers assigned to convex polytopes (Kouchnirenko [11]); second, characterizations of mixed volumes of convex bodies as the intersection index of certain varieties, and an alternative proof of the Alexandrov–Fenchel inequality (Bernstein [1], Burago and Zalgaller [4], Teissier [19]). In order to make the exposition comprehensible to nonspecialists, we restrict the discussion to elementary explanations.

1

We first recall some basic facts from the theory of convex bodies (see [3, 6, 7, 12, 14]). A compact convex subset K of $\mathbb{R}^n$ will be called a *convex body*. If $K = \text{conv}\,(M)$ is the convex hull of a finite set of points M (or equivalently, provided it is bounded, the intersection of finitely many closed half spaces) it is called a *convex polytope*. The faces of a convex polytope are again convex polytopes; they form a cell complex of dimension $n - 1$. The cells of dimension 0 are called *vertices* of K. If, in addition, all vertices are lattice points, K is said to be a *lattice polytope*.

Every convex body K is the limit, with respect to the Hausdorff metric, of a sequence $\{P_j\}$ of convex polytopes P_j. The surface measure (area) and volume of the P_j then lead to the surface measure and volume of K, respectively. Linear combinations $\lambda_1 K_1 + \cdots + \lambda_r K_r$ of convex bodies $K_1, \ldots, K_r$ are defined by

$$\lambda_1 K_1 + \cdots + \lambda_r K_r := \{\lambda_1 x_1 + \cdots + \lambda_r x_r | x_1 \in K_1, \ldots, x_r \in K_r\},$$

resulting again in convex bodies. We restrict our attention to nonnegative λ_i. For the volume of a linear combination, one shows:

$$V(\lambda_1 K_1 + \cdots + \lambda_r K_r) = \sum \lambda_{i_1}, \ldots, \lambda_{i_n} V(K_{i_1}, \ldots, K_{i_n}),$$

where the sum is taken over all nonnegative $i_1, \ldots, i_n$. The coefficient $V(K_{i_1}, \ldots, K_{i_n})$ is called the *mixed volume* of $K_{i_1}, \ldots, K_{i_n}$. In particular, $V(K, \ldots, K) = V(K)$ is the volume of K. The surface measure of K equals $nV(K, \ldots, K, B)$, where B is a unit ball in $\mathbb{R}^n$. Most mixed volumes, however, do not have an obvious geometrical meaning.

A number of inequalities for mixed volumes are known; for example, the *Minkowski inequalities*

$$V(K, \ldots, K, K')^n \geq V(K)^{n-1} V(K'),$$

$$V(K, \ldots, K, K')^2 \geq V(K) V(K, \ldots, K, K', K'),$$

which hold for arbitrary convex bodies K, K'. These inequalities play a decisive role in the solution of the isoperimetric problem. A generalization of the second inequality is the *Alexandrov–Fenchel inequality*:

$$V(K_1, \ldots, K_{n-1}, K_n)^2 \geq V(K_1, \ldots, K_{n-1}, K_{n-1}) V(K_1, \ldots, K_{n-2}, K_n, K_n).$$

2

We consider a ring R of functions $f\colon \mathbb{C}^n \to \mathbb{C}$ of the form

$$f(z) = \sum_{\rho \in I} c_\rho z^\rho,$$

where $z := (z_1, \ldots, z_n)$, $\rho := (\rho_1, \ldots, \rho_n)$, $z^\rho := z_1^{\rho_1} \ldots z_n^{\rho_n}$, $c_\rho \in \mathbb{C}$. ($\mathbb{C}$ can be replaced by any algebraically closed field.) If $I \subset \mathbb{Z}_+^n$ and if I is finite, R is the ring of polynomials $\mathbb{C}[z_1, \ldots, z_n]$ in n variables. If $I \subset \mathbb{Z}^n$ and I is finite, we obtain the ring $\mathbb{C}[z_1, z_1^{-1}, \ldots, z_n, z_n^{-1}]$ of Laurent polynomials. If only $I \subset \mathbb{Z}_+^n$ is required, $R = \mathbb{C}[[z_1, \ldots, z_n]]$ is the ring of formal power series.

The *support* supp f of f is given by $\{\rho \mid c_\rho \neq 0\}$. If I is finite, we call the lattice polytope

$$\mathcal{N}\omega(f) := \operatorname{conv}(\operatorname{supp} f)$$

the *Newton polytope* of f. One key to understanding what follows is the fact that many properties of the functions f or varieties defined by them depend only on $\mathcal{N}\omega(f)$ and not on supp f.

Let Δ be a lattice polytope such that $0 \notin \operatorname{aff} \Delta$ (affine hull). We consider the cone $\sigma_\Delta := \{tx \mid t \in \mathbb{R}_+ \backslash \{0\}, \mid x \in \Delta\}$ and set

$$P_\Delta := \sigma_\Delta \cap \mathbb{Z}^n.$$

Then P_Δ is a semigroup of lattice vectors (FIGURE 1). It represents the subring R_Δ of $\mathbb{C}[z_1, z_1^{-1}, \ldots, z_n, z_n^{-1}]$ consisting of all Laurent polynomials whose support lies in P_Δ. The use of semigroups P_Δ is a second key to understanding what follows: transformation of algebraic or algebraic-geometric facts into combinatorial-geometric ones.

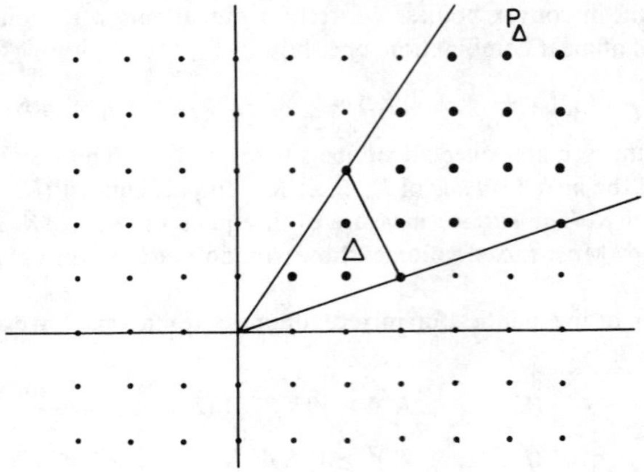

FIGURE 1

In a sense, the correspondence $R_\Delta \to P_\Delta$ is a logarithmic transformation. How it can be applied has been thoroughly studied by Hochster [8], Kempf *et al.* [9], Danilov [5], and others. Any time algebraic-geometric properties about singularities or blowups can be defined in local analytic coordinates by monomials, one may have a chance to apply the combinatorial-geometric technique of exponent semigroups. In [9], for example, this is done in a proof of a theorem about the existence of semistable reductions. The nucleus of the proof is a (55 page) proof of variants of the following combinatorial theorem about lattice points:

Let P be a lattice polytope of dimension n in $\mathbb{R}^n$. *Then there exist a* $v \in N$ *and a subdivision of P into finitely many simplices* T_j, *such that for all j*:

(a) *Every vertex of* T_j *lies in* $(1/v)\mathbb{Z}^n$.
(b) *The volume of* T_j *equals* $1/v^n n!$.

3

Let us discuss in more detail a geometric characterization of Milnor's number of an isolated critical point of an algebraic set defined by a polynomial equation $f(z) = 0$ (Kouchnirenko [11]). What is Milnor's number? In order to understand its geometric meaning let us first look at an example.

$$f(z_1, z_2) = z_1^p + z_2^q = 0, \qquad p, q \in N \backslash \{0, 1\}, \tag{1}$$

defines in $\mathbb{C}^2$ or $\mathbb{R}^4$ an algebraic set V. Both partial derivatives vanish at 0; f has an isolated critical point there. In order to investigate the critical point further, we consider the 3-sphere S_ε about 0 with radius $\varepsilon > 0$. The intersection $C := S_\varepsilon \cap V$ is a knotted curve or a link of several curves. $S_\varepsilon \backslash V$ can be considered a fiber space by setting

$$\phi\colon S_\varepsilon \backslash V \to \mathbb{C}; \qquad \phi(z) := \frac{f(z)}{|f(z)|}. \tag{2}$$

A fiber $F_\theta := \phi^{-1}(e^{i\Theta})$ is a surface homotopically equivalent to a bouquet $S^1 \vee \cdots \vee S^1$ of $\mu(f)$ circles. We now replace (**1**) by any polynomial equation

$$f(z) := f(z_1, \ldots, z_n) = 0, \qquad n \geq 2, \tag{3}$$

such that 0 is again an isolated critical point of the algebraic set V defined by (**3**). Let S_ε be a $(2n-1)$-sphere about 0 whose radius $\varepsilon > 0$ is so small that inside S_ε there is no critical point $\neq 0$ and such that $f^{-1}(0)$ is a transversal to $S_{\varepsilon'}$ for all $0 < \varepsilon' < \varepsilon$. The map (**2**) provides a fibration of $S_\varepsilon \backslash V$, the fiber being homotopic to a bouquet $S^{2n-3} \vee \cdots \vee S^{2n-3}$ of $\mu(f)$ spheres. $\mu(f)$ is called *Milnor's number* of the isolated critical point 0 of f (see [13]).

It is possible to calculate $\mu(f)$ directly from f (Palamodov [15]):

$$\mu(f) = \dim_{\mathbb{C}} \mathbb{C}[[z_1, \ldots, z_n]] \Big/ \left(\frac{\partial f}{\partial z}\right),$$

where $(\partial f/\partial z)$ denotes the ideal generated by all partial derivatives of f. In example (**1**), $\mu(f) = (p-1)(q-1)$ is easily obtained.

Following the work of Kouchnirenko [11], let us look at the Newton polytope $\mathcal{N}\!w(f)$ of f. We call f *permissible* if $\mathcal{N}\!w(f)$ touches all coordinate axes. The union of all faces Δ of $\mathcal{N}\!w(f)$ which are "visible" from 0, that is, for which $\mathcal{N}\!w(f) \cap \operatorname{conv}(\{0\} \cup \Delta) = \Delta$, 0 not in the affine hull of Δ, is called *Newton's boundary* $\Gamma(f)$ of f (FIGURE 2). f is called *nondegenerate* on $\Gamma(f)$ if $(z_1(\partial f/\partial z_1), \ldots, z_n(\partial f/\partial z_n))$ restricted to $\Gamma(f)$ is $\neq 0$ in $(\mathbb{C}\backslash\{0\})^n$.

We set $\Gamma_-(f) := \bigcup \operatorname{conv}(\{0\} \cup \Delta)$ for all $\Delta \subset \Gamma(f)$. Let V_n be the n-dimensional volume of $\Gamma_-(f)$, and let

$$V_k := \sum \operatorname{vol}_k (\Gamma_-(f) \cap U^k),$$

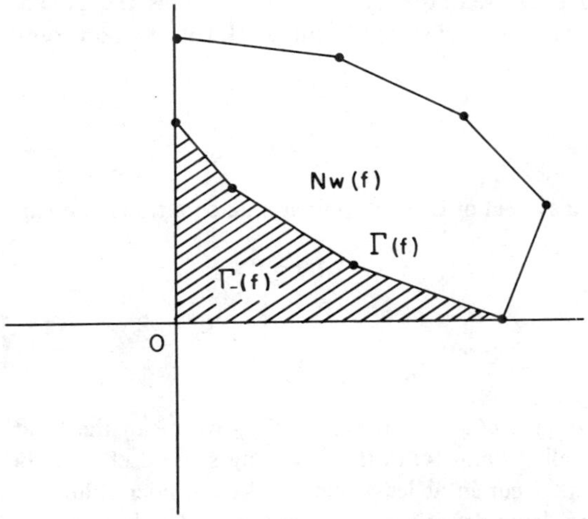

FIGURE 2

where the sum is taken over all k-dimensional coordinate subspaces U^k, $k < n$. We define

$$\nu(f) := n!\,V_n - (n-1)!\,V_{n-1} \pm \cdots + (-1)^{n-1} 1!\,V_1 + (-1)^n$$

to be *Newton's number.*

THEOREM 1 (Kouchnirenko [11]). *Let f be a permissible polynomial, and let* (3) *have 0 as an isolated critical point. Then*

(a) $\mu(f) \geq \nu(f)$,

(b) $\mu(f) = \nu(f)$ *iff f is nondegenerate on* $\Gamma(f)$.

It should be noted that (a) and (b) depend only on $\mathcal{N}\omega(f)$, not on supp f.

A related result is obtained as follows. Let $\tilde{\mu}(f)$ denote the sum of all Milnor numbers of f. Then

$$\tilde{\mu}(f) = \dim_{\mathbb{C}} \mathbb{C}[z_1, \ldots, z_n] \bigg/ \left(\frac{\partial f}{\partial z}\right).$$

We set $\tilde{\Gamma}_-(f) := \text{conv}(\{0\} \cup \mathcal{N}\omega(f))$ and denote by $\tilde{\Gamma}(f)$ the union of all faces of $\tilde{\Gamma}_-(f)$ that do not contain 0. If, in the definition of $\nu(f)$, we replace $\Gamma_-(f)$, $\Gamma(f)$ by $\tilde{\Gamma}_-(f)$, $\tilde{\Gamma}(f)$, respectively, we obtain a number $\tilde{\nu}(f)$.

THEOREM 2 (Kouchnirenko [11]). *Under the assumptions of* THEOREM 1 *we have*

(a) $\tilde{\mu}(f) \leq \tilde{\nu}(f)$,

(b) $\tilde{\mu}(f) = \tilde{\nu}(f)$ *iff f is nondegenerate on* $\tilde{\Gamma}(f)$.

In the above example (1), the Newton polytope $\mathcal{N}\omega(f)$ is the line segment joining $(p, 0)$ and $(0, q)$. We find $V_2 = \frac{1}{2}pq$, $V_1 = p + q$; hence, $\nu(f) = pq - p - q + 1 = (p-1)(q-1)$. Since f is nondegenerate on $\Gamma(f) = \tilde{\Gamma}(f)$, we obtain $\mu(f) = (p-1)(q-1) = \tilde{\mu}(f)$. ($f$ only has 0 as a critical point.)

Kouchnirenko has extended THEOREM 2 to Laurent polynomials. The proof of all three theorems makes use of the theory of graded rings, the Cohen–Macaulayness of the rings R_Δ (proved by Hochster [8]), and work with Koszul complexes.

4

We consider a system of Laurent polynomial equations in n variables

$$\begin{aligned} f_1(z) &= 0 \\ &\vdots \\ f_n(z) &= 0. \end{aligned} \tag{4}$$

By the *number of typical solutions* $\ell(f_1, \ldots, f_n)$ we mean the total number of solutions for almost all systems (4) in the following sense. Let n be the total number of lattice points that occur in at least one f_i. We consider solutions that occur in an open, everywhere dense set of the coefficient space $\mathbb{C}^n$; hereby $\mathcal{N}\omega(f_1), \ldots, \mathcal{N}\omega(f_n)$ is the same set of polytopes. Furthermore, we choose only solutions in $(\mathbb{C}\backslash\{0\})^n$.

THEOREM 3 (Bernstein [1]). *The typical number of solutions of* (4) *equals* $n!$ *times the mixed volume of the Newton polytopes of the* f_i:

$$\ell(f_1, \ldots, f_n) = n!\, V(\mathcal{N}\omega(f_1), \ldots, \mathcal{N}\omega(f_n)). \tag{5}$$

Again, there is a dependence only on $\mathcal{N}\omega(f_j)$, not on supp $f_j, j = 1, \ldots, n$.

EXAMPLE. Let two quadratic equations be given:

$$f_1(z_1, z_2) = a_1 z_1^2 + a_2 z_2^2 + a_3 = 0,$$
$$f_2(z_1, z_2) = b_1 z_1^2 + b_2 z_2^2 + b_3 = 0$$

all coefficients being $\neq 0$. The typical number of solutions is obviously 4 (FIGURE 3).

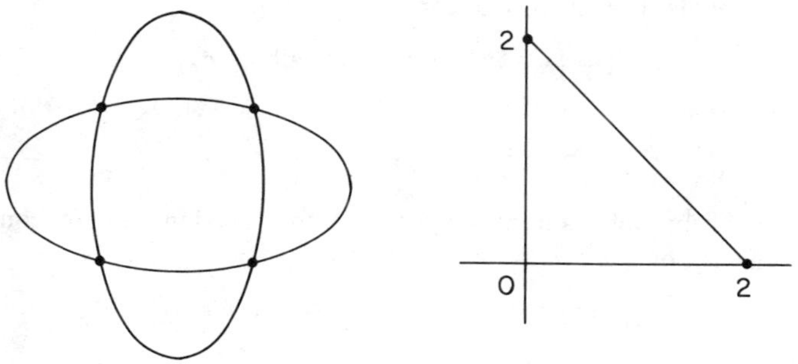

FIGURE 3

On the other hand, $\mathcal{N}\omega(f_1) = \mathcal{N}\omega(f_2)$ is the triangle with vertices (0, 0), (2, 0), and (0, 2). We have $V(\mathcal{N}\omega(f_1), \mathcal{N}\omega(f_2)) = V(\mathcal{N}\omega(f_1)) = 2$ (area of the triangle). Hence, $2!\, V(\mathcal{N}\omega(f_1), \mathcal{N}\omega(f_2)) = 4$.

Bernstein's proof of THEOREM 3 makes use of Puiseux series. Meanwhile, there is an alternative proof. Burago and Zalgaller mention in their book "Geometric Inequalities" [4] the possibility of proving (5) by characterizing the functionals on both sides of (5) uniquely by certain properties. Such properties have been found independently by W. J. Firey and by P. McMullen (unpublished).

When (5) *is established it is possible, by elementary arguments, to prove the Alexandrov–Fenchel inequality* (*see Section* 2) *for lattice polytopes and, by an approximation argument, extend the proof to arbitrary convex bodies.*

[**Note added in proof:** Meanwhile, the elementary proof by Phedotow published in [4] turned out to be false.]

5

A further characterization of mixed volumes and a proof of the Alexandrov–Fenchel inequality use Hodge Theory. Both have been achieved by Teissier [18, 19].

Let $K_1, \ldots, K_r$ be lattice polytopes in $\mathbb{R}^n$, $n \geq 2$. We consider the support function H of $K := K_1 + \cdots + K_r$ (Minkowski sum):

$$H(u) := \max_{x \in K} u \cdot x.$$

H is piecewise linear and convex. The same is true for the support functions H_i of $K_i, i = 1, \ldots, r$.

Let Δ be a face of K, and let p be a point in the relative interior of Δ. We denote the cone of "outer normals" in p by σ_Δ, that is,

$$\sigma_\Delta := \{u - p \mid \|u - p\| \leq \|u - q\| \text{ for all } q \in K\}.$$

σ_Δ only depends on Δ, not on p. It is readily seen that $\{\sigma_\Delta\}$, for $\Delta \neq \phi$, being a face of K, is a decomposition of $\mathbb{R}^n$ such that no two σ_Δ have a relative interior point in common. We denote the convex dual cone of σ_Δ by $\check{\sigma}_\Delta$:

$$\check{\sigma}_\Delta := \{x \in \mathbb{R}^n \mid u \cdot x \geq 0 \text{ for all } u \in \sigma_\Delta\}.$$

$\check{\sigma}_\Delta$ has dimension n.

$$R_\Delta := \mathbb{C}[\check{\sigma}_\Delta \cap \mathbb{Z}^n]$$

defines a subalgebra of Laurent polynomials (see Section 2). We consider the sub-R_Δ-module $L_{i\Delta}$ of $\mathbb{C}(\mathbb{Z}^n)$ (FIGURE 4) generated by

$$L'_{i\Delta} := \{x \in \mathbb{Z}^n \mid u \cdot x \geq H_i(u) \text{ for all } u \in \sigma_\Delta\}.$$

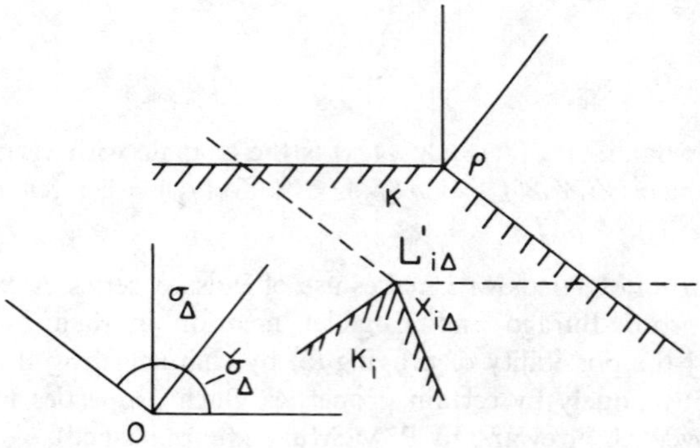

FIGURE 4

We can describe $L'_{i\Delta}$ geometrically as follows. An elementary property of convex polytopes is that each face Δ of K is the Minkowski sum of faces Δ_i of K_i,

$$\Delta = \Delta_1 + \cdots + \Delta_r.$$

We choose a vertex $m_{i\Delta}$ of Δ_i. Then $L'_{i\Delta}$ is obtained from $\check{\sigma}_\Delta \cap \mathbb{Z}^n$ by a translation: $L'_{i\Delta} = (\check{\sigma}_\Delta \cap \mathbb{Z}^n) + m_{i\Delta}$. Adding any vector of $\check{\sigma}_\Delta \cap \mathbb{Z}^n$ to $L'_{i\Delta}$ maps $L'_{i\Delta}$ into itself; therefore, $L_{i\Delta}$ is a sub-R_Δ-module of $\mathbb{C}(\mathbb{Z}^n)$. Furthermore, any vector of $L'_{i\Delta}$ is representable as $y = x + m_{i\Delta}$, where $x \in \check{\sigma}_\Delta \cap \mathbb{Z}^n$; so $z^{m_{i\Delta}}$ generates $L_{i\Delta}$.

Using sheaf-theoretic means one can glue up the affine varieties Spec $\mathbb{C}[\check{\sigma}_\Delta \cap \mathbb{Z}^n]$ (points, irreducible curves, irreducible surfaces, etc., represented by prime ideals) to a compact, normal, integral, and rational algebraic variety X. Hereby one takes advantage of the complex structure of $\{\sigma_\Delta\}$. Also the $L_{i\Delta}$ with any fixed i glue up together into an invertible sheaf of fractional ideals L_i. Let $L_j^{v_j}$ denote the v_j-fold tensor product of L_j. As is known from sheaf theory, there is a polynomial expression for the coherent Euler-characteristic of $L_1^{v_1} \otimes \cdots \otimes L_r^{v_r}$:

$$\chi(X, L_1^{v_1} \otimes \cdots \otimes L_r^{v_r}) = \sum_{\substack{\alpha \in N^r \\ |\alpha| = n}} \frac{1}{\alpha_1! \cdots \alpha_r!} S_\alpha v_1^{\alpha_1} \cdots v_r^{\alpha_r}$$

$$+ \text{ polynomial of degree } \leq n - 1.$$

It is now readily shown that

$$S_\alpha = n!\, V(\underbrace{K_1, \ldots, K_1}_{\alpha_1}, \ldots, \underbrace{K_r, \ldots, K_r}_{\alpha_r}).$$

This is another characterization of mixed volumes. *The Alexandrov–Fenchel inequality now follows from the Hodge Index Theorem* (Teissier [17]), according to which the (topological) index of a Kähler manifold M coincides with an alternating sum of complex cohomology group dimensions of M.

References

1. Bernstein, D. N. 1976. The number of roots of a system of equations. Functional Anal. Appl. **9:** 183–185.
2. Billera, L. J. & C. W. Lee. 1981. A proof of McMullen's conditions for f-vectors of simplicial convex polytopes. J. Combin. Theory Ser. A **31:** 237–255.
3. Bonnessen, T. & W. Fenchel. 1934. Theorie der Konvexen Körper. Ergebnisse der Mathematik Vol. 3. Springer-Verlag. Berlin, Germany.
4. Burago, D. M. & V. A. Zalgaller. 1980. Geometric Inequalities. Leningrad, USSR.
5. Danilov, V. I. 1978. The geometry of toric varieties. Russian Math. Surveys **33:** 97–154.
6. Ewald, G., P. Kleinschmidt, U. Pachner & C. Schulz. Recent results in combinatorial convexity. *In* Contributions to Geometry. Proceedings of the Geometry Symposium in Sigen 1978. J. Tölke & J. M. Wills, Eds.
7. Grünbaum, B. 1967. Convex Polytopes. London, England.
8. Hochster, M. 1972. Rings of invariants of tori, Cohen–Macaulay rings generated by monomials, and polytopes. Ann. of Math. **96:** 318–337.
9. Kempf, G., F. Knudsen, D. Mumford & B. Saint-Donat. 1973. Toroidal Embeddings. Lecture Notes in Mathematics No. 339. Springer-Verlag. West Berlin, FRG.
10. Kind, B. & P. Kleinschmidt. 1979. Schälbare Cohen–Macaulay-Komplexe und ihre Parametrisierung. Math. Z. **167:** 173–179.
11. Kouchnirenko, A. G. 1976. Polyèdres de Newton et nombres de Milnor. Invent. Math. **32:** 1–31.
12. Leichtweiss, K. 1980. Konvexe Mengen. Berlin.
13. Milnor, I. 1968. Singular Points of Complex Hypersurfaces. Annals of Mathematical Studies No. 61. Princeton, N.J.
14. McMullen, P. & G. C. Shephard. 1971. Convex Polytopes and the Upper Bound Conjecture. Cambridge.
15. Palamodov, V. P. 1967. On the multiplicity of a holomorphic mapping. Funkcional Anal. i Priložen **1:** 54–65.
16. Stanley, R. 1975. The upper bound conjecture and Cohen–Macaulay rings. Stud. Appl. Math. **54:** 135–142.

17. STANLEY, R. 1980. The number of faces of a simplicial convex polytope. Adv. in Math. **35:** 236–238.
18. TEISSIER, B. 1979. Du Theoreme de l'Index de Hodge aux inègalite's isoperimetriques. Note C. R. Acad. Sci. Paris **288:** 287–289.
19. TEISSIER, B. 1982. Bonnesen-type inequalities in algebraic geometry. I. Introduction to the problem. *In* Proceedings, Seminar on Differential Geometry. Pp. 85–105. Princeton, N.J.

TRIANGULATING THE d-CUBE

Carl W. Lee[a]

Department of Mathematics
University of Kentucky
Lexington, Kentucky 40506

Introduction

Let $I = [0, 1]$ be the unit interval. By a *triangulation* of the d-cube we mean a triangulation of I^d into (closed) simplices whose vertices are vertices of the cube itself. Mara [4, 5] described a triangulation of the 4-cube having 16 4-simplices and a triangulation of the 5-cube having 68 5-simplices, conjecturing that these triangulations are minimal and suggesting a formula for the minimum number of d-simplices in any triangulation of the d-cube. Sallee [9] disproved Mara's conjecture by exhibiting a new triangulation of I^d having fewer d-simplices than Mara's conjectured formula for $d \geq 5$. Both Sallee [8, 9] and Cottle [1] have demonstrated that Mara's triangulation of the 4-cube is minimal. In this paper we give an alternate proof of the minimality of Mara's 4-cube triangulation, rederive Sallee's method of triangulation by showing its relationship to pulling vertices of a polytope, and briefly mention another technique of triangulating polytopes.

1. Minimal Triangulations of the 4-Cube

We first give an alternate proof of the fact that any triangulation of I^4 requires at least 16 4-simplices.

Let Δ be any finite collection of simplices. For $F \in \Delta$ we call F a *face* of Δ and write dim F for the dimension of F. If dim $F = j$ we will call F a *j-face* of Δ. Faces of Δ of dimension 0 and 1 will be called *vertices* and *edges* of Δ, respectively. Taking $d = \max \{\dim F: F \in \Delta\}$, define

$$f(\Delta) = (f_{-1}(\Delta), f_0(\Delta), \ldots, f_d(\Delta)),$$

where

$$f_j(\Delta) = \text{card } \{F \in \Delta: \dim F = j\}, \qquad -1 \leq j \leq d$$

(regarding the empty set as a simplex of dimension -1), and

$$h(\Delta) = (h_0(\Delta), h_1(\Delta), \ldots, h_{d+1}(\Delta)),$$

where

$$h_i(\Delta) = \sum_{j=0}^{i} (-1)^{i-j} \binom{d-j+1}{d-i+1} f_{j-1}(\Delta), \qquad 0 \leq i \leq d+1.$$

[a] Research supported in part by NSF Grant MCS-8201653.

Set $h_i(\Delta) = 0$ if $i < 0$ or $i > d + 1$. It is known [6] that $f(\Delta)$ can be recovered from $h(\Delta)$ using the relations

$$f_j(\Delta) = \sum_{i=0}^{j+1} \binom{d-i+1}{d-j} h_i(\Delta), \qquad -1 \leq j \leq d.$$

PROPOSITION 1.1. *Suppose Δ is a triangulation of the d-cube, $\partial\Delta$ is the induced triangulation of ∂I^d, and Δ° is the set of those faces of Δ that are not in $\partial\Delta$. Then* (with the convention that $\phi \in \partial\Delta$):

(i) $h_i(\partial\Delta) = h_{d-i}(\partial\Delta)$, $0 \leq i \leq d$,
(ii) $h_i(\Delta^\circ) = h_{d-i+1}(\Delta)$, $0 \leq i \leq d + 1$,
(iii) $h_i(\partial\Delta) - h_{i-1}(\partial\Delta) = h_i(\Delta) - h_{d-i+1}(\Delta)$, $0 \leq i \leq d + 1$.

Proof. See [6]. ▮

We remark that PROPOSITION 1.1 holds more generally for all simplicial complexes Δ that are triangulations of homology d-balls.

EXAMPLE. Let Δ be the standard triangulation of the 4-cube having $4! = 24$ 4-simplices [5, 11]. Then,

$$f(\Delta) = (1, 16, 65, 110, 84, 24), \qquad h(\Delta) = (1, 11, 11, 1, 0, 0),$$

$$f(\Delta^\circ) = (0, 0, 1, 14, 36, 24), \qquad h(\Delta^\circ) = (0, 0, 1, 11, 11, 1),$$

$$f(\partial\Delta) = (1, 16, 64, 96, 48), \qquad h(\partial\Delta) = (1, 12, 22, 12, 1).$$

For any two 0-1 vectors v, w of the same length, we define $d(v, w)$ to be i if exactly i of the coordinates of v and w are different. For example, if $v = (0, 1, 0, 0)$ and $w = (0, 0, 1, 1)$, then $d(v, w) = 3$. Further, if $d(v, w) = 1$, we will say that v and w are *neighbors*.

Let Δ be a triangulation of I^4, and v, w be vertices of I^4. An edge of Δ joining v and w will be said to be of *type i* if $d(v, w) = i$.

LEMMA 1.2. *Suppose Δ is a triangulation of I^4. Let F be a 4-simplex of Δ and $V(F)$ be the set of the five vertices of I^4 contained in F. Assume that $d(v, w) \leq 2$ for all v, $w \in F$. Then $V(F)$ consists of a vertex of I^4 together with its four neighbors in I^4.*

Proof. Assume $d(v, w) = 2$ for all $v, w \in V(F)$. Then all of the vertices of F must be of the same parity. But then card $V(F) = 5$ forces $V(F)$ to contain an antipodal pair (v, w) of vertices, i.e., such that $d(v, w) = 4$, contradicting our assumption. Thus there must exist $u^1, u^2, u^3 \in V(F)$ such that $d(u^1, u^2) = 1$ and $d(u^2, u^3) = 2$. Without loss of generality, let us say

$$u^1 = (0, 0, 0, 0),$$

$$u^2 = (1, 0, 0, 0),$$

$$u^3 = (0, 1, 0, 0).$$

Now there must be a $u \in V(F)$ of the form $(*, *, 1, *)$. Without loss of generality, we may say

$$u^4 = (0, 0, 1, 0).$$

Again, there must be a $u \in V(F)$ of the form $(*, *, *, 1)$, necessarily

$$u^5 = (0, 0, 0, 1).$$

Thus, $V(F)$ consists of the vertex $(0, 0, 0, 0)$ together with its four neighbors. ▮

LEMMA 1.3. *Suppose Δ is a triangulation of I^4. Then Δ must contain at least one type 4 edge or three type 3 edges.*

Proof. The 4-simplices of Δ cannot include more than eight 4-simplices of the kind described in LEMMA 1.2, since each such simplex contains exactly 4 of the 32 original edges of I^4, and in no triangulation of I^4 may any of these original edges be contained in more than one such 4-simplex. Removing from Δ all such 4-simplices yields Σ, a simplicial complex with at least eight vertices. Every 4-simplex of Σ must contain at least one type 3 edge or one type 4 edge. If any 4-simplex of Σ contains a type 4 edge we are done, so assume that Σ has no type 4 edge. Let F be any 4-simplex of Σ. Without loss of generality, suppose $u^1, u^2 \in V(F)$, where

$$u^1 = (0, 0, 0, 0),$$

$$u^2 = (1, 1, 1, 0).$$

Now there must be a $u \in V(F)$ of the form $(*, *, *, 1)$, and it will be impossible to satisfy both $d(u, u^1) \leq 2$ and $d(u, u^2) \leq 2$. Thus we may, without loss of generality, take $u^3 = (1, 1, 0, 1)$, leaving $(0, 1, 0, 0)$, $(1, 0, 0, 0)$, and $(1, 1, 0, 0)$ as candidates for the remaining two vertices of F, assuming that F has no other type 3 edges. If (u^1, u^2) and (u^1, u^3) are the only two type 3 edges in Σ, then all 4-simplices of Σ must contain u^1, u^2, and u^3, forcing Σ to have at most six vertices, which is a contradiction. Therefore Δ must have at least three edges of type 3. ▮

THEOREM 1.4. *Suppose Δ is a triangulation of I^4. Then $f_4(\Delta) \geq 16$.*

Proof. First note that

$$f_{-1}(\partial\Delta) = 1 = f_{-1}(\Delta),$$

$$f_0(\partial\Delta) = 16 = f_0(\Delta),$$

$$f_{-1}(\Delta^\circ) = 0 = f_0(\Delta^\circ).$$

Now I^4 contains 32 edges (which are type 1 edges in Δ), and each of the 24 squares (2-faces) of I^4 must be triangulated by adding a diagonal (type 2 edge). Thus,

$$f_1(\partial\Delta) \geq 56.$$

We may then use PROPOSITION 1.1 to conclude

$$f(\partial\Delta) = (1, 16, a, *, *),$$

where $f_1(\partial\Delta) = a \geq 56$,

$$h(\partial\Delta) = (1, 12, b, 12, 1),$$

where $b = a - 42$,

$$h(\Delta) = (1, 11, c, d, 0, 0),$$

where $d = f_1(\Delta^\circ) \geq 0$ and $c - d = b - 12$, and

$$\begin{aligned} f_4(\Delta) &= \sum_{i=0}^{5} h_i(\Delta) \\ &= 12 + c + d \\ &= b + 2d \\ &= f_1(\partial\Delta) + 2f_1(\Delta^\circ) - 42. \end{aligned}$$

If Δ contains three type 3 edges, then $f_1(\partial\Delta) \geq 59$ and hence $f_4(\Delta) > 17$. If, on the other hand, Δ contains a type 4 edge, then $f_1(\Delta^\circ) \geq 1$ and $f_4(\Delta) \geq 16$. ∎

COROLLARY 1.5. *Every minimal triangulation of I^4 must contain a type 4 edge (i.e., an edge joining an antipodal pair of vertices of the 4-cube).*

We have as a consequence of arguments in [1] and the above corollary that there is no triangulation of the 4-cube having exactly nine 4-simplices of volume 1/24, six 4-simplices of volume 2/24, and one 4-simplex of volume 3/24.

2. TRIANGULATING ARBITRARY POLYTOPES

When a d-polytope P is a facet (d-face) of some $(d + 1)$-polytope Q, making Q simplicial by *pulling its vertices* [2, Section 5.2] will induce a triangulation of P. It is not necessary, however, to regard P as a facet of any $(d + 1)$-polytope to describe this triangulation of P directly: Let $V(P) = \{v^0, v^1, \ldots, v^n\}$ be the set of vertices of P. We obtain successive subdivisions $S_{-1}, S_0, \ldots, S_n$ of P by letting S_{-1} be P itself, and for $0 \leq k \leq n$, $0 \leq j \leq d$, letting the j-faces in S_k be

(1) the j-faces in S_{k-1} which do not contain v^k, and

(2) the j-faces of the form conv $(\{v^k\} \cup F)$, where F is a $(j - 1)$-face not containing v^k of a d-face in S_{k-1} which does contain v^k.

We denote S_n by $S(P)$, or by $S(P; v^0, v^1, \ldots, v^n)$ when we want to acknowledge the dependence of the subdivision upon the particular ordering of the vertices of P explicitly. Exactly as in the case of pulling vertices of a polytope, every j-face of $S(P)$ is a j-simplex, $0 \leq j \leq d$.

PROPOSITION 2.1. *Every d-simplex in $S(P; v^0, \ldots, v^n)$ is of the form* conv *$(\{v^0\} \cup G)$, where G is a $(d - 1)$-simplex in $S(F; v^{i_0}, \ldots, v^{i_m})$, for some $(d - 1)$-face F of P that does not contain v^0.*

Proof. It is immediate from the definition that for every j-face F of P, the triangulation of F induced by $S(P; v^0, \ldots, v^n)$ is precisely $S(F; v^{i_0}, \ldots, v^{i_m})$, regarding F itself as a j-polytope with vertex set $V(F) = \{v^{i_0}, \ldots, v^{i_m}\} \subset V(P)$, $i_0 < i_1 < \cdots < i_m$. Now, every d-face in S_0 is a pyramid with apex v^0 over some $(d - 1)$-face of P. Also, every d-dimensional face in any subdivision of such a pyramid must contain the apex of the pyramid. Therefore, every d-simplex in S_n contains v^0. ∎

For every j-face of P, $0 \leq j \leq d$, let $v(F)$ denote the vertex of P of smallest index that is in F. Let $P = F_d \supset F_{d-1} \supset \cdots \supset F_1 \supset F_0$ be a chain of faces of P such that F_j is a j-face of P and $v(F_j) \neq v(F_{j+1})$, $0 \leq j \leq d - 1$. Let $\mathscr{C}$ be the set of all such chains.

PROPOSITION 2.2. *W is the set of vertices of some d-simplex in* $S(P; v^0, \ldots, v^n)$ *if and only if* $W = \{v(F_d), v(F_{d-1}), \ldots, v(F_0)\}$ *for some chain* $F_d \supset F_{d-1} \supset \cdots \supset F_0$ *in* $\mathscr{C}$.

Proof. Use PROPOSITION 2.1 and induction on $d = \dim P$. ▌

PROPOSITION 2.2 implies that $S(P)$ is the same as the *full-flag triangulation* of P discussed by both Stanley [10] and Sallee, and its equivalence with the triangulation of Von Hohenbalken [12] is clear.

EXAMPLE. Consider the case $P = I^d$ with the vertices of the d-cube indexed according to the binary representation of the index; e.g., if $d = 4$, then $v^{11} = (1, 0, 1, 1)$. It is an easy exercise to see that $S(I^d)$ is the standard triangulation of the d-cube into $d!$ d-simplices [5, 11]. In fact, the number of d-simplices obtained is independent of the order in which the vertices of the d-cube are pulled.

3. SALLEE'S TRIANGULATION OF THE d-CUBE

We now turn to the d-cube and describe a triangulation due to Sallee and independently studied by the author.

For any 0-1 vector $v = (v_1, v_2, \ldots, v_d)$ we define the *parity* of v to be *odd* if v has an odd number of components equal to 1 and *even* if v has an even number of components equal to 1. Choose some $u \in V(I^d)$. The simplex whose vertex set is $\{u\} \cup \{v \in V(I^d)\colon v \text{ is a neighbor of } u\}$ will be called a simplex of *type I* with *apex u*. Let $d \geq 3$. We label the vertices of the d-cube according to the binary representation of the index exactly as in the example of Section 2. We begin our triangulation of I^d by "slicing off" from the d-cube each of the type I simplices with odd parity apex, leaving a d-polytope which we will denote Q_d; i.e., $Q_d = \operatorname{conv}\{v\colon v \text{ is an even vertex of } I^d\}$. (If $d = 3$, then Q_d is itself a 3-simplex.) Maintaining the original labeling of the vertices, we now obtain the triangulation $S(Q_d)$ of Q_d. The d-simplices of $S(Q_d)$ will be said to be of *type II*. Together the odd apex type I d-simplices and the type II d-simplices in $S(Q_d)$ constitute the d-simplices of a triangulation T_d of I^d.

Note that every $(d-1)$-face of Q_d is either (a) the convex hull of the set of neighbors of some odd vertex of I^d, or (b) congruent to Q_{d-1}, case (b) being possible only if $d \geq 4$.

Let $T(d)$ be the number of d-simplices in T_d and $\tilde{T}(d)$ be the number of d-simplices in $S(Q_d)$. Because T_d contains 2^{d-1} d-simplices of type I, it is clear that $T(d) = \tilde{T}(d) + 2^{d-1}$. Observe that $T(3) = 5$. Using PROPOSITION 2.1 and the above facts, we can determine a recursive formula for $T(d)$. Suppose $d \leq 4$. To begin, we have 2^{d-1} d-simplices of type I. Now Q_d has 2^{d-1} $(d-1)$-faces of type (a), each of which is a $(d-1)$-simplex, d of which contain v^0. This gives us $2^{d-1} - d$ d-simplices of $S(Q_d)$. On the other hand, there are $2d$ $(d-1)$-faces of type (b) of Q_d, half of which contain v^0. This gives us the remaining $d\tilde{T}(d-1)$ d-simplices of $S(Q_d)$. Therefore $T(d) = 2^{d-1} + 2^{d-1} - d + d\tilde{T}(d-1)$, implying:

THEOREM 3.1.

(i) $T(d) = dT(d-1) - d2^{d-2} + 2^d - d$,

(ii) $T(d) = d!\,p_d(2)/2 - d!\,p_d(1) + 2^{d-1} - d!/2 + 1$,

where $p_d(x) = \sum_{i=0}^{d} (x^i/i!)$.

Proof. Part (i) is immediate from the preceding comments, and part (ii) follows by induction on $d \geq 3$. ▮

Mara [5] conjectured that any triangulation of the d-cube has at least $M(d) = (2^{d-1} + d!)/2$ d-simplices, having constructed triangulations for $d = 4$ and $d = 5$ with these numbers of d-simplices; however, $T(d)$ is strictly smaller than $M(d)$ for $d \geq 5$, a fact that can be verified, for example, by comparing THEOREM 3.1(i) with the recursive formula $M(d) = dM(d-1) + 2^{d-2} - d2^{d-3}$. In particular, $T(5) = 67$. Todd [11] suggests that one comparison of a triangulation of the d-cube with the standard $d!$ triangulation (with respect to pivoting algorithms) is the extent to which the quantity $R(d) = (T(d)/d!)^{1/d}$ is less than 1. From THEOREM 3.1(ii), however, we see that $\lim_{d\to\infty} T(d)/d! = e^2/2 - e - \frac{1}{2} \simeq 0.476$ and so $\lim_{d\to\infty} R(d) = 1$.

Below are the values of $T(d)$ and $R(d)$ for $3 \leq d \leq 10$.

d	$d!$	$T(d)$	$R(d)$
3	6	5	0.941
4	24	16	0.904
5	120	67	0.890
6	720	364	0.893
7	5,040	2,445	0.902
8	40,320	19,296	0.912
9	362,880	173,015	0.921
10	3,628,800	1,728,604	0.929

An explicit description of the vertex sets of the d-simplices in T_d and a procedure for pivoting from simplex to simplex are informally discussed in [3].

4. ANOTHER TRIANGULATION METHOD: "PLACING" VERTICES

Another way to triangulate polytopes is described by Munson [7] in a more general context. The vertices of the polytope are successively "placed" into position, and at each stage the convex hull of the currently placed vertices is provided with a triangulation.

Again, let $V(P) = \{v^0, v^1, \ldots, v^n\}$. We obtain a sequence of polytopes $P_0, P_1, \ldots, P_n = P$ with respective dimensions $d_0, d_1, \ldots, d_n$ and respective triangulations $S'_0, S'_1, \ldots, S'_n$ as follows: Let $P_0 = \{v^0\}$, itself a simplex of dimension 0. For $0 < k \leq n$, let $P_k = \text{conv}\,(P_{k-1} \cup \{v^k\})$ and $d_k = \dim P_k$. If $d_k > d_{k-1}$ (i.e., $d_k = d_{k-1} + 1$), then for $0 \leq j \leq d_k$ the j-faces in S'_k are

(1) the j-faces in S'_{k-1}, and

(2) the j-faces of the form $\text{conv}\,(F \cup \{v^k\})$, where F is a $(j-1)$-face of S'_{k-1}.

If, on the other hand, we have $d_k = d_{k-1}$, then for $0 \leq j \leq d_k$ the j-faces in S'_k are

(1) the j-faces in S'_{k-1}, and

(2) the j-faces of the form $\text{conv}\,(F \cup \{v^k\})$, where F is a $(j-1)$-face of S'_{k-1} that is contained in a facet of P_{k-1} that v^k is "beyond" [relative to aff (P_{k-1})] [2, Section 5.2].

Intuitively, as each new vertex v^k is "placed," the triangulation S'_{k-1} is extended by constructing a pyramid with apex v^k and base F for every face $F \in S'_{k-1}$ that is "visible" from v^k. As with pulling vertices there is an explicit dependence upon the order in which the vertices are placed so we may sometimes wish to write $S'(P) = S'(P; v^0, \ldots, v^n)$. S' shares with S the property that for every j-face F of P, the triangulation of F induced by $S'(P; v^0, \ldots, v^n)$ is $S'(F; v^{i_0}, \ldots, v^{i_m})$, regarding F as a j-polytope with vertices $\{v^{i_0}, \ldots, v^{i_m}\}$, $i_0 < i_1 < \cdots < i_m$. Unlike S, however, the number of d-simplices in $S'(I^d)$ changes with the order in which the vertices of the d-cube are placed. For example, if the even parity vertices of the 3-cube are placed before the odd parity vertices, a minimal triangulation is obtained. Similarly, minimal triangulations of I^4 can be obtained directly by an appropriate placement order, without recourse to "slicing off" the odd parity vertices first. Whether the same holds for higher dimensions remains unexplored.

5. Remarks

That T_d is not a minimal triangulation of the d-cube in general has been shown by Sallee [9] with his triangulation of I^6 requiring 344 simplices. The nonminimality of T_d is not surprising in view of the fact that the only aspect of the triangulation that specifically depends upon the special structure of the cube is the "slicing off" of all of the odd parity vertices of I^d. The triangulation $S(P)$ can be applied to any d-polytope P, but of course the combinatorial properties of the d-cube enable us to use induction and easily compute the number of and describe the vertex sets of the d-simplices of T_d. So the size of a minimum triangulation of the d-cube remains unknown.

A more modest question to ask at this point is whether or not the 5-cube can be triangulated into fewer than 67 5-simplices. Using techniques similar to those of Section 1, one can see that it would be sufficient to prove that any minimal triangulation of I^5 must contain in its interior either no edge and at least 15 triangles (2-faces), or else 1 edge and at least 17 triangles.

References

1. Cottle, R. W. 1982. Minimal triangulations of the 4-cube. Disc. Math. **40:** 25–29.
2. Grünbaum, B. 1967. Convex Polytopes. John Wiley & Sons, Inc. London, England.
3. Lee, C. W. 1981. Triangulating the d-Cube. Research Report RC 8963. IBM Thomas J. Watson Research Center. Yorktown Heights, N.Y.
4. Mara, P. S. 1972. Triangulations of the Cube. M.S. Thesis. Colorado State University. Fort Collins, Colo.
5. Mara, P. S. 1976. Triangulations for the cube. J. Combin. Theory Ser. A **20:** 170–177.
6. McMullen, P. & D. W. Walkup. 1971. A generalized lower-bound conjecture for simplicial polytopes. Mathematika **18:** 264–273.
7. Munson, B. S. 1982. Triangulations of oriented matroids. *In* Proceedings of the Silver Jubilee Conference on Combinatorics, University of Waterloo, Waterloo. In press.
8. Sallee, J. F. 1982. A note on minimal triangulations of an n-cube. Disc. Appl. Math. **4:** 211–215.
9. Sallee, J. F. 1982. A triangulation of the n-cube. Disc. Math. **40:** 81–86.
10. Stanley, R. P. 1980. Decompositions of rational convex polytopes. Ann. Discrete Math. **6:** 333–342.
11. Todd, M. J. 1976. The Computation of Fixed Points and Applications. Lecture Notes in Economics and Mathematical Systems No. 124. Springer-Verlag. West Berlin, FRG.
12. Von Hohenbalken, B. August 1979. How to Simplicially Partition a Polytope. Research Paper No. 79-17. Department of Economics, University of Alberta. Edmonton, Alberta, Canada.

THE NUMBER OF FACES OF SIMPLICIAL POLYTOPES AND SPHERES[a]

Richard P. Stanley

Department of Mathematics
Massachusetts Institute of Technology
Cambridge, Massachusetts 02139

Let $\mathscr{P}$ be a d-polytope, i.e., a d-dimensional convex polytope [17, 30]. Let $f_i = f_i(\mathscr{P})$ denote the number of i-dimensional faces of $\mathscr{P}$. Thus, f_0 is the number of vertices of $\mathscr{P}$. The vector $f(\mathscr{P}) = (f_0, f_1, \ldots, f_{d-1})$ is called the *f-vector* of $\mathscr{P}$. What can be said about such a vector? In 1893 Poincaré proved (with an error corrected in 1899) the formula

$$f_0 - f_1 + f_2 - \cdots + (-1)^{d-1} f_{d-1} = 1 + (-1)^{d-1}, \tag{1}$$

generalizing Euler's famous formula $V - E + F = 2$. Proofs of (1) were published by several nineteenth century geometers, but they all contained unjustified assumptions about the structure of $\mathscr{P}$. It was not until 1955 that Hadwiger gave a nontopological proof of (1), and finally in 1971 Bruggesser and Mani [8] vindicated some of the nineteenth century efforts by showing that their assumptions were valid. For further information and references, see [17, Section 8.6; 19].

We can go on to ask, What more can be said about the f-vector of a d-polytope? The acquisition of any sort of definitive result for arbitrary d-polytopes seems hopeless. However, if we assume that $\mathscr{P}$ is *simplicial*, i.e., every proper face of $\mathscr{P}$ is a simplex, then much more can be said. For instance, if $\mathscr{P}$ is simplicial, then every $(d-2)$-face is contained in exactly two $(d-1)$-faces and every $(d-1)$-face contains d $(d-2)$-faces. Hence, $2f_{d-2} = df_{d-1}$, which is a linear relation among the f_i's independent from (1). Before we consider the most general linear relations that can hold among the f_i's, we first discuss a generalization of simplicial polytopes to which much of what we say applies.

Let $\Delta(\mathscr{P})$ denote the *boundary complex* of a simplicial d-polytope $\mathscr{P}$. By definition, it is the abstract simplicial complex whose vertices are the vertices of $\mathscr{P}$, and whose faces are those sets of vertices that span a proper face of $\mathscr{P}$. The geometric realization $|\Delta(\mathscr{P})|$ of $\mathscr{P}$ (as defined, e.g., in [34, pp. 110–111]) is therefore a $(d-1)$-sphere S^{d-1}. Thus we may consider more generally any (finite) simplicial complex Δ on a vertex set $V = \{x_1, \ldots, x_n\}$ whose geometric realization is S^{d-1}. We call Δ a *simplicial* $(d-1)$-*sphere*. It is by no means apparent that there exists a simplicial $(d-1)$-sphere Δ which is not the boundary complex of a simplicial d-polytope. Indeed, many earlier geometers simply assumed the two concepts were equivalent. It was not until 1965 that Grünbaum [16] gave the first example of such a Δ. The smallest possible example has $d = 4$ and $n = 8$. (It follows from Steinitz's theorem [17, Section 13.1] that an example cannot exist for $d = 3$.)

We can define the f-vector $f(\Delta)$ of a simplicial sphere (or of any simplicial complex) just as we did for polytopes. Thus $f_0 = n$, the number of vertices. The

[a] Research partially supported by National Science Foundation Grant MCF 81-04855.

Euler characteristic formula (**1**) continues to hold for simplicial $(d-1)$-spheres, as does the formula $2f_{d-2} = df_{d-1}$. To state the most general linear relation holding among the f_i's, it is convenient to introduce the numbers

$$h_i = h_i(\Delta) = \sum_{j=0}^{i} \binom{d-j}{d-i} (-1)^{i-j} f_{j-1}, \qquad 0 \le i \le d, \tag{2}$$

where we set $f_{-1} = 1$. Thus $h_0 = 1$, $h_1 = n - d$, and $h_d = (-1)^d(1 - f_0 + f_1 - \cdots + (-1)^d f_{d-1}) = 1$. The vector $h(\Delta) = (h_0, h_1, \ldots, h_d)$ is called the *h-vector* of Δ. It is easy to invert the equations (**2**) and solve for the f_i's, yielding

$$f_i = \sum_{j=0}^{d} \binom{d-j}{d-i-1} h_j.$$

Thus, knowing $h(\Delta)$ is equivalent to knowing $f(\Delta)$.

To compute the h-vector for particular examples, the following procedure can be used. We illustrate it for the four-dimensional cross-polytope, whose f-vector is (8, 24, 32, 16). Write down the f-vector on a diagonal, and put a 1 to the left of f_0.

$$\begin{array}{ccccc} 1 & 8 & & & \\ & & 24 & & \\ & & & 32 & \\ & & & & 16 \end{array}$$

Complete this array to construct a "difference table," by placing underneath a pair of consecutive entries their difference (and bringing down 1 on the left edge).

$$\begin{array}{ccccccccc} & & & 1 & & 8 & & & \\ & & 1 & & 7 & & 24 & & \\ & 1 & & 6 & & 17 & & 32 & \\ 1 & & 5 & & 11 & & 15 & & 16 \end{array}$$

The next row of differences will be $h(\Delta)$.

$$\begin{array}{ccccccccc} & & & 1 & & 8 & & & \\ & & 1 & & 7 & & 24 & & \\ & 1 & & 6 & & 17 & & 32 & \\ 1 & & 5 & & 11 & & 15 & & 16 \\ \hline \end{array}$$

$$h(\Delta) = (1, \quad 4, \quad 6, \quad 4, \quad 1).$$

Note that in this example we have $h_i = h_{d-i}$ $(d = 4)$. In general:

The Dehn–Sommerville equations: For any simplicial $(d-1)$-sphere Δ with h-vector $(h_0, h_1, \ldots, h_d)$, we have $h_i = h_{d-i}$.

This result was proved by Dehn in 1905 for $d \le 5$, and for general d by Sommerville in 1927. See [17, Sections 9.2 and 9.8; 30, Sections 2.4 and 5.1] for further details.

Little further study of the f-vectors of polytopes and spheres was undertaken until 1957, when Motzkin [32], motivated by the development of linear programming, asked how large $f_i(\mathscr{P})$ could be, given the dimension d and number of vertices n of $\mathscr{P}$. More precisely, for $n \geq d + 1$ let $C(n, d)$ denote the convex hull of any n points on the *moment curve* $\{(\tau, \tau^2, \ldots, \tau^d) \in \mathbb{R}^d : \tau \in \mathbb{R}\}$. The convex polytope $C(n, d)$ is known as a *cyclic polytope*, and its combinatorial type is independent of the choice of n extreme points. Cyclic polytopes were first investigated by Carathéodory in 1907 and 1911 and rediscovered by Gale in 1955 and Motzkin in 1957 (see [17, Section 7.4]). The following two properties concerning $C(n, d)$ can be proved without great difficulty:

(i) $C(n, d)$ is *simplicial*,

(ii) $f_i(C(n, d)) = \binom{n}{i+1}$, $0 \leq i \leq [\frac{1}{2}d] - 1$.

Clearly, no polytope or sphere with n vertices can have more than $\binom{n}{i+1}$ i-faces, so cyclic polytopes maximize $f_0, f_1, \ldots, f_{[d/2]-1}$ over all simplicial d-polytopes with n vertices [or simplicial $(d-1)$-spheres with n vertices]. On the other hand, the Dehn–Sommerville equations uniquely determine $f_{[d/2]}, \ldots, f_{d-1}$ once $f_0, \ldots, f_{[d/2]-1}$ are known. It is certainly reasonable to suppose that maximizing $f_0, \ldots, f_{[d/2]-1}$ will also maximize the other f_i's. We therefore say that a simplicial $(d-1)$-sphere Δ with n vertices satisfies the *Upper Bound Conjecture* (UBC) if

$$f_i(\Delta) \leq f_i(C(n, d)), \qquad 0 \leq i \leq d - 1.$$

In 1957 Motzkin [32] conjectured that all d-polytopes satisfy the UBC (and it is easy to reduce this conjecture to the simplicial case), and in 1964 Klee [25] pointed out that the UBC might as well be made for arbitrary simplicial spheres.

In 1970 McMullen [28] proved the UBC for simplicial polytopes. He first showed that the UBC for a $(d-1)$-sphere Δ with n vertices followed from the inequality

$$h_i \leq \binom{n-d+i-1}{i}, \qquad 0 \leq i \leq d. \tag{3}$$

He then used the recently proved result of Bruggesser and Mani [8] that the boundary complex of a convex polytope is shellable. Since simplicial spheres need not be shellable (the first example being given in [12], based on the fact that shellable spheres are combinatorial[b]), McMullen's techniques do not extend to simplicial spheres.

In order to extend the UBC for polytopes to arbitrary simplicial spheres, it is necessary to introduce the machinery of commutative algebra. Given an abstract simplicial complex Δ on a vertex set $V = \{x_1, \ldots, x_n\}$, let $A = k[x_1, \ldots, x_n]$ be the polynomial ring over the field k in the variables $x_1, \ldots, x_n$. Define I_Δ to be the ideal of A generated by all squarefree monomials $x_{i_1} x_{i_2} \cdots x_{i_r}$ such that $\{x_{i_1}, x_{i_2}, \ldots, x_{i_r}\} \notin \Delta$. Set $k[\Delta] = A/I_\Delta$. If we let $k[\Delta]_i$ denote the vector space of all homogeneous

[b] Danaraj and Klee [9] asked whether all combinatorial spheres are shellable. This was answered in the negative by Mandel [27, p. 197].

polynomials of degree i in $k[\Delta]$, then $k[\Delta]$ becomes what we will call a *standard graded k-algebra*, viz., a commutative, associative k-algebra R with identity, together with a collection of subspaces R_i, $i \geq 0$, satisfying:

(i) $R = \coprod_{i \geq 0} R_i$ (vector space direct sum),
(ii) $R_0 = k$,
(iii) $R_i R_j \subset R_{i+j}$,
(iv) R is generated as a k-algebra by R_1,
(v) R is a finitely generated k-algebra [which in view of (iv) is equivalent to $\dim_k R_1 < \infty$].

Given a standard graded k-algebra R, an element of R_i is said to be *homogeneous of degree i*. We also define the *Hilbert function* $H(R, \cdot): \mathbb{N} \to \mathbb{N}$ of R (where $\mathbb{N} = \{0, 1, 2, \ldots\}$) by

$$H(R, i) = \dim_k R_i.$$

[Condition (v) guarantees that $H(R, i) < \infty$.] An easy computation [36, Proposition 3.2] shows that

$$\begin{aligned} H(k[\Delta], i) &= 1, & i &= 0, \\ &= \sum_{j=0}^{d-1} f_j \binom{i-1}{j}, & i &> 0, \end{aligned} \tag{4}$$

where $(f_0, f_1, \ldots, f_{d-1})$ is the f-vector of Δ. The generating function

$$F(R, \lambda) = \sum_{i \geq 0} H(R, i)\lambda^i$$

is called the *Hilbert series* of R. A basic result of commutative algebra due to Hilbert (see [1, Chapter 11] for a simple proof) asserts that

$$F(R, \lambda) = (h_0 + h_1\lambda + \cdots + h_s\lambda^s)(1 - \lambda)^{-d}, \tag{5}$$

for certain integers $h_i = h_i(R)$ and d. We assume d has been chosen as small as possible (i.e., $h_0 + h_1 + \cdots + h_s \neq 0$), and we then call d the *Krull dimension* of R, denoted $d = \dim R$. It is not hard to show that $\dim R$ is equal to the largest number of elements of R (or of R_1) that are algebraically independent over k. It follows easily from (**2**) and (**4**) or otherwise [40, p. 63] that

$$\dim k[\Delta] = 1 + \dim \Delta$$

and

$$h_i(k[\Delta]) = h_i(\Delta).$$

A collection of d homogeneous elements $\theta_1, \ldots, \theta_d$ of R (where $d = \dim R$) of positive degree is called a *homogeneous system of parameters* (h.s.o.p.) if $\dim_k R/(\theta_1, \ldots, \theta_d) < \infty$. The Noether normalization lemma asserts that an h.s.o.p. $\theta_1, \ldots, \theta_d$ for R always exists, and each θ_i can be chosen to have degree one if k is infinite. (This graded version of the Noether normalization lemma is due to Hilbert.)

We are now in a position to make the crucial definition from commutative algebra. We will choose the most convenient definition for our purpose, though on the surface it does not seem to be related to the usual definition found in papers and textbooks. However, it is an easy matter to reconcile our definition with the "standard" one involving R-sequences [38, Corollary 3.2].

DEFINITION. A standard graded k-algebra R (where k is infinite) is called *Cohen–Macaulay* if for some (equivalently, every) h.s.o.p. $\theta_1, \ldots, \theta_d$ of degree one, we have

$$H(R/(\theta_1, \ldots, \theta_d), i) = h_i(R), \qquad i \geq 0.$$

In particular, $k[\Delta]$ is Cohen–Macaulay if for some h.s.o.p. $\theta_1, \ldots, \theta_d$ of degree one, we have

$$H(k[\Delta]/(\theta_1, \ldots, \theta_d), i) = h_i(\Delta), \qquad i \geq 0.$$

EXAMPLE. The algebra $k[\Delta] = k[x, y, z]/(xy, xz)$ is not Cohen–Macaulay, since $h(\Delta) = (1, 1, -1)$, and we can hardly have $H(k[\Delta]/(\theta_1, \theta_2), 2) = -1$.

THEOREM 1 [36, Corollary 4.4]. *If* $|\Delta| \approx S^{d-1}$ *and* $k[\Delta]$ *is Cohen–Macaulay, then the UBC holds for* Δ.

Proof. Let $\theta_1, \ldots, \theta_d$ be an h.s.o.p. for $k[\Delta]$ of degree one, and set $S = k[\Delta]/(\theta_1, \ldots, \theta_d)$. Since $H(k[\Delta], 1) = n$, we have $H(S, 1) = n - d$, so S is generated by $n - d$ linear forms. Hence, $H(S, i)$ cannot exceed the total number of monomials of degree i in $n - d$ variables. This number is $\binom{n-d+i-1}{i}$. Since $k[\Delta]$ is Cohen–Macaulay, we have $H(S, i) = h_i(\Delta)$. Thus $h_i(\Delta) \leq \binom{n-d+i-1}{i}$, which is (3). ∎

In order to prove the UBC for spheres, it remains to show that $k[\Delta]$ is Cohen–Macaulay whenever Δ is a sphere. It turns out that at about the same time that THEOREM 1 was proved, Reisner, working without knowledge of the UBC, found a complete characterization of those Δ for which $k[\Delta]$ is Cohen–Macaulay.[c] In order to state Reisner's result, recall that for any $F \in \Delta$ (including $F = \emptyset$) the *link* of F is defined by

$$\text{lk } F = \{G \in \Delta: G \cap F = \emptyset, G \cup F \in \Delta\}.$$

In particular, $\text{lk } \emptyset = \Delta$. Let $\tilde{H}_i(\Delta)$ denote reduced simplicial homology with coefficients in k.

THEOREM 2 [33]. *Let* Δ *be any finite simplicial complex. The algebra* $k[\Delta]$ *is Cohen–Macaulay if and only if for all* $F \in \Delta$, $\tilde{H}_i(\text{lk } F) = 0$ *unless* $i = \dim(\text{lk } F)$.

It is well known from topology that if $|\Delta| \approx S^{d-1}$, then for all $F \in \Delta$,

$$\begin{aligned} \tilde{H}_i(\text{lk } F) &\cong 0, \qquad i \neq \dim(\text{lk } F), \\ &\cong k, \qquad i = \dim(\text{lk } F). \end{aligned}$$

Hence, $k[\Delta]$ is Cohen–Macaulay for all $|\Delta| \approx S^{d-1}$, and we conclude:

THEOREM 3 [36, Corollary 5.3]. *The UBC is valid for all spheres* Δ.

[c] The simultaneity of THEOREMS 1 and 2 is not quite as coincidental as it may at first appear, since both Reisner and I were heavily influenced by Hochster's paper [20].

Let us mention for readers interested in understanding the proof of THEOREM 2 that in addition to Reisner's original argument, proofs also appear in [2, 21, 40]. All the proofs, except that in [2], use homological algebra. For readers familiar with homological algebra, the most straightforward proof is the one in [40]. (This proof is due to Hochster but is unpublished by him.)

While we now know that the UBC holds for spheres, we still can ask, What more can be said about the f-vector of simplicial polytopes or spheres? To this end, given integers $h, i > 0$, write

$$h = \binom{n_i}{i} + \binom{n_{i-1}}{i-1} + \cdots + \binom{n_j}{j},$$

where $n_i > n_{i-1} > \cdots > n_j \geq j \geq 1$. Such a representation always exists and is unique. For a nice discussion of the significance of this representation, see [13, Section 8]. Now define

$$h^{\langle i \rangle} = \binom{n_i + 1}{i+1} + \binom{n_{i-1} + 1}{i} + \cdots + \binom{n_j + 1}{j+1},$$

and set $0^{\langle i \rangle} = 0$. Call a vector $(h_0, \ldots, h_d) \in \mathbb{Z}^{d+1}$ an *M-vector* (after F. S. Macaulay) if $h_0 = 1$ and $0 \leq h_{i+1} \leq h_i^{\langle i \rangle}$, $1 \leq i \leq d-1$.

The next result is due essentially to Macaulay [26] but was first stated as such in [38, Theorem 2.2].

THEOREM 4. *Let k be a field. A vector $(h_0, \ldots, h_d) \in \mathbb{Z}^{d+1}$ is an M-vector if and only if there exists a standard graded k-algebra R such that $H(R, i) = h_i, 0 \leq i \leq d$.*

COROLLARY. *If $|\Delta| \approx S^{d-1}$, then $h(\Delta)$ is an M-vector.*

Proof. Follows from THEOREM 2 and our definition of Cohen–Macaulay. ∎

The conditions imposed on the h-vector of a sphere by the previous corollary and the Dehn–Sommerville equations are by no means sufficient to characterize such h-vectors. In 1971 McMullen [29], putting together all results concerning f-vectors known at that time, produced the following remarkable conjecture, called the "g-conjecture" after McMullen's notation g_i for our $h_i - h_{i-1}$.

Conjecture. A vector $(h_0, \ldots, h_d) \in \mathbb{Z}^{d+1}$ is the h-vector of some simplicial d-polytope $\mathscr{P}$ if and only if $h_i = h_{d-i}$ and $(h_0, h_1 - h_0, h_2 - h_1, \ldots, h_{[d/2]} - h_{[d/2]-1})$ is an M-vector.

As McMullen noted in [29], the above conjecture could conceivably be valid for simplicial $(d-1)$-spheres, but he felt he did not understand spheres well enough to have any confidence in such a generalization.

Note that the g-conjecture implies in particular that $h_0 \leq h_1 \leq \cdots \leq h_{[d/2]}$ for the h-vector of a simplicial polytope. This weaker result had earlier been conjectured by McMullen and Walkup [31] and had been called by them the "Generalized Lower-Bound Conjecture" (GLBC). This terminology was used because the GLBC implies the earlier "Lower-Bound Conjecture," which gives the minimum value of f_i for a simplicial d-polytope with n vertices. The Lower-Bound Conjecture was subsequently proved by Barnette [3, 4], but the GLBC remained open. The GLBC also included a condition as to when $h_i = h_{i+1}$. This part of the GLBC remains open at present.

It turns out that McMullen's g-conjecture (and therefore the inequality $h_0 \leq h_1 \leq \cdots \leq h_{[d/2]}$ of the GLBC) is indeed true (for polytopes). The "if" part of the conjecture was proved by Billera and Lee [5, 6] in 1979. They explicitly constructed the desired $\mathscr{P}$ by cleverly taking a sequence of stellar and inverse stellar subdivisions of a cyclic polytope. We will now sketch the proof of the "only if" part of the conjecture [39]. The proof consists basically of putting together some results from algebraic geometry. First note that we can assume that our simplicial polytope satisfies: (a) if dim $\mathscr{P} = d$, then $\mathscr{P} \subset \mathbb{R}^d$ (simply restrict one's attention to the affine span of $\mathscr{P}$), (b) the vertices of $\mathscr{P}$ lie in $\mathbb{Q}^d$, i.e., $\mathscr{P}$ is a *rational* polytope (because small perturbations of the vertices of $\mathscr{P}$ do not change the combinatorial type of $\mathscr{P}$ since $\mathscr{P}$ is simplicial), and (c) the origin of $\mathbb{R}^d$ lies in the interior of $\mathscr{P}$. We now note the following:

1. In 1970 Demazure [11] and in 1973 Mumford and co-workers [24] defined a certain class of complex varieties now known as *toric varieties*. Briefly, they may be described as follows [10]. Let Σ be a collection of convex polyhedral cones in a $\mathbb{Q}$-vector space V, satisfying:

(1) every cone in Σ has a vertex at the origin,
(2) if τ is a face of a cone $\sigma \in \Sigma$, then $\tau \in \Sigma$,
(3) if $\sigma, \sigma' \in \Sigma$, then $\sigma \cap \sigma'$ is a face of both σ and σ'.

We call Σ a *fan*. Σ is *complete* if $\bigcup_{\sigma \in \Sigma} \sigma$ is the whole space V, and *simplicial* if each $\sigma \in \Sigma$ is a simplicial cone. Let V^* be the dual space to V. For a cone $\sigma \in \Sigma$, define

$$\sigma^* = \{f \in V^*\colon f(\sigma) \geq 0\}.$$

Then σ^* is a convex polyhedral cone in V^*, called the *dual cone* to σ. Choose a lattice L in V, and let L^* be the dual lattice in V^*. Then $\sigma^* \cap L^*$ is an additive monoid in V^*; let $\mathbb{C}[\sigma^* \cap L^*]$ denote the monoid algebra of $\sigma^* \cap L^*$ over $\mathbb{C}$. Define an affine variety

$$X_{\sigma^*} = X_{(\sigma^*, L^*)} = \operatorname{Spec} \mathbb{C}[\sigma^* \cap L^*].$$

[Naively, view $\mathbb{C}[\sigma^* \cap L^*]$ as the quotient of a polynomial ring (whose variables correspond to the generators of $\sigma^* \cap L^*$, which are always finite in number) by some ideal I. Then $X_{(\sigma^*, L^*)}$ may be thought of as the variety of zeros of polynomials in I.] If τ is a face of σ, then X_{τ^*} can be identified with an open subvariety of X_{σ^*}. These identifications allow us to glue together from the X_{σ^*} (as σ ranges over Σ) a variety over $\mathbb{C}$, which is denoted by X_Σ and is called the *toric variety associated with* Σ. (Note that we have defined X_Σ as an *abstract variety*, in the sense of Weil. See, e.g., [18, p. 105].) The variety X_Σ does not a priori have an embedding into affine or projective space, and one of the main results stated below gives a criterion for X_Σ to be projective.

Now suppose $\mathscr{P}$ is a convex polytope (not necessarily simplicial) satisfying conditions (a)–(c) above. For any proper face F of $\mathscr{P}$, define a cone σ_F to consist of all rational points that lie on some ray with vertex 0 and passing through F. The set Σ of all such σ_F (together with 0) forms a complete fan. The corresponding toric variety X_Σ (say for the lattice $L = \mathbb{Z}^d$) is denoted $X(\mathscr{P})$.

2. A basic result of the theory of toric varieties [10, Proposition 6.9.1; 11; 24] asserts essentially that for any fan Σ, X is projective if and only if $X = X(\mathscr{P})$ for some convex polytope.

A second (easier) result states that when $\mathscr{P}$ is simplicial, $X(\mathscr{P})$, while not necessarily smooth, has very nice singularities. Namely, $X(\mathscr{P})$ looks locally like affine space $\mathbb{C}^d$ modulo a finite (Abelian) group of linear transformations. Varieties with these types of singularities are known as *V-varieties.*

3. In 1978 Danilov [10, Theorem 10.8 and Remark 10.9], extending work of Jurkiewicz, computed the cohomology ring $H^*(X(\mathscr{P}), \mathbb{C})$ when $\mathscr{P}$ is simplicial. All the cohomology occurs in even degree, i.e.,

$$H^*(X(\mathscr{P}), \mathbb{C}) = \coprod_{i=0}^{d} H^{2i}(X(\mathscr{P}), \mathbb{C}),$$

and there is an isomorphism

$$H^*(X(\mathscr{P}), \mathbb{C}) \xrightarrow{\cong} \mathbb{C}[\Delta(\mathscr{P})]/(\theta_1, \ldots, \theta_d) \tag{6}$$

for a certain h.s.o.p. of $\mathbb{C}[\Delta(\mathscr{P})]$ of degree one (which depends on how $\mathscr{P}$ is embedded in $\mathbb{R}^d$). In order for the isomorphism to be degree preserving one must consider elements of $H^{2i}(X, \mathbb{C})$ as having degree i, so $\theta_1, \ldots, \theta_d$ may be regarded as elements of $H^2(X, \mathbb{C})$. Note in particular that by THEOREM 2, the fact that $|\Delta(\mathscr{P})| \approx S^{d-1}$, and our definition of Cohen–Macaulay, we have

$$\dim_{\mathbb{C}} H^{2i}(X(\mathscr{P}), \mathbb{C}) = h_i(\mathscr{P}). \tag{7}$$

4. In 1976 Steenbrink [41] proved that projective V-varieties Y [in particular $Y = X(\mathscr{P})$ for $\mathscr{P}$ simplicial] satisfy the hard Lefschetz theorem. This means the following. Since Y is projective, we can embed it (as a closed subvariety) in some projective space $\mathbb{P}^N$. Let H be a generic hyperplane in $\mathbb{P}^N$. Then $Y \cap H$ is a closed subvariety of Y of complex codimension 1. With any closed subvariety Z of complex codimension i of an irreducible complex projective variety Y, a standard construction of algebraic geometry (e.g., [15] or [14, Chapter 5, Section 4]) associates a cohomology class $[Z] \in H^{2i}(Y, \mathbb{C})$. In particular, let $\omega \in H^2(Y, \mathbb{C})$ denote the class of the hyperplane section $Y \cap H$. The classical hard Lefschetz theorem (first proved completely by Hodge) states that if Y is smooth and irreducible of complex dimension d, and if $0 \le i \le d$, then the linear transformation

$$H^i(Y, \mathbb{C}) \xrightarrow{\omega^{d-i}} H^{2d-i}(Y, \mathbb{C}) \tag{8}$$

defined as multiplication by ω^{d-i} is a bijection (e.g., [15, Chapter 0, Section 7]). Steenbrink showed that this result continues to hold when Y is a projective V-variety, though Y need not be smooth. In general, the singularities of V-varieties are so "tame" that V-varieties behave almost like smooth varieties.

Note that if for all $0 \le i \le d$ the map (8) is a bijection, then it follows that for all $0 \le i < d$ the map $H^i(Y, \mathbb{C}) \xrightarrow{\omega} H^{i+2}(Y, \mathbb{C})$ is injective (and surjective for $d - 1 \le i \le 2d - 2$).

5. We now have at our disposal all the tools to prove the necessity of McMullen's g-conjecture. Let $\mathscr{P}$ be a simplicial d-polytope satisfying conditions (a)–(c)

above, and let $X = X(\mathscr{P})$ be the corresponding toric variety. Define the graded algebra

$$R = H^*(X, \mathbb{C})/(\omega),$$

where (ω) denotes the ideal of $H^*(X, \mathbb{C})$ generated by the class of a hyperplane section. The ith degree part of R is given by the quotient vector space

$$R_i = H^{2i}(X, \mathbb{C})/\omega H^{2i-2}(X, \mathbb{C}). \tag{9}$$

By (6), R is indeed a standard graded algebra and, in particular, is generated by R_1. From (9) we have

$$H(R, i) = \dim_{\mathbb{C}} R_i = \dim_{\mathbb{C}} H^{2i}(X, \mathbb{C}) - \dim_{\mathbb{C}} \omega H^{2i-2}(X, \mathbb{C}).$$

By (7), $\dim_{\mathbb{C}} H^{2i}(X, \mathbb{C}) = h_i(\mathscr{P})$. By the hard Lefschetz theorem, if $0 \le i \le [d/2]$, then

$$\dim_{\mathbb{C}} \omega H^{2i-2}(X, \mathbb{C}) = \dim_{\mathbb{C}} H^{2i-2}(X, \mathbb{C}) = h_{i-1}(\mathscr{P}).$$

Thus $H(R, i) = h_i(\mathscr{P}) - h_{i-1}(\mathscr{P})$, $0 \le i \le [d/2]$. Hence by THEOREM 4, $(h_0, h_1 - h_0, h_2 - h_1, \ldots, h_{[d/2]} - h_{[d/2]-1})$ is an M-vector, and the proof is complete. ▮

Two questions immediately suggest themselves, the second of which was alluded to earlier:

1. Is there a simpler proof? By the very nature of M-vectors it seems essential to consider graded algebras, in particular, the algebra $S = \mathbb{C}[\Delta(\mathscr{P})]/(\theta_1, \ldots, \theta_d)$. However, it might be possible to bypass the theory of toric varieties and prove directly the existence of a suitable $\omega \in S_1$. If one examines the "standard" proof of the hard Lefschetz theorem, the following can be seen. Define three linear transformations $\alpha, \beta, \gamma: S \to S$ as follows: $\alpha(x) = \omega x$, $\beta(x) = (2i - d)x$ if $x \in S_i$, and γ is the adjoint of α with respect to a certain scalar product on S. Then α, β, γ generate (under the bracket operation) the Lie algebra $\mathrm{sl}_2(\mathbb{C})$. This is equivalent in the present situation to saying $\alpha\gamma - \gamma\alpha = \beta$. Thus to prove the necessity of the g-conjecture, it would suffice to find α (i.e., $\omega \in S_1$), β (which is automatic), and γ (which involves a suitable choice of a scalar product on S). Some recent work of Kalai [23] might also lead to a simpler proof.

2. Does McMullen's g-conjecture hold for more general simplicial complexes? In particular, we have the following hierarchy:

$$\text{simplicial polytope} \Rightarrow P_1\text{-sphere} \Rightarrow \text{combinatorial sphere}$$
$$\Rightarrow P\text{-sphere} \Rightarrow \text{simplicial sphere} \Rightarrow \text{Gorenstein complex.} \tag{10}$$

The relevant definitions are as follows. A *P_1-sphere* is an abstract simplicial complex Δ which can be realized in $\mathbb{R}^d$ by a collection of rectilinear simplices whose union is the boundary of a convex polytope. In other words, triangulate the boundary of a convex polytope into (rectilinear) simplices. (Similarly define *P_1-ball.*) A *combinatorial sphere* is a triangulation of the sphere S^{d-1} which possesses a subdivision isomorphic to a subdivision of the boundary complex of a d-simplex. The notion of a P-sphere is due to J. Edmonds and seems to be the most general class of simplicial spheres avoiding certain pathologies and which lend themselves to proofs by induction. A simplicial complex Δ satisfying $|\Delta| \approx S^{d-1}$ is a *P-sphere* if it possesses a subdivision Δ' which is a P_1-sphere, such that the subdivision induced on each cell of Δ is itself a P_1-ball. (See [27, Chapter 5.VII].) Finally, a *Gorenstein complex* is an abstract simplicial complex Δ for which the ring $k[\Delta]$ is a Gorenstein ring (see [21, 40]).

It is known that all the implications in (**10**) are strict, and we can ask, For which of them does the g-conjecture hold? In the case of P_1-spheres Δ [possessing a realization $\mathscr{P}$ satisfying (a)–(c)] we still have a toric variety $X(\mathscr{P})$, but it need not be projective. Thus Steenbrink's theorem does not apply. Can one somehow "induce" the hard Lefschetz theorem on $X(\mathscr{P})$ from a subdivision $\mathscr{P}'$ of $\mathscr{P}$ for which $X(\mathscr{P}')$ is projective? (Such subdivisions always exist [10, Section 6.9.2; 24, p. 161].) Let us also note that Gorenstein complexes are precisely those simplicial complexes for which the ring $S = \mathbb{C}[\Delta]/(\theta_1, \ldots, \theta_d)$ satisfies (abstract) Poincaré duality. Since Poincaré duality is essentially a consequence of the hard Lefschetz theorem, it follows that Gorenstein complexes are the most general ones for which we could expect the g-conjecture to hold.

Historical Note. The following comments on how the proof of the necessity of the g-conjecture was found may be of interest. I had realized from my first work on the UBC (see the sentence after Conjecture 2 in [35]) that the necessity of the g-conjecture would follow from the existence of a suitable element $\omega \in S_1$, where $S = \mathbb{C}[\Delta(\mathscr{P})]/(\theta_1, \ldots, \theta_d)$. The proof of a rather uninteresting special case was announced in [37, Theorem 8]. In 1976, Tony Iarrobino brought the hard Lefschetz theorem to my attention. It was now apparent that one needed a smooth projective variety X whose cohomology ring was isomorphic to S [with the grading given by $S_i \cong H^{2i}(X, \mathbb{C})$]. I had been aware for some time of the theory of toroidal embeddings [24] and checked this reference to see whether the varieties $X(\mathscr{P})$ had the right properties. Three problems arose: (i) I could not understand [24] well enough to tell whether $X(\mathscr{P})$ was projective, (ii) in any case, no mention was made of $H^*(X(\mathscr{P}), \mathbb{C})$, and (iii) worst of all, $X(\mathscr{P})$ was not necessarily smooth, so the hard Lefschetz theorem (seemingly) did not apply. There matters rested until the spring or summer of 1979, when I stumbled upon the paper of Danilov [10] on the new journal shelf of the MIT library. Remark 3.8 immediately caught my attention. It asserted that for any triangulation Δ of the sphere S^{d-1}, the ring $S = \mathbb{C}[\Delta]/(\theta_1, \ldots, \theta_d)$ was the cohomology ring of a smooth variety $X(\mathscr{P})$! Thus, the necessity of the g-conjecture is valid for all spheres for which $X(\mathscr{P})$ is projective. But in reading [10] more carefully it became apparent that Remark 3.8 was stated rather carelessly. One needs to assume Δ is a P_1-sphere, not an arbitrary sphere, in order to define $X(\mathscr{P})$. Moreover, the corresponding variety $X(\mathscr{P})$ is not always smooth. I therefore asked some algebraic geometers whether $X(\mathscr{P})$, when projective, could satisfy the hard Lefschetz theorem, but none knew. Shortly thereafter I took the book [22] out of the library in order to look at a paper related to a completely different topic in which I was interested. In browsing through this book I discovered the paper of Steenbrink [41], with its proof of the hard Lefschetz theorem for projective V-varieties.[d] It remained only to ascertain that for convex polytopes the varieties $X(\mathscr{P})$ were projective. This was accomplished via a conversation with David Mumford on September 13, 1979, and the proof was complete.

[d] Danilov [10] refers to this paper of Steenbrink, but since he does not mention the hard Lefschetz theorem it did not occur to me to pursue this reference.

References

1. Atiyah, M. F. & I. G. Macdonald. 1969. Introduction to Commutative Algebra. Addison–Wesley. Reading, Mass.
2. Baclawski, K. & A. M. Garsia. 1981. Combinatorial decompositions of a class of rings. Adv. in Math. **39:** 155–184.
3. Barnette, D. W. 1971. The minimum number of vertices of a simple polytope. Israel J. Math. **9:** 121–125.
4. Barnette, D. W. 1973. A proof of the lower bound conjecture for convex polytopes. Pacific J. Math. **46:** 349–354.
5. Billera, L. J. & C. W. Lee. 1980. Sufficiency of McMullen's condition for f-vectors of simplicial polytopes. Bull. Amer. Math. Soc. **2:** 181–185.
6. Billera, L. J. & C. W. Lee. 1981. A proof of the sufficiency of McMullen's conditions for f-vectors of simplicial convex polytopes. J. Combin. Theory Ser. A **31:** 237–255.
7. Borel, A. & A. Haefliger. 1961. La class d'homologie fondamentale d'un espace analytique. Bull. Soc. Math. France **89:** 461–513.
8. Bruggesser, M. & P. Mani. 1971. Shellable decompositions of cells and spheres. Math. Scand. **29:** 197–205.
9. Danaraj, G. & V. Klee. 1978. Which spheres are shellable? Ann. Discrete Math. **2:** 33–52.
10. Danilov, V. I. 1978. The geometry of toric varieties. Russian Math. Surveys **33:** 97–154. [Translated from: Uspekhi Mat. Nauk **33** (1978): 85–134.]
11. Demazure, M. 1970. Sous-groupes algébriques de rang maximum du groupe de Cremona. Ann. Sci. École Norm. Sup. **3:** 507–588.
12. Edwards, R. D. 1975. The double suspension of a certain homology 3-sphere is S^5. Notices Amer. Math. Soc. **22:** A-334.
13. Greene, C. & D. J. Kleitman. 1978. Proof techniques in the theory of finite sets. *In* Studies in Combinatorics. G.-C. Rota, Ed.: 22–79. Mathematical Association of America. Washington, D.C.
14. Griffiths, P. & J. Adams. 1974. Topics in Algebraic and Analytic Geometry. Princeton University Press. Princeton, N.J.
15. Griffiths, P. & J. Harris. 1978. Principles of Algebraic Geometry. John Wiley & Sons, Inc. New York, N.Y.
16. Grünbaum, B. 1965. Diagrams and Schlegel diagrams. Notices Amer. Math. Soc. **12:** 578. (Abstract 625-112.)
17. Grünbaum, B. 1967. Convex Polytopes. John Wiley & Sons, Inc. New York, N.Y.
18. Hartshorne, R. 1977. Algebraic Geometry. Springer-Verlag. New York, N.Y.
19. Hilton, P. & J. Pedersen. 1981. Euler, Poincaré, Pólya- and polyhedra. Enseign. Math. **27:** 326–343.
20. Hochster, M. 1972. Rings of invariants of tori, Cohen–Macaulay rings generated by monomials, and polytopes. Ann. of Math. **96:** 318–337.
21. Hochster, M. 1977. Cohen–Macaulay rings, combinatorics, and simplicial complexes. *In* Ring Theory II. B. R. McDonald & R. Morris, Eds.: 171–223. Marcel Dekker, Inc. New York, N.Y.
22. Holm, P. (Ed.). 1977. Real and Complex Singularities, Oslo, 1976. Sijthoff & Noordhoff. Alphen aan den Rijn, the Netherlands.
23. Kalai, G. Characterization of f-vectors of families of convex sets in R^d. I. Necessity of Eckhoff's conditions. Israel J. Math. In press.
24. Kempf, G., F. Knudsen, D. Mumford & B. Saint-Donat. 1973. Toroidal Embeddings I. Lecture Notes in Mathematics No. 339. Springer-Verlag. West Berlin, FRG.
25. Klee, V. 1964. The number of vertices of a convex polytope. Canad. J. Math. **16:** 701–720.
26. Macaulay, F. S. 1927. Some properties of enumeration in the theory of modular systems. Proc. London Math. Soc. **26:** 531–555.
27. Mandel, A. 1982. Topology of Oriented Matroids. Ph.D. Thesis. University of Waterloo. Waterloo, Ontario, Canada.
28. McMullen, P. 1970. The maximum numbers of faces of a convex polytope. Mathematika **17:** 179–184.

29. McMullen, P. 1971. The numbers of faces of simplicial polytopes. Israel J. Math. **9:** 559–570.
30. McMullen, P. & G. C. Shephard. 1971. Convex Polytopes and the Upper Bound Conjecture. Cambridge University Press. Cambridge, England.
31. McMullen, P. & D. W. Walkup. 1971. A generalized lower-bound conjecture for simplicial polytopes. Mathematika **18:** 264–273.
32. Motzkin, T. S. 1957. Comonotone curves and polyhedra. Bull. Amer. Math. Soc. **63:** 35. (Abstract 111.)
33. Reisner, G. 1976. Cohen–Macaulay quotients of polynomial rings. Adv. in Math. **21:** 30–49.
34. Spanier, E. H. 1966. Algebraic Topology. McGraw–Hill Book Co. New York, N.Y.
35. Stanley, R. 1975. Cohen–Macaulay rings and constructible polytopes. Bull. Amer. Math. Soc. **81:** 133–135.
36. Stanley, R. 1975. The upper bound conjecture and Cohen–Macaulay rings. Stud. Appl. Math. **54:** 135–142.
37. Stanley, R. 1977. Cohen–Macaulay complexes. *In* Higher Combinatorics. M. Aigner, Ed.: 51–62. D. Reidel Publishing Co. Dordrecht, the Netherlands.
38. Stanley, R. 1978. Hilbert functions of graded algebras. Adv. in Math. **28:** 57–83.
39. Stanley, R. 1980. The number of faces of a simplicial convex polytope. Adv. in Math. **35:** 236–238.
40. Stanley, R. 1983. Commutative Algebra and Combinatorics. Progress in Mathematics Vol. 41. Birkhäuser Boston, Inc. Cambridge, Mass.
41. Steenbrink, J. H. M. 1977. Mixed Hodge structure on the vanishing cohomology. *In* Real and Complex Singularities, Oslo, 1976. P. Holm, Ed.: 525–563. Sijthoff & Noordhoff. Alphen aan den Rijn, the Netherlands.

POLYHEDRAL MANIFOLDS WITH PROPERTIES OF THE PLATONIC SOLIDS[a]

J. M. Wills

Mathematics Institute
University of Siegen
D-5900 Siegen, Federal Republic of Germany

1. Introduction

The five Platonic solids have nice algebraic, geometric, and combinatorial properties, so mathematicians have often tried to find analogues, which have some, but not all, properties in common with the Platonic solids. The 13 Archimedean solids have two or three types of regular faces. The four Kepler–Poinsot polyhedra and the uniform polyhedra described by Coxeter and others [5] have regular faces with self-intersections and "false" vertices. The five compound bodies are star-polyhedra with nonconvex vertices.

The regular tilings of the plane, which can be interpreted as degenerate regular polyhedra, are infinite and have no symmetry group of a Platonic solid. The apeirohedra or regular skew polyhedra described by Petrie, Coxeter [4], and Gott [7] are infinite polyhedra, again with symmetry group different from that of the Platonic solids.

There exist also higher-dimensional analogues, but for different and obvious reasons we are only interested in analogues in Euclidean 3-space E^3.

It seems to be appropriate to define compact oriented 2-manifolds after the mentioned analogues (convex polyhedra, star-polyhedra, regular tilings, manifolds with self-intersections, infinite manifolds). There have been some attempts before in this direction, but with different properties [13, 14]. The definition most like our own is that of Stewart [13], where the faces have to be regular but of different type (and with different valences of the vertices), whereas we require local combinatorial regularity. For classical and modern concepts of symmetry and regularity compare also [1–3, 6, 9].

2. Definitions

We use the notation of [8, 10, 11]. A polyhedral 2-manifold is a closed topological manifold in some Euclidean space E^d (here always in E^3) which is the underlying point-set of a geometric 2-complex. Its facets (i.e., 2-cells) are convex polygons. So a polyhedral 2-manifold is a geometric model of an abstract 2-manifold, made up of convex polygons and, in this paper, always embedded in E^3, and therefore necessarily oriented. We require further that adjacent faces, i.e., with a common edge, are not coplanar.

After this basic definition we call a polyhedral 2-manifold *equivelar* (i.e., with equal flags), if each of its faces has the same number p of edges and if each of its vertices belongs to the same number q of edges. We write a manifold of such type as $\{p, q; g\}$, where g denotes its genus (compare [10, 11, 15]).

[a] Manuscript received August 10, 1982.

If, as usual, $f_i = f_i(M)$, $i = 0, 1, 2$, denote the numbers of i-cells (vertices, edges, faces) of M, equivelarity implies $qf_0 = 2f_1 = pf_2$. From this and Euler's equation $f_0 - f_1 + f_2 = 2 - 2g$ it follows that

$$qf_0 = 2f_1 = pf_2 = 2(g-1)\left\{\frac{1}{2} - \frac{1}{p} - \frac{1}{q}\right\}^{-1} > 0. \qquad (*)$$

Equivelar manifolds have interesting extremal properties concerning the relations between the f_i, $i = 0, 1, 2$, and g [10, 11]. Infinitely many of them are realizable with the symmetry (respectively, rotation) group of one of the Platonic solids. But if one adds a rigidity condition and a suitable topological restriction, one obtains only finitely many of them.

In the following we use the fact that a manifold M together with its interior (bounded) component, ($M \cup \text{int } M$) for short, has the same homotopy type as a (possibly degenerate) graph with straight lines. So this graph is a deformation-retract of ($M \cup \text{int } M$) (compare, e.g., [12]), and we call it a retract-graph $R(M)$. That is, a retract-graph of a Platonic solid is a point. We remark that the introduction of a retract-graph leads to a certain asymmetry, because we require nothing about ($M \cup \text{ext } M$). We now come to the central definition. Compare also [15].

DEFINITION. A polyhedral 2-manifold $M \subset E^3$ of genus g is called a semi-Platonic manifold (abbreviated: SPM) if it has the following properties:

(A) Algebraic condition: M has the rotation group of a Platonic solid, i.e., the tetrahedral or octahedral or icosahedral group.

(B) Combinatorial condition: M is equivelar, i.e., all faces of M are convex p-gons, all vertices of M are q-valent.

(C) Geometric condition: The vertices of M lie on Platonic or Archimedean solids, and the rotation group of M acts transitively on the vertices of each solid.

(D) Topological condition: ($M \cup \text{int } M$) has the same homotopy type as a finite rectilinear graph $R(M)$ in E^3 with the following properties:

(1) $R(M)$ has the rotation group of M;

(2) all vertices of $R(M)$ have the same valence v;

(3) $R(M)$ contains the 1-skeleton of Platonic solids with the rotation group of M and at most one further vertex (i.e., a degenerate Platonic solid).

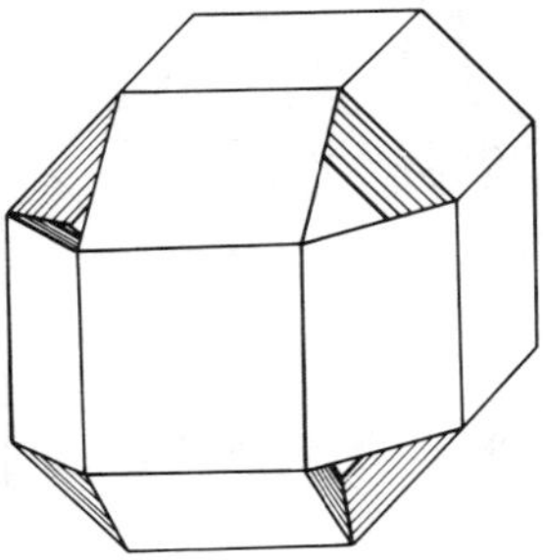

{4,5;7} f=12(4,10,5)

FIGURE 1

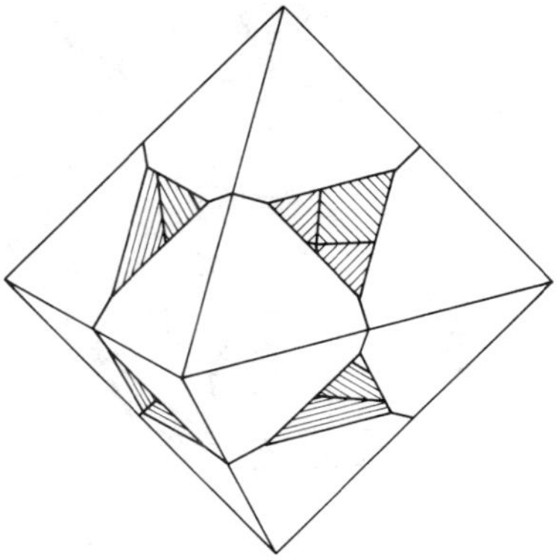

{5,4;7} f=12(5,10,4)

FIGURE 2

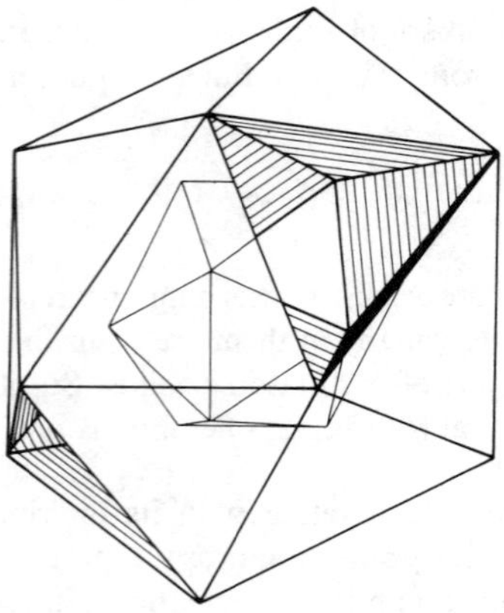

{3,7;3} f=4(6,21,14)

FIGURE 3

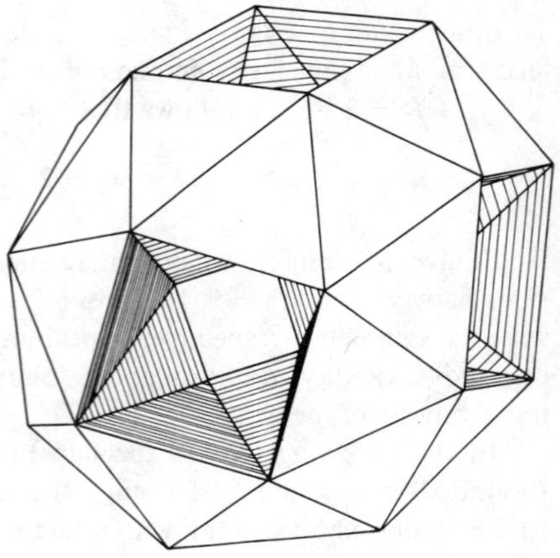

{3,7;5} f=8(6,21,14)

FIGURE 4

3. EXAMPLES

For $g = 0$ we obtain exactly the five Platonic solids. It can be shown that only finitely many SPMs exist. FIGURES 1–16 show 16 of the 19 known SPMs with $g > 0$. The three $\{3, 7; g\}$ exist in a right-hand and in a left-hand version. It is easy to check that (A), (B), and (D) are satisfied.

To make the figures easier to understand we have violated in some cases the condition in (C), that the vertices of M lie on Platonic or Archimedean solids. We

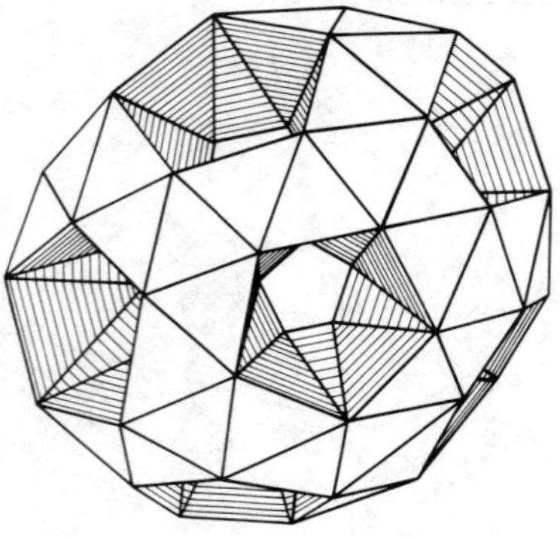

{3,7;11} f=20(6,21,14)

FIGURE 5

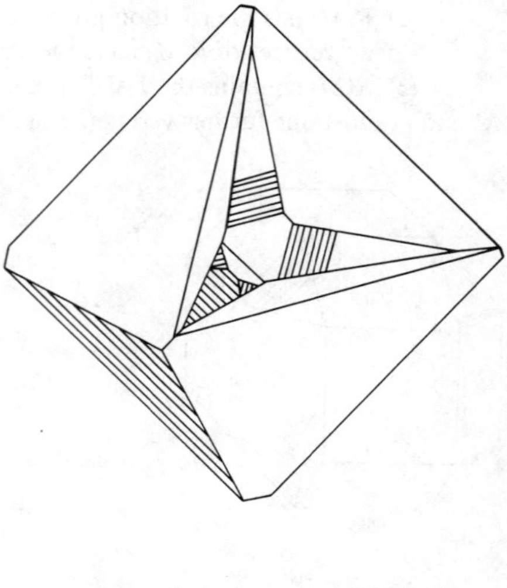

{6,4;6} f=10(3,6,2)

FIGURE 6

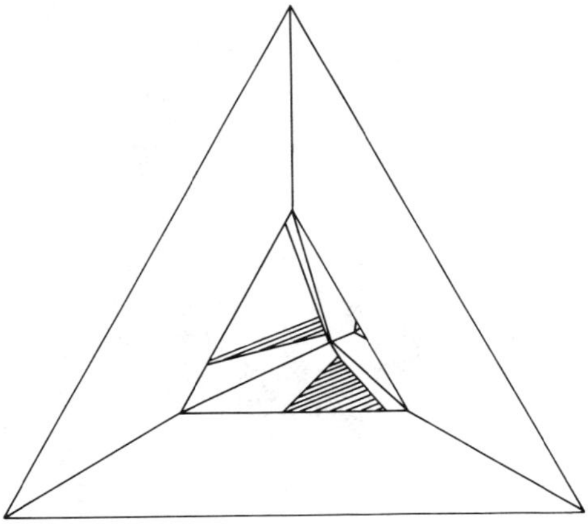

FIGURE 7

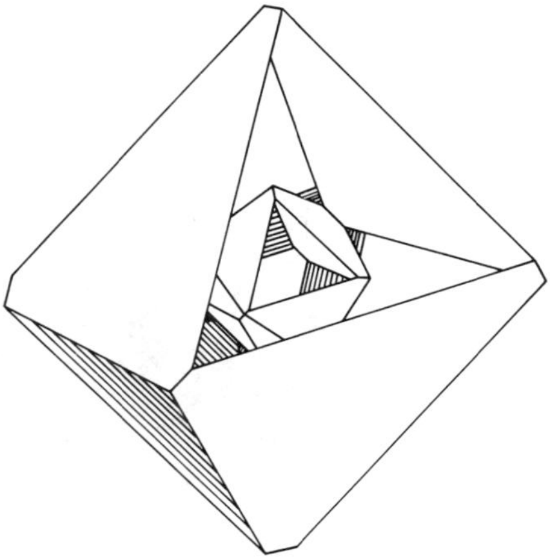

FIGURE 8

have also violated the condition that adjacent faces are not coplanar. There are some other SPMs, which exist only with some coplanar faces and which therefore are omitted. To each figure of an SPM we have added its f-vector

$$f = (f_0, f_1, f_2) = |\chi(M)| \cdot (f_0', f_1', f_2'),$$

where $\chi(M) = 2(1 - g)$ denotes its Euler–Poincaré characteristic.

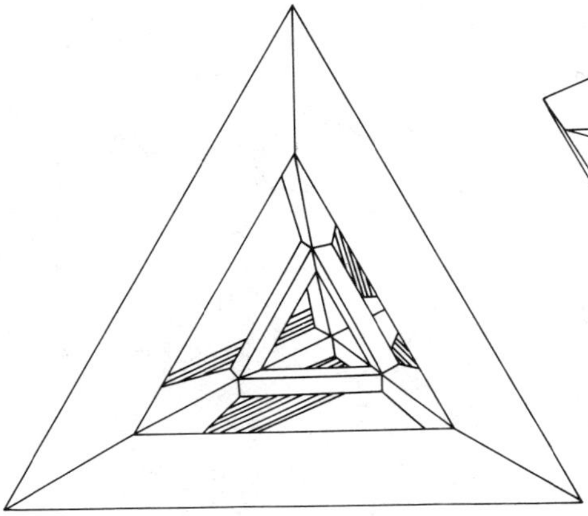

FIGURE 9

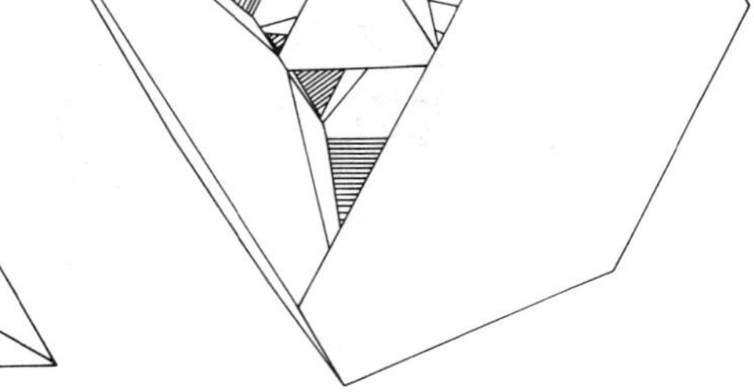

FIGURE 10

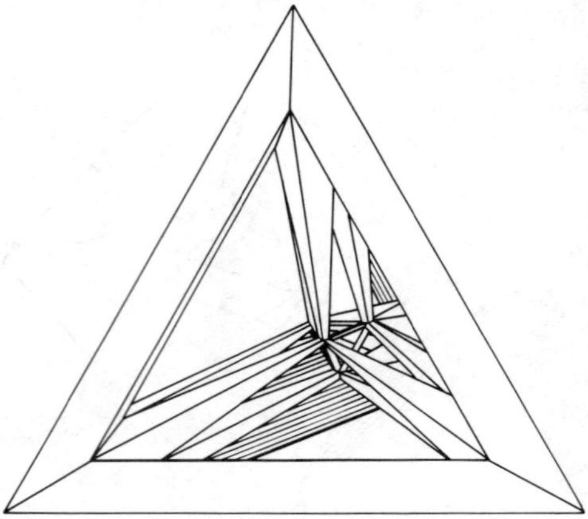

{4,6;17} f=32(2,6,3)

FIGURE 11

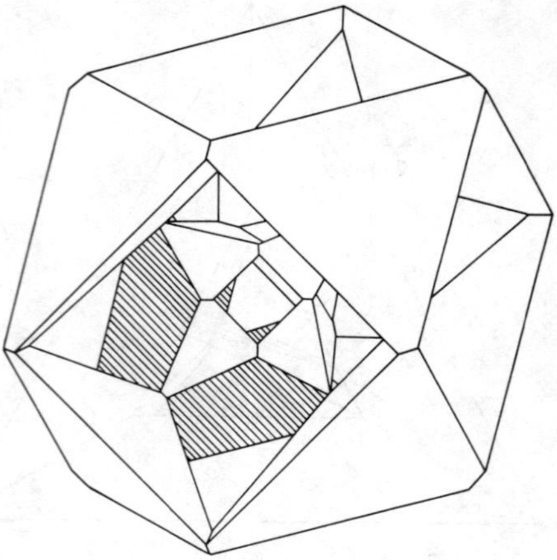

{6,4;17} f=32(3,6,2)

FIGURE 12

4. Additional Regularity Properties

In the cases of {3, 7; g}, $g = 3$, 5, 11, and {4, 5; 7} the vertices lie on two (Platonic or) Archimedean solids, which are (up to dilatation) either identical or enantiomorphic. So if we define a suitable 1–1 correspondence between the vertices of the two solids, we have the remarkable property that the corresponding symmetry group acts transitively on the vertices of this SPM. More precisely, if S_n, respectively A_n, denotes the symmetric group, respectively the alternating group, of order $n!$,

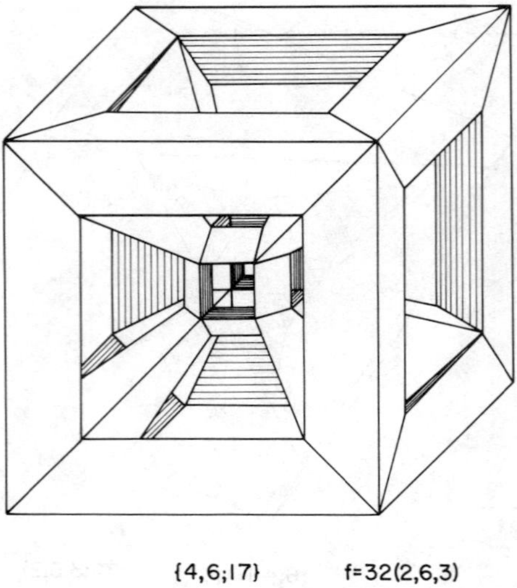

{4,6;17} f=32(2,6,3)

FIGURE 13

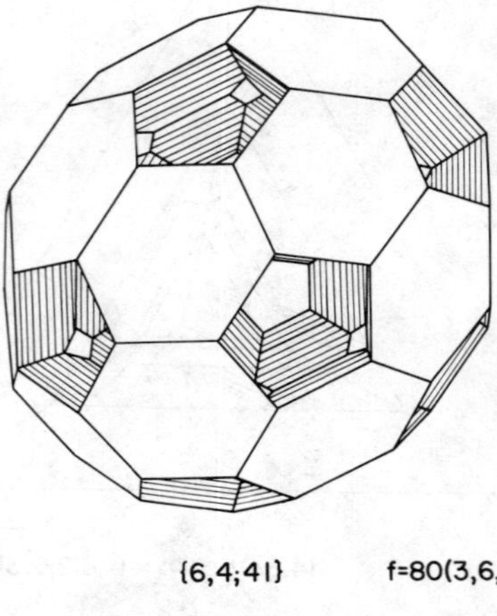

{6,4;41} f=80(3,6,2)

FIGURE 14

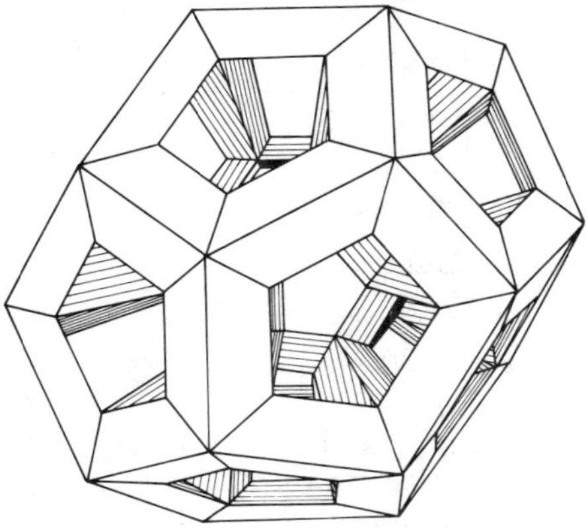

{4,6;41} f=80(2,6,3)

FIGURE 15

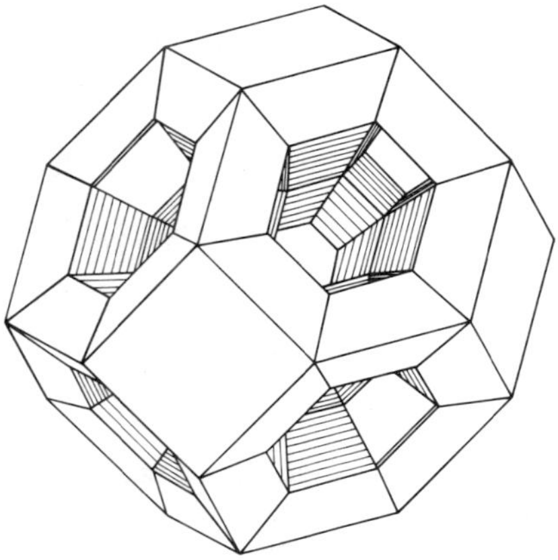

{4,5;19} f=36(4,10,5)

FIGURE 16

respectively $n!/2$: S_4 on {3, 7; 3}; $S_2 \times S_4$ on {3, 7; 5} and {4, 5; 7}; $S_2 \times A_5$ on {3, 7; 11}.

Analogously, $S_2 \times S_4$ acts transitively on the 48 pentagons on {5, 4; 7}. Obviously only finitely many SPMs with this property can exist.

REFERENCES

1. BRÜCKNER, M. 1900. Vielecke and Vielfläche. Teubner. Leipzig, Germany.
2. COXETER, H. S. M. 1973. Regular Polytopes. Dover. New York, N.Y.
3. COXETER, H. S. M. 1974. Regular Complex Polytopes. Cambridge University Press. Cambridge, England.
4. COXETER, H. S. M. 1937. Regular skew polyhedra in three and four dimensions, and their topological analogues. Proc. London Math. Soc. (2) **43:** 33–62. (Reprinted with minor changes in Twelve Geometric Essays, Southern Illinois University Press, Carbondale-Evansville, 1968.)
5. COXETER, H. S. M., M. S. LONGUET-HIGGINS & J. C. P. MILLER. 1954. Uniform polyhedra. Philos. Trans. Roy. Soc. London Ser. A **246:** 401–456.
6. FEJES TÓTH, L. 1964. Regular Figures. Pergamon. New York, N.Y.
7. GOTT, J. R. 1967. Pseudopolyhedrons. Amer. Math. Monthly **74:** 497–504.
8. GRÜNBAUM, B. 1967. Convex Polytopes. John Wiley & Sons, Inc. London, England.
9. MCMULLEN, P. 1967. Combinatorial regular polytopes. Mathematika **14:** 142–150.
10. MCMULLEN, P., CH. SCHULZ & J. M. WILLS. 1982. Equivelar polyhedral manifolds in E^3. Israel J. Math. **41:** 331–346.
11. MCMULLEN, P., CH. SCHULZ & J. M. WILLS. 1983. Polyhedral manifolds in E^3 with unusually large genus. Israel J. Math. **46:** 127–144.
12. SEIFERT, H. & W. THRELFALL. 1934. Lehrbuch der Topologie. (Reprint.) Chelsea Publishing Co. New York, N.Y.
13. STEWART, B. M. 1970. Adventures among the Toroids. Michigan State University. East Lansing, Mich. Private print.
14. WELLS, A. F. 1977. Three-Dimensional Nets and Polyhedra. Wiley–Interscience. London, England.
15. WILLS, J. M. 1983. Semi-Platonic manifolds. *In* Convexity and Its Applications. P. M. Gruber & J. M. Wills, Eds.: 413–421. Birkhäuser. Basel, Switzerland.

PART V. TILING, PACKING, COVERING, AND WEAVING

TILING EUCLIDEAN d-SPACE WITH CONGRUENT SIMPLEXES

Hans E. Debrunner

Mathematisches Institut
Universität Bern
CH 3000 Bern, Switzerland

1. INTRODUCTION

A tiling of Euclidean d-space E^d is a family $\mathscr{P} = \{P_\alpha \mid \alpha \in A\}$ of closed topological cells (called tiles) which fill E^d without gap and overlap, such that $E^d = \bigcup_{\alpha \in A} P_\alpha$, int $(P_\alpha \cap P_\beta) = \varnothing$ whenever $\alpha \neq \beta$. If each P_α is congruent to a fixed set $P \subset E^d$ (written $P_\alpha \cong P$), then the tiling is called *monohedral* and P is called its prototile, or in short P is said to admit the tiling $\mathscr{P}$. For this terminology and for a survey of known results and open problems on tilings cf. Grünbaum and Shephard [14] from where we have picked out the following: "Problem (E): Determine all simplexes ... which are prototiles of monohedral tilings of E^d." Of course, if P admits a tiling, then each P' similar to P in positive or negative ratio also does, so what we want is an enumeration of *types* of tiling simplexes, where the type of P is the set of all the P' similar to P. In dimension $d \leq 2$ any simplex obviously tiles, but for $d \geq 3$ the tiling property strongly restricts the possible types. For $d = 3$ Goldberg [12] gives a list of three one-parameter families and two additional special types ("Sommerville II and IV") of tiling simplexes. Pictures of them and the long history of the problem are compiled in [18]; we include the characterizing Coxeter diagrams (cf. Section 2.3) in FIGURE 1. For dimensions greater than 3 some special tiling types are described in [1, 2, 10] and Baumgartner [2, p. 223] conjectures that there are no more. But we will find new ones by the bushel. In fact, for E^d we will list $[\frac{1}{2}d] + 2$ one-parameter families (see THEOREMS I–III) of tiling types, and in addition several discrete series (see THEOREMS VII–X) with a total number of, roughly counted, $\frac{3}{4}d^2(1 + o(d))$ special types.

Now the reader is warned that the older literature and references on the subject are often obscure as to whether additional properties are to be required from tilings. In fact, most authors consider only monohedral tilings for which all tiles are *directly* congruent to the prototile; we call such tilings (and their prototiles) $(+)$-*monohedral.* The actual merit of Goldberg's aforementioned paper [12] is that it shows that in E^3 all known tiling types of simplexes admit $(+)$-monohedral tilings. Some authors consider only face-to-face tilings. Along this line, Sommerville [19, 20] (aside from considering spherical and hyberbolic 3-space) showed that there exist only four types of simplexes that admit $(+)$-monohedral face-to-face tilings of Euclidean E^3, but it seems that his proof is not complete, cf. [18]. Our investigation was first restricted to the listing of simplex types that admit a $(+)$-monohedral tiling of E^d. Now, if a prototile has in its symmetry group an orientation reversing isometry—we call such prototiles *amphicheiral* (="both-handed"), the remaining ones *monocheiral* (="enantiomorphous")—then of course each tiling is $(+)$-monohedral. But some of our methods for constructing tiling types yield simplexes which are amphicheiral or monocheiral according to the dimension d, and in the

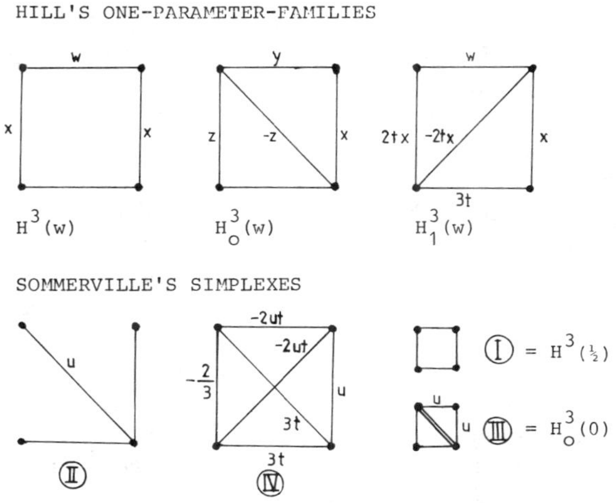

FIGURE 1. Coxeter diagrams of the tiling simplexes in E^3. $-\frac{1}{2} < w < 1$, $x = 2^{-1/2}(1 - w)^{1/2}$, $y = 2^{-1/2}(1 + w)^{1/2}$, $z = \frac{1}{2}(1 - w)^{1/2}(1 + w)^{-1/2}$, $u = 2^{-1/2}$, $t = 12^{-1/2}$

monocheiral case we are able to find (+)-monohedral tilings only for some of these values. For example (see THEOREMS V and VI), our $[\frac{1}{2}d] + 2$ one-parameter families of tiling types in E^d all admit (+)-monohedral tilings if $d \equiv 3 \bmod 4$, but if $d \equiv 0 \bmod 4$ we know (+)-monohedral tilings for only one of them. Furthermore, some types which tile, but not (+)-monohedrally, may well serve as bases for the construction of amphicheiral tiling types. Consequently, it seems appropriate not to restrict to (+)-monohedral tilings only.

There is such a wealth of tiling types that for $d > 4$ we cannot characterize each type explicitly by listing its dihedral angles or equivalently its Coxeter diagram; we can do this only for a stock of "central" types. But our theorems list all types by symbols which incorporate the precise way to construct them (together with corresponding tilings) starting from some "central" type, maybe in several steps, where for each step we give easy rules (LEMMAS 3, 4, and 8) which allow the reader to compute without difficulty the characterizing dihedral angles.

For $d = 3$ our enumeration reduces to Goldberg's list [12]. For $d = 4$ we list one amphicheiral and three monocheiral one-parameter families and in addition nine amphicheiral and five monocheiral special tiling types; we do not know whether in E^4 there is a monocheiral simplex which admits a (+)-monohedral tiling. Also, for no value $d \geq 3$ do we know whether our list is complete; we have exhausted only certain building principles and can assert that "most" simplexes do not admit tilings. In fact, each simplex that admits tilings of E^d with similar copies has a vanishing Dehn-functional, see [11]; but this condition is not sufficient for the tiling property.

2. DESCRIPTION AND MODIFICATION OF SIMPLICIAL TILINGS

2.1. *Notation*

We write $\mathbb{N}$ for $\{1, 2, \ldots\}$, $2\mathbb{N}$ for $\{2, 4, \ldots\}$, Z for the integers, and Z_d for the set $\{0, 1, \ldots, d - 1\}$ (sometimes with its additive structure modulo d); $\mathcal{S}_d$ stands for the symmetric group of the permutations of $(1, \ldots, d)$, and $\mathscr{A}_d$ for its subgroup of index 2, the alternating group; the characteristic ± 1 of $\pi \in \mathscr{S}_d$ is denoted $\chi(\pi)$. For Euclidean d-space E^d we take the vector space of real d-tuples $x = (x_1, \ldots, x_d)$ with scalar product $x \cdot y = \sum x_i y_i$ and canonical Euclidean distance; so we do not distinguish vectors and points in notation. $\mathscr{E}(d)$ stands for the full group of isometries of E^d, $\mathscr{E}^+(d)$ for the subgroup consisting of the orientation preserving isometries, and $\mathscr{E}^-(d)$ for $\mathscr{E}(d) - \mathscr{E}^+(d)$. $\mathscr{E}^\delta(d)$ means $\mathscr{E}^+(d)$ if $\delta = +1$ and $\mathscr{E}^-(d)$ if $\delta = -1$, and in analogy to the characteristic of permutations we write $\chi(\Phi) = \delta$ if $\Phi \in \mathscr{E}^\delta(d)$ and $\delta \in \{\pm 1\}$. Given $\Phi, \ldots, \Psi \in \mathscr{E}(d)$, Φ^i $(i = 0, 1, \ldots)$ will denote the ith iterate of Φ (in particular, $\Phi^0 = id_{E^d}$, $\Phi^1 = \Phi$), and $\langle \Phi, \ldots, \Psi \rangle$ the subgroup of $\mathscr{E}(d)$ generated by $\{\Phi, \ldots, \Psi\}$. $\Phi \in \mathscr{E}(d)$ is said to be a *symmetry* of $P \subset E^d$ if $\Phi(P) = P$, and it is said to be a *symmetry of the tiling* $\mathscr{P}$ if Φ permutes the tiles of $\mathscr{P}$. For polytopes and simplexes standard terminology is taken from Grünbaum [13]. Often we will find it convenient to consider a d-simplex $P \subset E^d$ as *ordered*, i.e., the vertex set $\{p_0, \ldots, p_d\}$ is provided with a fixed bijection onto Z_{d+1}; if this is not the order of the indexing, we will specify it by the order in a listing, say $p_{i_0}, \ldots, p_{i_d}$, where p_{i_j} will be the "jth vertex" in the order. It induces by oppositeness a simple ordering of the facets [$= (d-1)$-faces] of P. If a monohedral tiling of E^d is specified in the form $\{\Phi_\alpha(P) \mid \alpha \in A\}$ for suitable $\{\Phi_\alpha \mid \alpha \in A\} \subset \mathscr{E}(d)$, where P is a d-simplex, then an order on P induces via the Φ_α's an order on each tile. We say the d-polytope $P \subset E^d$ can be *dissected* into the d-polytopes $P_1, \ldots, P_r$ (in the sense of elementary geometry) and we write $P = P_1 + \cdots + P_r = \sum_1^r P_i$ if $P = \bigcup_1^r P_i$ and int $(P_j \cap P_k) = \varnothing$ for j, $k \in \{1, \ldots, r\}$, $j \neq k$; this last disjointedness condition is always understood if we write $\sum P_i$ in place of $\bigcup P_i$. In order to make the notation suggestive, certain frequently occurring monohedral tilings with prototile P^d or $Q^d, \ldots$ (specific d-polytopes) will be written as PP^d or $QQ^d, \ldots$, by doubling the capital.

2.2. *Simplex Normals*

Given a d-simplex $P \subset E^d$ with vertices $p_0, \ldots, p_d$, we will use the notation F_i for the facet opposite to p_i, Σ_i for the hyperplane aff (F_i) (which also is called a *wall* of P), and f_i for any nonzero vector orthogonal to Σ_i. A family $\{f_0, \ldots, f_d\}$ of facet normals to P is said to be *admissible* if all these normals point from the corresponding facet to the interior or all to the exterior of P; this happens if there are (uniquely up to a common factor) *positive* reals $\alpha_0, \ldots, \alpha_d$ such that $\sum \alpha_i f_i = 0$, cf. [9, p. 190]. An admissible family $\{f_0, \ldots, f_d\}$ carries even more geometrical information if it is *balanced*, i.e., if $\sum f_i = 0$; again by [9, p. 190] this means that the $|f_i|$ are proportional to the $(d-1)$-volume of the facets F_i $(i \in Z_{d+1})$. The type of the d-simplex $P \subset E^d$ is determined by any admissible family of facet normals of P or also by the $(d+1) \times (d+1)$-matrix $\|\cos \beta_{ij}\|$, where β_{ij} denotes the (interior) dihedral angle formed by the facets F_i, F_j at the $(d-2)$-face $F_i \cap F_j$. For an admissible family of facet normals, β_{ij} is supplementary to $\sphericalangle(f_i, f_j)$, so

$$\cos \beta_{ij} = -\cos \sphericalangle(f_i, f_j) = -\frac{f_i \cdot f_j}{|f_i|\,|f_j|}. \tag{1}$$

2.3. *Coxeter Diagrams*

To each d-simplex $P \subset E^d$ we associate a *labeled graph* $c(P)$, its Coxeter diagram, as follows: The $d + 1$ nodes of $c(P)$ represent the $d + 1$ facets of P. There is a branch marked with a label u_{ij} joining the ith and jth nodes whenever the dihedral angle β_{ij} formed by the corresponding facets is not a right angle, and for the label we choose $u_{ij} = \cos \beta_{ij} \neq 0$. In order to avoid overload of the Coxeter diagrams in our figures we will always omit the frequently occurring label $\frac{1}{2}$, so that an "unlabeled branch" •——• will correspond to a dihedral angle $\pi/3$; •—···—• will stand for a chain of several "unlabeled" branches. Furthermore, we will use •══• in place of a branch with label $-\frac{1}{2}$, corresponding to a dihedral angle $2\pi/3$.

We need also the slightly different notion of *complete Coxeter diagram* of a d-simplex $P \subset E^d$. This will be a *complete* graph with nodes $X_0, \ldots, X_d$ which correspond in this order to the facets $F_0, \ldots, F_d$ of P, together with the real function on the branch set which associates to the branch from X_i to X_j the label $\cos \beta_{ij}$; again β_{ij} is the dihedral angle of P at $F_i \cap F_j$. From here we fall back to our first notion by erasing branches with label 0. Occasionally we will draw diagrams in which only part of the branches with label 0 are erased; this occurs if some label is written as a variable in a one-parameter family and happens to assume the value 0.

A *symmetry* of a complete Coxeter diagram is a bijection ϕ of its set of nodes $\{X_0, \ldots, X_d\}$ onto itself such that the branch from $\phi(X_i)$ to $\phi(X_j)$ carries the same label as that from X_i to X_j for all $i, j \in Z_{d+1}$, $i \neq j$. Symmetries of $c(P)$ are defined similarly. Erasing all branches with label 0 often makes it easy to overlook all the symmetries of a diagram. This last point is important in view of:

LEMMA 1. *For each simplex $P \subset E^d$ the symmetries of P are in one-to-one correspondence with the symmetries of its Coxeter diagram $c(P)$.*

Proof. Each permutation ϕ of the node set $\{X_0, \ldots, X_d\}$ corresponds canonically to a permutation of the facet set $\{F_0, \ldots, F_d\}$ and by oppositeness to a permutation of the vertex set $\{p_0, \ldots, p_d\}$ of P. But this in turn corresponds uniquely to an affine mapping $\Phi: E^d \to E^d$ leaving P invariant as a set. Φ is an isometry if and only if ϕ preserves the labels on the branches. ∎

2.4. *The Halving Operation h_{ij}*

Dissection of tiles is an old technique to derive new tilings from a given one. The following case will occur frequently.

LEMMA 2. *Assume the d-polytope $P \subset E^d$ is the convex hull of the $d + 2$ points $a_1, \ldots, a_{d+2}$ and $\Phi \in \mathscr{E}(d)$ is a symmetry of P with "preferred" points a_{d+1}, a_{d+2} in the sense that Φ interchanges a_{d+1} with a_{d+2} and stabilizes Q, the convex hull of $\{a_1, \ldots, a_d\}$. Suppose $\frac{1}{2}(a_{d+1} + a_{d+2}) \in Q$. Then $P = P' + P''$, where $P' = \operatorname{conv}(Q \cup \{a_{d+1}\})$, $P'' = \Phi(P') = \operatorname{conv}(Q \cup \{a_{d+2}\})$, and both P' and P'' are d-simplexes.*

Proof. The conditions force Q to be $(d - 1)$-dimensional; $a_1, \ldots, a_d$ thus are the vertices of the $(d - 1)$-simplex Q and are permuted by Φ. Furthermore, a_{d+1} and a_{d+2} must lie on different sides of the hyperplane aff (Q) which therefore splits P into P' and P''. ∎

In a special case of the situation of LEMMA 2 we introduce a new useful notation. Namely, assume P is a d-simplex with vertex set $\{p_0, \dots, p_d\}$, Φ is a symmetry of P with $\Phi(p_i) = p_j$, $\Phi(p_j) = p_i$ for a fixed pair $i, j \in Z_{d+1}$, $i \neq j$. For the $d + 2$ points of LEMMA 2 we choose $p_0, \dots, p_d, \frac{1}{2}(p_i + p_j)$, where p_i, p_j play the role of "preferred" points. Then $P = P' + P''$ with $P' = \operatorname{conv}\{p_0, \dots, p_{j-1}, \frac{1}{2}(p_j + p_i), p_{j+1}, \dots, p_d\}$ and $P'' = \Phi(P')$. If for the ordered simplex P a symmetry Φ with these properties exists, we will say that P *admits operation* h_{ij} (or equally $h_{i,j}$) based on Φ and we set $h_{ij}(P) := P'$.

LEMMA 3. *Suppose the ordered d-simplex P admits the operation h_{ij} based on $\Phi \in \mathscr{E}(d)$. Then:*

(a) *from the vertex set $\{p_0, \dots, p_d\}$ of P we get that of $h_{ij}(P)$ by changing only p_j to $p'_j := \frac{1}{2}(p_j + p_i)$;*

(b) *from a balanced normal set $\{f_0 . \dots, f_d\}$ of P we get one of $h_{ij}(P)$ by changing only f_i to $f_i - f_j$ and f_j to $2f_j$;*

(c) *from an admissible normal set $\{f'_0, \dots, f'_d\}$ of P with $|f'_i| = |f'_j|$ we get one of $h_{ij}(P)$ by changing only f'_i to $f'_i - f'_j$;*

(d) *from the complete Coxeter diagram of P we get that of $h_{ij}(P)$ by changing labels on the branches which issue from the node X_i and lead to X_j or to a node connected to X_i or X_j with a branch of label $\neq 0$.*

Proof. (a) is clear; (c) and (d) follow from (b). More explicit rules on how to change labels should be given, but we do not need them. To prove (b) we remark that $h_{ij}(P) = P \cap H$, where H is the closed half space that contains p_i and is bounded by the bisector hyperplane Σ of the dihedral angle β_{ij}. Now Φ interchanges F_i with F_j and stabilizes Σ. So Σ dissects each facet of P except F_i, F_j into two congruent halves of which one is a facet of $h_{ij}(P)$. The facet F_j of P is also the jth facet of $h_{ij}(P)$, and the ith facet of $h_{ij}(P)$ is $\Sigma \cap P$. From $\Phi(f_i) = f_j$ it follows $|f_i| = |f_j|$; hence, $\pm(f_i - f_j)$ is normal to Σ. This means that from $f_0, \dots, f_d$ we get a normal set for $h_{ij}(P)$ with norms proportional to facet areas by leaving f_j unchanged, replacing f_i by $\alpha(f_i - f_j)$ for suitable $\alpha \in \mathbb{R} - \{0\}$, and changing f_k to $\frac{1}{2}f_k$ for each $k \neq i, j$. These vectors should add up to 0, so $\alpha = \frac{1}{2}$. We scale this set with the factor 2 to get (b). ▮

If in a tiling $\mathscr{P}$ of E^d by ordered simplexes each tile admits operation h_{ij} we will denote the resulting subdivision of $\mathscr{P}$ by $h_{ij}\mathscr{P}$. Of course, if a simplex P admits h_{ij} for suitable ordering of the vertices and admits a tiling $\mathscr{P}$ of E^d, then $h_{ij}P$ admits a tiling $h_{ij}\mathscr{P}$.

2.5. *The Doubling Operations Δ_i and Δ_{ij}*

Let a tiling $\mathscr{P} = \{P_\alpha \mid \alpha \in A\}$ of E^d with ordered simplexes be given such that for each tile there is an order preserving isometry onto the prototile P. We assume that there is a fixed $i \in Z_{d+1}$ such that to each tile P_α there is a tile $P_{\phi(\alpha)} \in \mathscr{P}$ such that the ith facets of P_α and $P_{\phi(\alpha)}$ are the same (as a set) and that the ith vertices of P_α and $P_{\phi(\alpha)}$ are mirror images of each other in the common ith wall. In this situation we set $\Delta_i(P_\alpha) := P_\alpha \cup P_{\phi(\alpha)}$, and there is $A' \subset A$ such that $\{\Delta_i(P_\alpha) \mid \alpha \in A'\}$ is a monohedral tiling with prototile $P \cup \sigma_i(P)$, where $\sigma_i \in \mathscr{E}(d)$ denotes reflection in the ith wall of P. Of course $\phi(\phi(\alpha)) = \alpha$, so ϕ is a permutation of order 2 on the index set A, and to get A' we have to choose one representative in A for each orbit of ϕ. The resulting tiling does not depend on this choice of $A' \subset A$ and will be written as $\Delta_i \mathscr{P}$. We are interested mainly in the case where the new prototile is again a simplex, so we note:

LEMMA 4. *If P is an ordered simplex and σ_i the reflection in its ith wall, then $P \cup \sigma_i P$ [written also $(1 + \sigma_i)P$] is a simplex if and only if the ith node of the Coxeter diagram $c(P)$ is an end node, i.e., X_i is connected with a branch to only one node, say X_j. In this case we say that P admits operation Δ_{ij} (or equally $\Delta_{i,j}$); we write $\Delta_{ij} P := P \cup \sigma_i P$ and*

(a) *from the ordered vertex set $\{p_0, \ldots, p_d\}$ of P we get that of $\Delta_{ij} P$ by changing only p_j to $p'_j := 2p_j - p_i$;*

(b) *from a balanced normal set $\{f_0, \ldots, f_d\}$ of P we get one of $\Delta_{ij} P$ by changing only f_i to $f_i + \frac{1}{2} f_j$ and f_j to $\frac{1}{2} f_j$;*

(c) *from an admissible normal set $\{f'_0, \ldots, f'_d\}$ of P we get one of $\Delta_{ij} P$ by changing only f'_i to $\sigma_i(f'_j)$;*

(d) *from the Coxeter diagram $c(P)$ we get that of $\Delta_{ij} P$ by changing the label* cos β_{ij} *on the branch joining X_i, X_j to* cos $2\beta_{ij}$ *and introducing a new branch labeled u_k from X_i to X_k for each node X_k which is joined to X_j with a branch labeled u_k;*

(e) *$\Delta_{ij} P$ admits the operations h_{ij} and h_{ji} based on σ_i and $h_{ij}(\Delta_{ij} P) = P$, $h_{ji}(\Delta_{ij} P) = \sigma_i P$.*

Proof. The 1-face $K = \operatorname{conv}(p_i, p_j)$ of P is orthogonal to the facet F_i if and only if all the hyperplanes containing K are orthogonal to F_i, or what amounts to the same: if some $d - 1$ hyperplanes which contain K and have linearly independent normals are orthogonal to F_i. As such we can choose the walls Σ_k ($k \neq i, j$) of P; this condition then reads in $c(P)$ that X_i is an end node joined to X_j. On the other hand, K is orthogonal to F_i if and only if p_j (which is fixed under σ_i as each point of F_i is) is the midpoint of the segment from p_i to $\sigma_i p_i$. This last condition states that $P \cup \sigma_i P$ is a simplex. Once the first part of the lemma is established we can easily verify (e). Finally (a), (b), and (c) will follow from LEMMA 3, and (d) from (c). ∎

We remark that for an ordered simplex $P \subset E^d$ it is easy to detect via the Coxeter diagram whether $P \cup \sigma_i P$ is a simplex $\Delta_{ij} P$; but whether a tiling $\mathscr{P}$ with this prototile P then admits operation Δ_i depends on the global structure of $\mathscr{P}$ and not on the prototile. If yes, the order on the simplexes of $\mathscr{P}$ and the choice of the index set $A' \subset A$ for $\Delta_i \mathscr{P}$ determine an order on each simplex of $\Delta_i \mathscr{P}$ via LEMMA 4(a).

2.6. *Determination of Simplexes*

Two ways of specifying a d-simplex in E^n (mostly $n = d$) will be particularly useful: Either we enumerate the vertices p_i (given by coordinates) or we describe a maximal simple edge path, leading from vertex to vertex, namely,

$$P = \operatorname{conv}\{p_0, \ldots, p_d\} = \langle p_0; b_1, \ldots, b_d \rangle, \tag{2}$$

where $b_i = p_i - p_{i-1}$ ($i = 1, \ldots, d$) or conversely $p_j = p_0 + \sum_{k \leq j} b_k$ ($j \in Z_{d+1}$). This is based on the definition

$$\langle p_0; b_1, \ldots, b_d \rangle := \left\{ p_0 + \sum_1^d \lambda_i b_i \,\middle|\, 1 \geq \lambda_1 \geq \cdots \geq \lambda_d \geq 0 \right\} \tag{3}$$

(cf. [16, p. 4]). In order for **(3)** to define a d-simplex (i.e., vertices are in general position) we have to start from a linearly independent d-tuple $b_1, \ldots, b_d$. Of course the notation **(3)** for a simplex P is in an essential way dependent on a chosen order $p_0, \ldots, p_d$ of the vertices. The facets of P then are given by $F_0 = \langle p_0 + b_1; b_2, \ldots, b_d \rangle$, $F_i = \langle p_0; b_1, \ldots, b_{i-1}, b_i + b_{i+1}, b_{i+2}, \ldots, b_d \rangle$ for $i = 1, \ldots, d-1$, and $F_d = \langle p_0; b_1, \ldots, b_{d-1} \rangle$. If the d-simplex is in E^d, we can obtain from these facet symbols a balanced family of facet normals by taking the cross product of the listed edge vectors. This results in:

$$f_0 = c_1, \qquad f_i = c_{i+1} - c_i \; (i = 1, \ldots, d-1), \qquad f_d = -c_d, \tag{4}$$

where for any fixed $\lambda \neq 0$ we put

$$c_i := (-1)^{i+1} \lambda (b_1 \times \cdots \times \hat{b}_i \times \cdots \times b_d) \qquad (i = 1, \ldots, d). \tag{5}$$

We recall the convention that in an array a symbol covered by ˆ is to be omitted. We also recall that the cross product $a_1 \times \cdots \times a_{d-1}$ of a $(d-1)$-tuple $a_1, \ldots, a_{d-1}$ of vectors of E^d is defined as that unique vector $c \in E^d$, for which $c \cdot v = \det(a_1, \ldots, a_{d-1}, v)$ for each $v \in E^d$. Equivalently (see [21, p. 84]): This cross product has absolute value equal to $(d-1)!$ times the $(d-1)$-volume of $\langle 0; a_1, \ldots, a_{d-1} \rangle$, if $a_1, \ldots, a_{d-1}$ are linearly independent it is orthogonal to the hyperplane spanned by $a_1, \ldots, a_{d-1}$; and the d-tuple $a_1, \ldots, a_{d-1}, c$ determines in this order the positive orientation of E^d. $a_1 \times \cdots \times a_{d-1}$ depends multilinearly and totally antisymmetrically on $a_1, \ldots, a_{d-1}$ and under $\Phi \in \mathscr{E}^\delta(d)$ $(\delta \in \{\pm 1\})$ we have $\Phi(a_1 \times \cdots \times a_{d-1}) = (-1)^\delta \Phi(a_1) \times \cdots \times \Phi(a_{d-1})$. The formulas **(1)**, **(2)**, **(4)**, **(5)** suffice to figure out the Coxeter diagram for any d-simplex $P \subset E^d$ which is given by its vertices. If we know a balanced family of normals of P, then **(1)** alone suffices to get the Coxeter diagrams of P, $h_{ij}P$, $\Delta_{ij}P$ using LEMMAS 3(b) and 4(b).

3. ONE-PARAMETER FAMILIES OF TILING d-SIMPLEXES

3.1. *The Basic Hadwiger–Hill Tiling $HH^d(w)$*

It is known (cf. [11]) that each 3-polytope that admits tilings of E^3 is "rectifiable," i.e., equivalent by dissection to some cube, so that in particular its volume can be determined from that of cubes without a limit process. This is the reason why all 3-simplex types which now are known to admit tilings of E^3 were described in 1896 by Hill [17] as solutions to the problem of exhibiting rectifiable simplexes (but as such Hill's list is not complete, cf. [5]!). Hadwiger [15] fully recognized the symmetry underlying Hill's basic construction; he reformulated and generalized it to arbitrary dimension to get a one-parameter family $H^d(w)$ of rectifiable d-simplexes for each d. These simplexes which we now will describe, stressing the tiling property, have been called basic Hadwiger–Hill simplexes in [3].

For fixed $\omega \in](1-d)^{-1}, 1[$ we choose an ordered basis $\mathbf{b}^\omega = \{b_1, \ldots, b_d\}$ of E^d such that

$$b_i \cdot b_i = 1, \; b_i \cdot b_j = \omega \qquad (1 \leq i, j \leq d, \; i \neq j). \tag{6}$$

The points $b_1, \ldots, b_d$ are the vertices of a regular spherical $(d-1)$-simplex of spherical side length arccos ω on the unit sphere $S^{d-1} \subset E^d$. Outer normals $c_1, \ldots, c_d$ to its facets are given by (**5**), and by symmetry we can normalize them in such a way that

$$c_i \cdot c_i = 1,\ c_i \cdot c_j = w \qquad (1 \le i, j \le d,\ i \ne j) \tag{7}$$

for some $w \in \mathbb{R}$. Then in analogy to (**1**) this spherical simplex has dihedral angles all equal to arccos $(-w)$. For the spherical side length arccos ω and the dihedral angle arccos $(-w)$ of any regular $(d-1)$-simplex on the unit sphere S^{d-1}

$$\text{either} \quad \omega = w = 0 \quad \text{or} \quad \omega^{-1} + w^{-1} = 2 - d \tag{8}$$

holds; see, e.g., [4, p. 278, (4)]. Note that not only does $\omega \in\ ](1-d)^{-1}, 1[$ uniquely determine $w \in\ ](1-d)^{-1}, 1[$, but the converse also holds.

With this basis $\mathbf{b}^{\omega}$, using the notation (**3**), we define the Hadwiger–Hill simplex

$$H^d(w) := \langle 0; b_1, \ldots, b_d \rangle = \operatorname{conv}\{p_0, \ldots, p_d\}, \tag{9}$$

where $p_0 = 0$ and $p_i = \sum_{k \le i} b_k$ for $i = 1, \ldots, d$. A balanced family of facet normals for $H^d(w)$ then is given by (**4**), and using (**7**) we see that the Coxeter diagram of $H^d(w)$, displayed in FIGURE 2, has labels $\cos \beta_{0,d} = w$, $\cos \beta_{0,1} = \cos \beta_{d-1,d} = (\frac{1}{2}(1-w))^{1/2}$ [corresponding to a dihedral angle $\frac{1}{2}(\pi - \beta_{0,d})$], $\cos \beta_{i,i+1} = \frac{1}{2}$ for $i = 1, \ldots, d-2$, and 0 otherwise.

In order to construct a tiling with prototile $H^d(w)$ we introduce the following isometries of E^d:

$$\text{for } \pi \in \mathscr{S}_d\text{: } S_\pi \in \mathscr{E}^{\chi(\pi)}(d) \text{ defined by } S_\pi(bb_i) = b_{\pi(i)}\ (i = 1, \ldots, d), \tag{10}$$

$$\text{for } n = (n_1, \ldots, n_d) \in Z^d\text{: } T_n \in \mathscr{E}^{+}(d) \text{ given by } T_n(x) = x + \textstyle\sum n_i b_i. \tag{11}$$

T_n is a translation, and S_π is an isometry because each $\pi \in \mathscr{S}_d$ is the product of transpositions (i, j), and $S_{(i,j)}$, which interchanges b_i and b_j and leaves b_k $(k \ne i, j)$ invariant, is a reflection of E^d in the $(d-1)$-subspace orthogonal to $b_i - b_j$. Obviously $S_\pi(H^d(w)) = \langle 0; b_{\pi(1)}, \ldots, b_{\pi(d)} \rangle$.

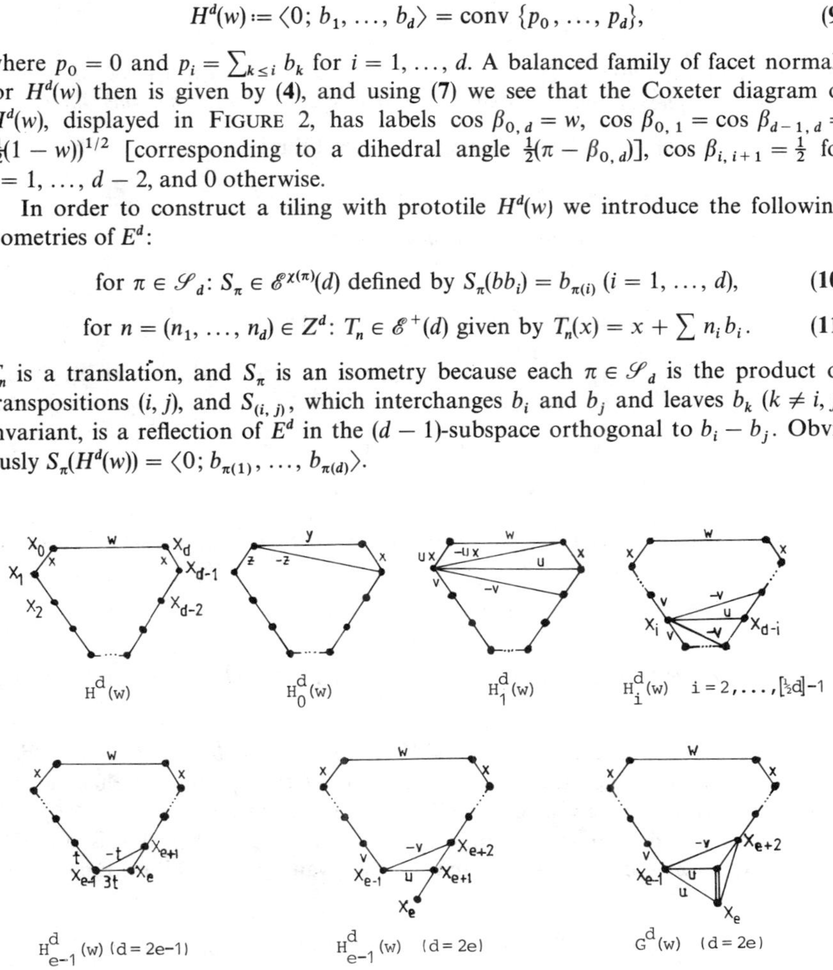

FIGURE 2. One-parameter families of tiling types in E^d. $(1-d)^{-1} < w < 1$, $x = 2^{-1/2}(1-w)^{1/2}$, $y = 2^{-1/2}(1+w)^{1/2}$, $z = \frac{1}{2}(1-w)^{1/2}(1+w)^{-1/2}$, $u = 2v = 2^{-1/2}$, $t = 12^{-1/2}$.

Now the rhombohedron

$$W_w = \left\{\sum_1^d \mu_i b_i \,|\, 0 \le \mu_i \le 1 \quad \text{for} \quad i = 1, \ldots, d\right\} \tag{12}$$

is the union of the pairwise congruent nonoverlapping simplexes $S_\pi(H^d(w))$ as π ranges over $\mathscr{S}_d$. In fact, according to (**3**), $\sum \mu_i b_i$ is in $\langle 0; b_{\pi(1)}, \ldots, b_{\pi(d)}\rangle$ exactly if (a) $1 \ge \mu_{\pi(1)} \ge \cdots \ge \mu_{\pi(d)} \ge 0$, and in its interior exactly if (b) $1 > \mu_{\pi(1)} > \cdots > \mu_{\pi(d)} > 0$; for each $\sum \mu_i b_i \in W_w$ there is at least one $\pi \in \mathscr{S}_d$ such that (a) holds and at most one such that (b) holds. If we carry over this dissection of W_w to each tile of the canonical lattice tiling $\{T_n W_w \,|\, n \in Z^d\}$ of E^d with prototile the rhombohedron (**12**), we eventually get the basic Hadwiger–Hill tiling of E^d. More precisely:

THEOREM I. *Given* $w \in \,]（1 - d)^{-1}, 1[$, *the family*

$$HH^d(w) := \{T_n S_\pi(H^d(w)) \,|\, n \in Z^d, \pi \in \mathscr{S}_d\}, \tag{13}$$

determined by the basis $\mathbf{b}^\omega$ *and by* (**9**), (**10**), (**11**), *is a monohedral tiling of* E^d *with prototile the simplex* $H^d(w)$. *The isometries* T_n *and* S_π *are symmetries of this tiling.*

Proof. It remains to check the last assertion. But this in turn follows readily from

$$T_{n'} S_{\pi'} \circ T_n S_\pi = T_m S_{\pi'\pi} \qquad \text{with } m_i = n'_i + n_{\pi'^{-1}(i)} \ (i = 1, \ldots, d), \tag{14}$$

a relation easy to check by letting both sides act on $\sum \lambda_i b_i$. ∎

3.2. *Further One-Parameter Families of Tilings*

We set $e = [\frac{1}{2}(d + 1)]$. The Coxeter diagram of $H^d(w)$ exhibits an involutory symmetry of $H^d(w)$. In fact, we have:

LEMMA 5. *The mapping* $U: E^d \to E^d$ *given by*

$$U\left(\sum \lambda_i b_i\right) = \sum (1 - \lambda_i) b_{d+1-i} \tag{15}$$

with respect to the basis $\mathbf{b}^\omega$ *is an involutory symmetry of* $H^d(w)$ *which induces the permutation*

$$\begin{pmatrix} p_0 & p_1 & \cdots p_d \\ p_d & p_{d-1} & \cdots p_0 \end{pmatrix}$$

on the vertex set. It is a symmetry of the tiling $HH^d(w)$, *and* $\chi(U) = +1$ *if* $d \equiv 0$ *or* 3 mod 4, $\chi(U) = -1$ *if* $d \equiv 1$ *or* 2 mod 4.

Proof. From (**15**) we deduce $U(p_i) = U(b_1 + \cdots + b_i) = b_{d-i} + \cdots + b_1 = p_{d-i}$, so U, being affine, maps the convex hull $H^d(w)$ of the p_i's onto itself and is an involution. From (**15**) we also see that $U = T_a \circ S_\alpha \circ I$, where $a = (1, 1, \ldots, 1) \in Z^d$, where $\alpha \in \mathscr{S}_d$ is determined by $\alpha(i) = d + 1 - i$ $(i = 1, \ldots, d)$, and I is given by $I(x) = -x$. So the relation $U \in \mathscr{E}(d)$ and the given value of $\chi(U)$ follow. Finally, with the same $\alpha \in \mathscr{S}_d$

$$U \circ T_n S_\pi = T_m S_{\alpha\pi\alpha} \circ U \qquad \text{for } m_i = -n_{\alpha(i)}. \tag{16}$$

This formula is easily checked by letting both sides act on $\sum \lambda_i b_i$, and it guarantees in view of $U(H^d(w)) = H^d(w)$ that U is a symmetry of $HH^d(w)$. ∎

From LEMMA 5 it follows that $H^d(w)$ admits the operations $h_{i,\,d-i}$ for $i = 0, \ldots, e - 1$ (this range for i is used in order to get each type just once). We set (see Section 2.4)

$$H_i^d(w) := h_{i,\,d-i}\, H^d(w) \qquad (i = 0, \ldots, e - 1). \tag{17}$$

We carry the dissection $H^d(w) = H_i^d(w) + UH_i^d(w)$ over to each tile of $HH^d(w)$ and get:

THEOREM II. *Set* $e = [\tfrac{1}{2}(d + 1)]$. *Given* $w \in\,](1 - d)^{-1}, 1[$ *and* $i \in \{0, 1, \ldots, e - 1\}$,

$$HH_i^d(w) := \{T_n S_\pi U^k(H_i^d(w)) \mid n \in Z^d,\ \pi \in \mathcal{S}_d,\ k \in Z_2\} \tag{18}$$

is a monohedral tiling of E^d *with prototile* $H_i^d(w)$ *given by* **(17)**.

The Coxeter diagrams of these "derived Hadwiger–Hill simplexes" are easy to construct from LEMMA 3 and are displayed in FIGURE 2.

Now let $d = 2e \in 2\mathbb{N}$. Then the Coxeter diagram of $H_{e-1}^d(w)$ has X_e as an end node joined to X_{e+1}, so $(1 + \sigma_e)H_{e-1}^d(w)$ is a simplex $\Delta_{e,\,e+1}H_{e-1}^d(w)$ according to LEMMA 4. To construct a tiling with this prototile we first observe that σ_e, the reflection of E^d in the eth wall of $H_{e-1}^d(w)$, is the isometry S_ε, where $\varepsilon = (e, e + 1) \in \mathcal{S}_d$. In fact, according to **(10)** we have $S_\varepsilon(p_i) = S_\varepsilon(\sum_{k \le i} b_k) = \sum_{k \le i} b_k = p_i$ for $i \neq e$ and $S_\varepsilon \in \mathcal{E}^-(d)$; this proves that S_ε is the reflection in the eth wall of $H^d(w)$ which also is the eth wall of $H_{e-1}^d(w)$. Then starting from the basic Hadwiger–Hill tiling $HH^d(w)$, the group decomposition $\mathcal{S}_d = \mathcal{A}_d + \mathcal{A}_d\varepsilon$ induces a monohedral tiling

$$\Delta_e HH^d(w) := \{T_n S_\pi(H^d(w) \cup S_\varepsilon H^d(w)) \mid n \in Z^d,\ \pi \in \mathcal{A}_d\}. \tag{19}$$

Its prototile $(1 + S_\varepsilon)H^d(w)$ equals conv $\{p_0, \ldots, p_d, p_e^*\}$, where we put $p_e^* := S_\varepsilon(p_e) = b_1 + \cdots + b_{e-1} + b_{e+1}$. We note that $U(p_e^*) = p_e^*$ and

$$\tfrac{1}{2}(p_e + p_e^*) = \tfrac{1}{2}(p_{e-1} + p_{e+1}), \tag{20}$$

so that according to LEMMA 2 $(1 + S_\varepsilon)H^d(w)$ has not only the symmetry S_ε with preferred points p_e, p_e^*, but also the symmetry U with preferred points p_{e-1}, p_{e+1}. So by LEMMA 2,

$$H^d(w) + S_\varepsilon H^d(w) = G^d(w) + U(G^d(w)),$$

where

$$G^d(w) := \text{conv}\,\{p_0, \ldots, p_e, p_e^*, p_{e+2}, \ldots, p_d\}.$$

We insert this into **(19)** and observe that from **(20)** and LEMMAS 3(a) and 4(a) it follows that $\Delta_{e,\,e+1}H_{e-1}^d(w) = \Delta_{e,\,e+1}h_{e-1,\,e+1}H^d(w) = G^d(w)$ because of the identical vertex sets on both sides. In summarizing we exclude the case $d = 2$ which gives nothing new:

THEOREM III. *Let* $4 \le d = 2e \in 2\mathbb{N}$. *Given* $w \in\,](1 - d)^{-1}, 1[$, *the family*

$$GG^d(w) := \{T_n S_\pi U^k(G^d(w)) \mid n \in Z^d,\ \pi \in \mathcal{A}_d,\ k \in Z_2\} \tag{21}$$

is a monohedral tiling of E^d *with prototile the simplex* $G^d(w) := \Delta_{e,\,e+1}H_{e-1}^d(w)$. *We have* $GG^d(w) = \Delta_e HH_{e-1}^d(w)$ *(cf. Section 2.5).*

The Coxeter diagram of $G^d(w)$, constructed with the help of LEMMA 4, is included in FIGURE 2.

3.3. *Discrimination of Types*

To construct new tilings based on $HH^d(w)$ we have exploited characteristic properties of the Coxeter diagrams such as symmetries and end nodes; the cases $w = 0$ and $w = \frac{1}{2}$, where we did not exhaust the possibilities, will be studied in Sections 4.5 and 4.8. Now it would be difficult to give complete characterizations of these Coxeter diagrams, but at least we can characterize each diagram within the class of Coxeter diagrams included in FIGURE 2, and thus we can show:

THEOREM IV. *For each $d \geq 3$ the infinitely many prototiles given by* THEOREMS I, II, *and* III *are of pairwise different type except that $H_i^{2e-1}(\frac{1}{2}) \cong H_{e-1-i}^{2e-1}(\frac{1}{2})$ for $0 \leq 2i < e - 1$.*

Proof. We assume $d > 4$; in the cases $d = 3, 4$ slight adjustments of the arguments are needed. Let a diagram $c(P)$ with $d + 1$ nodes (where $d = 2e$ or $d = 2e - 1$) be given which is isomorphic to a diagram in FIGURE 2. We first assume that $c(P)$ has some nontrivial symmetry ϕ. If ϕ fixes more than two nodes, inspection of FIGURE 2 shows that P is $G^{2e}(w)$. The value w can be recovered from the set L of all labeled branches (the omitted label $\frac{1}{2}$ included!) between the $d - 1$ nodes which are fixed under ϕ by the following fact:

The greatest label on L is w if it occurs once and is $(\frac{1}{2}(1 - w))^{1/2}$ if it occurs more than once. (*)

If ϕ fixes at most two nodes, P is $H_0^d(0)$ or $H_f^{4f+1}(\frac{1}{2})$ or $H^d(w)$, where the first two cases are characterized by a branch pair labeled $\frac{1}{2}$, $-\frac{1}{2}$ or $8^{-1/2}$, $-8^{-1/2}$ respectively, and, if this does not occur, w for $H^d(w)$ is recovered from the set L of all branches of $c(P)$ by (*). Now we restrict this discussion to diagrams without symmetries. Each such diagram in FIGURE 2 has exactly one node X_* from which a maximum number μ of branches issue, among which is a branch pair with labels m, $-m$ for some $m > 0$. First we consider the case (A) where $\mu = 3$. Then either (Aa) $m \neq 12^{-1/2}$ and $c(P)$ has end nodes, or (Ab) $m \neq 12^{-1/2}$ and $c(P)$ has no end nodes, or (Ac) $m = 12^{-1/2}$. In case (Aa), P is some $H_{e-1}^{2e}(w)$ (in fact, only $m = 8^{-1/2}$ occurs) and w is recovered from the set L of all branches not issuing from X_* by (*). In case (Ab), P is some $H_0^d(w)$ where we recover w from $m = \frac{1}{2}(1 - w)^{1/2}(1 + w)^{-1/2}$. In case (Ac), P is $H_0^d(\frac{1}{2})$ or some $H_{e-1}^{2e-1}(w)$. Then if more than four different labels occur in the diagram, P is $H_{e-1}^{2e-1}(w)$ for some $w \neq \frac{1}{2}$ which is recovered from (*) on the set L of all branches not issuing from X_*; if there are only four different labels, P is $H_0^d(\frac{1}{2})$ or $H_{e-1}^{2e-1}(\frac{1}{2})$ and for $d = 2e - 1$ the ambiguity remains, because $H_0^{2e-1}(\frac{1}{2}) \cong H_{e-1}^{2e-1}(\frac{1}{2})$. Now we come to the case (B) where $\mu > 3$. Then $\mu = 5$ and there are two branch pairs, labeled $8^{-1/2}$, $-8^{-1/2}$ and m, $-m$, issuing from X_*. Either (Ba) $m \neq 8^{-1/2}$ or (Bb) $m = 8^{-1/2}$. In case (Ba), P is $H_1^d(w)$ where w is recovered from $m = \frac{1}{2}(1 - w)^{1/2}$. In case (Bb), P is some $H_i^d(w)$ $(2 \leq i \leq e - 2)$ where first we recover w on the set L of all branches not issuing from X_* by (*). To recover $i \in \{2, \ldots, e - 2\}$ we remark that from X_* exactly one branch with label $2^{-1/2}$ issues and leads to a node X_{**}. If we now erase all the branches issuing from X_* and those issuing from X_{**}, the remaining graph is the disconnected union of strings. If $w = 0$, there are three strings, two with an equal number j of branches and the third on $d - 2j - 4$ branches; we have $i = j + 1$. If $w \neq 0$ and in addition $w \neq \frac{1}{2}$, there are two strings, one of which on an odd number (say k) of branches carries different labels; then

$i = \frac{1}{2}(k+1)$. Finally, if $w = \frac{1}{2}$ there are two strings, on k_1 and k_2 branches, all labeled $\frac{1}{2}$, where $k_1 + k_2 = d - 3$ and at least one of k_1, k_2 is odd. If k_2 is even, we have $i = \frac{1}{2}(k_1 + 1)$; if k_1 and k_2 are odd, we have $i = \frac{1}{2}(k_1 + 1)$ or $i = \frac{1}{2}(k_2 + 1)$, corresponding to the ambiguity $H_i^{2e-1}(\frac{1}{2}) \cong H_{e-1-i}^{2e-1}(\frac{1}{2})$. ▮

3.4. *One-Parameter Families of (+)-Monohedral Tilings*

Here we direct our attention to (+)-monohedral tilings. If $d \equiv 1$ or 2 modulo 4, $HH^d(w)$ is trivially (+)-monohedral because the prototile $H^d(w)$ is amphicheiral: It admits U as a symmetry and $\chi(U) = -1$ in this case, cf. LEMMA 5. Equally if $d = 2e \in 2\mathbb{N}$, $GG^d(w)$ is trivially (+)-monohedral because the prototile $G^d(w)$ is amphicheiral; it has the symmetry $S_\varepsilon = \sigma_e \in \mathscr{E}^-(d)$.

A glimpse of the Coxeter diagrams in FIGURE 2 shows that except for $w = 0$ or $\frac{1}{2}$, cases to be studied in Sections 4.5 and 4.8, $H^d(w)$ has no symmetry different from U and $H_i^d(w)$ ($i \in Z_e$) has no symmetry at all. So, if $w \neq 0, \frac{1}{2}$, $H^d(w)$ for $d \equiv 0$ or 3 mod 4 and $H_i^d(w)$ for $i \in Z_e$ are monocheiral and the tilings $HH^d(w)$, $HH_i^d(w)$ use both enantiomorphic forms of the prototile. Nevertheless, it is possible to modify these tilings in the case $d \equiv 3$ mod 4 in such a way that the resulting tilings $\bar{H}H^d(w)$, $\bar{H}H_i^d(w)$ are (+)-monohedral.

Aiming at this modification, we put (always with respect to the basis $\mathbf{b}^\omega$ introduced in Section 3.1) $b := b_1 + \cdots + b_d$ and denote by F_+ (F_-, respectively) the union of all those tiles $T_n S_\pi(H^d(w)) \in HH^d(w)$, for which $S_\pi \in \mathscr{E}^+(d)$ [$\mathscr{E}^-(d)$, respectively].

LEMMA 6. *Assume d is odd. Then from $x \in F_+$ (F_-, respectively) it follows $x + \lambda b \in F_+$ (F_-, respectively) for each $\lambda \in \mathbb{R}$. So F_+ and F_- are prismatic sets with prism axis parallel to $\mathbb{R}b$.*

Proof. Let $x = \sum \beta_i b_i$ be an interior point of $T_n S_\pi(H^d(w))$. According to (**3**), (**10**), and (**11**) we then have $\beta_{\pi(j)} = n_{\pi(j)} + \lambda_j$ ($j = 1, \ldots, d$) with $n_j = [\beta_j]$ and $1 > \lambda_1 > \cdots > \lambda_d > 0$, or equivalently, the mod 1 reductions $\beta_j^* := \beta_j - [\beta_j]$ of the components of x with respect to $\mathbf{b}^\omega$ satisfy $1 > \beta_{\pi(1)}^* > \cdots > \beta_{\pi(d)}^* > 0$. But addition of λ, which is a shift, induces a cyclic permutation of the reductions mod 1, so the components of $x + \lambda b$ satisfy $1 > (\beta_{\pi\zeta(1)} + \lambda)^* > \cdots > (\beta_{\pi\zeta(d)} + \lambda)^* \geq 0$ for some cyclic permutation $\zeta \in \mathscr{S}_d$. Putting $m_j := [\beta_j + \lambda]$ ($j = 1, \ldots, d$) and $m = (m_1, \ldots, m_d)$ it follows $x + \lambda b \in T_m S_{\pi\zeta}(H^d(w))$ where S_π and $S_{\pi\zeta}$ are both in $\mathscr{E}^+(d)$ or both in $\mathscr{E}^-(d)$, because for $d \equiv 1$ mod 2 cyclic permutations have characteristic $+1$. So $x + \lambda b \in F_+$ (F_-, respectively) if $x \in F_+$ (F_-, respectively). But F_+ and F_- are closed and $T_n S_\pi(H^d(w))$ is the closure of its interior, so by continuity of the transformation the result holds for general x. ▮

Now we define $W: E^d \to E^d$ by

$$W(x) = x - 2\,\frac{b \cdot x}{b \cdot b}\, b$$

as the reflection in the $(d-1)$-subspace Σ orthogonal to b, and we put

$$W^\pi := \begin{cases} id_{E^d} & \text{if } \chi(\pi) = +1, \\ W & \text{if } \chi(\pi) = -1. \end{cases}$$

Then we have:

THEOREM V. *For d odd and* $w \in \,](1-d)^{-1}, 1[$, *the family*

$$\bar{H}H^d(w) := \{W^\pi T_n S_\pi(H^d(w)) \mid n \in Z^d, \pi \in \mathscr{S}_d\} \tag{22}$$

is a $(+)$*-monohedral tiling of* E^d *with prototile* $H^d(w)$ *given by* **(9)**.

Proof. Obviously each $W^\pi T_n S_\pi$ is in $\mathscr{E}^+(d)$. $\bar{H}H^d(w)$ arises from $HH^d(w)$ by reflecting in Σ all tiles lying in F_-. Since F_- is a prismatic region stable under this reflection by LEMMA 6, the result is again a tiling of E^d. ▮

Now the prototile $H^d(w)$ admits operations $h_{i,\,d-i}$ $(i \in Z_e)$ based on U, and corresponding to THEOREM II we have:

THEOREM VI. *For* $d = 2e - 1 \equiv 3 \mod 4$, $w \in \,](1-d)^{-1}, 1[$, *and* $i \in \{0, \ldots, e-1\}$, *the family*

$$\bar{H}H_i^d(w) := \{W^\pi T_n S_\pi U^k(H_i^d(w)) \mid n \in Z^d, \pi \in \mathscr{S}_d, k \in Z_2\} \tag{23}$$

is a $(+)$*-monohedral tiling of* E^d *with prototile* $H_i^d(w)$ *given by* **(17)**.

Proof. For $d \equiv 3 \mod 4$ we have $W^\pi T_n S_\pi \in \mathscr{E}^+(d)$ and $U \in \mathscr{E}^+(d)$, cf. LEMMA 5. $\bar{H}H_i^d(w)$ is a subdivision of $\bar{H}H^d(w)$, so it is a tiling. ▮

We remark that $HH^d(w)$ is a face-to-face tiling, but $\bar{H}H^d(w)$, $HH_i^d(w)$, $\bar{H}H_i^d(w)$, and $GG^{2e}(w)$ are not.

4. TILINGS DERIVED FROM REFLECTION GROUPS

4.1. *Coxeter Tilings*

In each dimension d there exist a few simplicial tilings $\mathscr{P}$ of E^d having the following property: For a fixed tile $P^d \in \mathscr{P}$ the reflections $\sigma_0, \ldots, \sigma_d$ in the $d+1$ walls of the simplex P^d are symmetries of $\mathscr{P}$ but not of any tile. It follows easily that the group P_d^* generated by $\sigma_0, \ldots, \sigma_d$ acts transitively on the tiles of $\mathscr{P}$ and that P^d is a fundamental domain of this group P_d^*. To stress the dependence on the group we write in the above situation $P^d = C(P_d^*)$, $\mathscr{P} = CC(P_d^*)$ and call $CC(P_d^*)$ a Coxeter tiling. All these tilings and the associated groups P_d^* were determined by Coxeter [8] and are widely known because of their relations to other topics, e.g., classification of semisimple Lie groups (cf. [7]). The result can be summarized as follows:

LEMMA 7. *Let* P_d^* *denote an infinite discrete group of isometries* $\tau: E^d \to E^d$ *generated by finitely many reflections, and assume that no proper subspace of* E^d *is stable under* P_d^*. *Then*

(a) P_d^* *has as its fundamental domain a d-simplex* $C(P_d^*) = P^d \subset E^d$ *and is generated by the* $d+1$ *reflections* σ_i $(i = 0, \ldots, d)$ *in the walls of* P^d; *moreover,*

$$CC(P_d^*) := \{\tau(P^d) \mid \tau \in P_d^*\}$$

is a monohedral tiling of E^d *with the simplex* P^d *as prototile ("Coxeter tiling").*

(b) *The dihedral angles β_{ij} formed by the facets F_i, F_j of P^d are submultiples of π; more precisely,*

$$\beta_{ij} = \pi/n_{ij} \qquad \text{with } n_{ij} \in \{2, 3, 4, 6\},$$

and the group P_d^* has a corresponding presentation

$$P_d^* = \{\sigma_i\colon \sigma_i^2 = 1 = (\sigma_i \sigma_j)^{n_{ij}}\}_{i,\, j \in Z_{d+1},\, i \neq j}$$

which can be recovered from the Coxeter diagram of P^d and conversely (*in particular, σ_i and σ_j commute if the nodes X_i, X_j are not joined by a branch*).

(c) *The set of groups P_d^* (up to isomorphism) consists of four series, A_d^* ($d \geq 1$), B_d^* ($d \geq 2$), C_d^* ($d \geq 3$), D_d^* ($d \geq 4$), and five "exceptional" groups, F_4^*, G_2^*, E_6^*, E_7^*, E_8^*. The Coxeter diagrams of their fundamental regions are listed in* FIGURE 3 *for $d \geq 3$.*

We remark that our Coxeter diagrams are the "extended diagrams" of [7] and that our labels on branches are cos (π/n_{ij}) whereas other authors use n_{ij} as a label.

4.2. *The Dissection Operation $h_{i_1 \ldots i_r}$*

The various symmetries of Coxeter tilings and of their tiles give rise to many derived simplicial tilings. To demonstrate them we introduce the following generalization of operation h_{ij} (cf. Section 2.4). Assume for the d-simplex $P \subset E^d$ with vertex set $\{p_0, \ldots, p_d\}$ that the ordered r-tuple $(p_{i_1}, \ldots, p_{i_r})$ of vertices, where $r > 1$, is the orbit of the vertex p_{i_r} under a symmetry Φ of P; i.e., $p_{i_j} = \Phi^j(p_{i_r})$ for $j = 1, \ldots, r$ and $p_{i_j} \neq p_{i_r}$ for $j < r$. Then the barycenter $p^* = (1/r)(p_{i_1} + \cdots + p_{i_r})$ of the orbit is fixed under Φ and for the facets $F_{i_j} = \Phi^j(F_{i_r})$ holds. Now $p^* \in F_i$ if and only if $i \notin \{i_1, \ldots, i_r\}$, so the pyramid with apex p^* and base F_i is degenerate (its affine hull has dimension $< d$) if $i \notin \{i_1, \ldots, i_r\}$ and is a d-simplex P_i if $i \in \{i_1, \ldots, i_r\}$. It follows that we have a dissection $P = \sum_{k=1}^{r} P_k$ with $P_k = \operatorname{conv}(\{p^*\} \cup F_{i_k})$ and $P_k = \Phi^k(P_r)$ for $k = 1, \ldots, r$. If P has a symmetry Φ with these properties we will say that P *admits the operation $h_{i_1 \cdots i_r}$* (or equally, $h_{i_1, \ldots, i_r}$) *based on* Φ and we set $h_{i_1 \cdots i_r} P := P_r$.

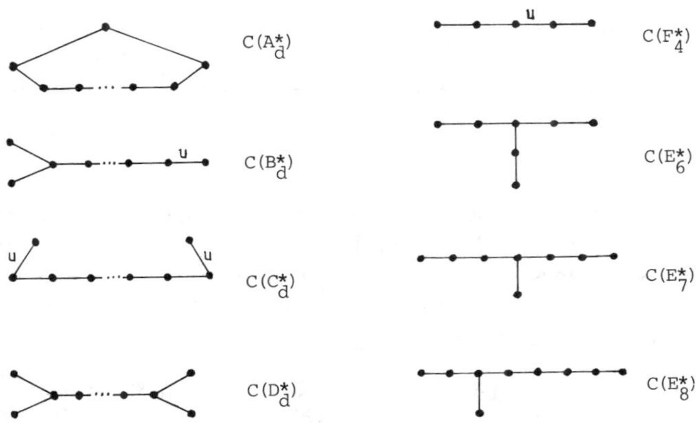

FIGURE 3. Types given as fundamental regions of reflection groups. $u = 2^{-1/2}$

LEMMA 8. *Suppose the ordered d-simplex P admits the operation* $h_{i_1 \cdots i_r}$ *based on* $\Phi \in \mathscr{E}(d)$. *Then*

(a) *from the vertex set* $\{p_0, \ldots, p_d\}$ *of P we get that of* $h_{i_1 \cdots i_r} P$ *by changing only* p_{i_r} *to* $p'_{i_r} := p^* = (1/r) \sum_{k=1}^{r} p_{i_k}$;

(b) *from a balanced normal set* $\{f_0, \ldots, f_d\}$ *of P we get one of* $h_{i_1 \cdots i_r} P$ *by changing* f_{i_k} *to* $f_{i_k} - f_{i_r}$ *for* $k = 1, \ldots, r-1$ *and* f_{i_r} *to* rf_{i_r};

(c) *from the complete Coxeter diagram of P we get that of* $h_{i_1 \cdots i_r} P$ *by changing the labels on the branches issuing from* $X_{i_1}, \ldots, X_{i_{r-1}}$.

Proof. (a) follows from the definition of $h_{i_1 \cdots i_r}$. To prove (b) we assume without loss of generality that $(i_1, i_2, \ldots, i_r) = (1, 2, \ldots, r-1, 0)$ and $P = \langle 0; b_1, \ldots, b_d \rangle$ for linearly independent $b_1, \ldots, b_d$. A balanced normal set $\{f_0, \ldots, f_d\}$ of P then is given by (4), (5). The analogous computation for $\bar{P} := h_{1, \ldots, r-1, 0} P = \langle p^*; b_1 - p^*, b_2, \ldots, b_d \rangle$, with $p^* = (1/r) \sum_{i=1}^{r-1} (r-i) b_i$ (= barycenter of the first r vertices), leads to normals $\bar{f}_0 = f_0$, $\bar{f}_i = (1/r)(f_i - f_0)$ $(i = 1, \ldots, r-1)$, and $\bar{f}_i = (1/r) f_i$ $(i = r, \ldots, d)$. Scaling with r gives the assertion (b). A weak corollary of (b) is assertion (c). ▮

Assume we have a tiling $\mathscr{P} = \{\tau(P) \mid \tau \in G\}$, where G is a subset of $\mathscr{E}(d)$. If P admits operation $h_{i_1 \cdots i_r}$ based on Φ so that we have a dissection $P = \sum_{k=0}^{r-1} \Phi^k(Q)$ with $Q = h_{i_1 \cdots i_r} P$, then we will denote the corresponding subdivision $\{\tau\Phi^k(Q) \mid \tau \in G, k \in Z_r\}$ of $\mathscr{P}$ by $h_{i_1 \cdots i_r} \mathscr{P}$.

4.3. *Forming Blocks*

The reflection group structure which governs each Coxeter tiling allows the operation Δ_i of Section 2.5 to be used with great efficacy and in an iterated manner. The group law allows the assembly of tiles globally into "blocks" without the need of checking geometric hypotheses. If we combine this with dissection we get:

LEMMA 9. *Assume that in the Coxeter tiling* $CC(P_d^*) = \{\tau(P^d) \mid \tau \in P_d^*\}$ *the prototile* P^d *has a dissection* $P^d = \sum_{\Phi \in \Omega} \Phi(Q)$, *where* Ω *is a set of symmetries of* P^d. *Let G be a finite subgroup of* P_d^* *and* $[P_d^*/G] \subset P_d^*$ *a system of representatives of the left cosets of G so that we have the group partition* $P_d^* = \sum \tau G$, *where the sum extends over all* $\tau \in [P_d^*/G]$. *Assume that* [*in* $\mathscr{E}(d)$] $\Phi G = G \Phi$ *holds for each* $\Phi \in \Omega$. *Then*

$$\{\tau\Phi(Q_G) \mid \tau \in [P_d^*/G], \Phi \in \Omega\} \tag{24}$$

is a monohedral tiling of E^d *with prototile the polyhedron*

$$Q_G := \bigcup_{\gamma \in G} \gamma(Q).$$

Remark. We will be concerned mainly with the special case where G is the subgroup $\langle \sigma_{j_1}, \ldots, \sigma_{j_s} \rangle$ of P_d^* generated by the reflections $\sigma_{j_1}, \ldots, \sigma_{j_s}$ in s pairwise orthogonal hyperplanes. Then G contains 2^s elements, namely, the summands of $(1 + \sigma_{j_1}) \cdots (1 + \sigma_{j_s})$ expanded in the group ring ZP_d^* of P_d^*. In this situation we write more explicitly $(1 + \sigma_{j_1}) \ldots (1 + \sigma_{j_s})Q$ in place of Q_G.

Proof. It is clear that $\{\tau(P_G) \mid \tau \in [P_d^*/G]\}$ is a tiling, where $P_G = \bigcup_{\gamma \in G} \gamma(P^d)$. Now we have a dissection $P_G = \sum_{\gamma \in G} \gamma(P^d) = \sum_{\gamma \in G} \sum_{\Phi \in \Omega} \gamma\Phi(Q) = \sum_{\Phi \in \Omega} \sum_{\gamma' \in G} \Phi\gamma'(Q) = \sum_{\Phi \in \Omega} \Phi(Q_G)$. Combination of these two facts gives the assertion. ▮

4.4. $CC(A_d^*)$ *is* $HH^d(\frac{1}{2})$

We insert the special value $\omega = -d^{-1}$ into (**8**) and obtain $w = \frac{1}{2}$. The resulting Coxeter diagram of $H^d(\frac{1}{2})$ is identical to that of $C(A_d^*)$ [cf. Lemma 7(c)]. So we assume that $C(A_d^*) = H^d(\frac{1}{2}) = \langle 0; b_1, \ldots, b_d \rangle$ with respect to the basis $\mathbf{b}^\omega$ with $\omega = -d^{-1}$. On the one hand, we now have the reflections σ_i in the ith wall of $H^d(\frac{1}{2})$ which are used to define $CC(A_d^*)$; on the other hand, the isometries $T_n S_\pi$ are used to define $HH^d(\frac{1}{2})$. For $i = 1, \ldots, d-1$, $\sigma_i = S_{(i, i+1)}$ holds. In fact, the right-hand transformation fixes the vertices $p_j = \sum_{k \le j} b_k$ $(j \ne i)$ of $H^d(\frac{1}{2})$ and is in $\mathscr{E}^-(d)$; but this characterizes σ_i. Along the dth wall of $\langle 0; b_1, \ldots, b_d \rangle$ we find in $HH^d(\frac{1}{2})$ the simplex $\langle -b_d; b_d, b_1, \ldots, b_{d-1} \rangle$. It has the d vertices $p_0, \ldots, p_{d-1}$ in common with $H^d(\frac{1}{2})$ and the remaining vertices $-b_d$, $b_1 + \cdots + b_d$ of these two simplexes are mirror images of each other in the common wall which contains the origin and is spanned by $b_1, \ldots, b_{d-1}$. In fact, the midpoint $\frac{1}{2}(-b_d + (b_1 + \cdots + b_d))$ is in this wall and the difference vector $(b_1 + \cdots + b_d) - (-b_d)$ is orthogonal to b_j $(j = 1, \ldots, d-1)$ in view of $b_j \cdot b_j = 1$, $b_i \cdot b_j = -d^{-1}$ $(j \ne i)$. So the mirror image $\sigma_d H^d(\frac{1}{2})$ of $H^d(\frac{1}{2})$ is $\langle -b_d; b_d, b_1, \ldots, b_{d-1} \rangle$, a tile of $HH^d(\frac{1}{2})$. Since the symmetry U of $HH^d(\frac{1}{2})$ (cf. Lemma 5) permutes the facets F_0, F_d of $H^d(\frac{1}{2})$, we have $\sigma_0 H^d(\frac{1}{2}) = U \sigma_d H^d(\frac{1}{2})$, so $\sigma_0 H^d(\frac{1}{2})$ too is a tile of $HH^d(\frac{1}{2})$. Since each $T_n S_\pi$ is a symmetry of $HH^d(\frac{1}{2})$ (cf. Theorem I), it follows that for each tile of $HH^d(\frac{1}{2})$ the mirror images of this tile in its $d+1$ walls are again tiles of $HH^d(\frac{1}{2})$. From this we conclude (in the manner of [9, pp. 80–81]) that the reflections $\sigma_0, \ldots, \sigma_d$ in the walls of $H^d(\frac{1}{2})$ are symmetries of $HH^d(\frac{1}{2})$, so that $HH^d(\frac{1}{2}) = CC(A_d^*)$. Note that we showed $\sigma_0 H^d(\frac{1}{2}) = \langle b_1; b_2, \ldots, b_d, b_1 \rangle = T_h S_\zeta H^d(\frac{1}{2})$ with $h = (1, 0, \ldots, 0) \in Z^d$, $\zeta = (1, 2, \ldots, d) \in \mathscr{S}_d$, but that $\sigma_0 \ne T_h S_\zeta$. In fact,

$$R := \sigma_0 T_h S_\zeta \tag{25}$$

is a nontrivial symmetry of the prototile $H^d(\frac{1}{2})$ which induces cyclic permutation of the vertices $p_0, \ldots, p_d$. Accordingly, we write $p_{d+1} := p_0$ and we have:

Lemma 10. *The symmetry group* Ξ *of the simplex* $C(A_d^*)$ *is isomorphic to the dihedral group* $\mathscr{D}_{d+1}$. *It has a presentation*

$$\Xi = (R, V \colon R^{d+1} = V^2 = 1, RV = VR^{-1}) \tag{26}$$

where the isometries R and V act according to

$$V(p_j) = p_{d+1-j}, \; R(p_j) = p_{j+1} \qquad (j = 0, \ldots, d)$$

on the vertex set. We have $R \in \mathscr{E}^-(d)$ *if* $d \equiv 1 \bmod 2$, $V \in \mathscr{E}^-(d)$ *if* $d \equiv 2$ *or* $3 \bmod 4$, *otherwise* $V, R \in \mathscr{E}^+(d)$. *R and V are symmetries of* $CC(A_d^*)$; *in fact,*

$$V\sigma_i = \sigma_{d+1-i} V, \; R\sigma_i = \sigma_{i+1} R \qquad (i = 0, \ldots, d) \tag{27}$$

hold.

Proof. For R we recall (**25**) and $C(A_d^*) = H^d(\frac{1}{2})$. For V, set $V := RU$ with U from Lemma 5, applied for $\omega = -d^{-1}$. $R, V \in \mathscr{E}(d)$ and the action of R, V on the vertices follow from this definition. $\chi(R)$ and $\chi(V)$ are the characteristics of the induced vertex permutations. That R and V generate Ξ can be seen from the Coxeter diagram of $C(A_d^*)$, which in its symmetric form is a regular $(d+1)$-gon. Now $V\sigma_i V$ fixes p_k $(k \ne d+1-i)$ and is in $\mathscr{E}^-(d)$, so it coincides with σ_{d+1-i}. Similarly, $R\sigma_i R^{-1}$ fixes p_k $(k \ne i+1)$ and is in $\mathscr{E}^-(d)$, so it coincides with σ_{i+1}. ∎

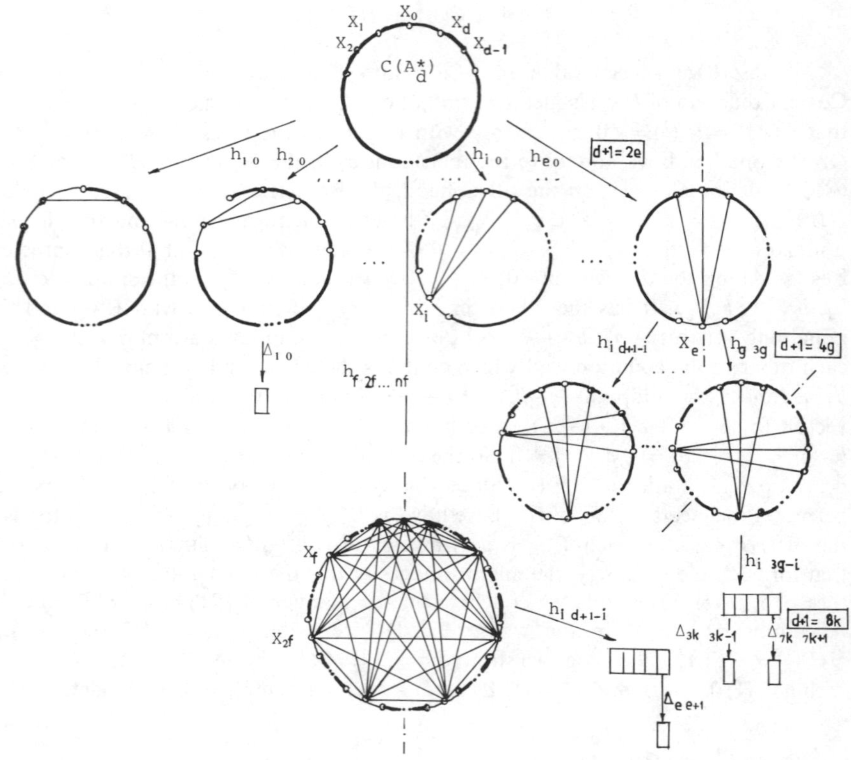

FIGURE 4. Pattern for modifications of $CC(A_d^*)$, cf. THEOREM VII. Heavy lines represent branches with label $\frac{1}{2}$; thin lines represent branches with labels $\neq \frac{1}{2}$.

4.5. *Tilings Derived from $CC(A_d^*)$*

In this section indices on vertices, facets, and corresponding operations are modulo $d + 1$. We exclude the case $d = 3$ to avoid double listing. FIGURE 4 may serve as a guide.

THEOREM VII. *Assume $d > 3$. The following simplexes are prototiles of monohedral tilings of E^d:*

(a) $A^d := C(A_d^*) = H^d(\frac{1}{2})$; *amphicheiral for $d \not\equiv 0$ mod 4.*

(b) $h_{i,\,0}A^d$ *for* $i = 1, \ldots, [\frac{1}{2}d]$; *(+)-monohedral if either* (a) $d \equiv 3$ mod 4 *and i odd, or* (b) $d \equiv 1$ mod 4 *and i even.*

(c) $\Delta_{1.\,0}h_{2.\,0}A^d$; *amphicheiral.*

(d) *If* $d + 1 = 4g \in 4\mathbb{N}$: $h_{i,\,3g-i}h_{g,\,3g}h_{2g,\,0}A^d$ *for* $(3/2)g < i < (7/2)g$, $i \neq 3g$.

(e) *If* $d + 1 = 8k \in 8\mathbb{N}$: $\Delta_{3k,\,3k-1}h_{3k+1,\,3k-1}h_{2k,\,6k}h_{4k,\,0}A^d$; *amphicheiral.*

(f) *If* $d + 1 = 8k \in 8\mathbb{N}$, $k > 1$: $\Delta_{7k,\,7k+1}h_{7k-1,\,7k+1}h_{2k,\,6k}h_{4k,\,0}A^d$; *amphicheiral.*

For each divisor $n > 1$ of $d + 1$, with $nf = d + 1$:

(g) $h_{f,\,2f,\,\ldots,\,nf}A^d$; *amphicheiral for* $d \equiv 2$ *or* 3 mod 4; *(+)-monohedral if* $d \equiv 1$ mod 4 *and f even.*

(h) $h_{i,d+1-i}h_{f,2f,\ldots,nf}A^d$ *for* $i = 1, \ldots, [\frac{1}{2}d]$; *amphicheiral for* $d + 1 = 4i = 2f \in 8\mathbb{N}$ (*i.e.*, $h_{g,3g}h_{2g,0}A^{4g-1}$ *for* $g = i =$ *even*); $(+)$-*monohedral if* $d \equiv 1$ mod 4 *and* f *even*.

(i) *If* $n > 1$ *is odd and* $d + 1 = nf = 2e \in 2\mathbb{N}$: $\Delta_{e,e+1}h_{e-1,e+1}h_{f,2f,\ldots,nf}A^d$; *amphicheiral.*

If not marked otherwise, these simplexes are monocheiral. They are of pairwise different types and each of the prototiles $H^d(w)$, $H_i^d(w)$ $(i = 0, \ldots, [\frac{1}{2}(d-1)])$, $G^d(w)$ *of* THEOREMS I, II, *and* III *belongs for* $w = \frac{1}{2}$ *and no other value of* w *to one of the types listed above.*

Proof. The assertions on amphicheirality or monocheirality are based on inspection of the symmetries in Coxeter diagrams. Such symmetries also demonstrate the applicability of dissection operations for the prototile and consequently for all the tiles of the corresponding tiling. On the other hand, end nodes in a Coxeter diagram demonstrate the applicability of operations Δ_{ij} for a prototile; we will be able to prove in each case that the corresponding operation Δ_i can be applied to some tiling with this prototile.

We start with cases (b), (d), (g), and (h) which involve only dissections of $CC(A_d^*)$. First we look for dissections based on involutions of A^d. The vertex pairs of A^d realize $[\frac{1}{2}(d+1)]$ different distances and any two pairs that realize the same distance are equivalent under some symmetry of A^d. The inequivalent pairs are represented by (p_i, p_0) $(i = 1, \ldots, [\frac{1}{2}(d+1)])$, and p_i, p_0 are interchanged by R^iV. On this symmetry we base the operation $h_{i,0}$ and we get the prototiles $h_{i,0}A^d$ of the tilings $h_{i,0}CC(A_d^*)$; see Section 2.4. Here we exclude the case $d = 2e - 1$, $i = e$ from the listing under (b) since it will occur under the next. Namely, we assume $d + 1 = nf$ with $n, f \in \mathbb{N}$, $n > 1$. Then A^d admits operation $h_{f,2f,\ldots,nf}$ based on R^f. This yields the tiling $h_{f,2f,\ldots,nf}CC(A_d^*)$ with prototile listed under (g). This prototile has V as a nontrivial symmetry, and based on V we perform the operations $h_{i,d+1-i}$ $(i = 1, \ldots, [\frac{1}{2}d])$. The result is tilings $h_{i,d+1-i}h_{f,\ldots,nf}CC(A_d^*)$ with prototiles listed under (h). Here there is one case where a symmetry is left, namely, if $d + 1 = 4g \in 4\mathbb{N}$, $f = 2g$, $i = g$. In this case $h_{g,3g}h_{2g,0}A^d$ admits operation $h_{i,3g-i}$ based on the involution R^gV, and a suitable range for $i \in \mathbb{N}$ is $(3/2)g < i < (7/2)g$. The resulting tilings $h_{i,3g-i}h_{g,3g}h_{2g,0}CC(A_d^*)$ have their prototile listed under (d), but we exclude $i = 3g$ from the listing since $h_{3g,0}h_{g,3g}h_{2g,0}A^d = h_{g,3g}h_{g,2g,3g,0}A^d$, where the right-hand side appears under (h) and the equality is easily checked by applying LEMMA 8(a) iteratively.

Now dissection based on symmetries is exhausted and we look for end nodes in the diagrams. Note that an end node X_k' always appears after an operation h_{ij} based on a symmetry that interchanges X_i, X_j and fixes X_k, where X_k is joined to X_i and X_j only.

In (b), $c(h_{2,0}A^d)$ has X_1 as an end node. Since $h_{2,0}A^d$ and A^d have their first wall in common, the reflection in this wall is $\sigma_1 \in A_d^*$. By (**27**), σ_1 commutes with R^2V, on which $h_{2,0}$ is based. Application of LEMMA 9 with $\Omega = \langle 1, R^2V\rangle$ and $G = \langle\sigma_1\rangle = \{1, \sigma_1\}$ then gives the monohedral tiling

$$\{\tau(R^2V)^j(Q_G) \mid \tau \in [A_d^*/\langle\sigma_1\rangle], j \in Z_2\}$$

with prototile the simplex $Q_G := (1 + \sigma_1)h_{2,0}A^d = \Delta_{1,0}h_{2,0}A^d$ listed under (c).

Next we start from (h). We assume that n is odd, f is even, and $d + 1 = 2e = nf$. Then the Coxeter diagram of $Q := h_{e-1,e+1} h_{f,\ldots,nf} A^d$ has X_e as an end node. By our choices f is not a divisor of e; hence, Q and A^d have the eth wall in common and $(1 + \sigma_e)Q$ is a simplex $\Delta_{e,e+1}Q$ [listed in (i)]. To get a corresponding tiling, we apply LEMMA 9 to the dissection $A^d = \sum_{\Phi \in \Omega} \Phi(Q)$, where $\Omega = \langle V, R^f \rangle = \{V^j R^{if} \mid j \in Z_2, i \in Z_n\}$. We choose for G the subgroup of A_d^* generated by the pairwise commuting reflections σ_{e+if} $(i \in Z_n)$. Then $\Phi G = G\Phi$ for each $\Phi \in \Omega$, since $R^f \sigma_i = \sigma_{i+f} R^f$ holds in view of (**27**). So LEMMA 9 provides us with a monohedral tiling with prototile the polyhedron

$$(1 + \sigma_{e+f}) \cdots (1 + \sigma_{e+(n-1)f})(1 + \sigma_e)Q$$

which obviously splits into 2^{n-1} congruent copies of the simplex $(1 + \sigma_e)Q$. Hence, we have a monohedral tiling

$$\{\tau V^j R^{if} \rho((1 + \sigma_e)Q) \mid \tau \in [A_d^*/G_1], j \in Z_2, i \in Z_n, \rho \in G_2\},$$

where $G_1 := \langle \sigma_e, \sigma_{e+f}, \ldots, \sigma_{e+(n-1)f} \rangle$ and $G_2 := \langle \sigma_{e+f}, \ldots, \sigma_{e+(n-1)f} \rangle$, and where the prototile is listed under (i).

Next we assume $d + 1 = 4g = 8k \in 8\mathbb{N}$. Then

$$Q' := h_{3k+1,\,3k-1} h_{2k,\,6k} h_{4k,\,0} A^d,$$

listed under (d), has an end node X_{3k} in its Coxeter diagram. The reflection in the $3k$th wall of Q' is $\sigma_{3k} \in A_d^*$, so $(1 + \sigma_{3k})Q'$ is a simplex $\Delta_{3k,\,3k-1}Q'$ listed under (e). We apply LEMMA 9 with the dissection $A^d = \sum_{\Phi \in \Omega} \Phi(Q')$, where $\Omega = \{R^{4ki_1} V^{i_2} (R^{2k}V)^{i_3} \mid i_1, i_2, i_3 \in Z_2\}$. If for G we choose the subgroup of A_d^* generated by the pairwise commuting reflections $\sigma_k, \sigma_{3k}, \sigma_{5k}, \sigma_{7k}$, then $R^{2k}G = GR^{2k}$ and $VG = GV$; hence, $\Phi G = G\Phi$ for each $\Phi \in \Omega$ holds. So LEMMA 9 provides us with a monohedral tiling with prototile the polyhedron

$$(1 + \sigma_k)(1 + \sigma_{3k})(1 + \sigma_{5k})(1 + \sigma_{5k})Q'$$

and this splits into 2^3 congruent copies of the simplex $(1 + \sigma_{3k})Q'$. The result is a monohedral tiling

$$\{\tau R^{4ki_1} V^{i_2} (R^{2k}V)^{i_3} \rho((1 + \sigma_{3k})Q') \mid \tau \in [A_d^*/G_1'], i_1, i_2, i_3, \in Z_2, \rho \in G_2'\},$$

where $G_1' = \langle \sigma_k, \sigma_{3k}, \sigma_{5k}, \sigma_{7k} \rangle$ and $G_2' = \langle \sigma_k, \sigma_{5k}, \sigma_{7k} \rangle$ and where the prototile is listed under (e).

Finally, we assume $d + 1 = 8k \in 8\mathbb{N}$, $k > 1$. Then

$$Q'' := h_{7k-1,\,7k+1} h_{2k,\,6k} h_{4k,\,0} A^d,$$

listed under (d), has in its Coxeter diagram X_{7k} as an end node connected with X_{7k+1}. The reflection in the $7k$th wall of Q'' is $\sigma_{7k} \in A_d^*$. We may apply LEMMA 9 with Ω and G the same as in the previous case; this gives us a monohedral tiling with prototile $(1 + \sigma_k)(1 + \sigma_{3k})(1 + \sigma_{5k})(1 + \sigma_{7k})Q''$ which splits into 2^3 congruent copies of $(1 + \sigma_{7k})Q''$. So we have a monohedral tiling

$$\{\tau R^{4ki_1} V^{i_2} (R^{2k}V)^{i_3} \rho((1 + \sigma_{7k})Q'') \mid \tau \in [A_d^*/G_1''], i_1, i_2, i_3 \in Z_2, \rho \in G_2''\},$$

where $G_1'' = \langle \sigma_k, \sigma_{3k}, \sigma_{5k}, \sigma_{7k} \rangle$, $G_2'' = \langle \sigma_k, \sigma_{3k}, \sigma_{5k} \rangle$, and where the prototile $(1 + \sigma_{7k})Q''$ is a simplex $\Delta_{7k,\,7k+1}Q''$ listed under (f).

Now our list of prototiles in THEOREM VII is complete. An inspection of the Coxeter diagrams, in particular counting nodes with valency >3 and labels $\neq \pm\frac{1}{2}$, $\pm 8^{-1/2}$, shows that the listed simplexes give new and distinct types except the mentioned cases. These last result from the relation $A^d = H^d(\frac{1}{2})$ by applying operations equivalent to $h_{i,d-i}$ and $\Delta_{e,e+1}h_{e-1,e+1}$ (where $d = 2e$) under a symmetry of A^d.

It remains to demonstrate $(+)$-monohedral tilings for some of the monocheiral prototiles. This is possible whenever these prototiles result from dissections of A^d based on symmetries $\Phi \in \mathscr{E}^+(d)$ and A^d itself is amphicheiral. Namely, we consider

$$h_{i,0}\bar{C}C(A_d^*) = \{(VR)^\sigma(R^iV)^j(h_{i,0}A^d) \mid \sigma \in A_d^*, j \in Z_2\},$$

$$h_{f,\ldots,nf}\bar{C}C(A_d^*) = \{(VR)^\sigma R^{jf}(h_{f,\ldots,nf}A^d) \mid \sigma \in A_d^*, j \in Z_n\},$$

$$h_{i,d+1-i}h_{f,\ldots,nf}\bar{C}C(A_d^*) = \{(VR)^\sigma R^{jf}V^k(h_{i,d+1-i}h_{f,\ldots,nf}A^d) \mid \sigma \in A_d^*, k \in Z_2, j \in Z_n\},$$

where

$$(VR)^\sigma := \begin{cases} \sigma & \text{if } \chi(\sigma) = +1, \\ \sigma VR & \text{if } \chi(\sigma) = -1. \end{cases}$$

Then $(VR)^\sigma \in \mathscr{E}^+(d)$ if $d \not\equiv 0 \bmod 4$ by LEMMA 10, and in the cases listed under (b), (g), and (h) all the isometries used to disperse the prototile in the three considered tilings are in $\mathscr{E}^+(d)$, so that we have $(+)$-monohedral tilings. ▮

4.6. *The Coxeter Groups B_d^*, C_d^*, and D_d^**

There is a well-known close connection among the groups B_d^*, C_d^*, and D_d^*, cf. [9, Chapter 9]. With our notation for tilings it is expressed by

$$CC(B_d^*) = \Delta_0\, CC(C_d^*), \qquad CC(D_d^*) = \Delta_d\, CC(B_d^*) = \Delta_d\Delta_0\, CC(C_d^*). \tag{28}$$

This is the consequence of short exact sequences of group homomorphisms

$$\tilde{B}_d^* \rightarrowtail C_d^* \overset{p'}{\twoheadrightarrow} \langle\sigma_0\rangle, \tag{29a}$$

$$\tilde{D}_d^* \rightarrowtail C_d^* \overset{p''}{\twoheadrightarrow} \langle\sigma_0, \sigma_d\rangle. \tag{29b}$$

Here $\sigma_0, \ldots, \sigma_d$ denote the reflections in the walls of the simplex $C(C_d^*) \subset E^d$. In **(29a)**, p' maps $\sigma_1, \ldots, \sigma_d$ onto 1 and σ_0 to itself; $\tilde{B}_d^*$ is the subgroup $\langle\bar{\sigma}_0, \sigma_1, \ldots, \sigma_d\rangle$ of C_d^*, where $\bar{\sigma}_0 := \sigma_0\sigma_1\sigma_0$. In **(29b)**, p'' maps $\sigma_1, \ldots, \sigma_{d-1}$ onto 1 and σ_0, σ_d each to itself; $\tilde{D}_d^*$ is the subgroup $\langle\bar{\sigma}_0, \sigma_1, \ldots, \sigma_{d-1}, \bar{\sigma}_d\rangle$ of C_d^*, where $\bar{\sigma}_0 := \sigma_0\sigma_1\sigma_0$ and $\bar{\sigma}_d := \sigma_d\sigma_{d-1}\sigma_d$. Using the group presentations from LEMMA 7(b) we easily check that these generator systems of $\tilde{B}_d^*$ and $\tilde{D}_d^*$ satisfy the relations required to identify $\tilde{B}_d^*$, $\tilde{D}_d^*$ with B_d^*, D_d^*. Application of LEMMA 9 with $\Omega = \{1\}$ and G equal to $\langle\sigma_0\rangle$ or $\langle\sigma_0, \sigma_d\rangle$ then gives **(28)**.

4.7. *A Series of Tilings from $HH^d(0)$ to $CC(C_d^*)$*

The Coxeter diagrams of $H_d(0)$ and $C(C_d^*)$ are identical; but the tilings $HH^d(0)$ and $CC(C_d^*)$ are essentially different. For example, the maximum number of tiles having a point in common is $(d+1)!$ for $HH^d(0)$ and $2^d d!$ for $CC(C_d^*)$. So it is appropriate to choose $H^d(0)$ and $C(C_d^*)$ not as identical, but merely homothetic: We set

$$H^d(0) = \langle 0; b_1, \dots, b_d \rangle, \qquad C(C_d^*) = \langle 0; \tfrac{1}{2}b_1, \dots, \tfrac{1}{2}b_d \rangle$$

with respect to the basis $\mathbf{b}^0$ of E^d consisting of the canonical orthonormal coordinate vectors b_i.

That way we will be able to construct a sequence $BB_d^d, \dots, BB_1^d, BB_0^d$ of monohedral simplicial tilings of E^d which starts from $BB_d^d := HH^d(0)$ and ends with $BB_1^d := CC(B_d^*)$ and $BB_0^d := CC(C_d^*)$, such that each BB_i^d results from BB_{i+1}^d by dissection of each tile into two congruent simplexes. Looking backwards in this sequence, each of these tilings, and in particular $HH^d(0)$, can be viewed as a "block tiling" in $CC(C_d^*)$, where the tiles (simplexes) are obtained by pasting together several tiles of $CC(C_d^*)$.

In order to define this sequence we introduce for $k = 1, \dots, d$ the isometries $U_k \in \mathscr{E}(d)$:

$$U_k(x_1, \dots, x_d) = (y_1, \dots, y_d), \quad \text{where } y_j = 1 - x_{k+1-j} \ (j \le k),\ y_j = x_j \ (j > k), \tag{30}$$

and for $k = 0, \dots, d$ the simplexes:

$$\begin{aligned} B_k^d &= \operatorname{conv}\{p_0, \dots, p_k, \tfrac{1}{2}p_{k+1}, \dots, \tfrac{1}{2}p_d\} \\ &= \langle 0; b_1, \dots, b_k, \tfrac{1}{2}(b_{k+1} - b_k - b_k - \cdots - b_1), \tfrac{1}{2}b_{k+2}, \dots, \tfrac{1}{2}b_d \rangle. \end{aligned} \tag{31}$$

In particular, U_d is the symmetry U of LEMMA 5 with respect to $\mathbf{b}^0$, and B_d^d is $H^d(0)$. Using that $p_k = (1, \dots, 1, 0, \dots, 0)$ with k of the d coordinates equal to 1, we check that U_k is a symmetry of B_k^d which reverses the vertex sequence $(p_0, \dots, p_k)$ and fixes all the remaining vertices of B_k^d. Hence, B_k^d admits operation $h_{j,k-j}$ ($j = 0, \dots, [\tfrac{1}{2}(k-1)]$) based on U_k, and in particular for $j = 0$ we have

$$h_{0,k} B_k^d = B_{k-1}^d \qquad \text{and} \qquad B_k^d = B_{k-1}^d + U_k(B_{k-1}^d) = \sum_{\delta_k = 0}^{1} U_k^{\delta_k}(B_{k-1}^d).$$

So starting with $BB_d^d := HH^d(0)$ we recursively get monohedral tilings

$$\begin{aligned} BB_i^d &:= h_{0,i+1} BB_{i+1}^d = h_{0,i+1} \cdots h_{0,d} HH^d(0) \\ &= \{T_n S_\pi U_d^{\delta_d} \cdots U_{i+1}^{\delta_{i+1}}(B_i^d) \mid n \in Z^d, \pi \in \mathscr{S}_d, \delta_j \in Z_2 \text{ for } j = i+1, \dots, d\} \end{aligned} \tag{32}$$

with prototile the simplex $B_i^d = h_{0,i+1} \cdots h_{0,d} H^d(0)$.

We show next that BB_0^d is $CC(C_d^*)$. From **(31)** we have $B_0^d = C(C_d^*)$, and our task is to prove the equality

$$\{T_n S_\pi U_d^{\delta_d} \cdots U_1^{\delta_1} \mid n \in Z^d, \pi \in \mathscr{S}_d, \delta_j \in Z_2 \text{ for } j = 1, \dots, d\} = C_d^* \tag{33}$$

of isometry sets. First we show that

$$\{S_\pi U_d^{\delta_d} \cdots U_1^{\delta_1} \mid \pi \in \mathscr{S}_d, \delta_j \in Z_2 \text{ for } j = 1, \dots, d\} = \langle \sigma_0, \dots, \sigma_{d-1} \rangle; \tag{34}$$

in fact, that both sides of **(34)** are the full symmetry group (of order $2^d d!$) of the d-cube $W_0 = \{0 \le x_j \le 1 \text{ for } j = 1, \dots, d\}$. This is easily shown for the left-hand side: Each S_π and each U_j obviously is a symmetry of W_0 and the $2^d d!$ words $S_\pi U_d^{\delta_d} \cdots U_1^{\delta_1}$ all represent different symmetries, since our recursive dissection process yields $W_0 = \sum_\pi S_\pi H^d(0) = \sum_\pi \sum_{\delta_d} S_\pi U_d^{\delta_d}(B_{d-1}^d) = \sum_\pi \sum_{\delta_d} \cdots \sum_{\delta_1} S_\pi U_d^{\delta_d} \cdots U_1^{\delta_1}(B_0^d)$. For the right-hand side of (**34**) we remark that for $i = 1, \ldots, d-1$ the ith wall of $C(C_d^*)$ is also the ith wall of $H^d(0)$ and the zeroth wall has equation $x_1 = \frac{1}{2}$. It follows first that $\sigma_0 = U_1$ and $\sigma_i = S_{(i,i+1)}$ for $i = 1, \ldots, d-1$, and from this that the group on the right side of (**34**) is contained in that on the left side and contains each S_π and for $j = 1, \ldots, d$ also $S_{(1,j)}\sigma_0 S_{(1,j)}$, which is the reflection in the hyperplane $x_j = \frac{1}{2}$. But the symmetry group of W_0 is generated by these reflections and the permutations S_π of the axes. Thus (**34**) is established. We have just seen that C_d^* contains the reflection in $x_d = \frac{1}{2}$; it also contains σ_d, which is the reflection in $x_d = 0$. So it contains the product which is the translation by $\pm(0, \ldots, 0, 1)$, furthermore its conjugates by $S_{(i,d)}$ $(i = 1, \ldots, d-1)$. But these elements generate the group $\{T_n \mid n \in Z^d\}$. It follows that the left-hand set in (**33**) is contained in C_d^*, and since both of these isometry sets produce a tiling of E^d with the same prototile, the equality (**33**) and $BB_0^d = CC(C_d^*)$ follow.

Since the Coxeter diagram of B_0^d has X_0 as an end node connected to X_1, $(1 + \sigma_0)B_0^d$ is a simplex $\Delta_{01}B_0^d$ and we have a tiling

$$\Delta_0 BB_0^d = \Delta_0 CC(C_d^*) = \{\tau((1 + \sigma_0)B_0^d) \mid \tau \in [C_d^*/\langle\sigma_0\rangle]\}$$

with this prototile. But we have seen that $\sigma_0 = U_1$ and that $[C_d^*/\langle\sigma_0\rangle]$ may be chosen as B_d^*, cf. (**29a**). So it follows that $\Delta_{01}B_0^d = (1 + U_1)B_0^d = B_1^d$ and $BB_1^d = CC(B_d^*)$, as asserted.

4.8. *Tilings Related to the Groups B_d^*, C_d^*, and D_d^**

Here we collect all possibilities of tilings arising from symmetries of the prototiles $H^d(0) = \langle 0; b_1, \ldots, b_d\rangle$ and $C(C_d^*) = \langle 0; \frac{1}{2}b_1, \ldots, \frac{1}{2}b_d\rangle$, where $b_1, \ldots, b_d$ is the canonical orthonormal base of E^d; cf. FIGURES 5, 6a, and 6b. How $C(B_d^*)$ and $C(D_d^*)$ enter the discussion was described in Sections 4.6 and 4.7. The determination of the Coxeter diagrams will start from the common balanced normal set $\{b_1, b_2 - b_1, b_3 - b_2, \ldots, b_d - b_{d-1}, -b_d\}$ of $H^d(0)$ and $C(C_d^*)$.

THEOREM VIII. *Let $d \geq 3$. The following simplexes are prototiles of monohedral tilings of E^d:*

(a) (FIGURE 5, *first column.*) *For $i \in \mathbb{N}$ with $0 \leq i \leq d-1$:*
$B_i^d := h_{0,i+1} \cdots h_{0,d} H^d(0)$, *in particular* $B_0^d = C(C_d^*)$, $B_1^d = C(B_d^*)$;
$h_{j,i-j} B_i^d$ $(j = 1, \ldots, [\frac{1}{2}(i-1)])$;
$\Delta_{e_i, e_i+1} h_{e_i-1, e_i+1} B_i^d$ *provided* $4 \leq i = 2e_i \in 2\mathbb{N}$.

(b) (FIGURE 5, *second column.*) *For $i \in \mathbb{N}$ with $1 \leq i \leq d-2$:*
$\Delta_{d,d-1} B_i^d$;
$\Delta_{d,d-1} h_{j,i-j} B_i^d$ $(j = 1, \ldots, [\frac{1}{2}(i-1)])$;
$\Delta_{d,d-1} \Delta_{e_i,e_i+1} h_{e_i-1,e_i+1} B_i^d$ *provided* $4 \leq i = 2e_i \in 2\mathbb{N}$.

(c) (FIGURE 5, *third column.*) *For $i \in \mathbb{N}$ with $2 \leq i \leq d-3$:*
$\Delta_{d,d-2} \Delta_{d,d-1} B_i^d$;
$\Delta_{d,d-2} \Delta_{d,d-1} h_{j,i-j} B_i^d$ $(j = 1, \ldots, [\frac{1}{2}(i-1)])$.

(d) (FIGURES 6a *and* 6b.) *With the abbreviations* $C^d := C(C_d^*) = B_0^d$, $D^d := C(D_d^*) = \Delta_{d,d-1}\Delta_{0,1}C^d = \Delta_{d,d-1}B_1^d$, *and for* $d \geq 5$ $K^d := \Delta_{d,d-2}\Delta_{0,2}D^d$:

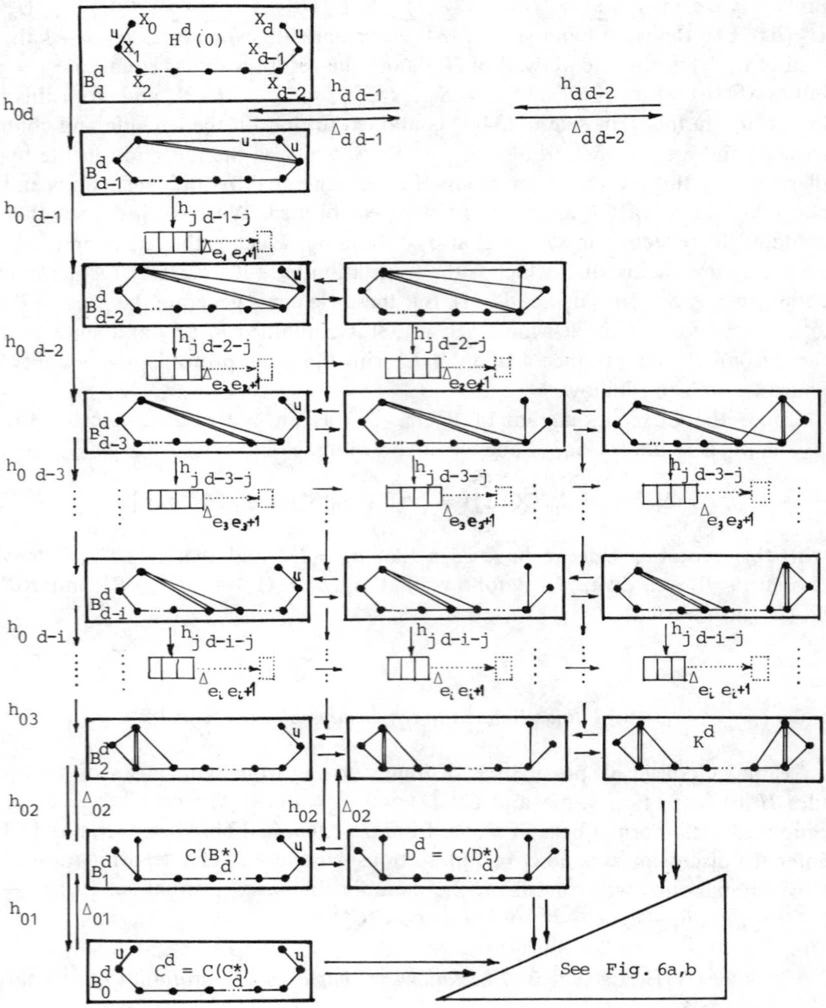

FIGURE 5. Three series related to the group C_d^*. $u = 2^{-1/2}$; dotted lines hold for $d - i = 2e_i \in 2\mathbb{N}$ only.

For $d \geq 3$*:*

$h_{0,\,d,\,1,\,d-1} D^d$;

$h_{j,\,d-j}\, C^d,\ h_{j,\,d-j}\, D^d\ (j = 1, \ldots, [\tfrac{1}{2}(d-1)])$.

For $d \geq 5$*:*

$h_{0,\,d,\,2,\,d-2}\, K^d,\ h_{d,\,d-1} h_{0,\,d,\,2,\,d-2}\, K^d$;

$h_{j,\,d-j} K^d\ (j = 1, \ldots, [\tfrac{1}{2}(d-1)])$;

$h_{0,\,1} h_{j,\,d-j}\, D^d,\ h_{d,\,d-1} h_{j,\,d-j}\, D^d\ (j = 2, \ldots, [\tfrac{1}{2}(d-1)])$.

For $d \geq 7$*:*

$h_{0,\,2}\, h_{j,\,d-j}\, K^d,\ h_{d,\,d-2}\, h_{j,\,d-j}\, K^d\ (j = 3, \ldots, [\tfrac{1}{2}(d-1)])$.

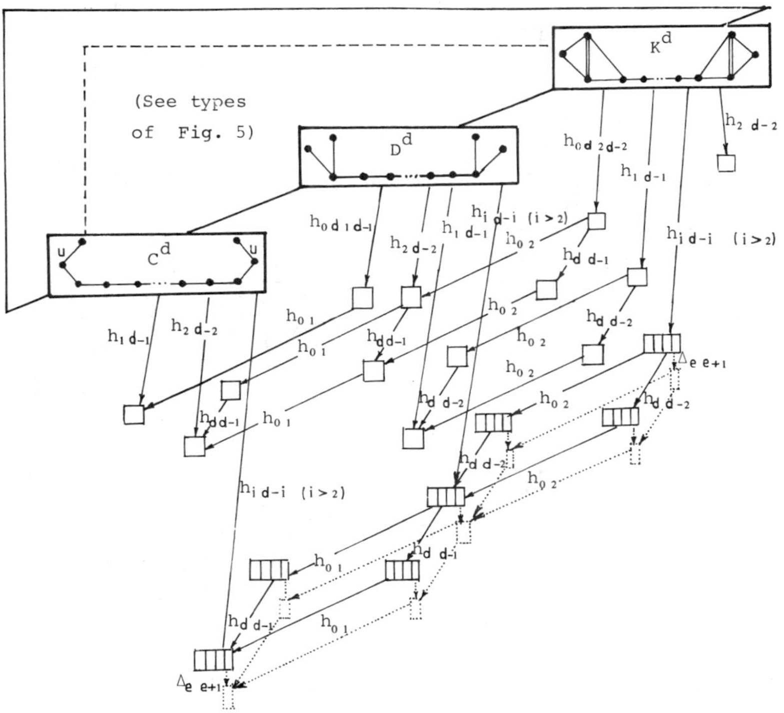

FIGURE 6a. $d > 4$: Types related to the groups B_d^*, C_d^*, D_d^*. $u = 2^{-1/2}$; dotted lines hold for $d = 2e \in 2\mathbb{N}$ only.

In addition, if $d = 2e \in 2\mathbb{N}$:

for $d = 4$: $\Delta_{23}\, h_{0413}\, D^4$;

for $d = 6$: $\Delta_{34}\, h_{0624}\, K^6$, $h_{65}\, \Delta_{34}\, h_{0624}\, K^6$;

for $d \geq 4$: $\Delta_{e,\,e+1} h_{e-1,\,e+1}\, C^d$;

for $d \geq 6$: $\Delta_{e,\,e+1} h_{e-1,\,e+1} D^d$, $h_{0,\,1} \Delta_{e,\,e+1} h_{e-1,\,e+1} D^d$, $h_{d,\,d-1} \Delta_{e,\,e+1} h_{e-1,\,e+1} D^d$;

for $d \geq 8$: $\Delta_{e,\,e+1} h_{e-1,\,e+1} K^d$, $h_{0,\,2}\, \Delta_{e,\,e+1} h_{e-1,\,e+1} K^d$, $h_{d,\,d-2}\, \Delta_{e,\,e+1} h_{e-1,\,e+1} K^d$.

The types of these simplexes are pairwise different and different from the types listed in THEOREMS I–VII, *with the following exceptions:* $H^d(0) \cong C^d = B_0^d$; $H_0^d(0) \cong B_{d-1}^d$; $H_j^d(0) \cong h_{j,\,d-j}\, C^d$ $(j = 1, \ldots, [\frac{1}{2}(d-1)])$; $G^d(0) \cong \Delta_{e,\,e+1} h_{e-1,\,e+1} C^d$ *for* $d = 2e \in 2\mathbb{N}$; $H^3(\frac{1}{2}) \cong C(A_3^*) \cong D^3 = \Delta_{32}\, B_1^3$; $H_0^3(\frac{1}{2}) \cong H_1^3(\frac{1}{2}) \cong h_{12}\, D^3$; $h_{13}\, D^4 \cong B_1^4$.

All listed simplexes are amphicheiral except:

In (a): B_0^d *if* $d \equiv 3$ *or* $0 \bmod 4$; B_i^d *if* $0 < i \equiv 3$ *or* $0 \bmod 4$; $h_{j,\,i-j} B_i^d$ $(j = 1, \ldots, [\frac{1}{2}(i-1)]$; $i = 3, \ldots, d-1)$.

In (d): $h_{1,\,d-1} D^d$ *if* $d \neq 4$; $h_{2,\,d-2}\, K^d$; $h_{j,\,d-j}\, C^d$ $(j = 1, \ldots, [\frac{1}{2}(d-1)])$.

For these monocheiral simplexes (+)*-monohedral tilings are known for* B_0^d *if* $d \equiv 3 \bmod 4$ [*namely*, $\bar{H}H^d(0)$, *cf.* THEOREM V], $h_{i,\,d-i}\, C^d$ *if* $d \equiv 3 \bmod 4$ [*namely*, $\bar{H}H_i^d(0)$, *cf.* THEOREM VI], *and* $h_{12}\, D^3$ [*namely*, $\bar{H}H_1^3(\frac{1}{2})$].

Proof. The symmetries in the Coxeter diagrams show that each listed dissection of prototiles is admitted, hence admitted by each corresponding tiling. The harder

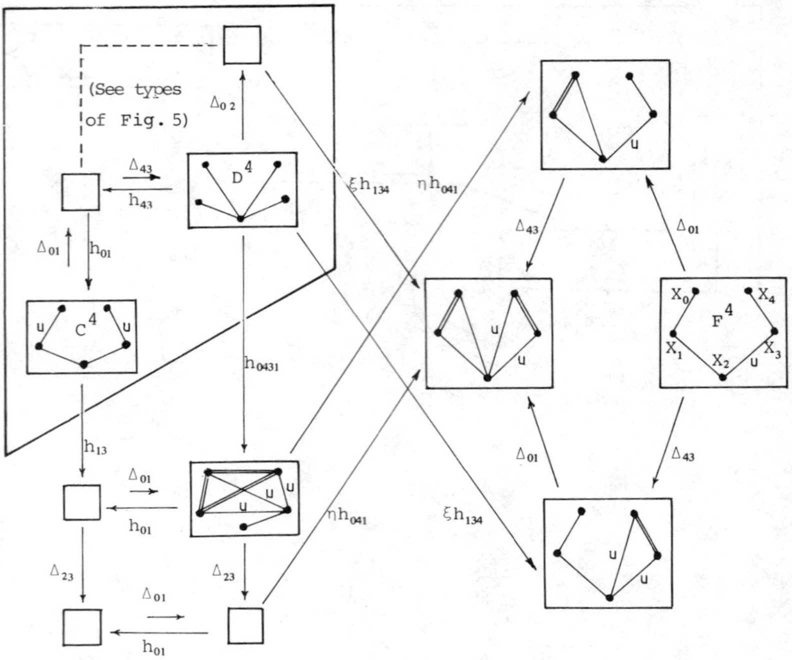

FIGURE 6b. $d = 4$: Types related to the groups B_4^*, C_4^*, D_4^*, F_4^*. $u = 2^{-1/2}$, $\xi = \eta^{-1} = (X_0)(X_3)(X_1, X_2, X_4)$.

part of the proof is to show that suitable tilings admit the required pasting operations Δ_i. We proceed stepwise.

ad (a). The tiling BB_i^d with prototile B_i^d is given by (**32**); but we add another description of BB_i^d which will demonstrate tilings for all the prototiles under (a). Observe that the operation h_{0k} performed on BB_k^d introduces new vertices which are the centers of the k-faces of the cubical cell complex $\mathscr{C}^d = \{4, 3^{d-1}\}$ whose d-cells are the cubes $T_n W_0$ ($n \in Z^d$). In fact, this holds for the prototile given by (**31**) and is, according to (**32**), propagated by the symmetries $T_n S_\pi U_d^{\delta_d} \cdots U_k^{\delta_k}$ since they are symmetries of $\mathscr{C}^d$. So by recursion we see that we can get BB_i^d in the following way: We start with the cubical complex $\mathscr{C}^d$. As the first modification we subdivide the i-skeleton of $\mathscr{C}^d$ by the Hadwiger–Hill process $HH^i(0)$. This means that the tiling $HH^i(0)$ of $E^i \subset E^d$ (where E^i is spanned by the first i base vectors $b_1, \ldots, b_i$) is isometrically carried to the i-skeleton of $\mathscr{C}^d$ by the isometries $T_n S_\pi$ ($n \in Z^d, \pi \in \mathscr{S}_d$); the dissection of each i-face of $\mathscr{C}^d$ and the vertex order $q_0, \ldots, q_i$ for each of the i-simplexes in the subdivision are well defined. As the final step we associate to each chain $F_d \supset F_{d-1} \supset \cdots \supset F_i$ of incident j-faces F_j of $\mathscr{C}^d$ ($j \geq i$) and to each of the $i!$ simplexes, say conv $\{q_0, \ldots, q_i\}$, into which F_i is subdivided, the d-simplex conv $\{q_0, \ldots, q_i, \ldots, q_d\}$ where q_j, for $j = i + 1, \ldots, d$, is the center of F_j. The set of all these simplexes is BB_i^d. In fact, any two of these d-simplexes are congruent, since any two of the i-simplexes into which F_i was dissected are congruent under a symmetry of F_i which extends to a symmetry of F_d fixing $q_{i+1}, \ldots, q_d$; furthermore, any two chains in $\mathscr{C}^d$ of the form $F_d \supset F_{d-1} \supset \cdots \supset F_i$ are equivalent under an isometry $T_n S_\pi$.

Now exactly the same reasoning applies if for our first modification in place of $HH^i(0)$ we take one of the tilings $HH^i_j(0)$ ($j = 1, \ldots, [\frac{1}{2}(i-1)]$) or $GG^i(0)$ (provided $4 \leq i = 2e_i \in 2\mathbb{N}$) from THEOREMS II and III to subdivide the i-skeleton of $\mathscr{C}^d$. The resulting tilings are $h_{j,i-j}BB^d_i$ and $\Delta_{e_i}h_{e_i-1,e_i+1}BB^d_i$ with the prototiles listed under (a).

ad (b). We color the d-cubes $T_n W_0$ of $\mathscr{C}^d$ checkerboardlike black and white according to whether $\sum n_i$ is even or odd. Let $\mathscr{P}_i$ be one of the tilings BB^d_i, $h_{j,i-j}BB^d_i$, or $\Delta_{e_i}h_{e_i-1,e_i+1}BB^d_i$ under (a) and P_i its prototile. Each tile of $\mathscr{P}_i$ has its dth vertex in the center of a d-cube of $\mathscr{C}^d$ and its dth facet on a $(d-1)$-face of this d-cube; moreover, each $(d-1)$-face of $\mathscr{C}^d$ belongs with the same dissection to a black and to an adjacent white d-cube. This proves that the tiling $\mathscr{P}_i$ admits operation Δ_d, yielding a tiling $\Delta_d\mathscr{P}_i$ with prototile $(1+\sigma_d)P_i$. For $i < d-1$ this prototile is a simplex $\Delta_{d,d-1}P_i$ according to LEMMA 4, so we get the list under (b), where the case $i = 0$ is discarded in order to avoid double occurrence; in fact, $\Delta_{d,d-1}B^d_0 \cong \Delta_{0,1}B^d_0 = B^d_1$. In each remaining tiling $\Delta_d\mathscr{P}_i$ an arbitrary tile is the union of a black simplex $P = \operatorname{conv}\{q_0, \ldots, q_d\}$ and a white simplex $\operatorname{conv}\{q_0, \ldots, q_{d-1}, q^*_d\}$, where $q_{d-1} = \frac{1}{2}(q_d + q^*_d)$. We choose for this simplicial tile the vertex order $q_0, \ldots, q_{d-2}, q^*_d, q_d$.

ad (c). The centers of the black cubes in $\mathscr{C}^d$ form a point lattice. We write $\mathscr{D}^d$ for the tessellation of E^d consisting of the Voronoi regions to this lattice. Each d-cell in $\mathscr{D}^d$ is the union of a black cube of $\mathscr{C}^d$ and the $2d$ pyramids which have as their base a facet of this black cube and as their apex the center of the white cube adjoining this facet. In particular, $\mathscr{D}^3$ is the tessellation of E^3 by rhombic dodecahedra and $\mathscr{D}^4$ the tessellation of E^4 by regular 24-cells, cf. [9, p. 156]. Each tiling $\Delta_d\mathscr{P}_i$ constructed under (b) consists of tiles that have the dth vertex in the center of a d-cell of $\mathscr{D}^d$ and the dth facet on a $(d-1)$-face of this d-cell; moreover, each $(d-1)$-face of $\mathscr{D}^d$ belongs with the same dissection to two adjacent d-cells of D^d. This proves that the tiling $\Delta_d\mathscr{P}_i$ admits operation Δ_d. The result is a tiling $\Delta_d\Delta_d\mathscr{P}_i$ with prototile $(1+\bar{\sigma}_d)(1+\sigma_d)P_i$. The case $i = d-2$ has to be discarded from the listing under (c) since the prototile is not a simplex; the case $i = 1$ is excluded to avoid double occurrence of types.

ad (d). Here we start from the tilings $CC(C^*_d)$ with prototile C^d, $CC(D^*_d) = \Delta_d\Delta_0 CC(C^*_d)$ with prototile D^d, and $\{\tau(K^d) \mid \tau \in [D^*_d/\langle\bar{\sigma}_0, \bar{\sigma}_d\rangle]\}$ for K^d. Note that $U' \in \mathscr{E}(d)$ given by $U'(x_1, \ldots, x_d) = (\frac{1}{2} - x_d, \ldots, \frac{1}{2} - x_1)$ is a symmetry of C^d which reverses the order of the vertices. It follows $U'\sigma_i = \sigma_{d-i}U'$, $U'\bar{\sigma}_j = \bar{\sigma}_{d-j}U'$ for $i = 0, \ldots, d$ and $j = 0, d$. Hence, U' is also an order reversing symmetry of D^d and of K^d. Moreover, D^d has the additional symmetries σ_0 and σ_d which induce the transpositions $(0, 1)$ and $(d, d-1)$, respectively, of the vertex indices, and similarly K^d has the symmetries $\bar{\sigma}_0$ and $\bar{\sigma}_d$ which induce $(0, 2)$ and $(d, d-2)$, respectively. On the basis of these symmetries and their products we can apply LEMMAS 2 and 8 to get dissections of tilings for all the prototiles listed in the first half of (d). It remains to display block tilings in even dimensions $d = 2e$, namely, for (α) $\Delta_{23}h_{0413}D^4$, (β) $\Delta_{34}h_{0624}K^6$, and (γ) $\Delta_{e,e+1}h_{e-1,e+1}$ applied to C^d, D^d, K^d; these block tilings will then allow some further dissections. For (α) we first use LEMMA 9 with $d = 2e \geq 4$, $P^*_d := D^*_d$, $P^d := D^d$, $\psi := \sigma_0 U'$, $\Omega := \{\psi^j \mid j = 1, 2, 3, 4\}$, $Q := h_{0,d,1,d-1}D^d$, $G = \langle\sigma_e\rangle$. The result is a tiling **(24)** with prototile $(1+\sigma_e)h_{0,d,1,d-1}D^d$. It appears in our listing for $d = 4$ only, since it is a bipyramid in general and a simplex only for $d = 4$. In fact, from the vertices $p_i := \frac{1}{2}\sum_{k \le i} b_k$ $(i = 0, \ldots, d)$ of C^d we get those of D^d by changing p_1 to $2p_1 - p_0$ and p_{d-1} to $2p_{d-1} - p_d$. But the latter changes in $h_{0,d,1,d-1}D^d$ to the barycenter $v := \frac{1}{4}(p_0 + (2p_1 - p_0) + (2p_{d-1} - p_d) + p_d) = \frac{1}{2}(p_1 + p_{d-1})$ of its orbit under ψ. In the case $d = 2e = 4$, (**20**), scaled by $\frac{1}{2}$, reveals that this vertex v of $h_{0,d,1,d-1}D^d$ is also the barycenter of the two vertices $p_e, \sigma_e p_e$ of the bipyramid $(1 + \sigma_e)h_{0,d,1,d-1}D^d$, so in this case this prototile is a simplex $\Delta_{34} h_{0413} D^4$ according to LEMMA 4. For (β) we will construct similarly for $6 \le d = 2e \in 2\mathbb{N}$ a tiling with prototile the bipyramid $(1 + \sigma_e)h_{0,d,2,d-2}K^d$; it appears in our listing for $d = 6$ only, where the bipyramid is actually a simplex. We start from LEMMA 9 with P_d^*, P^d, G, Ω given by D_d^*, D^d, $\langle \bar\sigma_0, \bar\sigma_d, \sigma_e \rangle$, $\{1\}$, respectively, to get a tiling

$$\{\tau(S) \mid \tau \in [D_d^*/\langle \bar\sigma_0, \bar\sigma_d, \sigma_e \rangle]\} \tag{35}$$

with prototile

$$S := (1 + \sigma_e)(1 + \bar\sigma_0)(1 + \bar\sigma_d)D^d = (1 + \sigma_e)K^d. \tag{35'}$$

The symmetry $\bar\psi := \bar\sigma_0 U'$ of K^d of order 4 commutes with σ_e and with $\langle \bar\sigma_0, \bar\sigma_d \rangle$, and we have $K^d = \sum_{j=1}^4 \bar\psi^j(h_{0,d,2,d-2}K^d)$; hence,

$$S = (1 + \sigma_e)K^d = \sum_{j=1}^4 \bar\psi^j(1 + \sigma_e)h_{0,d,2,d-2}K^d.$$

If we insert this into (**35**) we get the tiling

$$\{\tau\bar\psi^j((1 + \sigma_e)h_{0,d,2,d-2}K^d) \mid \tau \in D_d^*/\langle \bar\sigma_0, \bar\sigma_d, \sigma_e \rangle, j \in Z_4\}$$

with prototile $(1 + \sigma_e)h_{0,d,2,d-2}K^d$. But $h_{0,d,2,d-2}K^d$ has as a vertex the barycenter $\bar v$ of the vertex orbit $\{p_0, p_d, 2p_2 - p_0, 2p_{d-2} - p_d\}$ of K^d under $\bar\psi$, so $\bar v = \frac{1}{2}(p_2 + p_{d-2})$. In the case $d = 2e = 6$, (**20**) shows again that $\bar v$ is also the barycenter of the vertex pair $p_e, \sigma_e p_e$ of $(1 + \sigma_e)h_{0,d,2,d-2}K^d$, so this bipyramid is actually a simplex $\Delta_{34} h_{0624} K^4$, as asserted. For (γ) we apply LEMMA 9 with P_d^*, P^d, G, Ω, Q given a first time by C_d^*, C^d, $\langle \sigma_e \rangle$, $\{1, U'\}$, $h_{e-1,e+1}C^d$ and a second time by D_d^*, D^d, $\langle \sigma_e \rangle$, $\{1, U'\}$, $h_{e-1,e+1}D^d$. Then (**24**) gives tilings for $\Delta_{e,e+1}h_{e-1,e+1}C^d$ and $\Delta_{e,e+1}h_{e-1,e+1}D^d$. For K^d we insert $K^d = (1 + U')h_{e-1,e+1}K^d$ into (**35′**) and since U' commutes with σ_e we have

$$S = (1 + U')(1 + \sigma_e)h_{e-1,e+1}K^d = (1 + U')\Delta_{e,e+1}h_{e-1,e+1}K^d.$$

The tiling (**35**) then subdivides to the tiling

$$\{\tau U'^j(\Delta_{e,e+1}h_{e-1,e+1}K^d) \mid \tau \in [D^*/\langle \bar\sigma_0, \bar\sigma_d, \sigma_e \rangle], j \in Z_2\}$$

with prototile the simplex $\Delta_{e,e+1}h_{e-1,e+1}K^d$.

We have now exhausted the occurrence of end nodes and of symmetries except in dimension 4, where a casual symmetry of order 3 remains and will be exploited in the next section. The assertion on the distinctness of types is verified by inspection of the Coxeter diagrams; only a very few distinct labels occur. ▮

4.9. *Tilings Related to the Exceptional Group F_4^**

The fundamental region of the group F_4^* is a simplex F^4 with balanced normal set $\{b_1 + b_2, 2(b_3 - b_2), 3(b_4 - b_3), -4b_4, b_2 + b_3 + b_4 - b_1\}$, cf. [7].

THEOREM IX. *The following simplex types admit monohedral tilings of* E^4 *based on the Coxeter group* F_4^*:

$$F^4, \qquad \Delta_{01}F^4, \qquad \Delta_{43}F^4, \qquad \Delta_{01}\Delta_{43}F^4.$$

F^4 *is monocheiral, the other three are amphicheiral. For the interrelations see* FIGURE 6b.

COROLLARY. *Outside the one-parameter families* $H^4(w)$, $H_0^4(w)$, $H_1^4(w)$, $G^4(w)$ $(-\frac{1}{3} < w < 1)$, *there are in* E^4:

three monocheiral types given in THEOREM VII(g, h);

one monocheiral and six amphicheiral types given in THEOREM VIII(a, b, d);

one monocheiral and three amphicheiral types given in THEOREM IX

of tiling simplexes.

Proof. Tilings corresponding to the types of THEOREM IX are easily provided by LEMMA 9. Up to similarity the same tilings can be obtained by subdivision of some tilings described in Section 4.8, as is shown in FIGURE 6b. ▮

4.10. *Tilings Related to the Exceptional Groups* E_6^*, E_7^*, E_8^*

For the prototiles $C(E_d^*)$ ($d = 6, 7, 8$) the ordering of the vertices and balanced normal sets are taken from [7, pp. 213–219]. A simple geometrical account to determine these normals can be found in [6].

THEOREM X. *On the basis of the Coxeter group* E_d^* ($d = 6, 7, 8$) *and the dissecting and pasting operations of Sections* 2.4, 2.5, *and* 4.3, *monohedral tilings of* E^d *can be constructed:*

(a) *if* $d = 6$ *for* 32 *types* E_j^6 ($j = 1, \ldots, 32$) *of simplexes of which* 19 *are amphicheiral;* 1 *of the monocheiral types admits also a* (+)*-monohedral tiling;*

(b) *if* $d = 7$ *for* 22 *types* E_j^7 ($j = 1, \ldots, 22$) *of simplexes of which* 18 *are amphicheiral;*

(c) *if* $d = 8$ *for* 8 *types* E_j^8 ($j = 1, \ldots, 8$) *of simplexes of which* 7 *are amphicheiral. Interrelations of these types are shown in* FIGURES 7, 8, *and* 9, *where monocheiral types are marked with an* m *and the monocheiral type admitting a* (+)*-monohedral tiling with* m_+.

Proof. Again all the dissection operations shown in the figures follow from the Coxeter diagrams which in turn are based on modification of balanced normal sets along LEMMA 8. The same holds for the operations Δ_{ij} on prototiles, but here we have to show that they extend to the operation Δ_i on a tiling. In the figures we omit many Coxeter diagrams for reasons of space.

$d = 6$: It suffices to construct tilings for the types E_4^6, E_{10}^6, and E_{23}^6. All other tilings arise from these by suitable subdivision; to see this we note that Δ_{ij}, if admitted, has the dissection h_{ij} as an inverse, cf. LEMMA 4(e).

First, $E_4^6 := \Delta_{13}\Delta_{65}\Delta_{02}C(E_6^*) = (1 + \sigma_1)(1 + \sigma_6)(1 + \sigma_0)C(E_6^*)$ is the prototile of $\{\tau(E_4^6) \mid \tau \in [E_6^*/\langle\sigma_0, \sigma_1, \sigma_6\rangle]\}$. Moreover, we can derive from this tiling a (+)-monohedral tiling for the monocheiral prototile $E_{13}^6 := h_{352}E_4^6$. In fact, h_{352} is based on a symmetry χ of order 3 of E_4^6 which belongs to $\mathscr{E}^+(4)$. This tripartition by h_{352} gives a tiling $\{\tau\chi^i(h_{352}E_4^6) \mid \tau \in [E_6^*/\langle\sigma_0, \sigma_1, \sigma_6\rangle], i \in Z_3\}$ for E_{13}^6, and we may well choose our system of representatives $[E_6^*/\langle\sigma_0, \sigma_1, \sigma_6\rangle]$ in such a way that all its members are direct congruences. Next we consider $E_{10}^6 := \Delta_{13}\Delta_{02}h_{35}C(E_6^*) = (1 + \sigma_1)(1 + \sigma_0)h_{35}C(E_6^*)$. The operation h_{35} is based on a symmetry Ψ of $C(E_6^*)$

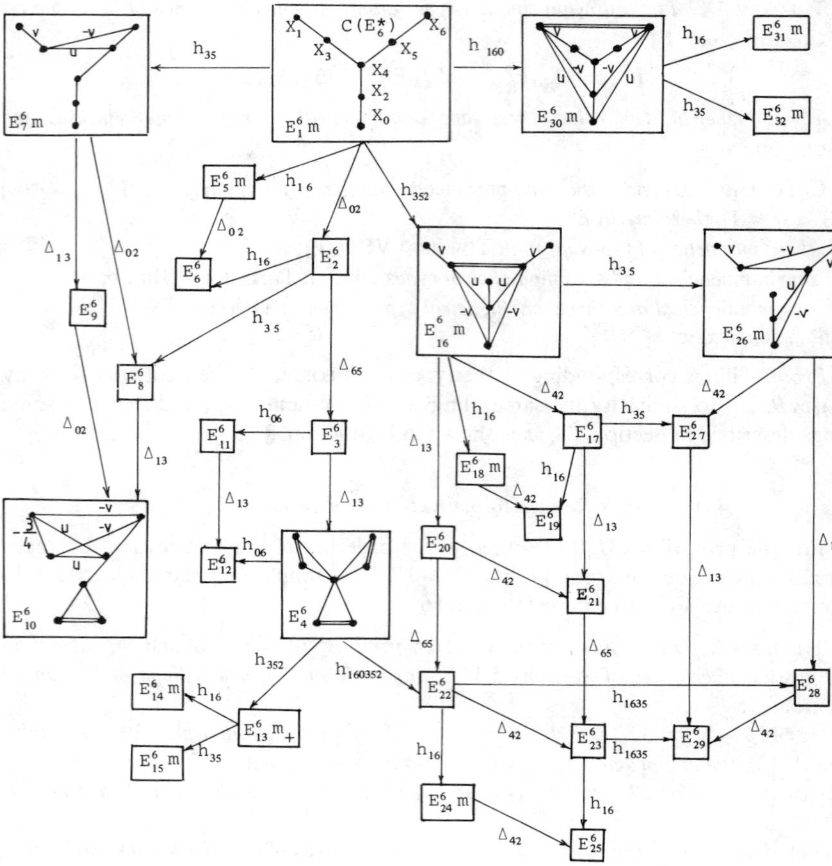

FIGURE 7. $d = 6$: Types related to the group E_6^*. $u = 2v = 2^{-1/2}$.

given by the vertex permutation $(p_1, p_6)(p_3, p_5)$, so $\Psi\langle\sigma_1, \sigma_0, \sigma_6\rangle = \langle\sigma_1, \sigma_0, \sigma_6\rangle\Psi$ and LEMMA 9 with $\Omega = \{1, \Psi\}$, $G = \langle\sigma_0, \sigma_1, \sigma_6\rangle$ provides us with a tiling with prototile the polytope $(1 + \sigma_6)(1 + \sigma_1)(1 + \sigma_0)h_{35}\, C(E_6^*)$ which splits obviously into two congruent parts E_{10}^6, $\sigma_6(E_{10}^6)$; if each tile is equally split, we get the desired tiling

$$\{\tau\Psi^j\sigma_6^i(E_{10}^6) \mid \tau \in [E_6^*/\langle\sigma_0, \sigma_1, \sigma_6\rangle], i \in Z_2, j \in Z_2\}.$$

Finally we consider $E_{23}^6 := \Delta_{42}\Delta_{65}\Delta_{13}h_{352}\, C(E_6^*)$. The operation h_{352} is based on a symmetry Θ of $C(E_6^*)$ given by the vertex permutation $(p_0, p_6, p_1)(p_2, p_5, p_3)$, so we may apply LEMMA 9 with $\Omega = \{1, \Theta, \Theta^2\}$ and $G = \langle\sigma_0, \sigma_1, \sigma_4, \sigma_6\rangle$. This provides us with a tiling with prototile $(1 + \sigma_0)(1 + \sigma_4)(1 + \sigma_6)(1 + \sigma_1)h_{352}\, C(E^*)$; the prototile (and similarly each tile) splits into two congruent simplexes $E_{23}^6, \sigma_0 E_{23}^6$.

$d = 7$: For the same reason as before it suffices to give tilings for the types E_6^7, E_{12}^7. By definition $E_6^7 = (1 + \sigma_7)(1 + \sigma_2)(1 + \sigma_0)C(E_7^*)$ and LEMMA 9 with $\Omega = \{1\}$

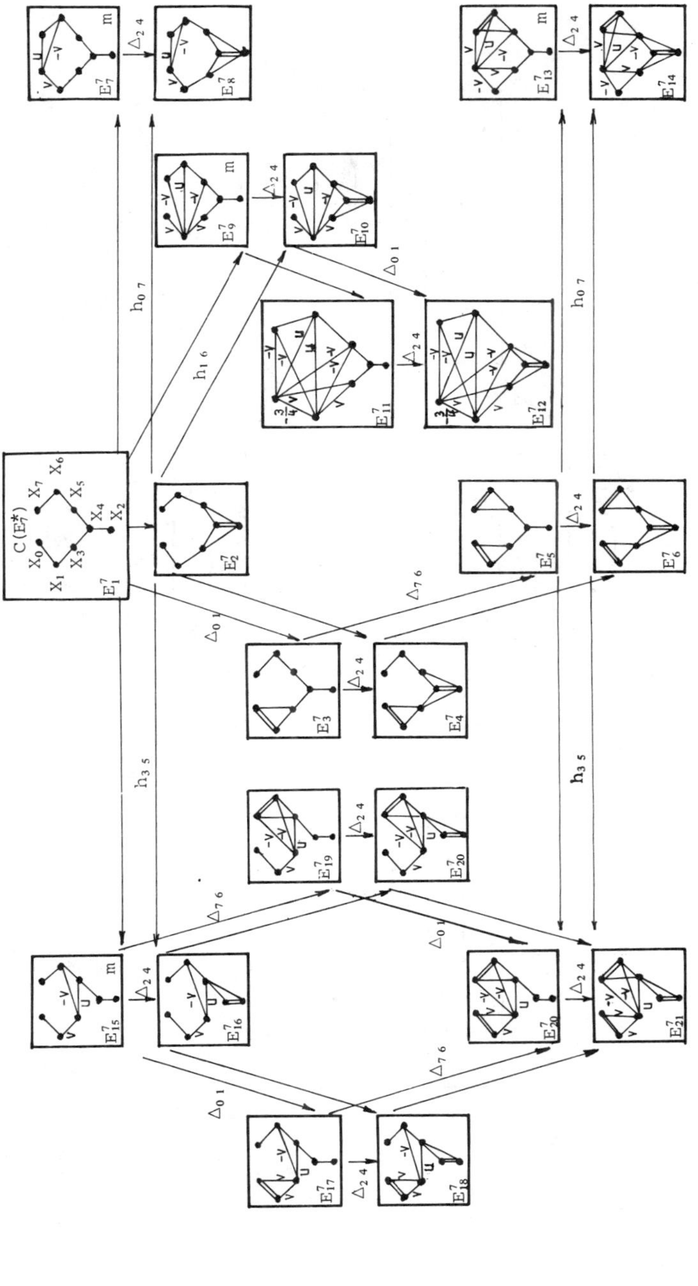

FIGURE 8. $d = 7$: Types related to the group E_7^*. $u = 2v = 2^{-1/2}$.

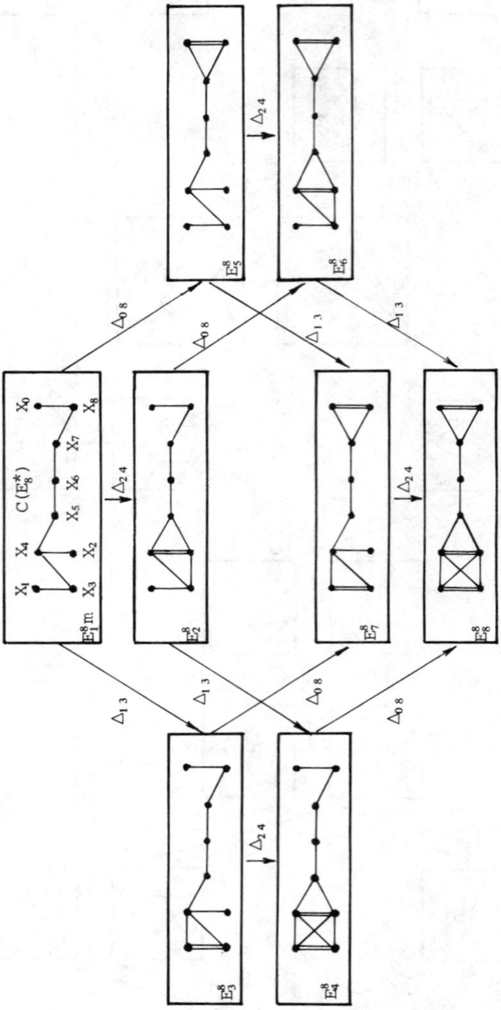

FIGURE 9. $d = 8$: Types related to the group E_8^*.

and $G = \langle \sigma_0, \sigma_2, \sigma_7 \rangle$ gives the desired tiling. Next we have

$$E_{12}^7 := (1 + \sigma_2)(1 + \sigma_0)h_{16}\, C(E_7^*).$$

The operation h_{16} is based on a symmetry Φ which induces the vertex permutation $(p_0, p_7)(p_1, p_6)(p_3, p_5)$ of $C(E_7^*)$; hence, $\Phi\langle \sigma_0, \sigma_2, \sigma_7 \rangle = \langle \sigma_0, \sigma_2, \sigma_7 \rangle \Phi$, and LEMMA 9 with $\Omega = \{1, \Phi\}$ and $G = \langle \sigma_0, \sigma_2, \sigma_7 \rangle$ gives a tiling with prototile $(1 + \sigma_7)(1 + \sigma_2)(1 + \sigma_0)h_{16}\, C(E_7^*)$. This prototile (and similarly each tile) splits into two halves congruent to E_{12}^7, yielding the tiling

$$\{\tau \Phi^j \sigma_7^i (E_{12}^7) \mid \tau \in [E_7^* / \langle \sigma_0, \sigma_2, \sigma_7 \rangle], j \in Z_2, i \in Z_2\}.$$

$d = 8$: Here it suffices to give a tiling for $E_8^8 := \Delta_{13} \Delta_{24} \Delta_{08}\, C(E_8^*)$. It is given by LEMMA 9 with $\Omega = \{1\}$ and $G = \langle \sigma_0, \sigma_1, \sigma_2 \rangle$. ∎

REFERENCES

1. BAUMGARTNER, L. 1968. Zerlegung des vierdimensionalen Raumes in kongruente Fünfzelle. Math.-Phys. Semesterber. **15:** 76–86.
2. BAUMGARTNER, L. 1971. Zerlegung des *n*-dimensionalen Raumes in kongruente Simplexe. Math. Nachr. **48:** 213–224.
3. BERGER, J. 1978. Rektifizierbare vierdimensionale Simplices. Elem. Math. **33:** 107–115.
4. BÖHM, J. & E. HERTEL. 1981. Polyedergeometrie in *n*-dimensionalen Räumen konstanter Krümmung. Birkhäuser Verlag. Basel, Switzerland.
5. BOLTIANSKII, V. G. 1978. Hilbert's Third Problem. Translated edit. V. H. Winston & Sons. Washington, D.C.
6. BOS, W. 1962. Zur Klassifikation der einfachen Lie-Gruppen. Indag. Math. **24:** 240–244.
7. BOURBAKI, N. 1981. Groupes et algèbres de Lie. Chapters 4–6. Masson. Paris, France.
8. COXETER, H. S. M. 1934. Discrete groups generated by reflections. Ann. of Math. **35:** 588–621.
9. COXETER, H. S. M. 1973. Regular Polytopes. 3rd edit. Dover Publications. New York, N.Y.
10. DANZER, L. 1968. Zerlegung endlich-dimensionaler Räume in kongruente Simplices. Math.-Phys. Semesterber. **15:** 87.
11. DEBRUNNER, H. E. 1980. Ueber Zerlegungsgleichheit von Pflasterpolyedern mit Würfeln. Arch. Math. **35:** 583–587.
12. GOLDBERG, M. 1974. Three infinite families of tetrahedral space-fillers. J. Combin. Theory **16:** 348–354.
13. GRÜNBAUM, B. 1967. Convex Polytopes. John Wiley & Sons, Inc. London, England.
14. GRÜNBAUM, B. & G. C. SHEPHARD. 1980. Tilings with congruent tiles. Bull. Amer. Math. Soc. (New Series) **3:** 951–973.
15. HADWIGER, H. 1951. Hillsche Hypertetraeder. Gaz. Mat. (Lisbon) **12**(No. 50): 47–48.
16. HADWIGER, H. 1957. Vorlesungen über Inhalt, Oberfläche und Isoperimetrie. Springer-Verlag. West Berlin, FRG.
17. HILL, M. J. M. 1896. Determination of the volume of certain species of tetrahedrons. Proc. London Math. Soc. **27:** 39–52.
18. SENECHAL, M. 1981. Which tetrahedra fill space? Math. Mag. **54:** 227–243.
19. SOMMERVILLE, D. M. Y. 1923. Space-filling tetrahedra in Euclidean space. Proc. Edinburgh Math. Soc. **41:** 49–57.
20. SOMMERVILLE, D. M. Y. 1923. Division of space by congruent triangles and tetrahedra. Proc. Roy. Soc. Edinburgh **43:** 85–116.
21. SPIVAK, M. 1965. Calculus on Manifolds. W. A. Benjamin, Inc. New York, N.Y.

COVERINGS AND PACKINGS BY SEQUENCES OF CONVEX SETS[a]

H. Groemer

Department of Mathematics
The University of Arizona
Tucson, Arizona 85721

1. Introduction

This paper is a survey of results dealing with a particular class of covering and packing problems. Most of these results have been found in recent times, and there are some unsolved problems that appear to be of interest.

We are primarily concerned with the following question: If K is a convex subset of n-dimensional Euclidean space E^n, and if (C_i) is a finite or infinite sequence of closed convex subsets of E^n, is it possible to apply to the sets C_i rigid motions (or perhaps only translations) so that the resulting sets cover K, or are contained in K and have mutually disjoint interiors? Moreover, if such coverings or packings exist the problem of estimating the possible efficiency (density) arises.

For most problems of this kind it is immaterial in which order the terms C_i appear in the sequence (C_i). However, we prefer to deal with sequences, rather than sets $\{C_i\}$, since sequences provide a natural possibility of including some of the sets C_i repeatedly. The principal difference between the subject of this survey and the classical sort of covering and packing problems, as presented, for example, in the books of L. Fejes Tóth [13], Rogers [48], and the recent survey of G. Fejes Tóth [11], is the absence of any strong conditions on the mutual relationship between the sets C_i. In general, these sets are not supposed to be translation equivalent, homothetic, congruent, or similar. Possible restrictions that are imposed on the closed convex sets C_i concern their boundedness or the boundedness of their diameter set. Also, some theorems deal with particular cases, like sequences of strips, slabs, or boxes.

It appears that the first result dealing with a packing problem of the type being discussed here was stated (without proof) by Auerbach, Banach, Mazur, and Ulam in the well-known "Scottish Book" (see Mauldin [40, Problem 10.1]). Other references to early contributions are Tarski [52], Hlawka [31], and L. Fejes Tóth [12]; but these articles are only very loosely connected with our subject. A proof of the Auerbach–Banach–Mazur–Ulam theorem has been published by Kosiński [34]. The first more comprehensive study of covering and packing properties of sequences of compact convex sets, whose corresponding diameter set is bounded, was made by Moon and Moser [42]. The investigation of coverings of all E^n by sequences of compact convex sets without any restrictions on the diameters has been initiated by Chakerian [7].

[a] Supported by National Science Foundation Research Grants MCS 8001578 and MCS 8300825.

Our principal aim here is to describe the further developments in this area. We divide the material into sections headed by the following titles: 1. Introduction; 2. Definitions and Notations; 3. Approximation by Boxes; 4. Isometric Coverings and Packings by Bounded Sequences of Convex Bodies; 5. Translative Coverings and Packings by Bounded Sequences of Convex Bodies; 6. Isometric Coverings by Unbounded Sequences of Convex Bodies; 7. Translative Coverings by Unbounded Sequences of Convex Bodies; 8. Coverings by Sequences of Unbounded Convex Sets; 9. Coverings by Sequences of Strips and Slabs; and 10. Coverings of the Sphere.

2. Definitions and Notations

We define a *convex body* as a nonempty compact convex subset of E^n (it is not required that the interior be nonempty). The class of all convex bodies of E^n will be denoted by $\mathbf{C}^n$. Whenever we talk about sequences of convex sets the sequence is permitted to be finite or infinite, unless specified otherwise.

Let us now assume that $K \subset E^n$ and that (M_i) is a sequence of closed convex subsets of E^n. The sequence (M_i) is called a *covering of* K if $K \subset \bigcup M_i$, and a *packing in* K if $\bigcup M_i \subset K$ and $\operatorname{int} M_i \cap \operatorname{int} M_k = \varnothing$ (for all $i \neq k$). We say that (M_i) *permits an isometric covering of* K, or *packing in* K, if there are rigid motions σ_i so that $(\sigma_i M_i)$ is, respectively, a covering of K or a packing in K. If there are not only rigid motions but translations τ_i so that $(\tau_i M_i)$ is a covering of K, or a packing in K, we express this by saying, respectively, that (M_i) *permits a translative covering of* K, *or packing in* K. The diameter of a set $X \subset E^n$ will be denoted by $d(X)$, and the volume of a set $M \in \mathbf{C}^n$ by $v(M)$. For $d(X)$ the value ∞ is permitted. (M_i) is said to be a *bounded sequence* if the corresponding diameter set $\{d(M_i)\}$ is bounded. We let $B^n(r)$ denote the closed n-dimensional ball of radius r centered at the origin of E^n. Frequently we have to deal with the subsequence consisting of all terms of (M_i) with $d(M_i) \leq 1$. This subsequence will be denoted by $(M_i)_{d(M_i) \leqq 1}$, and similar notations will be used if other conditions are imposed on (M_i). If $C_i \in \mathbf{C}^n$ and (C_i) is bounded and a covering of E^n we define the *covering density* of (C_i) by

$$\varlimsup_{r \to \infty} \frac{1}{v(B^n(r))} \sum_{C_i \cap B^n(r) \neq \varnothing} v(C_i).$$

Similarly, if (C_i) is a packing in E^n the *packing density* of (C_i) is defined by

$$\varliminf_{r \to \infty} \frac{1}{v(B^n(r))} \sum_{C_i \subset B^n(r)} v(C_i).$$

In addition to coverings of all E^n we also have to consider coverings of almost all points of E^n. We say that a sequence (M_i) of closed convex subsets of E^n *permits an isometric or translative covering of almost all* E^n if there are, respectively, rigid motions or translations σ_i such that $E^n \backslash \bigcup \sigma_i M_i$ has Lebesgue measure 0. Finally, we let $I^n(s)$ denote the cube defined (in the usual Cartesian coordinate system of E^n) by $I^n(s) = \{(x_1, \ldots, x_n): -\frac{1}{2}s \leq x_i \leq \frac{1}{2}s\}$.

3. Approximation by Boxes

By a box in E^n we mean an orthogonal parallelepiped, i.e., a set of the form $L_1 + L_2 + \cdots + L_n$ with mutually orthogonal line segments L_i. Many covering and packing theorems of the type to be discussed here are proved by reducing the problem to the case of boxes. For this purpose it is essential to show that every convex body contains a box of "large" volume and is contained in a box of "small" volume. The principal result in this direction is the following theorem:

For any $K \in \mathbf{C}^n$ there exist two boxes, say Q_1, Q_2, such that

$$Q_1 \subset K, \qquad v(Q_1) \geq n^{-n} v(K)$$

and

$$K \subset Q_2, \qquad v(Q_2) \leq n!\, v(K).$$

Both parts of this theorem have been proved by Macbeath [38] and subsequently by Hadwiger [29]. The second part has also been proved by Kosiński [34], and a more recent proof of the first statement, based on a mapping theorem of Rattray, has been given by Chakerian [7]. Already Macbeath has noted that it is unlikely that the above inequalities are best possible for any $n \geq 2$ (except the second one when $n = 2$). Moon and Moser [42] have conjectured that the factor n^{-n} can be replaced by $n!\, n^{-n}$. If $n = 2$ this has been shown to be true by Süss [51] and Radziszewski [47]. Thus, for $n = 2$ one has

$$Q_1 \subset K, \qquad v(Q_1) \geq \tfrac{1}{2} v(K).$$

The latter article and the joint paper of Bielecki and Radziszewski [5] contain also some information concerning the case $n = 3$ (but not a complete solution).

A very comprehensive survey of results concerning the approximation of a convex body by polytopes or other special sets has been published by Gruber [27].

4. Isometric Coverings and Packings by Bounded Sequences of Convex Bodies

Remember that we call a sequence (C_i) of convex bodies bounded if $d(C_i) \leq d_o$ for all i and some $d_o < \infty$. It is often convenient to assume that $d_o = 1$. The general case is then obtained by an obvious similarity transformation. If some box P is given, the principal problem is to decide how large (or how small) $\sum v(C_i)$ has to be so that (C_i) permits an isometric covering of P (packing in P).

Moon and Moser [42] have outlined a proof of the theorem that (C_i) permits an isometric covering of the cube $I^n(1)$ if $d(C_i) \leq 1$ and $\sum v(C_i) \geq n^n 2^n c_n$, where c_n is defined by $c_n = ((2 \cdot 4 \cdots 2^n)/(1 \cdot 3 \cdots (2^n - 1))) - 1 < 2.463$. If all the sets C_i are boxes the factor n^n can be removed and it suffices to assume that the edge lengths (rather than the diameters) of the boxes C_i are at most 1. A detailed proof that the conditions $d(C_i) \leq 1$ and $\sum v(C_i) \geq n^n(2 \cdot 3^{n-1} - 1)$ imply that (C_i) permits an isometric covering of $I^n(1)$ (without the factor n^n in the case of boxes) has been given in [17]. The factor n^n stems of course from the approximation by boxes (cf. Section 3) and can be replaced by 2 if $n = 2$.

All these results are improved and generalized by the following theorem of Groemer [23].

Let (C_i) *be a sequence of convex bodies whose diameters are at most* 1. *If for some* $s > 0$

$$\sum v(C_i) \geq n^n((s + 1)^n - 1),$$

then (C_i) *permits an isometric covering of* $I^n(s)$. (*If* $n = 2$ *the factor* n^n *can be replaced by* 2.) *Moreover, if all the sets* C_i *are boxes whose edge lengths are at most* 1 *and if*

$$\sum v(C_i) \geq (s + 1)^n - 1,$$

then (C_i) *permits an isometric covering of* $I^n(s)$.

The case $n = 2$ of this theorem is already contained in the paper of Moon and Moser [42]. In the general case the proof uses an induction argument with respect to dimension. The basic ideas can already be found in the proof of a packing theorem of Kosiński [34]. If $n = 2$ coverings of the desired kind can be obtained in the following way: One may assume that the sets C_i are rectangles and ordered so that the lengths of the largest sides decrease. Then the rectangles are placed on the square, side by side from left to right along the base. If the right-hand side of the square is reached, a new layer of rectangles is started above the first so that there are no uncovered portions between the first and the second layer, and so on. If $n = 2$ the necessary estimates are not difficult (see also [8]), but for the general case the situation becomes more involved.

It is unlikely that the inequalities in the above theorem are best possible for any value of s. However, in the case of boxes the expression $(s + 1)^n - 1$ is "asymptotically" best possible since the inequality $\sum v(C_i) \geq s^n$ is obviously necessary. As shown in [23] one can subdivide E^n into a sequence of increasing cubes and cover each cube by a suitable collection of sets from the given sequence (C_i) to prove the following theorem concerning coverings of all E^n. In the case of boxes it yields an optimal result.

Let (C_i) *be a bounded sequence of convex bodies so that* $\sum v(C_i) = \infty$. *Then,* (C_i) *permits an isometric covering of* E^n *of density at most* n^n (*at most* 2 *if* $n = 2$). *Moreover, if all the sets* C_i *are boxes, then* (C_i) *permits an isometric covering of* E^n *of density* 1.

We turn now to the corresponding packing theorems. First we state analogues of the above covering theorems. These are also proved in [23].

Let (C_i) *be a sequence of convex bodies whose diameters are at most* 1. *If for some* $s \geq 3$

$$\sum v(C_i) \leq \frac{1}{n!}\left((s - 1)^n - \frac{s - 1}{s - 2}((s - 1)^{n-2} - 1)\right),$$

then (C_i) *permits an isometric packing in* $I^n(s)$. *Moreover, if all the sets* C_i *are boxes whose edge lengths are at most* 1 *and*

$$\sum v(C_i) \leq (s - 1)^n - \frac{s - 1}{s - 2}((s - 1)^{n-2} - 1),$$

then (C_i) *permits an isometric packing in* $I^n(s)$.

The corresponding statement for packings in E^n is:

Let (C_i) *be a bounded sequence of convex bodies with the property that* $\sum v(C_i) = \infty$. *Then,* (C_i) *permits an isometric packing in* E^n *of density at least* $1/n!$. *Moreover, if all the sets* C_i *are boxes, then* (C_i) *permits an isometric packing in* E^n *of density* 1.

As already mentioned in the Introduction, one of the first results in this area was a theorem of Kosiński [34]. It states that a sequence (C_i) of boxes, all of whose edge lengths are at most 1, permits an isometric packing in a box of edge lengths 3, 3, ..., 3, $(v_o + 1)$, where $v_o = \sum v(C_i)$. This was later improved by Moon and Moser [42] who showed that a box of edge lengths 2, 2, ..., $2(v_o + 1)$ is already sufficient. This result has been established by a proof that is quite different from that of Kosiński. In the simplest case $n = 2$ it begins by noting that it suffices to consider rectangles of side lengths $2^{-i}, 2^{-j}$ $(i = 0, 1, \ldots; j = 0, 1, \ldots)$ and proceeds by giving rules on how to combine smaller rectangles to form larger ones and where to place leftover rectangles in another rectangle of side lengths 1 and 2.

Actually, if $n = 2$ better results are known. For example, Meir and Moser [41] have outlined a proof of the fact that a sequence (C_i) of rectangles with side lengths at most 1 and $\sum v(C_i) = v_o$ permits an isometric packing in any rectangle of side lengths a, b with $1 \leq a$ and of area $2v_o + a^2/8$. Although this result is in a certain sense best possible, the above general theorem for packings in $I^n(s)$ yields better estimates if s is large. This is again due to the fact that the condition on $\sum v(C_i)$ is asymptotically best possible. The problem of finding for all n and s the smallest number $g(s, n)$ such that any sequence (C_i) with $C_i \in \mathbf{C}^n$, $d(C_i) \leq 1$, and $\sum v(C_i) \leq g(n, s)$ permits an isometric packing in $I^n(s)$ appears to be extremely difficult; it is listed as one of the problems in [40] (Problem 10.1).

Packing results of the kind described in this section have practical applications concerning "bin packings" (cf. Kleitman and Krieger [33]). From the general theorem stated before, one can deduce that it is always possible to pack a bin with a bounded sequence of boxes so that one achieves nearly 100% efficiency if the bin is large enough.

There exist several subtle investigations dealing with packings of squares in rectangles; see Moon and Moser [42], Meir and Moser [41], Erdös and Graham [10], and Kleitman and Krieger [32, 33]. Further references can be found in the survey of Göbel [16]. Some unsolved problems in this area are discussed in the problem collection of W. Moser [44, Problem 23]. Concerning the analogous problem for cubes in boxes see Meir and Moser [41], Kotljar [35], and the comments of Meir and Bezdek regarding a packing problem of D. J. Newman in *SIAM Review* **25** (1983): 99–101.

The question of whether a bounded sequence of boxes of infinite total volume permits an isometric covering of E_n of density 1 has been raised in [18]. To a certain extent the answer has been anticipated in [17]. However, the pertinent result in that article is based on the assumption that the set of all edge lengths of the boxes has a positive lower bound. This assumption enables one to use compactness methods to obtain a subsequence of nearly congruent boxes which can then be arranged to provide a covering of the desired kind. In this connection it should be noted that already Hlawka [31] and L. Fejes Tóth [12; 13, pp. 106–109] have used compactness arguments for the solution of certain packing and covering problems. The disadvantage of this method is that rather unnatural restrictions are necessary.

5. Translative Coverings and Packings by Bounded Sequences of Convex Bodies

The idea of considering coverings and packings obtained by translations of the members of a given sequence of convex bodies can already be found in the investigations of Hlawka [31]. Independently, S. K. Stein has informally raised the question of whether there are translative analogues of the results concerning isometric coverings and packings.

We deal first with the case $n = 2$. The following proposition, proved in [23], is the starting point for almost all quantitative results in this area.

Let (R_i) be a sequence of rectangles in E^2, and let β_i denote the angle between the longer side of R_i and some fixed line $(0 \le \beta_i < \pi/2)$. Then, the conditions $\beta_i \le \beta < \pi/4$, $d(R_i) \le 1$, $s > 0$, and $\sum v(R_i) > (s^2 + 2(\tan\beta + 1)s + 2\tan\beta)/(1 - \tan\beta)$ imply that (R_i) permits a translative covering of a square of side length s. Similarly, the conditions $\beta_i \le \beta \le \pi/4$, $d(R_i) \le 1$, $s > 3$, and $\sum v(R_i) < ((s-1)(s - 2\tan\beta - 1))/(1 + \tan\beta)$ imply that (R_i) permits a translative packing in a square of side length s.

Using the approximation results of Section 3 and subsequences satisfying the pertinent angle conditions (for suitable β) one can easily derive theorems of the following kind (see [23] for details).

Let (C_i) be a sequence of convex bodies in E^2, and assume that $d(C_i) \le 1$. Then (C_i) permits a translative covering of a unit square if $\sum v(C_i) > \frac{3}{2}(13 + 7\sqrt{3})$. Furthermore, if $s > 3$ and $\sum v(C_i) \le \frac{1}{4}(s-1)(s-3)$, then (C_i) permits a translative packing in a rectangle of side lengths s and $2s$.

Perhaps more interesting is the application of the above general proposition to coverings and packings with respect to all E^2. By a suitable subdivision of E^2 into a sequence of increasing squares and a selection of rectangles with small β_i the following theorem has been shown (see [23]).

Let (C_i) be a bounded sequence of convex bodies in E^2, and assume that $\sum v(C_i) = \infty$. Then (C_i) permits a translative covering of E^2 of density 2, and also a translative packing in E^2 of density $\frac{1}{2}$. Moreover, if all the sets C_i are rectangles, then (C_i) permits such coverings and packings of density 1.

The fact that under the above assumptions (C_i) permits a translative covering of E^2 has also been proved by Makai and Pach [39], but no density estimates are given.

For $n > 2$ it is one of the major unsolved problems in this area whether any bounded sequence (C_i) with $\sum v(C_i) = \infty$ permits a translative covering of E^n. Actually, a translative covering of $I^n(1)$ would suffice since (C_i) can always be decomposed into infinitely many subsequences, each of which has a divergent sum of the corresponding volumes.

If one does not require covering of all E^n but only of almost all E^n the situation is much easier to deal with. The following theorem, which is proved in [19], shows that one does not even need convexity.

If (M_i) is a bounded sequence of Lebesgue measurable subsets of E^n, then (M_i) permits a translative covering of almost all E^n if and only if $\sum \lambda(M_i) = \infty$, where λ denotes Lebesgue measure.

The proof is a fairly simple consequence of the well-known integral formula $\int_M \lambda(M \cap (M_i + p))d\lambda(p) = \lambda(M)\lambda(M_i)$, where M is assumed to be bounded and Lebesgue measurable (see, for example, Halmos [30, p. 261]).

6. Isometric Coverings by Unbounded Sequences of Convex Bodies

In this section we consider again sequences (C_i) with $C_i \in \mathbf{C}^n$, but permit the possibility that the diameter set $\{d(C_i)\}$ is unbounded. Most results concern coverings of all E^n. Covering problems of this kind were first discussed by Chakerian [7]. The remarkable feature of these problems is that even if $n = 2$ a condition of the kind $\sum v(C_i) = \infty$ is no longer sufficient for (C_i) to permit an isometric covering of a unit square. For example, if we let C_i denote a rectangle of side lengths i and $1/2i^2$, then $\sum v(C_i) = \infty$ but (C_i) does not permit an isometric covering of $I^2(1)$. The latter statement follows from the solution of Tarski's "plank problem" (see Section 9), since it implies that not even the strips of widths $1/2i^2$ permit such a covering.

Regarding the case $n = 2$ it has been shown by Chakerian and Groemer [8] that a sequence (C_i) with $C_i \in \mathbf{C}^2$ permits an isometric covering of E^2 if and only if either $\sum_{d(C_i)\le 1} v(C_i) = \infty$ or $\sum_{d(C_i)>1} \Delta(C_i) = \infty$, where $\Delta(C_i)$ denotes the minimal width of C_i. As a consequence of this theorem it is shown that each of the following conditions is necessary and sufficient for (C_i) to permit an isometric covering of E^2:

$$\sum \Delta(C_i) \min \{1, d(C_i)\} = \infty,$$

$$\sum \frac{v(C_i)}{1 + d(C_i)} = \infty,$$

and

$$\sum \frac{v(C_i)}{1 + p(C_i)} = \infty,$$

where $p(C_i)$ denotes the perimeter of C_i.

The situation is more complicated if there is no restriction on the dimension. The following definition has turned out to be essential for dealing with this problem. If $C \in \mathbf{C}^n$ and if p, q are points in C so that the length of the line segment $[p, q]$ equals $d(C)$, we call the orthogonal projection of C onto a hyperplane orthogonal to $[p, q]$ a *normal projection* of C. It will be denoted by $N(C)$. [Of course, $N(C)$ is not necessarily unique; but if it is not, we assume that one of these projections has been selected arbitrarily and is kept fixed.]

Using this definition Chakerian and Groemer [9] have proved the following general theorem:

A sequence (C_i) of convex bodies in E^n permits an isometric covering of E^n if and only if either the sequence $(C_i)_{d(C_i)\le 1}$ permits an isometric covering of E^n or the sequence $(N(C_i))_{d(C_i)>1}$ permits an isometric covering of E^{n-1}.

In this connection the sets $N(C_i)$ are of course viewed as subsets of E^{n-1}. This theorem provides an inductive procedure for deciding whether a given sequence (C_i) permits an isometric covering of E^n. Note that according to the results discussed in Section 4 the first alternative in the theorem is equivalent to $\sum_{d(C_i)\le 1} v(C_i) = \infty$.

To derive more explicit results it is convenient to introduce the following notations. For any $C \in \mathbf{C}^n$ we set

$$N^0(C) = C, \qquad N^m(C) = N(N^{m-1}(C)) \qquad (m = 1, 2, \ldots, n),$$

$$d^m(C) = d(N^m(C)) \qquad (m = 0, 1, \ldots, n),$$

$$v^m(C) = v(N^m(C)) \qquad (m = 0, 1, \ldots, n;\ v^n(C) = 1;\ v \text{ denotes the volume in } E^{n-m}).$$

Furthermore, if $C_i \in \mathbf{C}^n$ we define $(n + 1)$ subsequences of (C_i) by setting

$$S^0 = (C_i)_{d^0(C_i) \leq 1}, \qquad S^m = (C_i)_{d^m(C_i) \leq 1,\, d^{m-1}(C_i) > 1} \qquad (m = 1, 2, \ldots, n).$$

It is shown in [9] that a sequence (C_i) of convex bodies in E^n permits an isometric covering of E^n if and only if for at least one $m \in \{0, 1, \ldots, n\}$ we have

$$\sum_{C_i \in S^m} v^m(C_i) = \infty.$$

An equivalent, and sometimes more practical, condition is

$$\sum_{C_i \in S^m} d^m(C_i) d^{m+1}(C_i), \ldots, d^{n-1}(C_i) = \infty,$$

where the terms of the series are assumed to be 1 if $m = n$.

If each C_i is a box (or ellipsoid) with respective edge lengths (or lengths of principal axes) $a_i^1 \geq a_i^2 \geq \cdots \geq a_i^n$, the above theorem yields that (C_i) permits an isometric covering of E^n if and only if at least one of the series

$$\sum_{a_i^1 \leq 1} a_i^1, \ldots, a_i^n, \quad \sum_{\substack{a_i^1 > 1 \\ a_i^2 \leq 1}} a_i^2, \ldots, a_i^n, \quad \ldots, \quad \sum_{\substack{a_i^{n-2} > 1 \\ a_i^{n-1} \leq 1}} a_i^{n-1} a_i^n, \quad \sum_{a_i^{n-1} > 1} a_i^n$$

diverges.

Also proved in [9] is the result that (C_i) permits an isometric covering of E^n if and only if

$$\sum \left(\frac{1}{v^0(C_i)} + \frac{1}{v^1(C_i)} + \cdots + \frac{1}{v^{n-1}(C_i)} \right)^{-1} = \infty.$$

Furthermore, if we let $W_k(C_i)$ denote the kth mean projection measure (Quermassintegral) of C_i, then each of the following conditions is also necessary and sufficient:

$$\sum \frac{d^0(C_i) d^1(C_i), \ldots, d^{n-1}(C_i)}{1 + d^0(C_i) + d^0(C_i) d^1(C_i) + \cdots + d^0(C_i) d^1(C_i), \ldots, d^{n-2}(C_i)} = \infty,$$

$$\sum \frac{W_0(C_i)}{W_1(C_i) + W_2(C_i) + \cdots + W_n(C_i)} = \infty,$$

$$\sum \frac{v(C_i)}{v(C_i + B^n(1)) - v(C_i)} = \infty.$$

The third condition has the remarkable feature that it involves only ordinary volume in E^n.

One of the basic lemmas in the proof of these results concerns the approximation of convex bodies by cylinders. It is shown that for any $C \in \mathbf{C}^n$ the right cylinder of height $d(C)$ and base $N(C)$ that contains C has the property that it is contained in a translate of $3C$.

If one is concerned with coverings of a convex body in E^n, rather than all E^n, then most of the concepts and methods developed for coverings of all E^n are still applicable but the results that can be obtained are less succinct. For example, Makai and Pach [39] have shown that a sequence (C_i) of convex bodies in E^n permits a covering of a set $C \in \mathbf{C}^n$ with int $C \neq \varnothing$ if

$$\sum \tilde{d}^0(C_i)\tilde{d}^1(C_i), \ldots, \tilde{d}^{n-1}(C_i) \geq kd^0(C)d^1(C), \ldots, d^{n-1}(C),$$

where $\tilde{d}^j(C_i) = \min\{d^j(C_i), d^j(C)\}$ and k is a constant depending only on n. It is also shown that a condition of the same type, but with a different constant, is necessary. Moreover, a similar condition (with the inequality sign reversed) is given for packings in C, but the additional conditions $d^j(C_i) \leq hd^j(C)$ $(j = 0, 1, \ldots, n-1)$, with some suitable constant h, have to be added. The remark made for coverings with respect to the necessity of these conditions also applies for packings.

7. Translative Coverings by Unbounded Sequences of Convex Bodies

Most results dealing with translative coverings of E^n by possibly unbounded sequences of convex bodies are types of transference theorems relating translative and isometric coverings. A typical example of this kind of result is the following theorem which is proved in [19].

Let (C_i) be a sequence of convex bodies in E^n. Then, (C_i) permits a translative covering of almost all E^n if and only if it permits an isometric covering of E^n.

The proof given in [19] proceeds by establishing inductive covering criteria for translative coverings of almost all E^n which are of the same type as those discussed in the preceding section. However, it is much easier to use the following procedure. If $C_i \in \mathbf{C}^n$ and if there are translations or rigid motions σ_i such that $(\sigma_i C_i)$ covers all or almost all E^n, then it is obviously possible to find translation vectors p_k and corresponding unit balls $B_k = B^n(1) + p_k$ $(k = 1, 2, \ldots)$ such that each C_i meets at most one B_k. Setting $C_i' = C_i \cap B_k$ if there is a B_k with $C_i \cap B_k \neq \varnothing$, and $C_i' = \varnothing$ if no such B_k exists, we find immediately that (C_i) permits an isometric or translative covering of all or almost all E^n if and only if (C_i') permits a covering of the same kind. But (C_i') is bounded and it follows from the results stated in Sections 4 and 5 that in all cases (with the possible exception of translative coverings of all E^n if $n > 2$) the sequence (C_i') permits a covering of the respective type if and only if $\sum v(C_i') = \infty$. Thus, the following statements are equivalent:

(i) (C_i) permits an isometric covering of almost all E^n,
(ii) (C_i) permits an isometric covering of all E^n,
(iii) (C_i) permits a translative covering of almost all E^n.

In the case $n = 2$ one may add:

(iv) (C_i) permits a translative covering of all E^n.

The restriction $n = 2$ in the last statement is certainly a rather regrettable blemish. The idea of using the balls B_k in proofs of this kind originally emerged, in a somewhat different context, in [24].

8. Coverings by Sequences of Unbounded Convex Sets

We consider now sequences (C_i) of closed convex sets C_i, but no longer impose the condition that these sets be bounded. First we need a definition and some simple facts on inscribed cones (see [20] or [28] for proofs). If C is a closed convex subset of E^n and $p \in C$ there exists a cone K in C with apex p such that every other cone in C with apex p is contained in K. It is easily seen that for every $p \in C$ there exists exactly one such cone and that it is closed and convex. Furthermore, any two cones of this kind belonging to different points in C are translation equivalent. Each of these cones will be called a *maximal cone* of C.

Disregarding coverings by special sets, like strips or slabs, one of the first results in this area concerns coverings by finite sequences. To formulate this result we say that a finite sequence of closed convex sets C_i with maximal cones K_i permits an isometric or translative *central covering* of E^n if there are, respectively, rigid motions or translations σ_i so that the cones $\sigma_i K_i$ cover E^n and have the origin of E^n as their apex. Central coverings appear to be a natural but very special possibility for covering E^n. Therefore the following theorem, proved in [20], may be somewhat surprising.

Let (C_i) be a finite sequence of closed convex sets. Then (C_i) permits an isometric or translative covering of E^n if and only if it permits, respectively, an isometric or translative central covering of E^n.

Actually, only the case of isometric coverings is considered in [20], but the proof for translative coverings is practically the same. One can also prove a generalization of this theorem that concerns coverings of some convex cone, rather than all E^n. If the cones of a central covering are intersected with the unit sphere bdr $B(1)$, one sees that for finite sequences the isometric covering problem of E^n can be reduced to a sphere covering problem. For example, if $n = 2$ one obtains immediately that a finite sequence of closed convex subsets of E^2 permits an isometric covering of E^2 if and only if the sum of the angles of the respective maximal cones is at least 2π. (See Section 10 for results concerning sphere coverings without dimension restrictions.)

Another consequence of the theorem is the fact that a finite sequence of closed convex sets can be altered by adding or removing finitely many bounded sets without changing the covering behavior with regard to all E^n.

Let us now turn to the general situation when (C_i) is permitted to be infinite. We say that a closed convex set C in E^n is of *finite width* if it is contained between two parallel hyperplanes. If this is not the case we say that C is of *infinite width.* Note that a convex set of infinite width always contains a unit ball. Therefore, if (C_i) contains infinitely many sets of infinite width it permits trivially a translative covering of all E^n. If all the sets C_i are of finite width the following theorem, which is proved in [24], enables one to reduce the covering problem to the case of bounded sequences.

Let (C_i) be a sequence of closed convex sets of finite width. Then (C_i) permits an isometric or translative covering of E^n if and only if there are unit balls B_i such that the sequence $(C_i \cap B_i)$ permits, respectively, an isometric or translative covering of E^n. Moreover, the corresponding statement with "E^n" replaced by "almost all E^n" is also true.

If the theorem is combined with the pertinent results stated in Sections 4 and 5 it can be concluded that (C_i) permits an isometric covering of E^n, or a translative covering of almost all E^n (or of all E^n if $n = 2$), if and only if $\sum \tilde{v}_i = \infty$, where $\tilde{v}_i = \sup \{v(C_i \cap (B(1) + p)) : p \in E^n\}$. In particular, we obtain that statements (i)–(iv) of Section 7 are equivalent even if the sets C_i are unbounded, provided they are closed, convex, and of finite width [and $n = 2$ as far as (iv) is concerned].

It is also worth noting that the relationship between (C_i) and $(C_i \cap B_i)$, as expressed by the above theorem, remains valid even for sequences of arbitrary sets of finite widths if attention is restricted to locally finite coverings, i.e., coverings with the property that each ball intersects only finitely many sets. The proof of the theorem follows the pattern outlined in Section 7 [immediately before statements (i)–(iv)]. The basic ideas of this proof can also be used to derive a result of Makai and Pach [39] which essentially states that a sequence (C_i) of closed convex sets of finite width permits an isometric covering of E^n, or a translative covering of almost all E^n (or all E^n if $n = 2$), if and only if there are, respectively, rigid motions or translations σ_i so that $\overline{\lim}_{r\to\infty} v((\bigcup \sigma_i C_i) \cap B^n(r))/v(B^n(r)) > 0$.

Similarly as in Section 6 one can define a normal projection $N(C)$ of a closed convex set C even if it is unbounded. In this case $N(C)$ is defined as the orthogonal projection of C onto a hyperplane that is orthogonal to a half-line contained in C. With this extension of the definition of $N(C)$ it has been shown in [24] that the inductive covering theorem of Section 6 can be generalized to arbitrary sequences of closed convex sets of finite width. In other words, such a sequence (C_i) permits an isometric covering of E^n if and only if either $(C_i)_{d(C_i)\leq 1}$ permits an isometric covering of E^n, or the sequence $(N(C_i))_{d(C_i)>1}$ permits an isometric covering of E^{n-1}. Makai and Pach [39] have shown that with a natural extension of the diameters $d^j(C_i)$ and the sets S^m (see Section 6) one can again prove that (C_i) permits an isometric covering of E^n if and only if for $m = 0, 1, \ldots, n$ at least one of the series $\sum_{C_i \in S^m} d^m(C_i)d^{m+1}(C_i), \ldots, d^{n-1}(C_i)$ diverges.

Up to this point we have only considered coverings by finite sequences of arbitrary closed convex sets, and coverings by sequences of closed convex sets of finite width. It might appear that there are many other cases that have to be dealt with. But the following theorem shows that in a certain sense we have already exhausted all possibilities for coverings of E^n by closed convex sets.

Let (C_i) be a sequence of closed convex subsets of E^n. Then, (C_i) permits an isometric covering of E^n (or almost all E^n), or a translative covering of almost all E^n (or all E^n if $n = 2$), if and only if one of the following conditions is satisfied:

(a) *(C_i) contains infinitely many terms of infinite width.*

(b) *There exists a subsequence of (C_i) that consists entirely of sets of finite widths and permits an isometric covering of E^n (or almost all E^n), or a translative covering of almost all E^n (or all E^n if $n = 2$).*

(c) *There exists a finite subsequence of (C_i) that permits, respectively, an isometric or translative central covering of E^n.*

As discussed earlier in this section, the conditions stated in (b), which concern only sets of finite width, are actually all equivalent. Although it is not known if the restriction $n = 2$ is necessary, the theorem remains true without this limitation if only locally finite sequences are considered. More precisely, a sequence (C_i) permits a locally finite translative covering of E^n if and only if either (a) or (c) holds (the latter with "translative"), or (C_i) has a subsequence that consists of sets of finite width and permits a locally finite translative covering of E^n.

A proof of the above theorem, together with the additions about locally finite sequences, has been given in [24]. Independently, Makai and Pach [39] have arrived at essentially the same conclusions but have formulated their results differently. They also have considered, similarly as in [20], coverings of convex cones in E^n.

9. Coverings by Sequences of Strips and Slabs

A *slab* in E^n is defined as the closed set consisting of all points between two parallel hyperplanes. The distance between the two boundary planes is called the *width* of the slab. If $n = 2$ we also talk about a strip instead of a slab.

The question of when a finite sequence of slabs S_i of widths w_i permits an isometric covering of a convex body K is completely answered by Bang's solution of Tarski's "plank problem." For any $K \in \mathbf{C}^n$ of minimal width $\Delta(K)$ the condition $\sum w_i \geq \Delta(K)$ is necessary and sufficient. This result, conjectured by Tarski [52], was proved first in full generality by Bang [2, 3]; see also Fenchel [15] and Bognár [6]. An affine-invariant version, involving the widths of the strips measured by the Minkowskian metric induced by K, has been formulated and discussed by Bang [3, 4]. The following articles concern themselves with proofs of this "relative plank problem": Ohmann [45] (this paper contains a serious gap, as noted by Fenchel in *Mathematical Reviews* **14:** 788), Lee *et al.* [36] (the authors use a definition of width of a convex body that differs from the usual definition, so their claim that they have proved Bang's conjecture is therefore not justified), and Ohmann [46] [see the reviews of Santaló in *Mathematical Reviews* **20:** #1948, and Rogers in *Zentralblatt für Mathematik und ihre Grenzgebiete* **81:** 165 for some critical remarks]. In view of the shortcomings of these efforts to prove the relative plank problem some authors consider the problem unsolved; see, for example, Alexander [1] and Rogers [49]. For the special case of the plank problem when only two strips are involved we refer to the articles of Straszewicz [50] and, regarding the relative version, Bang [4], W. Moser [43], and Alexander [1]. It is easy to extend the solution of the ordinary plank problem to infinite sequences of strips. A kind of dual of the plank problem has been introduced by L. Fejes Tóth [14] [Makai, in a colloquium on discrete geometry (Salzburg, 1980) announced further results concerning this dual plank problem].

We now turn to the principal subject of this section, namely, translative coverings by sequences of slabs or strips. We will first deal with the case $n = 2$. Thus, we assume that we are given a sequence of strips S_i of respective widths w_i. For some time it was not known whether the condition $\sum w_i = \infty$ implies that (S_i) permits a translative covering of the unit square in E^2. This problem was eventually solved by P. Erdös and E. Straus (unpublished). Their basic idea of ordering first the strips so that their angle with a given line increases was later used to prove theorems of a more quantitative nature. In [23] it has been shown that (S_i) permits a translative covering of a given $K \in \mathbf{C}^2$ if $\sum w_i \geq 6d(K)$. Makai and Pach [39] have proved that even the condition $\sum w_i > p(K)$, where $p(K)$ denotes the perimeter of K, is sufficient. The proof of the latter result depends also on an ordering of the Erdös–Straus type, but proceeds by constructing first a particular kind of covering of the boundary of K. Then it is observed that such coverings have the property that all K is covered if the boundary is. Using this method together with the device of circumscribing a rectangle or triangle of proper shape and size, Groemer [26] has found that each of the following conditions is sufficient for (S_i) to permit a covering of a given $K \in \mathbf{C}^2$:

$$\sum w_i \geq \frac{3}{\pi} p(K),$$

$$\sum w_i \geq 2\sqrt{2}\, d(K),$$

$$\sum w_i \geq d(K) + 2\Delta(K).$$

[As before, $p(K)$ denotes the perimeter, $d(K)$ the diameter, and $\Delta(K)$ the minimal width of K.] Moreover, if (S_i) is finite and consists of m strips, the condition

$$\sum w_i \geq \frac{m}{m+1} p(K)$$

is also sufficient. It is unlikely that any of these inequalities is best possible (with the trivial exception of the last one in the case $m = 1$), but at present no better results seem to be known. A simple consequence of each of these conditions is the fact that (S_i) permits a translative covering of E^2 if and only if $\sum w_i = \infty$. The "only if" part of this statement follows, for example, from Bang's theorem.

None of the proofs of these results can be generalized to the case $n > 2$. It is not even known whether a sequence of slabs S_i in E^n $(n > 2)$ of respective widths w_i that satisfies the condition $\sum w_i = \infty$ permits a translative covering of a unit cube in E^n. If this were true one could easily deduce that (S_i) also permits a translative covering of all E^n. At present the only result in this direction which has been shown to hold for all $n \geq 1$ is that the condition $\sum w_i^{(n+1)/2} = \infty$ implies that (S_i) permits a translative covering of E^n. This result has been found in a slightly different context which we are now going to describe.

For any $n \geq 1$ let K be a convex body and (S_i) a sequence of slabs in E^n. We say that (S_i) permits a *nondissective* translative covering of K if there are translates S_i' of S_i so that $K \subset \bigcup S_i'$ and that the sets $K \backslash S_1'$, $K \backslash (S_1' \cup S_2')$, $K \backslash (S_1' \cup S_2' \cup S_3')$, ... are convex. Changing the ordering of the sequence (S_i) is not permitted. As before, we let w_i denote the width of S_i. For this covering problem it is not $\sum w_i$ but $\sum w_i^{(n+1)/2}$ which is of crucial importance. This is clearly demonstrated by the following theorem which is proved in [22]:

Let (S_i) be a sequence of slabs in E^n of respective widths w_i, where $w_i \leq 1$, and let K be a convex body with mean projection measures W_j. Furthermore, let κ_{n-1} denote the volume of the $(n-1)$-dimensional unit ball. Then, (S_i) permits a nondissective translative covering of K if

$$\sum w_i^{(n+1)/2} > \frac{n}{\kappa_{n-1}} \left(W_0(K) + \binom{n}{1} W_1(K) + \cdots + \binom{n}{n-1} W_{n-1}(K)\right).$$

Moreover, for every $n \geq 1$ there exists a sequence of slabs which does not permit a nondissective translative covering of an n-dimensional unit ball, but whose widths w_i have the property that $w_i \leq 1$ and $\sum w_i^\alpha = \infty$ for all $\alpha < (n+1)/2$.

The proof of this theorem is obtained by investigating the decrease of the volume in the transition from the parallel body $(K \backslash (S_1' \cup S_2' \cup \cdots \cup S_i')) + B^n(1)$ to the parallel body $(K \backslash (S_1' \cup S_2' \cup \cdots \cup S_i' \cup S_{i+1}')) + B^n(1)$. If $n = 2$ the above sufficient condition becomes

$$\sum w_i^{3/2} > a(K) + p(K),$$

with $a(K)$ being the area and $p(K)$ the perimeter of K. A similar, but slightly weaker, result has been proved in [21] by showing that for any $C \in \mathbf{C}^n$ a strip of width w can always be translated so that either it covers C or it does not dissect C and decreases the function $\frac{4}{3}a(C) + p(C)$ by an amount at least $w^{3/2}$.

As a corollary of the above theorem one obtains the already mentioned statement that a sequence of slabs of respective widths w_i permits a translative covering of E^n if $\sum w_i^{(n+1)/2}$ diverges.

Regarding coverings of almost all E^n (instead of all E^n), it follows of course from the results discussed in Section 8 that the condition $\sum w_i = \infty$ implies that (S_i) permits a translative covering of almost all E^n.

10. Coverings of the Sphere

Let $S^n = \operatorname{bdr} B^{n+1}(1)$ denote the n-dimensional unit sphere embedded in E^{n+1} and centered at its origin o. A subset of S^n will be called a *cap* if it is of the form $K \cap S^n$, where K is a closed convex cone in E^{n+1} with apex o and nonempty interior. If C is a cap we let $m(C)$ denote its (n-dimensional) measure. Similarly as in the Euclidean case, we say that a sequence of caps C_i permits a covering of S^n if there are rotations ρ_i (about o) such that $S^n \subset \bigcup \rho_i C_i$.

It is intuitively quite clear that (C_i) will permit a covering of S^n if $\sum m(C_i)$ is sufficiently large. However, a proof of this statement is not easily obtained. The main obstacle is that in S^n there is no simple analogue of a box, and consequently it is impossible to transfer to the sphere the method of approximation by boxes and their orderly arrangement. The only approach that has led to some success begins by radially projecting the caps onto the tangent planes of S^n determined by the facets of a circumscribed cube, and proceeds by using the theorem on isometric coverings in these planes and projecting back onto the sphere. However, it is not possible to obtain good estimates due to the various distortions under radial projections. In [25] methods of this kind are used to prove the following theorem [as before κ_n denotes the volume of $B^n(1)$]:

For any dimension $n \geq 1$ let (C_i) be a sequence of caps on S^n. If

$$\sum m(C_i) \geq 2\kappa_n(2n)^n(n+1)((\sqrt{n}+1)^n - 1) + \kappa_{n+1}2^{n-1}(n+1)(2n+1),$$

then (C_i) permits a covering of S^n. If $n = 2$ this inequality can be replaced by the weaker condition

$$\sum m(C_i) \geq 144\sqrt{3}\,(\sqrt{2}+1) + 40\pi = (57{,}916, \ldots)4\pi.$$

Moreover, if all the sets C_i *have (geodesic) diameter at most* t*, where* $t \leq \pi/2$*, then* (C_i) *permits a covering of* S^n *if*

$$\sum m(C_i) \geq 2\kappa_n n^n(n+1)\left(\left(\sqrt{n} + \tan\frac{t}{2}\right)^n - \tan^n\frac{t}{2}\right) + \kappa_{n+1}\frac{(n+1)(2n+1)}{2}$$

and in the case $n = 2$ *if*

$$\sum m(C_i) \geq 36\sqrt{3}\left(\sqrt{2}\tan\frac{t}{2} + 1\right) + 10\pi.$$

It may be worth noting that the last inequality implies that (C_i) permits a covering of S^2 if the diameters of the caps are sufficiently small and their total area is less than 7.5 times the area of S^2. An examination of the proof of the above theorem shows also that for particular kinds of caps improvements are possible. For example, if all the caps are circular or elliptic (i.e., intersections of S^n with elliptic cones having apex o) the factor n^n can be removed from both inequalities where it now appears.

Instead of dealing with coverings of S^n one may consider packings in S^n. It is obvious what is meant by saying that a sequence of caps permits a packing in S^n. The principal problem is to find for each n a constant k_n such that any sequence of caps of total measure at most k_n permits a packing in S^n. At present it has not even been proved that such a k_n exists. Both the covering and packing problems are mentioned in the collection [18] of unsolved problems. One can also consider a kind of plank problem on the sphere where the "strips" are defined as parallel sets of great circles. See Linhart [37] for results concerning this problem.

References

1. Alexander, R. 1968. A problem about lines and ovals. Amer. Math. Monthly **75:** 484–487.
2. Bang, Th. 1950. On covering by parallel-strips. Mat. Tidsskr. B **1950:** 49–53.
3. Bang, Th. 1951. A solution of the "plank problem." Proc. Amer. Math. Soc. **2:** 990–993.
4. Bang, Th. 1954. Some remarks on the union of convex bodies. *In* Proceedings, Tolfte Skandinaviska Matematikerkongressen, Lund, 1953. Pp. 5–11. Lunds Universitets Matematiska Institution.
5. Bielecki, A. & K. Radziszewski. 1954. Sur les parallelépipèdes inscrits dans les corps convexes. Ann. Univ. Mariae Curie–Sklodowska Sect. A **8:** 101–103.
6. Bognár, N. 1961. On W. Fenchel's solution of the plank problem. Acta Math. Acad. Sci. Hungar. **12:** 269–270.
7. Chakerian, G. D. 1975. Covering space with convex bodies. *In* The Geometry of Metric and Linear Spaces (Michigan, 1974). L. M. Kelly, Ed.: 187–193. Lecture Notes in Mathematics No. 490. Springer-Verlag. West Berlin, FRG.
8. Chakerian, G. D. & H. Groemer. 1974. On classes of convex sets that permit plane coverings. Israel J. Math. **19:** 305–311.
9. Chakerian, G. D. & H. Groemer. 1978. On coverings of Euclidean space by convex sets. Pacific J. Math. **75:** 77–86.
10. Erdös, P. & R. L. Graham. 1975. On packing squares with equal squares. J. Combin. Theory Ser. A **19:** 119–123.
11. Fejes Tóth, G. 1983. New results in the theory of packing and covering. *In* Convexity and Its Applications. P. M. Gruber & J. M. Wills, Eds.: 318–359. Birkhäuser Verlag. Basel, Switzerland.

12. FEJES TÓTH, L. 1950. Covering with dismembered discs. Proc. Amer. Math. Soc. **1:** 806–812.
13. FEJES TÓTH, L. 1972. Lagerungen in der Ebene, auf der Kugel und im Raum. 2. Auflage. Springer-Verlag. West Berlin, FRG.
14. FEJES TÓTH, L. 1974. Remarks on the dual of Tarski's plank problem. Mat. Lapok **25:** 13–20. (In Hungarian with an English summary.)
15. FENCHEL, W. 1951. On Th. Bang's solution of the plank problem. Mat. Tidsskr. B **1951:** 49–51.
16. GÖBEL, F. 1979. Geometrical packing and covering problems. *In* Packing and Covering in Combinatorics. A. Schrijver, Ed.: 179–199. Mathematical Centre Tracts 106. Mathematisch Centrum. Amsterdam, the Netherlands.
17. GROEMER, H. 1976. On a covering property of convex sets. Proc. Amer. Math. Soc. **59:** 346–352.
18. GROEMER, H. 1976. Some covering and packing problems. Amer. Math. Monthly **83:** 726–727.
19. GROEMER, H. 1979. Space coverings by translates of convex sets. Pacific J. Math. **82:** 379–386.
20. GROEMER, H. 1980. On finite classes of convex sets that permit space coverings. Amer. Math. Monthly **84:** 286–288.
21. GROEMER, H. 1981. On coverings of plane convex sets by translates of strips. Aequationes Math. **22:** 215–222.
22. GROEMER, H. 1981. On coverings of convex sets by translates of slabs. Proc. Amer. Math. Soc. **82:** 261–266.
23. GROEMER, H. 1982. Covering and packing properties of bounded sequences of convex sets. Mathematika **29:** 18–31.
24. GROEMER, H. 1983. On space coverings by unbounded convex sets. J. Combin. Theory Ser. A **34:** 71–79.
25. GROEMER, H. 1983. On coverings of spheres by convex sets. Arch. Math. (Basel) **41:** 89–96.
26. GROEMER, H. 1984. Some remarks on translative coverings of convex domains by strips. Canad. Math. Bull. **27:** 233–237.
27. GRUBER, P. 1983. Approximation of convex bodies. *In* Convexity and Its Applications. P. M. Gruber & J. M. Wills, Eds.: 131–162. Birkhäuser Verlag. Basel, Switzerland.
28. GRÜNBAUM, B. 1967. Convex Polytopes. Interscience. London, England.
29. HADWIGER, H. 1955. Volumschätzung für die einem Eikörper überdeckenden und unterdeckenden Parallelotope. Elem. Math. **10:** 122–124.
30. HALMOS, P. R. 1950. Measure Theory. D. Van Nostrand. Princeton, N.J.
31. HLAWKA, E. 1949. Ausfüllung und Überdeckung konvexer Körper durch konvexe Körper. Monatsh. Math. **53:** 81–131.
32. KLEITMAN, D. & M. KRIEGER. 1970. Packing squares in rectangles, I. Ann. N.Y. Acad. Sci. **175:** 253–262.
33. KLEITMAN, D. & M. KRIEGER. 1975. An optimal bound for two dimensional bin packing. *In* Proceedings, 16th Annual Symposium on Foundations of Computer Science, Berkeley, California, 1975. Pp. 163–168. IEEE Computer Society. Long Beach, Calif.
34. KOSIŃSKI, A. 1956–1957. A proof of an Auerbach–Banach–Mazur–Ulam theorem on convex bodies. Colloq. Math. **4:** 216–218.
35. KOTLJAR, B. D. 1978. Packings of rectangular parallelotopes. Cybernetics **14**(No. 2): 294–296.
36. LEE, T. Y., J. S. LIN, K. C. TONG & M. Y. ZHANG. 1953. A solution of Bang's "plank problem." J. Chinese Math. Soc. **2:** 139–143. (In Chinese with an English summary.)
37. LINHART, J. 1974. Eine extremale Verteilung von Grosskreisen. Elem. Math. **29:** 57–59.
38. MACBEATH, A. M. 1951. A compactness theorem for affine equivalence classes of convex regions. Canad. J. Math. **3:** 54–61.
39. MAKAI, E., JR. & J. PACH. Controlling of function classes and covering of Euclidean space. To appear.
40. MAULDIN, R. D. (Ed.). 1981. The Scottish Book. Birkhäuser Verlag. Basel, Switzerland.
41. MEIR, A. & L. MOSER. 1968. On packing of squares and cubes. J. Combin. Theory **5:** 126–134.

42. MOON, J. W. & L. MOSER. 1967. Some packing and covering theorems. Colloq. Math. **17:** 103–110.
43. MOSER, W. O. J. 1958. On the relative widths of coverings by convex bodies. Canad. Math. Bull. **1:** 154, 168, 174.
44. MOSER, W. O. J. 1981. Research Problems in Discrete Geometry. (Privately published collection of problems.)
45. OHMANN, D. 1952. Eine Abschätzung für die Dicke bei Überdeckung durch konvexe Körper. J. Reine Angew. Math. **190:** 125–128.
46. OHMANN, D. 1957. Kurzer Beweis einer Abschätzung für die Breite bei Überdeckung durch konvexe Körper. Arch. Math. (Basel) **8:** 150–152.
47. RADZISZEWSKI, K. 1952. Sur un problème extrémal relative aux figures inscrites et circonscrites aux figures convexes. Ann. Univ. Mariae Curie–Sklodowska Sect. A **6:** 5–18.
48. ROGERS, C. A. 1964. Packing and Covering. Cambridge University Press. Cambridge, England.
49. ROGERS, C. A. 1981. Some problems in the geometry of convex sets. *In* The Geometric Vein. C. Davis, B. Grünbaum & F. A. Scherk, Eds.: 279–284. Springer-Verlag. New York, N.Y.
50. STRASZEWICZ, S. 1948. Un theoreme sur la largeur des ensembles convexes. Ann. Soc. Math. Polon. **21:** 90–93.
51. SÜSS, W. 1955. Über Parallelogramme und Rechtecke, die sich ebenen Eibereichen einschreiben lassen. Rend. Mat. e Appl. (5) **14:** 338–341.
52. TARSKI, A. 1932. Further remarks about the degree of equivalence of polygons. Odbitka z Parametru **2:** 310–314.

A CATALOGUE OF ISONEMAL FABRICS

Branko Grünbaum[a]

University of Washington, GN-50
Seattle, Washington 98195

G. C. Shephard

University of East Anglia
Norwich, NR4 7TJ, England

1. Introduction

Many interesting problems in discrete geometry and combinatorics relate to familiar structures in the real world. This assertion is illustrated by the topic of this paper—the geometry of woven fabrics. In spite of the fact that people have used weaving techniques for several millennia, and the design of fabrics presents many challenges to the combinatorialist, the mathematical literature on this subject is extremely meager. This is particularly surprising when one realizes that practical weavers have, over the centuries, evolved many fabric designs that have interesting mathematical properties. Here we shall be concerned with one of these properties known as *isonemality*.

Some problems in the geometry of fabrics are very challenging; that of determining the possible kinds of isonemal fabrics seems, at present, to be intractable by theoretical methods. In order to provide some data, we wrote a search program for the ICL 1904S computer at the University of East Anglia, Norwich, England, and here we report on some of the results obtained. In particular we determined all kinds of isonemal fabrics with periods up to 13, and information about these is given in Table 1. Due to restrictions on space we can illustrate *all* fabrics of periods up to and including 8; for larger periods we can show only a small number of kinds that seem to us to be visually attractive. In the last section of the paper we make some general comments about the results and state some open problems.

In order to make the paper self-contained we begin with a brief review of the necessary terminology, and then explain the theoretical considerations that were necessary before the computer program could be written.

2. Definitions

A (2-way 2-fold) *fabric* $\mathscr{F}$ consists of two *layers* of *strands* which are interwoven. The strands of one layer (the *warp*) are perpendicular to those of the other layer (the *weft*). A *design* $D(\mathscr{F})$ for a fabric $\mathscr{F}$ consists—following weavers' practice—of an array of squares colored either black or white. The rows of squares correspond to the weft strands, and the columns to the warp strands. A black square means that the warp strand (corresponding to the column containing the square) *passes over* the

[a] Research supported by National Science Foundation grant MCS-8001570 and by a Fellowship from the Guggenheim Foundation.

TABLE 1
THE ISONEMAL FABRICS OF PERIODS UP TO 13[a]

n	*b*	*T*	*S*	*I*	Totals
2	1	1	0	0	1
3	1	1	0	0	1
4	1	1	0	1	2
	3	1	0	1	2
	Totals	2	0	2	4
5	1	1	1	0	2
	3	1	0	0	1
	5	1	0	0	1
	Totals	3	1	0	4
6	1	1	0	1	2
	3	1	0	1	2
	5	1	0	1	2
	7	1	0	0	1
	11	1	0	1	2
	Totals	5	0	4	9
7	The only isonemal fabrics are 8 twills corresponding to sequences with binary indices $b = 1, 3, 5, 7, 9, 11, 19$ and 21.				
	Totals	8	0	0	8
8	1	1	1	1	3
	3	1	0	1	2
	5	1	0	2	3
	7	1	0	2	3
	9	1	0	1	2
	11	1	0	9	10
	15	1	0	1	2
	19	1	0	9	10
	21	1	0	2	3
	23	1	0	2	3
	27	1	0	5	6
	37	1	0	2	3
	43	1	0	2	3
	45	1	0	1	2
	Totals	14	1	40	55
9	The only isonemal fabrics are 21 twills corresponding to sequences with binary indices $b = 1, 3, 5, 7, 9, 11, 15, 17, 19, 21, 23, 27, 35, 37, 39, 43, 45, 51, 75, 83$ and 85.				
	Totals	21	0	0	21
10	1	1	1	2	4
	3	1	0	2	3
	5	1	0	2	3
	7	1	0	0	1
	9	1	0	2	3
	11	1	0	2	3
	15	1	0	1	2
	17	1	0	2	3
	19	1	0	2	3
	21	1	0	2	3

TABLE 1. *(Continued)*

n	b	T	S	I	Totals
10	23	1	0	2	3
(Continued)	27	1	0	1	2
	31	1	0	0	1
	35	1	0	2	3
	37	1	0	2	3
	39	1	0	4	5
	43	1	0	2	3
	45	1	0	1	2
	47	1	0	1	2
	51	1	0	1	2
	55	1	0	4	5
	69	1	0	2	3
	71	1	0	0	1
	73	1	0	0	1
	75	1	0	4	5
	77	1	0	2	3
	83	1	0	2	3
	85	1	0	3	4
	87	1	0	2	3
	91	1	0	4	5
	93	1	0	1	2
	103	1	0	1	2
	107	1	0	1	2
	147	1	0	1	2
	149	1	0	0	1
	155	1	0	0	1
	171	1	0	2	3
	173	1	0	1	2
	179	1	0	1	2
Totals		39	1	64	104
11	The only isonemal fabrics are 62 twills corresponding to sequences with binary indices b = 1, 3, 5, 7, 9, 11, 15, 17, 19, 21, 23, 27, 31, 33, 35, 37, 39, 43, 45, 47, 51, 55, 67, 69, 71, 73, 75, 77, 79, 83, 85, 87, 91, 93, 99, 103, 107, 119, 137, 139, 147, 149, 151, 153, 155, 157, 163, 165, 167, 171, 173, 179, 199, 203, 219, 293, 299, 301, 307, 331, 339 and 341.				
Totals		62	0	0	62
12	There are 112 nonequivalent sequences with binary indices b = 1, 3, 5, 7, 9, 11, 15, 17, 19, 21, 23, 27, 31, 33, 35, 37, 39, 43, 45, 47, 51, 55, 63, 67, 69, 71, 73, 75, 77, 79, 83, 85, 87, 91, 93, 95, 99, 103, 107, 111, 119, 133, 135, 137, 139, 141, 143, 147, 149, 151, 153, 155, 157, 163, 165, 167, 171, 173, 175, 179, 183, 187, 189, 199, 203, 207, 215, 219, 231, 275, 277, 279, 283, 291, 293, 295, 297, 299, 301, 307, 309, 311, 313, 315, 327, 331, 333, 339, 341, 343, 347, 349, 355, 359, 363, 365, 371, 403, 411, 423, 427, 587, 595, 597, 603, 619, 661, 683, 685, 691, 693 and 717. The numbers of isonemal fabrics are as follows:				
Totals		112	1	316	429

TABLE 1. (*Continued*)

n	b	T	S	I	Totals
13	1	1	1	0	2
	45	1	0	1	2
	99	1	0	1	2
	107	1	0	1	2
	557	1	0	1	2
	581	1	0	1	2
	583	1	0	1	2
	There are also 182 other sequences each of which leads only to a twill.				
	Totals	189	1	6	196

[a] For each (exact) period n from 2 to 13 we give the binary indices b of the $t(n)$ nonequivalent sequences of n squares, as explained in the text. For each such sequence we list the corresponding number of isonemal fabrics under the following headings:

T is the number of twills,

S is the number of satins,

I is the number of isonemal fabrics other than twills and satins.

weft strand (corresponding to the row containing the square) and a white square means that the weft strand passes over the warp strand. Formal definitions of the italicized words can be found in Grünbaum & Shephard [5], but we believe that examination of FIGURE 1 will enable the concepts to be understood without difficulty. One further essential feature of a fabric is that it *hangs together*. This means that it is impossible to partition the set of all strands into two nonempty subsets so that each strand of the first subset passes over every strand of the second subset, see FIGURE 2. It is not at all easy to determine by inspection whether a given design corresponds to a genuine fabric or to a "fabric" that does not hang together, although recently Clapham [2] and Enns [3] have given useful criteria for deciding this question. We used a simpler algorithm which is explained below, in Section 3. In this connection we would like to correct an error in [5]. The design of Figure

FIGURE 1. A fabric of exact period 8, together with its design. A fundamental 8 × 8 block is outlined.

11(b) of [5] does not represent a color-alternate twill, as stated in the caption to that diagram, since—as can be easily verified—it does not hang together. (Fabrics with strands in more than two directions are possible; they are discussed in [8] and are not considered here.)

Associated with any fabric $\mathcal{F}$ is its symmetry group $S(\mathcal{F})$. Envisaging the fabric physically as made up of thin layers of strands in three-dimensional space, we may interpret the symmetries of $S(\mathcal{F})$ as consisting of those isometries of the space which map $\mathcal{F}$ onto itself. One can determine $S(\mathcal{F})$ from inspection of the design $D(\mathcal{F})$; every element of $S(\mathcal{F})$ corresponds to a symmetry of $D(\mathcal{F})$ or to such a symmetry combined with an interchange of the colors black and white. For example, for the fabric shown in FIGURE 1 there is a symmetry which maps the upper row (weft strand) onto the third row from top—it is a translation through two units to the left and two units down. (The *unit* is the edge-length of one of the squares in the design; for brevity we shall in the sequel suppress the mention of the obviously necessary vertical component of the translations and of the other symmetries which map rows onto rows.) There is also a symmetry of the design which maps the upper row onto the second row—it is a translation through one unit to the left combined with an interchange of the colors, and it corresponds in the fabric to a translation combined with a reversal of the sides of the fabric. The top row of the diagram is equivalent to its rightmost column by reflection in the diagonal—this corresponds in the fabric to the equivalence of the topmost weft strand to the rightmost warp strand under reflection in the diagonal combined with a reversal of the sides of the fabric.

If $S(\mathcal{F})$ contains translations through n units in both horizontal and vertical directions, then we say $\mathcal{F}$ has *period n*. If n is the smallest integer with this property then n is called the *exact period* of the fabric. The design of any fabric of period n can be obtained from a *fundamental* $n \times n$ *block* of squares, suitably colored, translated in both horizontal and vertical directions through multiples of n units. The fabric of FIGURE 1 has exact period 8 and a fundamental 8×8 block is outlined. The fabrics whose designs appear in FIGURES 3(a) and 3(b) have exact periods 6 and 8 respectively.

A fabric $\mathcal{F}$ is *isonemal* if the symmetry group $S(\mathcal{F})$ is transitive on the strands of $\mathcal{F}$; it is easily verified, for example, that the fabrics of FIGURES 1 and 3 are isonemal. The combinatorial problem considered in this paper is that of determining designs of all essentially distinct isonemal fabrics of exact period n, for various small values of n. In Grünbaum & Shephard [5] methods of constructing isonemal fabrics were explained, and many examples were given. However, these methods are difficult to apply systematically and so do not lend themselves to a complete enumeration. In fact, by empirical means we determined all isonemal fabrics of periods $n \leq 7$, but for larger values of n the task became practically impossible to carry out by hand.

For this reason a computer program was written to carry on the search, and some of the designs it produced appear in FIGURE 5.

One kind of isonemal fabric which lends itself to a theoretical investigation is that of the *twills*. Examples are shown in FIGURE 3 from which the characteristic diagonal "stripe" of a twill will be observed. The essential feature of a twill is that each row in its design can be obtained from the row above by a translation of one unit to the left (or to the right). For a twill of period n we may take any row of n black and white squares to be the upper row in a fundamental $n \times n$ block, and use this to generate a design by applying unit translations to obtain successive rows.

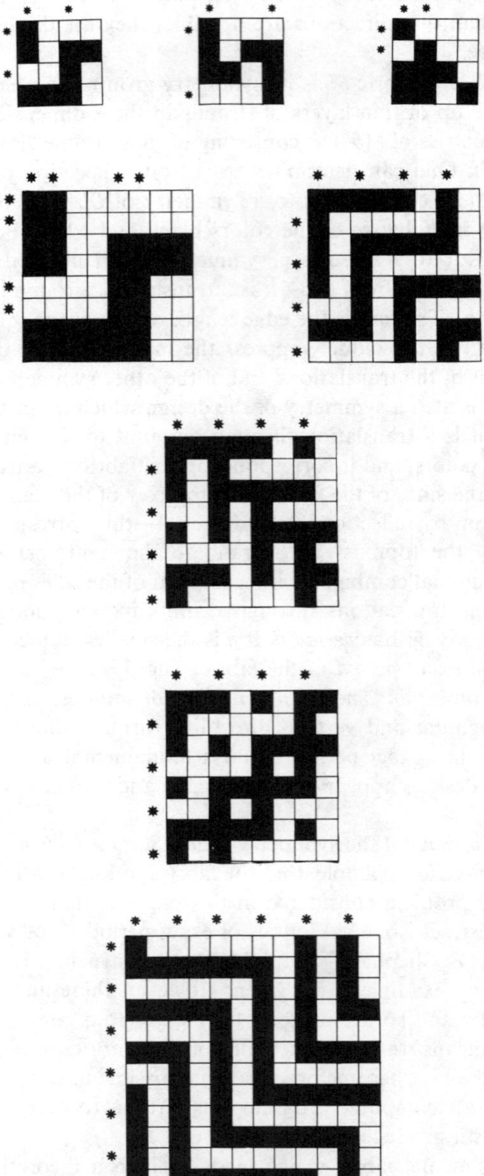

FIGURE 2. These designs do not represent fabrics because they violate the condition that a fabric must hang together. The stars indicate subsets of strands that can be "lifted off", since they all pass over all the strands of the complementary subsets. The designs shown are all the isonemal designs with period up to 13 which do not represent fabrics.

FIGURE 2. (*Continued.*)

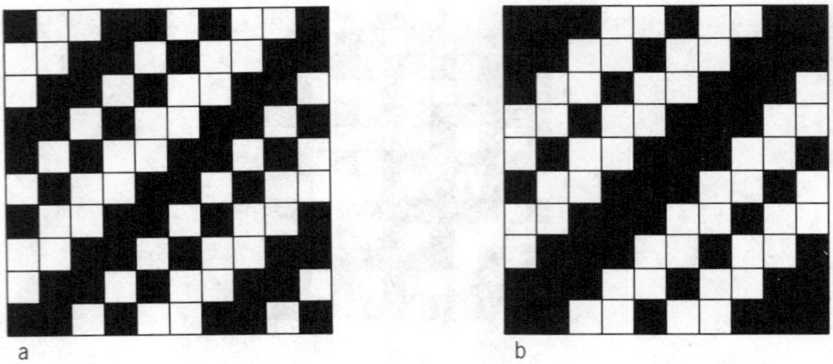

FIGURE 3. Examples for designs of twills of exact periods (a) 6 and (b) 8.

Assuming that *both* black and white squares occur in the given row then we always obtain the design of a genuine fabric (in the sense that it hangs together); moreover it is necessarily isonemal.

As twills are produced so easily it was originally intended that they should be excluded from the computer search—however, it turned out to be convenient to list them as a first step in determining the non-twilly isonemal fabrics (see Section 3).

The problem of enumerating the twills is equivalent to that of finding colorings of the original row of n squares that lead to essentially distinct twills. Here, and earlier, we have used the words "essentially distinct" without explanation. The precise meaning is as follows: two fabrics are *essentially distinct* if a design for one cannot be made to coincide with any design for the other by a rigid motion, possibly combined with an interchange of the colors black and white. In terms of the more general classification of geometric objects with respect to symmetry (see Grünbaum & Shephard [6], [7]), "essentially distinct" means "of different homeomeric types".

3. The Enumeration Procedure

The procedure that was used is in four stages, which will be described in this section. They will be stated in terms of black and white squares in the design; in the computer program they were coded using binary digits 0 and 1. Fabrics of period 4 exhibit some exceptional features, but since they can be enumerated empirically without difficulty they were excluded from the search program. However, for completeness they are included in our catalogue (TABLE 1).

(I) *Distinct Sequences*

Two sequences, each containing n black or white squares, are called *equivalent* if one can be obtained from the other by a cyclic permutation, possibly combined with a reversal of order or an interchange of the colors black and white. As we shall be interpreting each such row of n squares as the top row of a fundamental $n \times n$ block, it is clear that in enumerating essentially distinct fabrics it will suffice to choose one sequence out of each equivalence class. We denote the number of equivalence classes by $t(n)$, where n is the exact period of the sequence.

From the remarks at the end of the previous section we see that $t(n)$ is the number of essentially distinct twills of exact period n—hence determining a maximal set of nonequivalent sequences is the same problem as finding the essentially distinct twills of a given period. The value of $t(n)$ can be determined using an extension of Pólya's enumeration theorem, see [5, p. 149]. (Notice, however, that the formula given in this reference is for the number of sequences of period n, while here $t(n)$ is defined as the number of sequences of *exact* period n.) This theoretical enumeration was used to check the first stage of the program, and numerical values for small values of n appear in TABLE 1.

It is convenient to associate with each sequence of n black and white squares its binary index. This is the decimal equivalent of the n-digit binary number obtained by interpreting each black square as 1 and each white square as 0. For an equivalence class of sequences, the *binary index* is defined as the minimum of the binary indices of all the sequences in the class. The binary indices of the $t(n)$ equivalence classes of sequences enable the twills of exact period n to be listed in a compact and convenient manner.

(II) *Weft-Isonemal Fabrics*

A fabric $\mathscr{F}$ is *weft-isonemal* if the symmetry group $S(\mathscr{F})$ is transitive on the weft strands. The second stage in the enumeration consists of constructing the designs of all weft-isonemal fabrics of the given period n from the $t(n)$ nonequivalent sequences determined in the first stage. To do this we consider all the ways in which weft strands can be mapped onto one another. Clearly there are eight possible mappings: the four isometries (translations, rotations (half-turns), reflections and glide-reflections) and four more resulting from each of these followed by an interchange of the colors black and white. It is necessary to consider how these mappings can be combined in the diagram of a weft-isonemal fabric, and it turns out that for fabrics with exact period greater than 4 there are just five possibilities. These are shown diagrammatically in FIGURE 4, where they are referred to as *genera*; each genus depends upon one or two parameters, as indicated. In FIGURE 4 a black arrow is used for the initial sequence of n squares, and a white arrow represents the same sequence with the colors black and white interchanged.

In some of the genera the permissible values of the parameters can be restricted without loss of generality. For example, in the case of Genus I, the parameter s is the magnitude of the horizontal component of the translation that maps the first row onto the second, and we need only consider those values which satisfy $2 \leq s \leq [\frac{1}{2}n]$ and s is prime to n. The value $s = 1$ is excluded since this only produces twills, and the values of s in the range $[\frac{1}{2}n] < s < n$ clearly lead to designs which are not essentially distinct from those that arise from smaller values of s. Further, if s is not prime to n then the rows will repeat with period strictly less than n and such designs are clearly inadmissible in our search for fabrics of exact period n.

Notice that none of the five genera in FIGURE 4 involve reflections or glide-reflections. The reason for this is that (as can be easily shown) the presence of either of these types of isometries necessarily leads to a design in which the rows repeat with period 1, 2 or 4; but these are inadmissible since, as stated above, we restrict our search to fabrics with exact period n for $n > 4$.

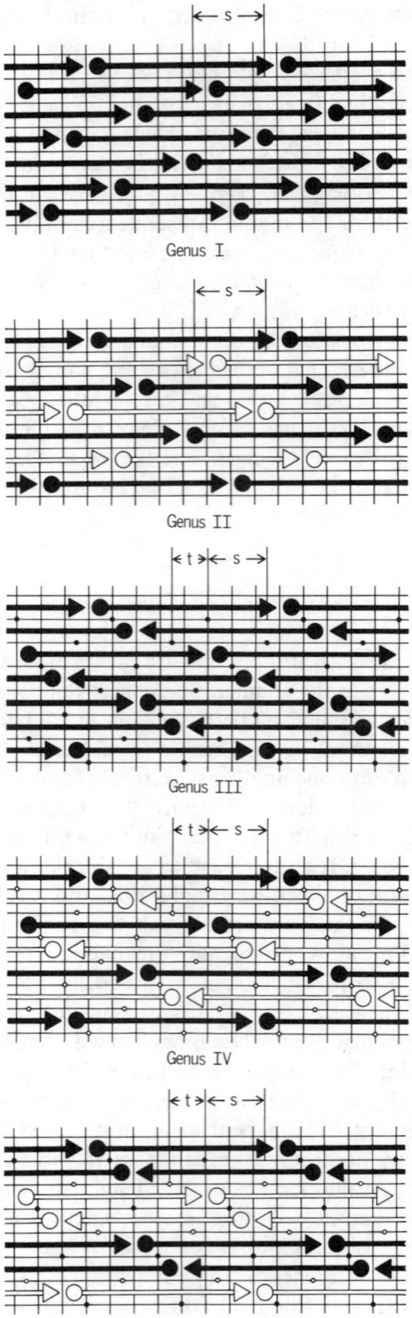

FIGURE 4. The five genera of weft-isonemal fabrics of exact period greater than 4. A black arrow represents a sequence of n black and white squares, and a white arrow represents the same sequence with the colors black and white interchanged. A small black dot denotes a center of 2-fold rotation that maps a row into the following row; if there is a color change then a small white dot is used instead. Definitions of the parameters s and t are also indicated for each of the five genera.

(III) *Testing for Isonemality*

Each of the weft-isonemal designs produced in the second stage is then tested for isonemality. This is done by "turning" the given design through 90° (in all possible ways), possibly combining this with an interchange of the colors black and white, and then seeing if some translation will "fit" exactly over the original design. Whenever such a fit is found, the design is printed out, together with a symbol n-b-ℓ defined as follows. The integer n is the exact period of the fabric, the integer b is the binary index of the row of n black and white squares from which the design was generated, and ℓ is a list number. In all the diagrams we give the symbol below the design, and display the designs in lexicographic order of their symbols. Conventionally the list number $\ell = 0$ is reserved for the twills.

(IV) *Testing whether the Fabric Hangs Together*

Finally, we check all the isonemal designs to see whether they correspond to fabrics that hang together. The following procedure applies to all fabrics, not only the isonemal ones. It is very simple to implement on the computer but—in the range of n considered—also easy to carry out manually.

An arbitrary column of a fundamental n by n block of squares of the design is chosen, and we attempt to "lift it off"; then we have to "lift off" all rows meeting the chosen column in *white* squares, and also all columns meeting at least one of these rows in a *black* square, as well as all the rows meeting at least one of the latter columns in a *white* square, etc. The process clearly terminates after a (small) finite number of steps; at this stage, *either* a proper subset of the strands "lifts off" indicating that the design does not correspond to a fabric, *or else* we "lifted off" all the strands. In that case we turn to the second half of the procedure, namely we try to "push down" the originally chosen column and—in analogy to the above—determine the set of all strands necessarily "pushed down" with it. If this is a proper subset of the set of all strands—we have a decomposition of the "fabric"; otherwise, the design indeed represents a genuine isonemal fabric. In the case of isonemal fabrics, a further drastic reduction of effort is possible. In a design which represents a "fabric" that does not hang together, consider a column that can be "pushed down". Then each row which meets that column in a black square is also "pushed down" and therefore must meet every column which is *not* "pushed down" in a black square. Thus, since the design cannot contain completely black rows, it follows that the number of black squares in columns "lifted off" is strictly greater than the number of black squares in the columns "pushed down". It follows that an isonemal design of Genus I or III necessarily hangs together. For the other three genera, the design represents a fabric which hangs together *unless* the columns contain unequal numbers of black squares, and the columns containing the larger numbers of black squares can be "lifted off" together with the rows containing fewer black squares. Therefore we can forgo the testing of designs of Genus I or III, and have only to test one definite set of strands in case of a design of Genus II, IV or V.

4. The Catalogue

The procedure used to generate the designs of isonemal fabrics of a given period necessarily produces repetitions—for example, this occurs whenever a fabric belongs

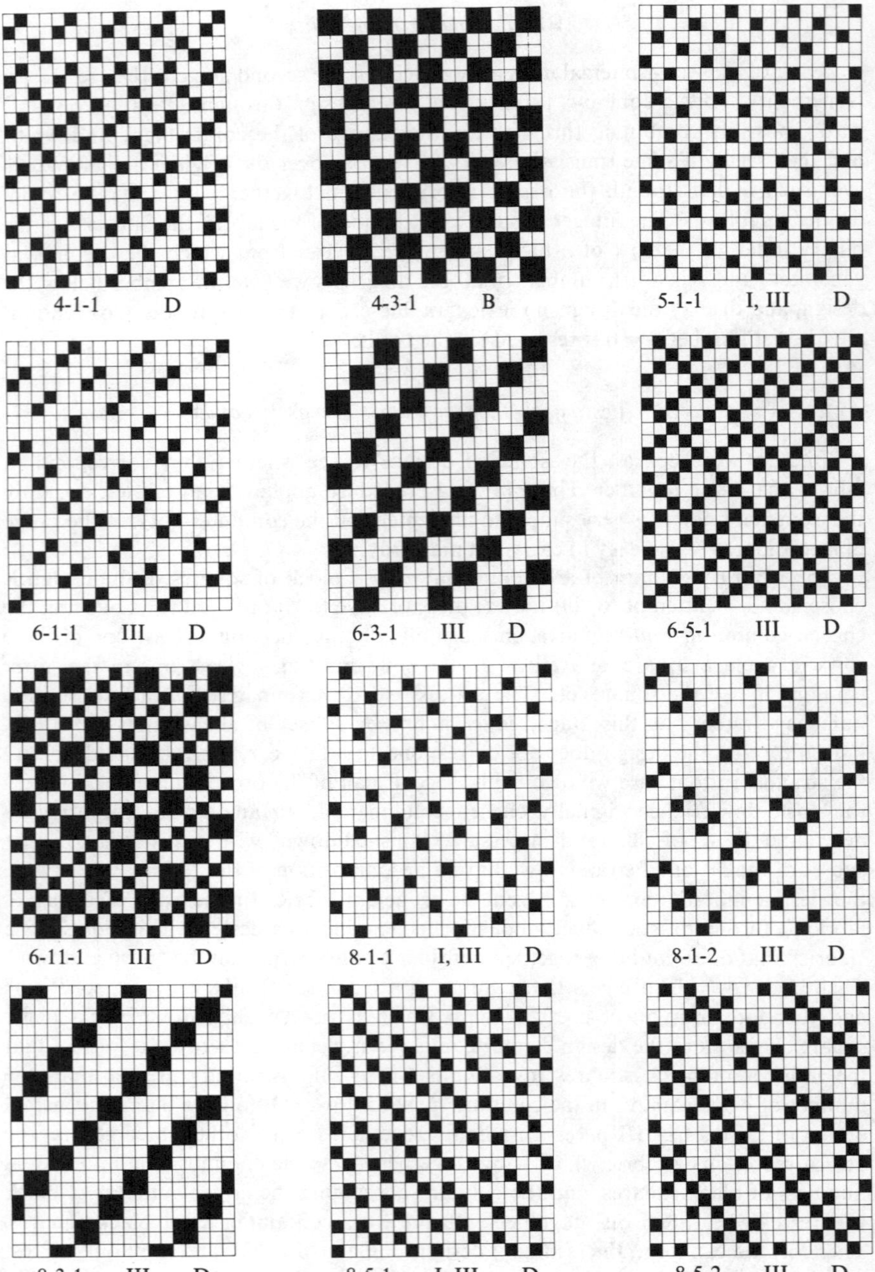

FIGURE 5. Examples of designs for isonemal fabrics which are not twills. All such fabrics are shown for exact period $n \leq 8$ (see TABLE 1) and also for $n = 13$. For $n = 10$ and $n = 12$ a few examples of fabrics are shown. There are no isonemal fabrics with exact periods 7, 9 or 11 other than the twills. The symbol below each design is explained in Section 3 of the text. Roman numerals (I, II, III, IV or V) list all the genera to which the design belongs. The letters A, B, C, D indicate the class of the fabric, as explained at the end of Section 4.

8-7-1 II, IV A

8-7-2 V A

8-9-1 III D

8-11-1 II A

8-11-2 II A

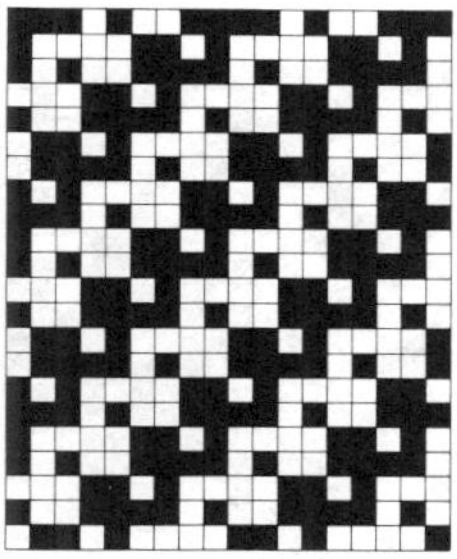

8-11-3 IV A

8-11-4 IV A

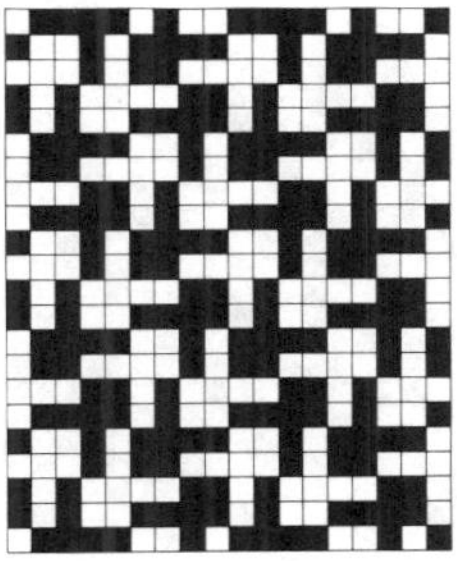

8-11-5 V A

8-11-6 I D

8-11-7 III D

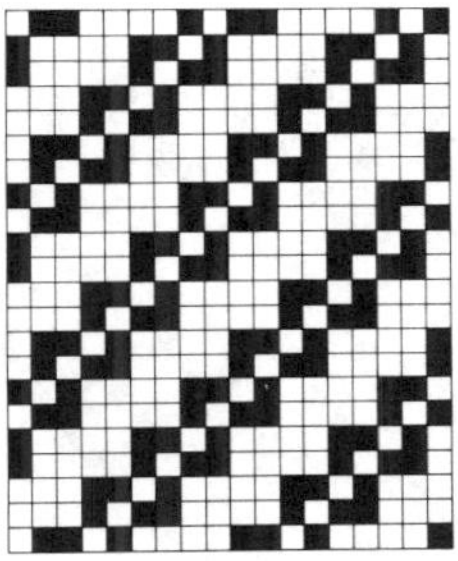

8-11-8 III D

8-15-1 III, IV, V B

FIGURE 5. *(Continued.)*

8-19-1 II A

8-19-2 II A

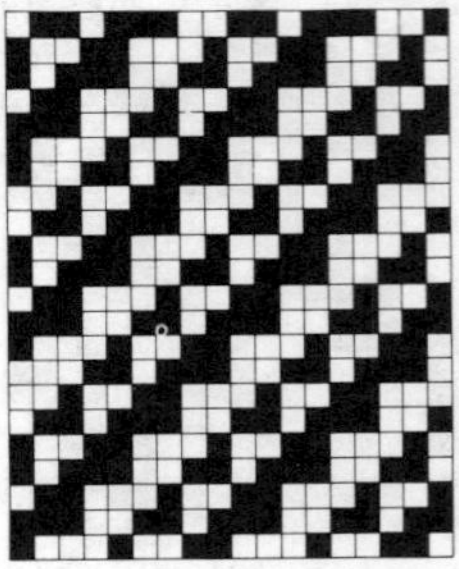

8-19-3 IV A

8-19-4 IV A

8-19-5 I D

8-19-6 III D

8-19-7 III D

8-19-8 III D

8-19-9 III D

8-21-1 I, III D

8-21-2 III D

8-23-1 III D

FIGURE 5. (*Continued.*)

8-27-1 II, IV A

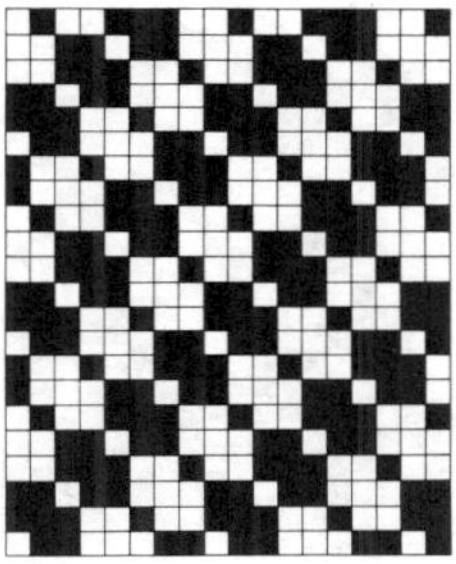

8-27-2 II, IV A

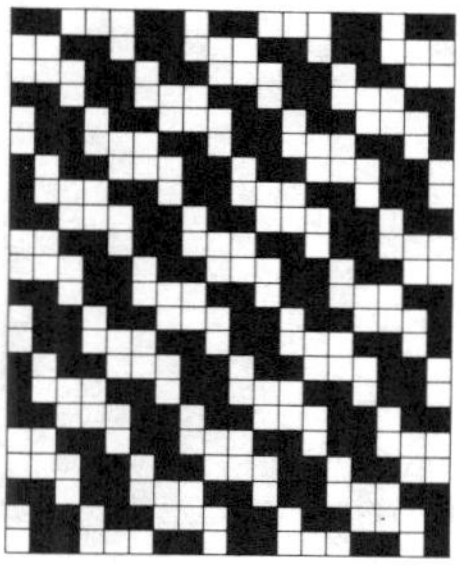

8-27-3 III C

8-27-4 IV A

8-27-5 I, III D

8-37-1 II, IV A

8-37-2 V A

8-43-1 V A

8-43-2 III D

8-45-1 III, IV, V B

10-39-1 III D

10-39-2 III D

FIGURE 5. (*Continued.*)

10-39-3 III D

10-39-4 III D

10-87-1 III C

10-171-2 III C

12-75-5 III D

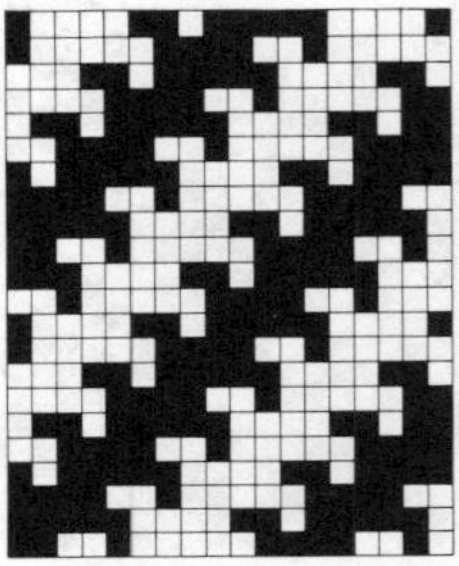

12-79-1 II A

12-179-1 II A

12-231-1 I, III C

12-275-10 IV A

12-371-1 I, IV B

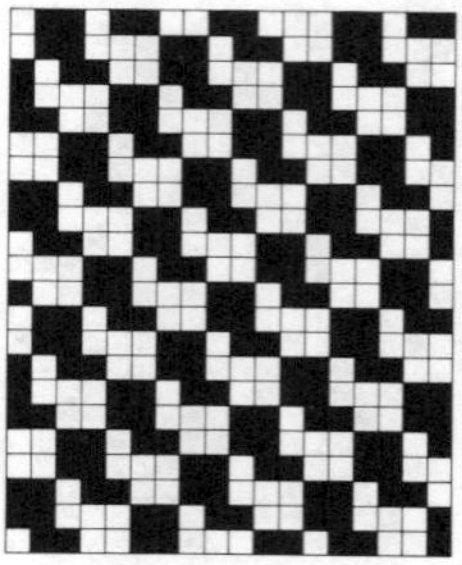

12-411-1 I C

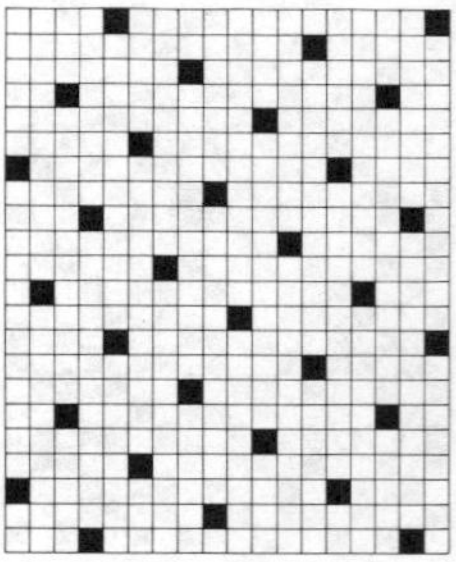

13-1-1 I, III D

FIGURE 5. (*Continued.*)

13-45-1 I, III D

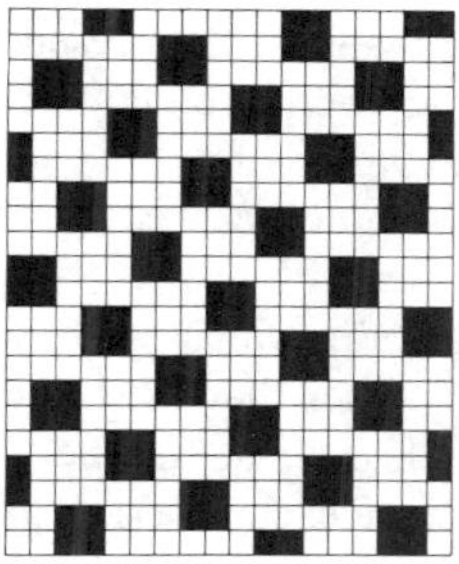

13-99-1 I, III D

13-107-1 I, III D

13-557-1 I, III D

13-581-1 I, III D

13-583-1 I, III D

FIGURE 5. (*Continued.*)

to more than one genus. These repetitions were removed by inspection since, for the small values of n considered here this seemed to be more convenient than modifying the computer program to carry out this task.

Although the twills are included in TABLE 1, none of these are shown in the diagrams (except for the two of FIGURE 3). Fabrics of Genus I with binary index $b = 1$ are called *satins* [5, p. 150], and a list of all possible satins with periods up to 100 is given in [5, p. 153]. Earlier enumerations of satins occur in: Lucas [9] in 1867 (for periods $n \leq 50$), Gand [4] in 1871 (for $n \leq 82$), Lucas [10] in 1880 (for $n \leq 95$), Aubry [1] in 1911 (for n up to 100, but only for certain types of isonemal satins). All these enumerations, with the exception of [9], are incomplete in the ranges under consideration. It should be noted that if a satin is not isonemal it has two distinct representations: an (n, p)-satin is identical with an (n, q)-satin if $pq \equiv \pm 1 \pmod{n}$. In [5, Table 3] each satin occurs once; only as (n, p) with the smaller value of p.

The data in TABLE 1 reveal some interesting features. The most surprising is the small number of isonemal fabrics of odd periods. In fact, apart from the twills and the satin of period 5, there are no isonemal fabrics of any odd period $n < 13$. Clearly, isonemal fabrics of odd period must be of Genus I (or *twillins* as they are called in [5, p. 154]), and we conjecture that these only exist for those odd values of n which admit an (isonemal) satin. If this is so, then it seems possible that a theoretical enumeration of isonemal fabrics may be possible in this case.

From the expression for $t(n)$ we see that the number of twills (or of essentially distinct sequences of n squares) is of the order $2^{n-2}/n$. The number of weft-isonemal fabrics of even period is then of order $2^{n-2} \cdot 3n$, but we have no idea how many of these are isonemal. The data available at present suggests, however, that this number increases so rapidly that even enumeration by computer methods necessarily fails already for quite small values of n.

From the point of view of the practical weaver, the isonemal fabrics can conveniently be classified into four *classes* with respect to their "appearance".

Class A consists of those fabrics in which establishing the mutual equivalence of weft strands necessitates in some cases that the fabric be reversed; for example, the fabric shown in FIGURE 1 is of Class *A*, as may be seen by considering the two top rows. Clearly, the warp strands of each Class *A* fabric have the same property. Hence Class *A* fabrics "look the same" in the warp and in the weft directions, and on both sides of the fabric—although not all strands are equivalent on the same side.

Class B consists of fabrics in which all the weft strands are equivalent to each other under symmetries which do not reverse the sides of the fabric, as well as under symmetries which do reverse the sides; clearly, the warp strands then have the same property. The fabric whose design is shown in FIGURE 3(a) is of Class *B*. Fabrics of Class *B* not only "look the same" in the warp as in the weft directions and on both sides of the fabric, but each strand is equivalent to every other strand, both by symmetries which interchange the sides of the fabric and by symmetries which preserve the sides.

Classes C and D consist of fabrics in which all the weft strands are equivalent under symmetries which do not reverse the fabric, and no two of these strands are equivalent under any fabric-reversing symmetry. Clearly, the warp strands have the same property. For fabrics in Class *C* the warp strands are equivalent to the weft strands under symmetries which do not reverse the sides of the fabric, while for fabrics in Class *D* the equivalence of the two kinds of strands can be established only by symmetries which reverse the sides of the fabric. Thus the fabrics in Class *C* "look the same" in the warp and in the weft directions of the same side of the fabric, but the two sides "look different". The fabrics of Class *D* "look the same" in the weft direction of one side as in the warp direction of the other side—but on the same side the appearance in the two directions is different. The twill in FIGURE 3(b) is of Class *D*, while the fabric with design 8-27-3 in FIGURE 5 is an example of Class *C*.

* * * * *

Since the completion of our manuscript, Dr. A. P. Street has kindly drawn our attention to several other papers devoted to the enumeration of fabrics of certain special kinds. In particular we must mention "Enumerating the compound twillins" by J. A. Hoskins, R. G. Stanton and A. P. Street [*Congressus Numerantium* **38** (1983), 3–22] and "Some elementary isonemal binary matrices" by J. A. Hoskins, W. D. Hoskins, A. P. Street and R. G. Stanton [*Ars Combinatoria* **13** (1982), 3–38]. It is unfortunate that in these and other publications the authors have completely ignored the condition that a *fabric* (both in the practical and mathematical sense) must hang together. In particular, as we pointed out in [5], the design of Figure 4 of the first of the above mentioned papers does not represent a genuine fabric, yet, in spite of this fact, it is given as an example of a "compound twillin with single complementation". In the second paper mentioned, nine of the designs given in the diagrams at the end do not represent fabrics for the same reason. It appears that despite introductory remarks to the contrary, these enumerations concern certain binary arrays and not fabrics. The comparison of their results with ours is made harder by the inclusion of arrays of smaller periods in their enumeration of $n \times n$ arrays with composite n.

We are indebted to Prof. P. J. Campbell for several corrections of an earlier version of this paper, and also for bringing to our attention a manuscript "Conditions for isonemal arrays on a Cartesian grid" by W. D. Hoskins and R. S. D. Thomas (which has since then appeared in *Linear Algebra and Its Applications* **57** (1984), 87–103). This paper presents some general results on enumeration of "isonemal arrays", develops a criterion for an array to represent a fabric that hangs together (in our terminology, it fails to hang together if and only if it is of Genus II, IV or V, and precisely one quarter of the squares in some row are black), and enumerates the 6×6 and 8×8 arrays. From this enumeration we find that the two fabrics of Genus V whose designs are shown in FIGURE 6 have been missed in our enumeration. The entries in TABLE 1 have been corrected to account for these two fabrics.

8-11-9 V A

8-23-2 V A

FIGURE 6. Designs for two isonemal fabrics of exact period 8 found by Hoskins and Thomas, which are missing from FIGURE 5.

We plan to publish, in the near future, an extension of the catalogue presented here, to designs of all isonemal fabrics with periods 10 and 12.

ACKNOWLEDGMENT

We are grateful to the director and staff of the Computer Centre at the University of East Anglia for the facilities they provided and for their help in dealing with the difficulties that arose in carrying out this project.

REFERENCES

1. AUBRY, A. 1911. Les principes de la géométrie des quinconces. Enseign. Math. **13:** 187–203.
2. CLAPHAM, C. R. J. 1980. When a fabric hangs together. Bull. London Math. Soc. **12:** 161–164.

3. Enns, T. C. 1984. An efficient algorithm determining when a fabric hangs together. Geom. Dedicata **15:** 259–260.
4. Gand, E. 1871. Le transpositeur ou l'improvisateur de tissus. iii + 47 pages. J. Baudry. Paris, France.
5. Grünbaum, B. & G. C. Shephard. 1980. Satins and twills: an introduction to the geometry of fabrics. Math. Mag. **53:** 139–161 and 313.
6. Grünbaum, B. & G. C. Shephard. 1981. A hierarchy of classification methods for patterns. Z. Krist. **154:** 163–187.
7. Grünbaum, B. & G. C. Shephard. 1983. Tilings, patterns, fabrics and related topics in discrete geometry. Jahresber. Deutsch. Math.-Verein. **85:** 1–32.
8. Grünbaum, B. & G. C. Shephard. Isonemal fabrics. To be published.
9. Lucas, E. 1867. Application de l'arithmétique a la construction de l'armure des satins réguliers. 16 pp. G. Retaux. Paris, France.
10. Lucas, E. 1880. Principii fondamentali della geometria dei tessuti. L'ingegneria civile e le arti industriali **6:** 104–115.

TILING CONVEX POLYGONS WITH EQUILATERAL TRIANGLES AND SQUARES

Joseph Malkevitch

Department of Mathematics
York College
Jamaica, New York 11451

Given a supply of equilateral triangles and squares of side length one, for which values of n is there a convex n-gon P so that the triangles and squares tile P edge-to-edge? We will sometimes refer to such a polygon as an ETS polygon. For examples of ETS polygons see FIGURE 1. The problem above was treated in [6], where it was stated as a dissection problem and where it was shown that n can only be an integer from 3 to 12. The purpose of this paper is to show how a slight variant of the above question leads to an Eberhard-type problem and to raise other related problems. For a discussion of Eberhard problems the reader is referred to [2] or [3].

If equilateral triangles and squares of side one tile a convex polygon P edge-to-edge, the possible interior angles of P can be 60, 90, 120, 150, and 180°, and thus the only exterior angles for P can be 120, 90, 60, 30, and 0°. If we denote by v_{120}, v_{90}, v_{60}, v_{30}, and v_0 the number of vertices of P having exterior angle the value of the subscript, then it follows that

$$120v_{120} + 90v_{90} + 60v_{60} + 30v_{30} + 0v_0 = 360 \qquad (*)$$

since the sum of the exterior angles of a convex polygon is 360°. Equation $(*)$ simplifies to the diophantine equation $(**)$, whose solutions can easily be found by inspection:

$$4v_{120} + 3v_{90} + 2v_{60} + v_{30} = 12. \qquad (**)$$

It is perhaps noteworthy that essentially the same diophantine equation arises from two other problems in geometry: the problem of determining the combinatorial type of three-dimensional convex three-valent polytopes with no face having more than six sides (see [7]) and the problem of determining the deltahedra—convex three-dimensional polytopes all of whose faces are triangles (see [1]).

Equation $(**)$ suggests several questions:

(1) For which solutions of $(**)$ are there convex polygons P which can be tiled edge-to-edge with equilateral triangles and squares of side one?

(2) Can $(**)$ be used to determine the "shape" of all such polygons P, and the number of triangles and squares in such tilings?

(3) (Eberhard type.) Since v_0 appears in $(*)$ with a zero coefficient, is it possible to select a v_0 so that for a solution of $(**)$ there is a convex polygon P' that has exterior angles corresponding to this solution and v_0 exterior angles of 0° as well? (The value of v_0 is assumed not to affect the count n of the number of sides of P'.)

To begin the investigation of these questions, TABLE 1 shows the 34 solutions to the diophantine equation $(**)$, where the rows of the table have not been written in the "natural" lexicographic order, but in order of increasing n. Note that for each line of the table, one can ask not only if there is a convex n-gon corresponding to

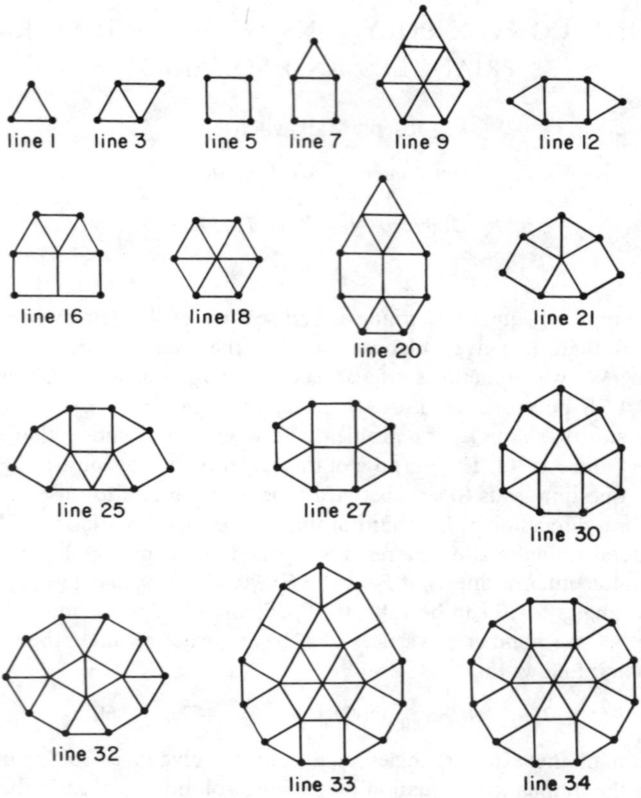

FIGURE 1

this solution that can be tiled edge-to-edge with equilateral triangles and squares of side one, but also which values of v_0 can go with this solution. FIGURE 1 shows an ETS polygon with a minimal value of v_0 for each line in TABLE 1 for which an ETS polygon exists. The determination of all values of v_0 which might occur for a given "feasible" line of the table would also be of interest. For 18 lines of the table no value of v_0 works, so no Eberhard-type theorem exists in this situation. Note, however, that for some lines of TABLE 1, not only is there a solution corresponding to the given line, but also one can have different orders in which the angles appear in a cycle around the polygon. This is illustrated in FIGURE 2 where the cyclic orders for the angles of the two quadrilaterals for line 3 of TABLE 1 are 60°, 120°, 60°, 120°, and 120°, 120°, 60°, 60°, respectively. It is noteworthy, as FIGURE 1 shows, that for each n, between 3 and 12 except 11, there is an ETS polygon for that n with v_0 equal

FIGURE 2

TABLE 1
POSSIBLE EXTERIOR ANGLES FOR ETS POLYGONS WITH n SIDES

	v_{120}	v_{90}	v_{60}	v_{30}	n	Minimal v_0
1.	3	0	0	0	3	0
2.	2	1	0	1	4	None
3.	2	0	2	0	4	0
4.	1	2	1	0	4	None
5.	0	4	0	0	4	0
6.	2	0	1	2	5	None
7.	1	2	0	2	5	0
8.	1	1	2	1	5	None
9.	1	0	4	0	5	2
10.	0	3	1	1	5	None
11.	0	2	3	0	5	None
12.	2	0	0	4	6	0
13.	1	1	1	3	6	None
14.	1	0	3	2	6	None
15.	0	3	0	3	6	None
16.	0	2	2	2	6	1
17.	0	1	4	1	6	None
18.	0	0	6	0	6	0
19.	1	1	0	5	7	None
20.	1	0	2	4	7	2
21.	0	2	1	4	7	0
22.	0	1	3	3	7	None
23.	0	0	5	2	7	None
24.	1	0	1	6	8	None
25.	0	2	0	6	8	1
26.	0	1	2	5	8	None
27.	0	0	4	4	8	0
28.	1	0	0	8	9	None
29.	0	1	1	7	9	None
30.	0	0	3	6	9	0
31.	0	1	0	9	10	None
32.	0	0	2	8	10	0
33.	0	0	1	10	11	2
34.	0	0	0	12	12	0

to zero. This is related to the fact that for each n other than 11, there exist equilateral ETS polygons (see [6]).

The following lemmas help in showing that no value of v_0 allows one to construct an ETS polygon for several rows in TABLE 1.

LEMMA 1. *If a convex polygon P is ETS and if AB is a side of P (disregarding* 180° *angles), then the interior angles at A and B cannot be*:

(i) 60 *and* 90°,

(ii) 90 *and* 120°.

The lemma can be verified using simple arguments, in case (i) applied to the diagram in FIGURE 3, where two subcases arise according to whether or not $Q = Q'$. Using the notation M–N to indicate angles of M and N degrees having a side of an ETS polygon in common (disregarding 180° angles), we have:

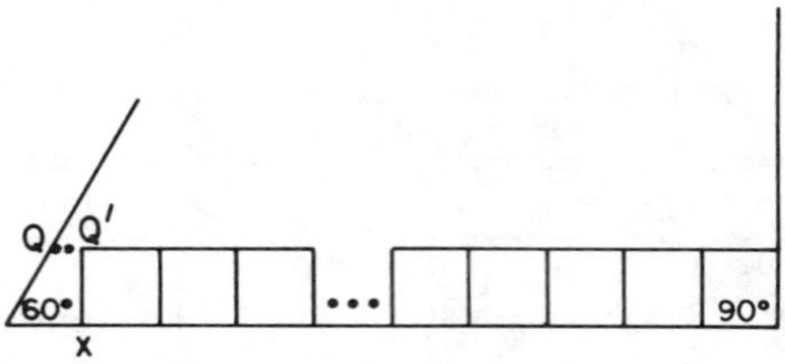

FIGURE 3

COROLLARY. *In an ETS polygon the following are impossible:*

$$60°\text{–}150°\text{–}60°,$$

$$60°\text{–}150°\text{–}120°,$$

$$120°\text{–}150°\text{–}120°.$$

LEMMA 2. *In an ETS polygon,* 90°–150°–90° *is impossible.*

LEMMAS 1 and 2 together with ad hoc arguments allow verification of TABLE 1 and thus answer questions (1) and (3) raised above. This leaves question (2), concerning the possible "shapes" of ETS polygons. Partial results on this problem are easy to obtain. For example, for $n = 4$, there are only three shapes possible: rectangles, isosceles trapezoids, and parallelograms. If the sides of these are $\{a, b, a, b\}$, $\{a, b, a - b, b\}$ $(a > b)$, and $\{a, b, a, b\}$, respectively, we see (FIGURE 4) that we can construct 4-sided ETS polygons with ab squares, $2ab - b^2$ triangles, and $2ab$ triangles, where a, b vary independently and in the case of the isosceles trapezoid $a > b$. However, a complete analysis for all the other allowable values of n is not complete. The difficult case is $n = 12$ where it may not be possible to find a finite number of classes into which all 12-sided ETS polygons can be placed. Starting with a "patch" [4, 5] of equilateral triangles and squares that are edge-to-edge one can often complete the configuration to a 12-sided ETS in several ways.

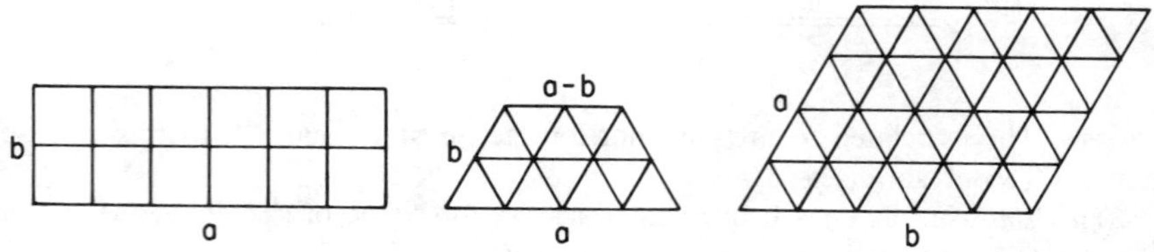

FIGURE 4

It is tempting to think that connected configurations of equilateral triangles and squares can be completed to be a subset of an ETS polygon, or a tiling of the plane, by the addition of triangles and squares. That this is not so is shown by the configuration in FIGURE 5. This patch can be completed locally by additional triangles and/or squares at any vertex, but cannot be completed globally. For a related observation see [4]. The difficulty in resolving the "shape" problem for ETS polygons with 12 sides may be related to the fact that the classification of all edge-to-

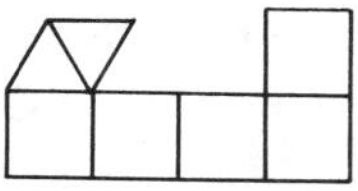

FIGURE 5

edge tessellations of the plane with equilateral triangles and squares is still open [4, 5]. In fact, resolving the $n = 12$ case may be useful in answering this tiling problem.

Acknowledgments

The author is grateful to the referees for their helpful suggestions. Thanks also go to Professor Robert Coen for a useful conversation.

References

1. Beck, A., M. Bleicher & D. Crowe. 1967. Excursions into Mathematics. Worth. New York, N.Y.
2. Grünbaum, B. 1968. Some analogues of Eberhard's theorem on convex polytopes. Israel J. Math. **6:** 398–411.
3. Grünbaum, B. 1970. Polytopes, graphs, and complexes. Bull. Amer. Math. Soc. **76:** 1131–1201.
4. Grünbaum, B. & G. C. Shephard. 1977. Tilings by regular polygons. Math. Mag. **50:** 227–247.
5. Grünbaum, B. & G. C. Shephard. 1984. Tilings and Patterns. Freeman. San Francisco, Calif.
6. Malkevitch, J. 1969. Problem 707, solution by M. Goldberg. Math. Mag. **42:** 158.
7. Malkevitch, J. 1970. 3-Valent, 3-polytopes with faces having fewer than 7 edges. Ann. N.Y. Acad. Sci. **175:** 285–286.

TILING BOUNDED OPEN SETS WITH SQUARES THAT TOUCH THE BOUNDARY[a]

William T. Trotter, Jr.

Department of Mathematics and Statistics
University of South Carolina
Columbia, South Carolina 29208

1. Introduction

In this paper, we discuss the problem of tiling a bounded open subset S of the Euclidean plane with a (possibly infinite) family of squares each of which touches the boundary of S. The sides of the squares are required to be parallel to the coordinate axes. With these restrictions, we show that if S is an open subset of $\mathbb{R}^2$ and S is either a homeomorph of the unit disk or a homeomorph of an annulus, then S can be tiled with a family of squares each of which touches the boundary of S. We construct an open subset S which is a homeomorph of a disk with two holes which cannot be tiled with a family of squares each of which touches the boundary of S. We then show that every bounded open subset $S \subseteq \mathbb{R}^2$ can be tiled with a family of rectangles each of which touches the boundary of S. We also consider the obvious generalization of these problems to $\mathbb{R}^n$ and show that for every $n \geq 3$, there exists a homeomorph S of the open unit ball in $\mathbb{R}^n$ which cannot be tiled by a family of cubes each of which touches the boundary of S. On the other hand, for $n \geq 2$ every bounded open subset S of $\mathbb{R}^n$ can be tiled with a family of boxes each of which touches the boundary of S.

We denote the interior, boundary, and closure of a subset S of n-dimensional Euclidean space $\mathbb{R}^n$ by Int (S), Bdry (S), and $\bar{S}$, respectively. An *n-box* in $\mathbb{R}^n$ is a subset R of the form $[a_1, b_1] \times [a_2, b_2] \times \cdots \times [a_n, b_n]$, where $b_i > a_i$ for $i = 1, 2, 3, \ldots, n$. An n-box is called an *n-cube* when $b_i - a_i = b_j - a_j$ for all i, j with $1 \leq i < j \leq n$. A 2-box is called a *rectangle* and a 2-cube is called a *square*. Note that the sides of an n-box are required to be parallel to the coordinate axes in $\mathbb{R}^n$.

A family $\mathscr{T}$ of sets is called a *partial tiling* of a subset $S \subseteq \mathbb{R}^n$ when it satisfies the following two properties:

P_1: $R \subseteq \bar{S}$ for every $R \in \mathscr{T}$.

P_2: Int $(R) \cap$ Int $(R') = \varnothing$ for every distinct $R, R' \in \mathscr{T}$.

A partial tiling $\mathscr{T}$ of a subset $S \subseteq \mathbb{R}^n$ is called a *tiling* of S when it also satisfies:

P_3: Int $(S) \subseteq \bigcup\{\bar{R}\colon R \in \mathscr{T}\}$.

In this paper, we will be concerned with the following problem:

If S is a bounded open subset of $\mathbb{R}^n$, under what conditions does there exist a tiling $\mathscr{T}$ of S so that every set in $\mathscr{T}$ is an n-box (alternately, an n-cube) that touches the boundary of S? More formally, we will require that the tiling also satisfy the following property:

P_4: $\bar{R} \cap$ Bdry $(S) \neq \varnothing$ for every $R \in \mathscr{T}$.

[a] Research supported in part by National Science Foundation Research Grants ISP 80110451 and MCS-8202172.

Every open subset S of $\mathbb{R}^n$ can be tiled with a family of n-cubes, although it takes a little effort to show this. However, when we require that the tiling also satisfy property P_4, some rather surprising results are encountered.

2. Summary of Results

In this section, we summarize the principal results of the paper. We defer the proofs of these theorems, as well as the development of supporting lemmas, until Section 3. We begin with a positive result.

THEOREM 1. *If $S \subseteq \mathbb{R}^2$ and S is homeomorphic to the open disk $\{(x, y) \in \mathbb{R}^2: x^2 + y^2 < 1\}$, then S can be tiled with a family of squares each of which touches the boundary of S.*

THEOREM 1 is somewhat surprising since several intuitive approaches to defining an algorithm for producing a tiling fail when the boundary of S is irregular. In particular, the obvious scheme of choosing a square of maximum size in the region not already tiled is defective. To see that this is the case, consider the open set S bounded by the curve shown in FIGURE 1. The first three squares chosen under this scheme will have side lengths of 10, 8, and 6, respectively. However, these three squares will then leave a 3×2 rectangle that cannot be suitably tiled.

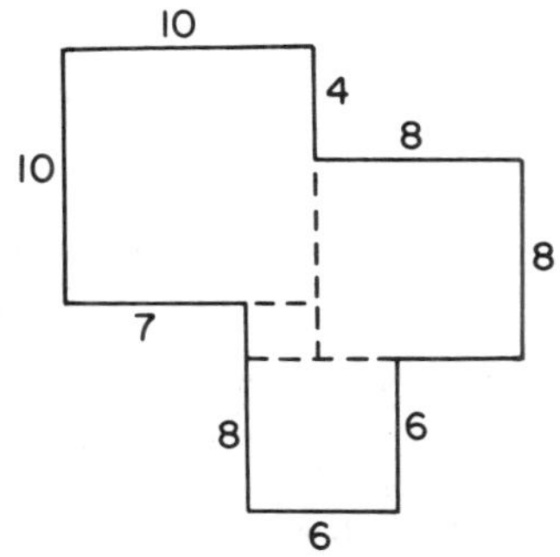

FIGURE 1

In view of THEOREM 1, it is natural to expect that bounded open regions that are not simply connected may present difficulties. However, we will show that the argument given for THEOREM 1 can be modified to settle the question in the affirmative when S is an annulus, i.e., an open disk with one hole.

THEOREM 2. *If $S \subseteq \mathbb{R}^n$ and S is homeomorphic to the open disk with one hole $\{(x, y) \in \mathbb{R}^2: 1 < x^2 + y^2 < 2\}$, then S can be tiled with a family of squares each of which touches the boundary of S.*

A careful examination of the argument used to prove THEOREM 2 would lead one to conjecture that regions with more than one hole may fail to have a tiling with squares each of which touches the boundary.

To show that the conjecture is true and to confirm that THEOREM 2 is best possible, we establish the following result.

THEOREM 3. *There exists a homeomorph S of the open disk with two holes* $\{(x, y) \in \mathbb{R}^2\colon 1 < x^2 + y^2 < 5$ *and* $(x-3)^2 + y^2 > 1\}$ *which cannot be tiled with squares each of which touches the boundary of S.*

In view of THEOREM 3, it is reasonable to conjecture that there exist more complicated regions which cannot be tiled with rectangles (each of which touches the boundary) either. In fact, it is not too difficult to construct a bounded open set S in the plane so that for every $\varepsilon > 0$, there exists a partial tiling $\mathscr{T}(\varepsilon)$ of S so that: (1) if R is a rectangle in $\mathscr{T}(\varepsilon)$ and R is properly contained in a rectangle R', then Int (R') is not a subset of S; (2) the sum of the areas of the rectangles in $\mathscr{T}(\varepsilon)$ is less than ε; and (3) Int $(S - \bigcup \mathscr{T}(\varepsilon)) = \varnothing$. In other words, we have covered an arbitrarily small portion of S but cannot augment the partial tiling $\mathscr{T}(\varepsilon)$ to complete the task of tiling S. Despite these apparent difficulties, we will establish the following result.

THEOREM 4. *If S is a bounded open subset of the plane, then S can be tiled with rectangles each of which touches the boundary of S.*

The proof of THEOREM 4 can be extended immediately to higher dimensions.

THEOREM 5. *If* $n \geq 2$ *and S is a bounded open subset of* $\mathbb{R}^n$, *then S can be tiled with n-boxes each of which touches the boundary of S.*

Although THEOREM 4 extends immediately to higher dimensions, we will show that THEOREMS 1 and 2 do not.

THEOREM 6. *If* $n \geq 3$, *there exists a homeomorph S of the open ball* $\{(x_1, x_2, \ldots, x_n)\colon x_1^2 + x_2^2 + \cdots + x_n^2 < 1\}$ *which cannot be tiled with n-cubes each of which touches the boundary of S.*

3. DETAILS OF THE ARGUMENTS

Let S be a bounded open subset of $\mathbb{R}^2$ and let $\mathscr{T}$ be a tiling of S with rectangles each of which touches the boundary of S. If $\mathscr{T}'$ is a finite subcollection of $\mathscr{T}$, then $\mathscr{T} - \mathscr{T}'$ is a tiling of the open set $S' = S - \bigcup \mathscr{T}'$ and every rectangle in $\mathscr{T} - \mathscr{T}'$ touches Bdry $(S') \cap$ Bdry (S). It is therefore natural to consider the following problem.

Let S be a bounded open subset of the plane and let Bdry $(S) = B_a(S) \cup B_r(S)$ be a partition of the boundary of S into two sets which we call the *artificial boundary* of S and the *real boundary* of S, respectively. Under what conditions can S be tiled with a family of squares each of which touches the real boundary of S? More formally, we require that the tiling satisfy the following variant of property P_4.

P_4': $\bar{R} \cap B_r(S) \neq \varnothing$ for every $R \in \mathscr{T}$.

EXAMPLE 7. Let S be the interior of the rectangle $[a, b] \times [c, d]$ in $\mathbb{R}^2$, and let $B_r(S)$ consist of the points on the top and right sides of this rectangle, i.e., $B_r(S) = \{(x, d)\colon a \leq x \leq b\} \cup \{(b, y)\colon c \leq y \leq d\}$. Then S can be tiled with a family of squares each of which touches the real boundary of S.

Proof. Let $S_0 = S$. We then define S_k and Q_k inductively for every $k \geq 0$ by the following rule. If $S_k = (a, b) \times (c, d)$, let $m = \min\{b - a,\ d - c\}$, $Q_k = [a, a + m] \times [c, c + m]$, and $S_{k+1} = S_k - Q_k$. If this definition results in $S_{k+1} = \varnothing$ for some $k \geq 0$, then the finite collection $\mathscr{T} = \{Q_0, Q_1, \ldots, Q_k\}$ is a tiling of S by squares each of which touches the real boundary of S. On the other hand, if $S_{k+1} \neq \emptyset$ for all $k \geq 0$, then the collection $\mathscr{T} = \{Q_k: k \geq 0\}$ is the desired tiling. ∎

We have included this elementary example for two reasons. First, the algorithm used in the example will be an essential device in subsequent arguments, and second, we want to emphasize the fact that we allow infinite tilings. Tiling problems for finite families of squares and rectangles have been studied extensively, and we refer the reader to [2–4] for details.

In order to generalize Example 7, we consider an open set $S \subseteq \mathbb{R}^2$ whose artificial boundary forms a pair of perpendicular line segments which are parallel to the coordinate axes and form a right angle that is included in S.

We call S a *quarter-region* and refer to the point $p_0(S)$ where the perpendicular line segments meet as the *corner point* of S. We will require that the real boundary $B_r(S)$ of a quarter-region S be a closed connected subset of Bdry (S). We denote the endpoints of $B_r(S)$ by $p_0'(S)$ and $p_0''(S)$ and call the two line segments determined by $\{p_0(S), p_0'(S)\}$ and $\{p_0(S), p_0''(S)\}$ the *sides* of S. We will assume that these points have been labeled so that if we begin at $p_0'(S)$ and proceed clockwise around Bdry (S), then we will encounter the corner point $p_0(S)$ before $p_0''(S)$. (See FIGURE 2.)

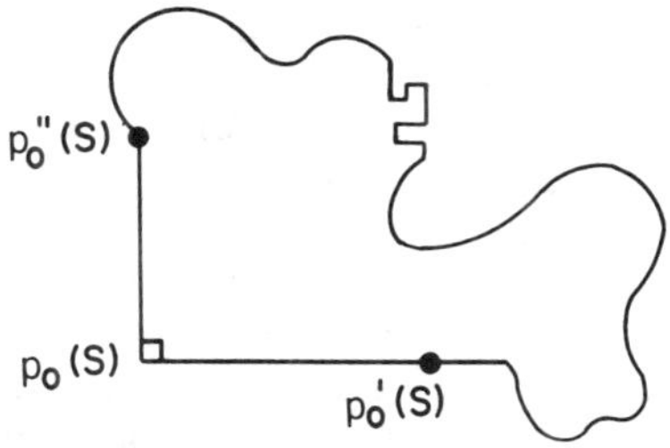

FIGURE 2

We will also need to consider an open subset $S \subseteq \mathbb{R}^2$ bounded by a simple closed curve for which the real boundary is a closed connected subset of Bdry (S) and whose artificial boundary is a straight line segment which is parallel to one of the coordinate axes. We call such a set a *half-region*. The tiling problem for a half-region S can be transformed into two separate tiling problems for quarter-regions by choosing an arbitrary point $p \in B_a(S)$ and drawing a line through p which is perpendicular to $B_a(S)$. This line determines a tiling $\mathscr{T}(S, p) = \{S_1, S_2\}$ of S, where S_1 and S_2 are quarter-regions and each has p as its corner point. Note that $B_r(S_1) \subseteq B_r(S)$ and $B_r(S_2) \subseteq B_r(S)$. (See FIGURE 3.)

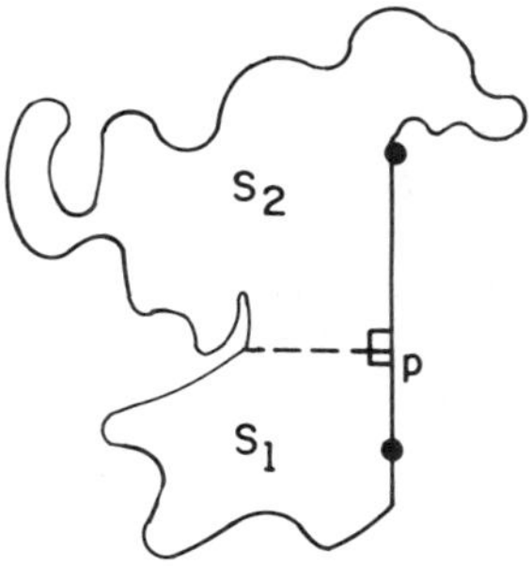

FIGURE 3

Next we describe an algorithm for tiling a quarter-region. Let S_0 be an arbitrary quarter-region. Then there is a unique square $Q(S_0)$ of maximum size satisfying the following conditions: $Q(S_0) \subseteq \bar{S}_0$ and $p_0(S_0)$ is one of the corners of $Q(S_0)$. (See FIGURE 4.)

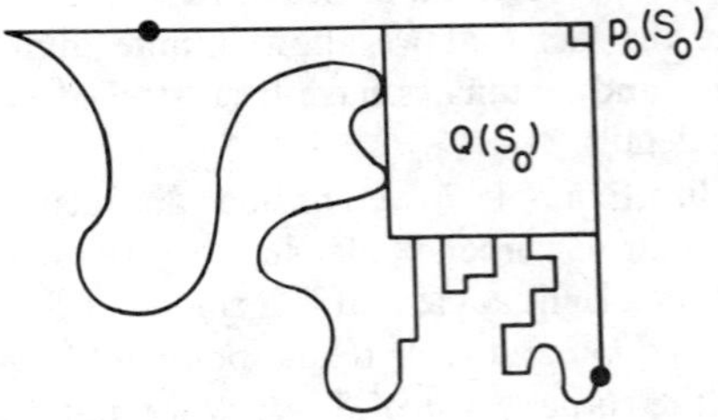

FIGURE 4

Note that $Q(S_0)$ touches the real boundary of S_0. We assume that $Q(S_0) \neq \bar{S}_0$ and observe that Bdry $(Q(S_0)) \cap S_0$ is a nonempty open subset of Bdry $Q(S_0)$ so that Bdry $(Q(S_0)) \cap S_0$ can be uniquely partitioned into a countable (and possibly infinite) collection of pairwise disjoint connected sets each of which is an open subset of Bdry $(Q(S_0))$. Let Bdry $(Q(S_0)) \cap S_0 = \bigcup\{U_\alpha: \alpha \in \mathscr{A}(S_0)\}$ be this partition.

We now describe a process for determining a tiling $\mathscr{F}(S_0)$ of $S - Q(S_0)$ so that each set $S \in \mathscr{F}(S_0)$ is a quarter-region and $B_r(S) \subseteq B_r(S_0)$. Beginning at the corner point $p_0(S_0)$ and proceeding clockwise around $Q(S_0)$, we label the sides s_1, s_2, s_3, and s_4; i.e., s_1 and s_4 meet at $p_0(S_0)$. For $i = 1, 2, 3$, label the corner where s_i and s_{i+1} meet as $p_i(S_0)$. We then observe that the partition Bdry $(Q(S_0)) \cap S_0 = \bigcup\{U_\alpha: \alpha \in \mathscr{A}(S_0)\}$ induces a natural tiling $\mathscr{U}(S_0) = \{S_\alpha: \alpha \in \mathscr{A}(S_0)\}$ of the open set $S_0 - Q(S_0)$ where each S_α is an open subset of S_0 whose artificial boundary contains U_α and whose real boundary is a subset of the real boundary of S_0. Note that each S_α is the interior of a simple closed curve. At this point, we must divide the construction of $\mathscr{F}(S_0)$ into two cases. (See FIGURE 5.)

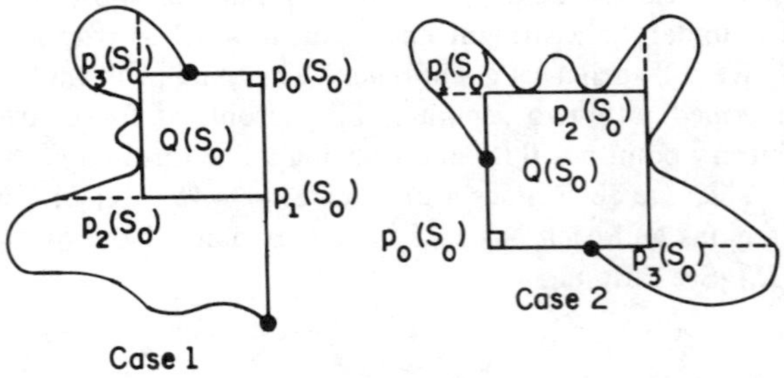

FIGURE 5

Case 1. There exists $\alpha \in \mathscr{A}(S_0)$ for which $p_1(S_0)$ is an endpoint of U_α, $p_1(S_0) \in B_a(S_0)$, and $p_2(S_0) \in U_\alpha$.

Let $\mathscr{U}_1(S_0) = \{S_\alpha \in \mathscr{U}(S_0): U_\alpha$ contains one or more corners of $Q(S_0)\}$. For each corner $p_i(S_0)$ that belongs to some $U_\alpha \in U_1(S_0)$, extend side s_i through $p_i(S_0)$ until it touches $B_r(S_0)$.

Case 2. There does not exist $\alpha \in \mathscr{A}(S_0)$ in which $p_1(S_0)$ is an endpoint of U_α, $p_1(S_0) \in B_a(S_0)$, and $p_2(S_0) \in U_\alpha$.

In this case, for each corner $p_i(S_0)$ that belongs to some $U_\alpha \in \mathscr{U}_1(S_0)$, we extend side s_{i+1} through $p_i(S_0)$ until it touches $B_r(S_0)$.

It is easy to see that this two-part definition determines a tiling $\mathscr{F}'(S_0)$ of $S_0 - Q(S_0)$ where each region in $\mathscr{F}'(S_0)$ is either a quarter-region or a half-region whose real boundary is a subset of $B_r(S_0)$.

Let $\mathscr{F}''(S_0) = \{S \in \mathscr{F}'(S_0)\colon S \text{ is a half-region}\}$. Then for each $S \in \mathscr{F}'(S_0)$, choose a point $p(S) \in B_a(S) \cap \text{Bdry}(Q(S_0))$ and define $\mathscr{F}(S_0) = (\mathscr{F}' - \mathscr{F}''(S_0)) \cup (\bigcup\{\mathscr{T}(p(S), S)\colon S \in \mathscr{F}''(S_0)\})$. It is easy to see that $\mathscr{F}(S_0)$ is a tiling of $S_0 - Q(S_0)$ with quarter-regions and $B_r(S) \leq B_r(S_0)$ for every $S \in \mathscr{F}(S_0)$.

The procedure described above can now be repeated on each quarter-region in $\mathscr{F}(S_0)$. We set $\mathscr{G}_0(S_0) = \{S_0\}$ and $\mathscr{T}_0(S_0) = \{Q(S_0)\}$. If $\mathscr{G}_k(S_0)$ and $\mathscr{T}_k(S_k)$ have been defined for some $k \geq 0$, we define $\mathscr{G}_{k+1}(S_0) = \bigcup\{\mathscr{F}(S)\colon S \in \mathscr{G}_k(S_0)\}$ and $\mathscr{T}_{k+1}(S_0) = \mathscr{T}_k(S_0) \cup \{Q(S)\colon S \in \mathscr{G}_{k+1}(S_0)\}$. We then define $\mathscr{G}(S_0) = \bigcup\{\mathscr{G}_n(S_0)\colon n \geq 0\}$ and $\mathscr{T}(S_0) = \bigcup\{\mathscr{T}_n(S_0)\colon n \geq 0\}$.

Hereafter, we will refer to this recursive construction as *Algorithm* 1 and call $\mathscr{G}(S_0)$ and $\mathscr{T}(S_0)$, respectively, the family of quarter-regions and the family of squares obtained by applying Algorithm 1 to S_0. Strictly speaking, $G(S_0)$ and $\mathscr{T}(S_0)$ are not uniquely determined since each time that a half-region S is encountered in the construction, the choice of the point $p(S) \in B_a(S)$ is arbitrary. We will now proceed to show that regardless of the manner in which the half-regions are tiled by quarter-regions, the collection $\mathscr{T}(S_0)$ is a tiling of the quarter-region S_0 by squares, and every square in $\mathscr{T}(S_0)$ touches the real boundary of S_0.

We begin by observing that it is clear that $\mathscr{T}(S_0)$ satisfies properties P_1, P_2, and P'_4. In order to complete the proof of our claim, we need only show that $\mathscr{T}(S_0)$ satisfies property P_3; i.e., we must show that $S_0 \subseteq \bigcup\mathscr{T}(S_0)$. To accomplish this will require some preliminary work. For a point $x \in \mathbb{R}^2$ and a positive number ε we let $U(x, \varepsilon)$ denote the open ball with center at x and radius ε.

LEMMA 8. *Let S_0 be a quarter-region and let $\mathscr{T}(S_0)$ be the family of squares obtained by applying Algorithm* 1 *to S_0. Also let $\{Q_\alpha\colon \alpha \in \mathscr{A}\}$ be an infinite subfamily of $\mathscr{T}(S_0)$ and for each $\alpha \in \mathscr{A}$, let $p_\alpha \in Q_\alpha$. Then every limit point of $\{p_\alpha\colon \alpha \in \mathscr{A}\}$ is an element of $B_r(S_0)$.*

Proof. Let y_0 be a limit point of $\{p_\alpha\colon \alpha \in \mathscr{A}\}$. Then choose an infinite sequence $\{\alpha_n\colon n \geq 1\} \subseteq \mathscr{A}$ so that $\{p_{\alpha_n}\colon n \geq 1\}$ converges to y_0. Now let ε be an arbitrary positive number. We show that $U(y_0, \varepsilon) \cap B_r(S_0) \neq \varnothing$.

Since $\{p_{\alpha_n}\colon n \geq 1\}$ converges to y_0 and $\varepsilon/10 > 0$, there exists an integer n so that $p_{\alpha_n} \in U(y_0, \varepsilon/10)$ for all $n \geq n_1$. However, there are at most four squares in $\mathscr{T}(S_0)$ which intersect both $U(y_0, \varepsilon/10)$ and $\bar{S}_0 - U(y_0, \varepsilon)$. Thus there exists an integer n_2 so that $Q_{\alpha_n} \subseteq U(y_0, \varepsilon)$ for all $n \geq n_2$. Since each Q_α touches the real boundary of S_0, it follows that $U(y_0, \varepsilon) \cap B_r(S_0) \neq \varnothing$. Since $B_r(S_0)$ is closed and ε is arbitrary, we conclude that $y_0 \in B_r(S_0)$. And since y_0 was arbitrary, the proof is complete. ∎

LEMMA 9. *Let S_0 be a quarter-region and let $\mathscr{T}(S_0)$ be the family of squares obtained by applying Algorithm* 1 *to S_0. If $x_0 \in B_a(S_0)$, then there exists an integer $n \geq 0$ and a square $Q \in \mathscr{T}_n(S_0)$ so that $x_0 \in Q$.*

Proof. Suppose to the contrary that $x_0 \in B_a(S_0)$ but that $x_0 \notin \bigcup\mathscr{T}_n(S_0)$ for every $n \geq 0$. Since $p_0(S_0) \in Q(S_0) \in \mathscr{T}_0(S_0)$, we know that $x_0 \neq p_0(S_0)$ and $x_0 \notin Q(S_0)$. Now suppose that for some integer $k \geq 0$, we have $x_0 \in B_a(S_k)$ and $x_0 \notin Q(S_k)$. Then there is a unique $S_{k+1} \in \mathcal{F}(S_k)$ for which $x_0 \in B_a(S_{k+1})$. Hence $x_0 \notin Q(S_{k+1})$. This inductive definition determines an infinite sequence $\{S_n: n \geq 0\}$ of quarter-regions for which the sequence $\{p_0(Q(S_n)): n \geq 0\}$ converges to a point y_0 on the side of $B_a(S_0)$ that contains x_0. Furthermore, y_0 is between $p_0(S_0)$ and x_0 on this side. However, in view of LEMMA 8, we must also have $y_0 \in B_r(S_0)$ which is a contradiction. ∎

COROLLARY 10. *Let S_0 be a quarter-region and let $\mathcal{T}(S_0)$ be the collection of squares obtained by applying Algorithm 1 to S_0. If $x_0 \in B_a(S_0)$, then there exists a positive number ε and a subcollection $\mathcal{T}' \subseteq \mathcal{T}(S_0)$ so that $\mathcal{T}'$ contains at most two squares and $U(x_0, \varepsilon) \cap \bar{S}_0 \subseteq \bigcup \mathcal{T}'$.*

Proof. If $x = p_0(S_0)$, the result is trivial since we may choose $\mathcal{T}' = \{Q(S_0)\}$ and ε as the length of a side of the square $Q(S_0)$. Now suppose that $x_0 \neq p_0(S_0)$. Then it follows from the algorithm given in LEMMA 9 that there is a least integer $n \geq 0$ for which $x_0 \in \bigcup \mathcal{T}_n(S_0)$. Then let S_1 be the unique quarter-region in $\mathcal{G}_n(S_0)$ for which $x_0 \in B_a(S_1)$. Note that $x_0 \in Q(S_1)$ and that $x_0 \neq p_0(Q(S_1))$. If x_0 is not a corner of $Q(S_1)$, the conclusion follows immediately since we may take $\mathcal{T}' = \{Q(S_1)\}$ and ε as the minimum distance from x_0 to a corner of $Q(S_1)$. On the other hand, if x_0 is a corner of $Q(S_1)$, then there exists a unique quarter-region $S_2 \in \mathcal{F}(S_1)$ for which $x_0 = p_0(S_2)$. Then set $\mathcal{T}' = \{Q(S_1), Q(S_2)\}$ and ε as the minimum of the side lengths of these two squares. ∎

LEMMA 11. *Let S_0 be a quarter-region and let $\mathcal{T}(S_0)$ be the family of squares obtained by applying Algorithm 1 to S_0. If $x_0 \in S_0$, $S^* \in \mathcal{G}(S_0)$, and $x_0 \in B_a(S^*)$, then there exists $\varepsilon > 0$ so that $U(x_0, \varepsilon) \subseteq \bigcup \mathcal{T}(S_0)$.*

Proof. From the recursive definitions for $\mathcal{G}_n(S_0)$ and $\mathcal{T}_n(S_0)$, it follows that for each $n \geq 0$, the family $\mathcal{C}_n(S_0) = \mathcal{T}_n(S_0) \cup \{\bar{S}: S \in \mathcal{G}_{n+1}(S_0)\}$ is a tiling of S_0. For each $n \geq 0$, let $\mathcal{C}_n(S_0, x_0) = \{A \in \mathcal{C}_n(S_0): x_0 \in A\}$. If $n \geq 0$ and there exists $A \in \mathcal{C}_n(S_0, x_0)$ for which $x \in \text{Int}(A)$, then $|\mathcal{C}_n(S_0, x_0)| = 1$. On the other hand, if $n \geq 0$ and $x \in \text{Bdry}(A)$ for every $A \in \mathcal{C}_n(S_0, x_0)$, then it follows that $2 \leq |\mathcal{C}_n(S_0, x_0)| \leq 4$.

From the hypotheses we know that $x_0 \in B_a(S^*)$ so we may choose the unique integer n_0 for which $\bar{S}^* \in \mathcal{C}_{n_0}(S_0, x_0)$. If $A \in \mathcal{C}_{n_0}(S_0, x_0)$ and $A \in \mathcal{T}_{n_0}(S_0)$, set $\mathcal{T}_A = \{A\}$ and let ε be the length of a side of the square A. If $A \in \mathcal{C}_{n_0}(S_0 x_0)$ and $A = \bar{S}$ for some $S \in \mathcal{G}_{n+1}(S_0)$, then we know that $x_0 \in B_a(S)$. We may therefore choose a finite subcollection $\mathcal{T}_A \subseteq \mathcal{T}(S_0)$ and a positive number ε_A so that $U(x_0, \varepsilon_A) \cap \bar{S} \subseteq \bigcup \mathcal{T}_A$.

Finally we set $\varepsilon = \min \{\varepsilon_A: A \in \mathcal{C}_n'(S_0, x_0)\}$. Then

$$\begin{aligned} U(x_0, \varepsilon) &= \bigcup \{U(x_0, \varepsilon) \cap A: A \in \mathcal{C}_n'(S_0, x_0)\} \\ &\subseteq \bigcup \{U(x_0, \varepsilon_A) \cap A: A \in \mathcal{C}_n'(S_0, x_0)\} \\ &\subseteq \bigcup \{\bigcup \mathcal{T}_A: A \in \mathcal{C}_n'(S_0, x_0)\} \\ &\subseteq \bigcup \mathcal{T}(S_0). \quad ∎ \end{aligned}$$

Recall that a *path* from a point x to a point y in $\mathbb{R}^2$ is a continuous injection f: $[0, 1] \to \mathbb{R}^2$ so that $f(0) = x$ and $f(1) = y$.

LEMMA 12. *Let S_0 be a quarter-region and let $\mathcal{T}(S_0)$ be the family of squares obtained by applying Algorithm 1 to S_0. Then $B_a(S_0) \cup S_0 \subseteq \bigcup \mathcal{T}(S_0)$.*

Proof. Choose an arbitrary point $x_0 \in B_a(S_0) \cup S_0$. We show that $x_0 \in \bigcup\mathscr{T}(S_0)$. The result follows from LEMMA 9 if $x_0 \in B_a(S_0)$, so we may assume that $x_0 \in S_0$. Choose a path f: $[0, 1] \rightarrow \mathbb{R}^2$ from $p_0(S_0)$ to x_0 so that $f(t) \in S_0$ for all $t \in (0, 1]$.

Now suppose that $x_0 \notin \bigcup\mathscr{T}(S_0)$ and let $T = \{t \in [0, 1]: f([0, t]) \subseteq \bigcup\mathscr{T}(S_0)\}$ and let $t_0 = \text{l.u.b. } (T)$. Also, let $y_0 = f(t_0)$. Now suppose that $y_0 \in \bigcup\mathscr{T}(S_0)$. Then $y_0 \neq x_0$ and $t_0 < 1$. Choose a square $Q \in \mathscr{T}(S_0)$ for which $y_0 \in Q$. If $y_0 \in \text{Int } (Q)$, then there exists $\varepsilon_1 > 0$ so that $y_0 \in U(y_0, \varepsilon_1) \subseteq \text{Int } (Q) \subseteq \bigcup\mathscr{T}(S_0)$. On the other hand, if $y_0 \in \text{Bdry } (Q)$, then there exists $S \in G(S_0)$ so that $y_0 \in B_a(S)$. It follows from LEMMA 11 that there exists $\varepsilon_2 > 0$ so that $U(y_0, \varepsilon_2) \subseteq \bigcup\mathscr{T}(S_0)$. In either case, there exists an open set U so that $y_0 \in U \subseteq \bigcup\mathscr{T}(S_0)$. Since f is continuous, there exists $\varepsilon_3 > 0$ so that $t_0 + \varepsilon_3 \leq 1$ and $f([t_0, t_0 + \varepsilon_3]) \subseteq U \subseteq \bigcup\mathscr{T}(S_0)$. Thus $f([0, t_0 + \varepsilon_3]) \subseteq \bigcup\mathscr{T}(S_0)$ and $t_0 + \varepsilon_3 \in T$. This contradicts the assumption that $t_0 = \text{l.u.b. } (T)$. We may therefore assume that $y_0 \notin \bigcup\mathscr{T}(S_0)$. In this case, we note that $t_0 > 0$.

We will now show that $y_0 \in B_r(S_0)$. The argument is essentially the same as the proof of LEMMA 8.

We first show that y_0 is a limit point of $B = \{p_0(Q): Q \in \mathscr{T}(S_0)\}$. To the contrary, suppose that y_0 is not a limit point of B and choose $\varepsilon_1 > 0$ so that $U(y_0, \varepsilon_1) \cap B = \varnothing$ and $U(y_0, \varepsilon_1) \subseteq S_0$.

Since each square in $\mathscr{T}(S_0)$ touches the real boundary of S_0, it follows that $\mathscr{A} \cap (\bar{S}_0 - U(y_0, \varepsilon_1)) \neq \varnothing$ for every $Q \in \mathscr{T}(S_0)$. Now let $\mathscr{S} = \{Q \in \mathscr{T}(S_0): Q \cap U(y_0, \varepsilon_1/10) \neq \varnothing\}$. Then $|S| \leq 4$. Since $y_0 \notin \bigcup\mathscr{T}(S_0)$, we may then choose ε_2 so that $\varepsilon_2 < \varepsilon_1/10$ and $U(y_0, \varepsilon_2) \cap (\bigcup\mathscr{S}) = \varnothing$. Thus $U(y_0, \varepsilon_2) \cap (\bigcup\mathscr{T}(S_0)) = \varnothing$. Since f is continuous, there exists $\varepsilon_3 > 0$ so that $0 \leq t_0 - \varepsilon_3$ and $f((t_0 - \varepsilon_3, t_0]) \subseteq U(y_0, \varepsilon_2)$. This contradicts the assumption that $t_0 = \text{l.u.b. } (T)$.

We may thus conclude that y_0 is a limit point of B, and thus by LEMMA 8, $y_0 \in B_r(S_0)$. The contradiction completes the proof. ▮

COROLLARY 13. *Let S_0 be a half-region. Then S_0 can be tiled with a family of squares each of which touches the real boundary of S_0.*

Proof. Choose an arbitrary point $p \in B_a(S_0)$ and let $\{S_1, S_2\} = \mathscr{T}(S_0, p)$. Then $\mathscr{T} = \mathscr{T}(S_1) \cup \mathscr{T}(S_2)$ is the desired tiling. ▮

We are now ready to present the proof of THEOREM 1.

THEOREM 1. *Let S be a homeomorph of the open disk $\{(x, y) \in \mathbb{R}^2: x^2 + y^2 < 1\}$. Then S can be tiled with squares each of which touches the boundary of S.*

Proof. Choose an arbitrary point $p \in S$ and draw a horizontal line through p. This determines a tiling $\{S_1, S_2\}$ of S where each S_i is a half-region containing p on its artificial boundary. The desired tiling is then found by applying COROLLARY 13 to each of these half-regions. ▮

Before proceeding to the proof of THEOREM 2, we pause to develop some notation and conventions concerning lines and curves. When we speak of a line or line segment, it will be assumed that it is parallel to one of the coordinate axes. If x and y are distinct points on a line (or line segment) L, then we let $L(x, y)$ denote the set of points on L that are (strictly) between x and y. Note that $x, y \notin L(x, y)$ and that $L(x, y) = L(y, x)$. We then define $L[x, y] = L[y, x] = L(x, y) \cup \{x, y\}$, $L(x, x) = \varnothing$ and $L[x, x] = \{x\}$.

On the other hand, when we try to extend this notation to curves, it is necessary to specify the direction in which the curve is to be traversed. For distinct points x, y on a simple closed curve C in $\mathbb{R}^2$, we let $C(x, y)$ be the points on C that are between x and y when we traverse C from x to y in the clockwise direction. Note that $C(x, y) \neq C(y, x)$. We then define $C[x, y] = C(x, y) \cup \{x, y\}$, $C(x, x) = \emptyset$ and $C[x, x] = \{x\}$.

Let π_1 and π_2 denote the projection maps from $\mathbb{R}^2$ to $\mathbb{R}$ defined by $\pi_1(x, y) = x$ and $\pi_2(x, y) = y$. Now let x be a point on a horizontal line L. Then we refer to the sets $\{y \in L: \pi_1(y) \geq \pi_1(x)\}$ and $\{y \in L: \pi_1(y) \leq \pi_1(x)\}$ as *half-lines emanating from* x. Half-lines are defined similarly for vertical lines.

Now let S be an open subset of the plane, with the boundary of S being a simple closed curve C containing a closed line segment $L[x, y]$. Now consider the collection $\mathscr{L}$ of all half-lines that are perpendicular to $L[x, y]$ and emanate from a point in $L[x, y]$. There is a natural partition $\mathscr{L} = \mathscr{L}_1 \cup \mathscr{L}_2$ of this collection into two subcollections depending on the direction taken by the line segments. For example, if $L[x, y]$ is vertical, let $\mathscr{L}$ denote the half-lines that emanate from a point in $L[x, y]$ and satisfy $\pi_1(z) \leq \pi_1(x)$ for every $z \in M$ and every $M \in \mathscr{L}_1$. For exactly one of these subcollections, the following statement holds: There exists $p \in L[x, y]$, $M \in \mathscr{L}_i$, and $\varepsilon > 0$ so that M emanates from p and $M \cap U(p, \varepsilon) \subseteq \bar{S}$. In this case, we say that the half-lines in $\mathscr{L}_i$ are *interior to S*.

We are now ready to establish THEOREM 2. The argument will involve the construction of a tiling $\mathscr{H}(S)$ of an annulus S by a collection of quarter-regions and half-regions so that $B_r(S') \leq \text{Bdry}(S)$ for every $S' \in \mathscr{H}(S)$. At a crucial stage of the argument, we will need to take advantage of the fact that when we tile a half-region S, we choose an arbitrary point $p \in B_r(S)$ and begin by setting $\{S_1, S_2\} = \mathscr{T}(S, p)$.

THEOREM 2. *Let S_0 be a homeomorph of the open disk with one hole $\{(x, y) \in \mathbb{R}^2: 1 < x^2 + y^2 < 2\}$. Then S_0 can be tiled with squares each of which touches the boundary of S_0.*

Proof. Let C and C' denote the interior and exterior boundary curves of S_0; i.e., S_0 consists of those points in $\mathbb{R}^2$ that are outside the curve C and inside the curve C'. Let $t_0 = \text{l.u.b. } \{\pi_2(p): p \in C\}$. Since C is compact, there exists a nonempty compact subset $A \subseteq C$ so that $\pi_2(p) = t_0$ for every $p \in A$. Then let $t_1 = \text{l.u.b. } \{\pi_1(p): p \in A\}$. It follows that there exists a unique point $q_0 \in A$ so that $\pi_1(q_0) = t_1$.

Let L_0 be the horizontal line containing q_0. Then there exists a unique point $q_1 \in C'$ with $\pi_1(q_0) < \pi_1(q_1)$ for which $L_0(q_0, q_1) \subseteq S_0$. Similarly there exists a unique point $q'_0 \in L_0 \cap C$ so that $\pi_1(q'_0) = \text{g.l.b. } \{\pi_1(p): p \in A\}$, and a unique point $q'_1 \in L_0 \cap C'$ so that $\pi_1(q'_1) < \pi_1(q'_0)$ and $L_0(q'_0, q'_1) \subseteq S$. Note that $\pi_1(q'_1) < \pi_1(q'_0) \leq \pi_1(q_0) < \pi_1(q_1)$. Then let D_0 denote the half-region whose real boundary is $C'[q'_1, q_1]$ and whose artificial boundary is $L_0(q'_1, q_1)$. (See FIGURE 6.)

If $L_0(q'_0, q_0) \cap S_0 = \emptyset$, we set $\mathscr{E}_0(S_0) = \emptyset$. If $L_0(q'_0, q_0) \neq \emptyset$, we let $L_0(q'_0, q_0) \cap S_0 = \{U_\alpha: \alpha \in \mathscr{A}\}$ be the partition into connected open intervals. Then for each $\alpha \in \mathscr{A}$, we choose $q_\alpha, q'_\alpha \in L_0$ so that $U_\alpha = L_0(q_\alpha, q'_\alpha)$ where $\pi_1(q_\alpha) < \pi_1(q'_\alpha)$. Next, for each $\alpha \in \mathscr{A}$, we let E_α be the half-region whose artificial boundary is $L_0(q_\alpha, q'_\alpha)$ and whose real boundary is $C[q'_\alpha, q_\alpha]$. Then let $\mathscr{E}_0(S_0) = \{E_\alpha: \alpha \in \mathscr{A}\}$.

Let M_0 be the half-line that emanates from q'_1 and contains q'_0, q_0, and q_1. Then let $B_0 = M_0[q_0, q_1]$ and let S_1 be the open set whose artificial boundary is $M_0(q'_1, q'_0) \cup M_0(q_0, q_1)$ and whose real boundary is $C[q_0, q'_0] \cup C'[q_1, q'_1]$. Now consider the following recursive definition. For an integer $k \geq 0$, suppose we have a set of

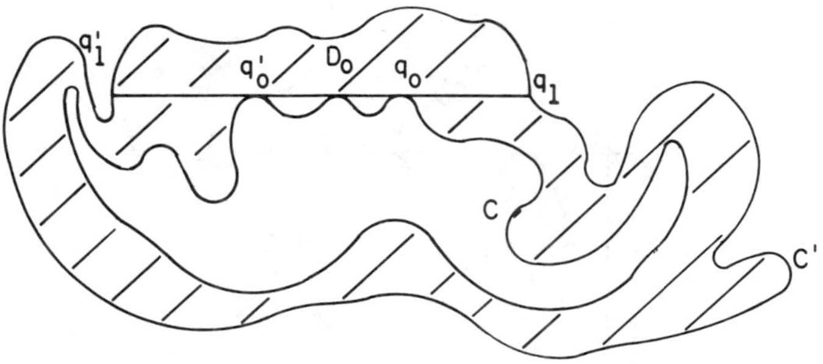

FIGURE 6

points $\{q_i: 0 \le i \le 2k+1\}$, a set of half-lines $\{M_i: 0 \le i \le k\}$, a set of closed line segments $\{B_i: 0 \le i \le k\}$, a set of open sets $\{S_i: 0 < i \le k+1\}$, and a set of points $\{p_i: 0 \le i \le 2k+1\}$ which satisfy the following properties:

1. If $0 \le i \le k$, then $q_{2i} \in C \cap M_k$ and $q_{2i+1} \in C' \cap M_k$.
2. If $0 \le i \le k$, then $B_i = M_i[q_{2i}, q_{2i+1}]$.
3. If $0 < i \le k$, then $q_{2i} \in C[q_{2i-2}, q_0']$ and $q_{2i+1} \in C'[q_{2i-1}, q_1']$.
4. If $0 < i \le k+1$, S_i is the open subset of S_{i-1} whose artificial boundary is $M_0(q_1', q_0') \cup M_i(q_{2i}, q_{2i+1})$ and whose real boundary is $C[q_{2i}, q_0'] \cup C'[q_{2i+1}, q_1']$.
5. If $0 < i \le k$, then $p_{2i-2} \in B_{i-1}$.
6. If $0 < i \le k$, then M_i is perpendicular to M_{i-1} and M_i emanates from p_{2i-2} into the interior of S_i.
7. If $0 < i \le k$, then the component of $M_i \cap \bar{S}_i$ that contains p_{2i-2} is $M_i[p_{2i-2}, p_{2i-1}]$.
8. If $0 < i \le k$, then $p_{2i-1} \in B_i \subseteq M_i[p_{2i-2}, p_{2i-1}]$.

We now describe a procedure which either will determine q_{2k+2}, q_{2k+3}, M_{k+1}, B_{k+1}, S_{k+2}, p_{2k}, and p_{2k+1} so that these eight properties still hold, or will tell us that the construction is to be halted.

For each point $p \in B_k$, let $M(p)$ be the half-line that is perpendicular to B_k and emanates from p into the interior of S_{k+1}. Then let $N(p)$ be the component of $M(p) \cap \bar{S}_{k+1}$ that contains p. It is easy to see that there exists a unique point $p_{2k} \in B_k$ which satisfies the following property:

$$N(p_{2k}) \cap C \neq \varnothing \neq N(p_{2k}) \cap C'.$$

Then let $M_{k+1} = M(p_{2k})$ and $N(p_{2k}) = M_{k+1}[p_{2k}, p_{2k+1}]$. If $p_{2k+1} \in C'$, we let $q_{2k+3} = p_{2k+1}$ and let q_{2k+2} be the unique point of $N(p_{2k}) \cap C$ which is closest to q_{2k+3}. Dually, if $p_{2k+1} \in C$, we let $q_{2k+2} = p_{2k+1}$ and let q_{2k+3} be the unique point of $N(p_{2k}) \cap C'$ which is closest to q_{2k+2}. Next, we let $B_{k+1} = M_{k+1}[q_{2k+2}, q_{2k+3}]$.

At this step in the construction, there are two cases.

Case 1. Either $k+1 \le 2$, or $k+1 \ge 3$ and $B_{k+1} \cap M_0[q_1', q_0'] = \varnothing$.

In this case, we note that $q_{2k+2} \in C[q_{2k}, q_0']$ and $q_{2k+3} \in C'[q_{2k+1}, q_1']$. We then define S_{k+2} as the open subset of S_{k+1} whose artificial boundary is $M_0(q_1', q_0') \cup M_{k+1}(q_{2k+2}, q_{2k+3})$ and whose real boundary is $C[q_{2k+2}, q_0'] \cup C'[q_{2k+3}, q_1']$. At this point, each of the eight properties listed previously holds.

Case 2. $k+1 \ge 3$ and $B_{k+1} \cap M_0[q_1', q_0'] \neq \varnothing$.

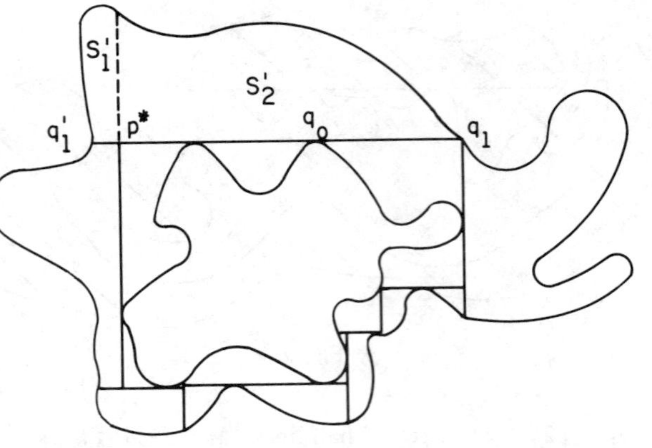

FIGURE 7

In this case, it is clear that B_{k+1} is a vertical line segment. Furthermore, for each i with $1 \leq i \leq k$, it is clear that there is a natural tiling $\mathscr{H}_i$ of $S_i - \bar{S}_{i+1}$ by a family of quarter-regions and half-regions so that $B_r(S) \subseteq \text{Bdry}(S_0)$ and $B_a(S) \subseteq B_{i-1} \cup M_i[p_{2i-2}, p_{2i-1}]$ for every $S \in \mathscr{H}_i$.

Now suppose that $B_{k+1} \cap M_0[q'_1, q'_0] = \{p^*\}$ and that $p^* \neq q'_1$. (See FIGURE 7.) Then let $\{S'_1, S'_2\} = \mathscr{T}(D_0, p^*)$ where $q'_1 \in \bar{S}'_1$. Then it follows that there is a natural tiling $\mathscr{H}_{k+1}$ of the open set $S'_1 \cup S_{k+1} \cup M_0(q'_1, p^*)$ where each set in $\mathscr{H}_{k+1}$ is either a quarter-region or half-region with $B_r(S) \subseteq \text{Bdry}(S_0)$ and $B_a(S) \subseteq B_k \cup M_{k+1}[p_{2k}, p_{2k+1}] \cup M_0(p^*, q'_0)$ for every $S \in \mathscr{H}_{k+1}$. However, we may then conclude that $\mathscr{H}(S_0) = \{S'_2\} \cup \mathscr{E}_0(S_0) \cup \mathscr{H}_1 \cup \mathscr{H}_2 \cup \cdots \cup \mathscr{H}_{k+1}$ is the tiling of S_0 by a collection of quarter-regions and half-regions so that $B_r(S) \subseteq \text{Bdry}(S_0)$ for every $S \in \mathscr{H}(S_0)$. The desired tiling of S_0 by squares each of which touches the boundary of S_0 then follows from LEMMA 12 and COROLLARY 13.

Now suppose that $B_{k+1} \cap M_0[q'_1, q'_0] = \{p^*\}$ but $p^* = q'_1$. Then there is a natural tiling $\mathscr{H}'_{k+1}$ of the open set S_{k+1} by a family of quarter-regions and half-regions so that $B_a(S) \subseteq B_k \cup M_{k+1}[p_{2k}, p_{2k+1}] \cup M_0[q'_1, q'_0]$ and $B_r(S) \subseteq \text{Bdry}(S_0)$ for every $S \in \mathscr{H}'_{k+1}$. We may then conclude that $\mathscr{H}'(S_0) = \{D_0\} \cup \mathscr{E}_0(S_0) \cup \mathscr{H}_1 \cup \mathscr{H}_2 \cup \cdots \cup \mathscr{H}_k \cup \mathscr{H}'_{k+1}$ is a tiling of S_0 by a family of quarter-regions and half-regions with $B_r(S) \subseteq \text{Bdry}(S_0)$ for every $S \in \mathscr{H}'(S_0)$.

We may summarize the argument presented thus far by stating that if the recursive definition ever halts because the criteria for Case 2 are satisfied, then we obtain a tiling of S_0 by a family of squares each of which touches the boundary of S_0. It remains only to consider the case where this recursive construction never terminates. If this occurs, then we obtain an infinite sequence of open subsets $\{S_n : n \geq 1\}$ and an infinite sequence of line segments $\{B_n : n \geq 1\}$ in which $B_n \subseteq \text{Bdry}(S_{n+1})$, $S_{n+1} \subseteq S_n$, and B_n is perpendicular to B_{n+1} for every $n \geq 1$. Now let d denote the usual metric in $\mathbb{R}^2$ and then let $t_2 = \text{g.l.b.}\{d(x, y): x \in C, y \in C'\} = d(C, C')$. Since C and C' are compact, $t_2 > 0$. Also let μ denote the usual measure function for subsets of $\mathbb{R}^2$. Then let $x_n = \{q_{2n}, q_{2n+1}\} - \{p_{2n-1}\}$ for every $n \geq 1$. It follows that $\mu(S_n - S_{n+1}) \geq (\pi/4)(t_2^2)$ for every $n \geq 1$. To see that this is true consider the open disk $U(x_n, t_2)$ and observe that one of the quadrants of this disk must be a subset of $S_n - S_{n+1}$.

However, we observe that it is not possible to have $S_{n+1} \subseteq S_n \subseteq S_0$ and $\mu(S_n - S_{n+1}) \geq (\pi/4)(t_2^2)$ for every $n \geq 1$ since S_0 is bounded. We therefore conclude that the recursive construction described previously must halt in a finite number of steps and with this observation, the proof of the theorem is complete. ∎

We now proceed immediately to show that the previous theorem cannot be improved.

THEOREM 3. *There exists a homeomorph S of the open disk with two holes* $\{(x, y) \in \mathbb{R}^2: 1 < x^2 + y^2 < 5, (x-3)^2 + y^2 > 1\}$ *which cannot be tiled with squares each of which touches the boundary of S.*

Proof. Let $M = 100$, $\varepsilon = 0.01$, and $\delta = 0.0001$. Then consider the open set S consisting of those points which belong to the interior of the large square Q_3 with corner points at $(\pm M, \pm M)$ and are in the exterior of the small squares Q_1 and Q_2 which have side length ε and have their centers respectively at $(-1/2 - \varepsilon/2, 0)$ and $(+1/2 + \varepsilon/2, 0)$.

We suppose that the open set S can be tiled with a family $\mathcal{T}$ of squares each of which touches the boundary of S and proceed to obtain a contradiction.

We refer to a closed line segment B as a *fault line* of $\mathcal{T}$ when there exists a finite subcollection $\mathcal{T}' \subseteq \mathcal{T}$ so that $B \subseteq \bigcup\{\text{Bdry }(Q): Q \in \mathcal{T}'\}$. We now establish the following property of $\mathcal{T}$ which we call the "Tunnel Property."

1. Let $R = [a, b] \times [c, d]$ be a rectangle with $R \subseteq S$. Let B_1 and B_2 be opposite sides of R and let d_1 be the distance from B_1 to B_2. Then let d_2 be the length of the other two sides of R. If B_1 and B_2 are fault lines of $\mathcal{T}$, then $d_2 \leq 2d_1$.

To see that this inequality must hold, we observe that if $d_2 > 2d_1$ and Q is any square in $\mathcal{T}$ that contains the midpoint $p = ((a+c)/2, (b+d)/2)$ of R, then $Q \subseteq R \subseteq S$ and $Q \cap \text{Bdry }(S) = \varnothing$.

We now show that $\mathcal{T}$ satisfies the following property which we call the "Tangent Property."

2. Let $B = L[x, y]$ be a fault line of $\mathcal{T}$ and let Q_i be one of the square holes in S. If $L[z, w]$ is a side of Q_i, $z \in L(x, w)$, and $w \in L(z, y)$, then either $d(x, z) < 5\varepsilon$ or $d(w, y) < 5\varepsilon$.

Before giving the argument for this property, we pause to emphasize that this property holds for each of the four sides of the square holes in S. Now suppose that the Tangent Property fails to hold, i.e., $d(x, z) \geq 5\varepsilon$ and $d(w, y) \geq 5\varepsilon$. If necessary, we rotate the coordinate system so that the configuration formed by B and Q_i is aligned as shown in FIGURE 8.

It follows from the Tunnel Property that there exist squares Q_4 and Q_5 in $\mathcal{T}$ so that (1) Q_4 and Q_5 are tangent to B; (2) Q_4 and Q_5 are above B; (3) Q_4 contains a point p_1 which is δ above B and δ to the right of x; and (4) Q_5 contains a point p_2 which is δ above B and δ to the left of y. Suppose first that $Q_4 \cap Q_i \neq \varnothing \neq Q_5 \cap Q_i$. Then it

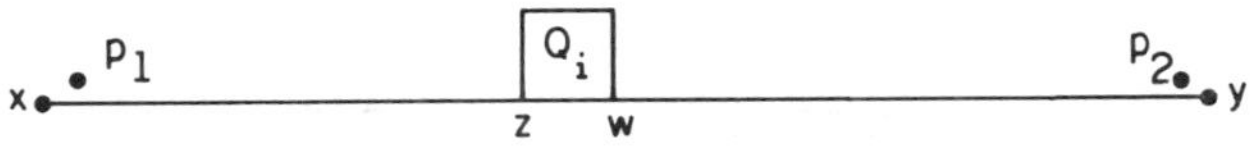

FIGURE 8

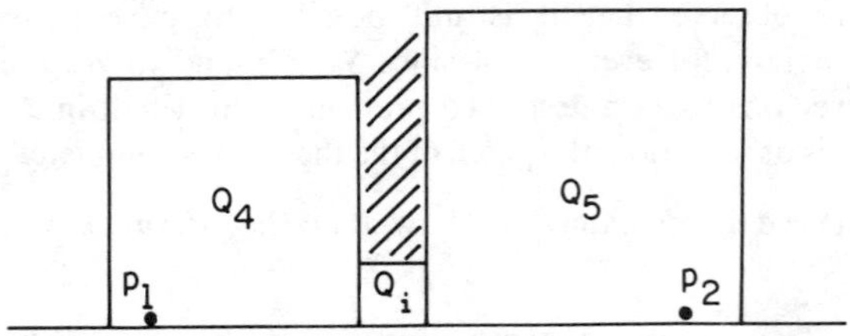

FIGURE 9

follows that Q_4 and Q_5 are tangent to Q_i and each has side length of at least $5\varepsilon - \delta$. However, it is clear that this implies that $\mathscr{T}$ violates the Tunnel Property between Q_4 and Q_5. (See FIGURE 9.)

Now suppose that neither Q_4 nor Q_5 touches Q_i. Then Q_4 and Q_5 are tangent to B, and each has side length of at least $1 - 5\varepsilon$. Again this implies that $\mathscr{T}$ violates the Tunnel Property between Q_4 and Q_5.

We may therefore assume without loss of generality that $Q_4 \cap Q_i \neq \varnothing = Q_5 \cap Q_i$. Then Q_4 has side length at least $5\varepsilon - \delta$ and Q_5 has side length at least $1 - 5\varepsilon$. (See FIGURE 10.)

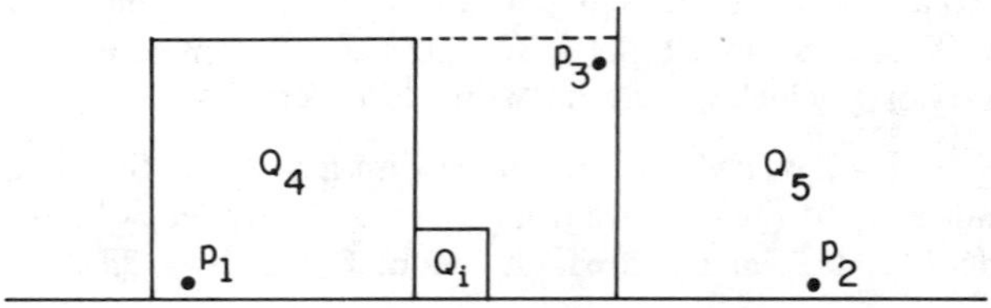

FIGURE 10

Let B_1 be the left boundary of Q_5 and consider the point p_3 which is δ to the left of B_1 and $5\varepsilon - 2\delta$ above B. If $p_3 \in Q_6 \in \mathscr{T}$, then Q_6 has side length at most 5ε and thus $Q_6 \cap Q_i \neq \varnothing$. Thus Q_6 has side length at least 4ε. If Q_6 is not tangent to Q_4, then $\mathscr{T}$ violates the Tunnel Property between Q_4 and Q_6. On the other hand, if Q_6 is not tangent to B, then $\mathscr{T}$ violates the Tunnel Property between Q_6 and B. Since Q_6 cannot be tangent to Q_4 and B, the argument for the Tangent Property is complete.

Next, we consider the square $Q_0 \in \mathscr{T}$ which contains the origin $(0, 0)$. Q_0 must touch the boundary of one or both of the square holes in S. Thus Q_0 has side length of at least $\frac{1}{2}$ and at most 1. It follows from the Tangent Property that $\mathscr{T}$ has a horizontal fault line B_0 (found by either the top or the bottom of Q_0) which is between the square holes and at least $\frac{1}{2} - 6\varepsilon$ away from the origin. Without loss of generality we assume that B_0 is beneath the origin. Now consider the point p which is δ beneath the center of B and let $p \in Q \in \mathscr{T}$. Then it follows that Q does not touch Q_1 or Q_2 so Q must touch Q_3 and hence has side length at least $M - 1$. Let B_1 denote the top side of Q. We may assume without loss of generality that B_1 extends at least $M - 2$ to the right of Q_2. Let $d_1 = d(Q_2, B_1)$. Then $\frac{1}{2} - 6\varepsilon < d_1 < 1$. (See FIGURE 11.)

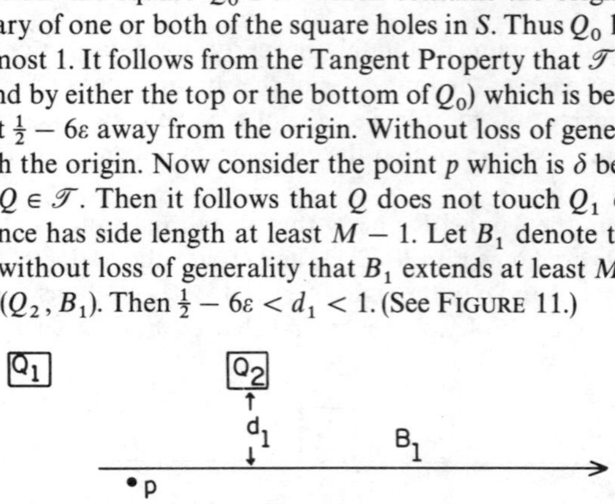

FIGURE 11

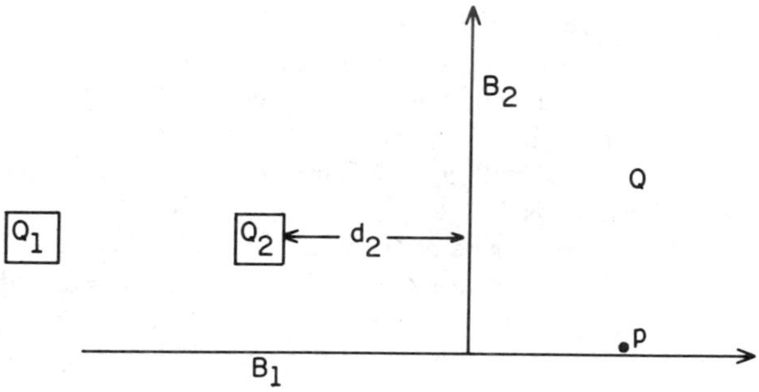

FIGURE 12

Now consider a point p which is δ above B_1 and 2 to the right of Q_2. Choose $Q \in \mathcal{T}$ so that $p \in Q$. Then Q is tangent to B_1. If Q touches Q_2, then $\mathcal{T}$ violates the Tangent Property. Therefore Q touches Q_3 and has side length at least $M - 3$. The left side of Q forms a fault line B_2 which is perpendicular to B_1 and extends at least $M - 2$ above Q_2. Let $d_2 = d(Q_2, B_2)$; then $0 < d_2 \leq 2$. (See FIGURE 12.)

Now consider a point p to the left of B_2 with $d(p, B_2) = \delta$ and $d(p, B_1) = 5$ and choose a square $Q \in \mathcal{T}$ with $p \in Q$. If Q touches Q_2 then it also touches Q_1 and violates the Tangent Property on the top of Q_1. Therefore Q is tangent to B_2, touches Q_3, and has side length at least $M - 5$. Let B_3 be the bottom edge of Q and let $d_3 = d(Q_1, B_3)$. Then $0 < d_3 \leq 5$.

Now consider a point p under B_3 with $d(p, B_3) = \delta$ and $d(p, B_2) = 15$ and choose a square $Q \in \mathcal{T}$ with $p \in Q$. It is clear that Q is tangent to B_2.

At this point, we consider two cases.

Case 1. $Q \cap Q_1 \neq \varnothing$.

We let B_4 be the right side of Q and $d_4 = d(B_4, Q_1) = 0$. Then let p' be a point to the right of B_4 with $d(p', B_4) = \delta$ and $d(p', B_3) = 8$. Choose a square $Q' \in \mathcal{T}$ with $p' \in Q'$; then Q' is tangent to B_4 and Q' touches Q_3. Furthermore, the top side of Q is a fault line which contains B_0.

Case 2. $Q \cap Q_1 = \varnothing$.

We let B_4 be the right side of Q and $d_4 = d(B_4, Q_1)$. Then $0 < d_4 \leq 14$. Let B_4 be the right side of Q_4 and choose a point p' to the right of B_4 with $d(p', B_4) = \delta$ and $d(p', B_3) = 25$. Then choose a square $Q' \in \mathcal{T}$ with $p' \in Q'$. It follows that Q' is tangent to B_4 and that Q' touches Q_3. Furthermore, the top side of Q is a fault line which contains B_0.

It follows that in either case we have determined a rectangle R which contains Q_1 and Q_2 and whose four sides are all fault lines. We denote the bottom, right, top, and left sides of R by T_1, T_2, T_3, and T_4, respectively. Then we may weaken the inequalities obtained previously to the following set:

$$\tfrac{1}{3} < d_1 = d(Q_1, T_1) = d(Q_2, T_1) \leq 1,$$

$$0 \leq d_2 = d(Q_2, T_2) \leq 30,$$

$$0 \leq d_3 = d(Q_1, T_3) = d(Q_2, T_3) \leq 30,$$

$$0 \leq d_4 = d(Q_1, T_4) \leq 30.$$

Now we recall that we chose $Q_0 \in \mathcal{T}$ so that $(0, 0) \in Q_0$. Furthermore, the bottom side of Q_0 is contained in T_1. In view of the symmetric form of the weakened inequalities given above, we may assume without loss of generality that Q_0 touches Q_2. Now consider a point p_4 to the right of Q_0 and above T_1 with $d(p_4, Q_0) = d(p_4, T_1) = \delta$. Then choose a square $Q_4 \in \mathcal{T}$ with $p_4 \in Q$. Since $Q_4 \subseteq R$, Q_4 must be tangent to Q_2 and thus the side length of Q_4 is d_1.

Now suppose that the right side of Q_4 is not contained in T_2. Then consider a point p_5 which is above T_1 and between Q_4 and T_2 with $d(p_5, T_1) = \delta$ and $d(p_5, Q_4) = d(p_5, T_2)$. It follows easily that no square in $\mathcal{T}$ that contains p_5 can touch the boundary of S. The contradiction shows that the right side of Q_4 is contained in T_2, and thus $d_2 = d_1 - \varepsilon$.

Now consider the point p_6 which is above Q_4 and to the left of T_2 with $d(p_6, Q_4) = d(p_6, T_2) = \delta$. Then choose a square $Q_6 \in \mathcal{T}$ with $p_6 \in Q_6$. Then it follows that Q_6 is tangent to Q_2 and has side length $d_1 - \varepsilon$. As before, it follows that the top side of Q_6 is contained in T_3 and thus $d_3 = d_1 - 2\varepsilon$.

Next consider the point p_7 which is below T_3 and to the left of Q_6 with $d(p_7, T_3) = d(p_7, Q_2) = \delta$. Choose a square $Q_7 \in \mathcal{T}$. Then Q_7 is tangent to Q_2 and has side length $d_1 - 2\varepsilon$. In view of the Tunnel Property, we may also conclude that Q_0 has side length $d_1 + \varepsilon$. Since $d_1 + \varepsilon < 1$, we know that $d_1 - 2\varepsilon < 1$ so that Q_7 does not touch Q_1. Next consider the point p_8 which is to the left of Q_7 and under T_3 with $d(p_8, Q_7) = d(p_8, T_3) = \delta$ and choose $Q_8 \in \mathcal{T}$ with $p_8 \in Q_8$. Then Q_8 must touch Q_1 and thus Q_8 has side length $d_1 - 2\varepsilon$. The requirement that Q_8 touch Q_1 implies that $2(d_1 - 2\varepsilon) \geq 1 + \varepsilon$ and thus $d_1 \geq 1/2 + 5\varepsilon/2 > 1/2$. However, in order that Q_8 and Q_1 not violate the Tangent Property, we must have $2(d_1 - 2\varepsilon) \leq \varepsilon + 1 + \varepsilon + 5\varepsilon = 1 + 8\varepsilon$ and thus $d_1 \leq 1/2 + 6\varepsilon$. (Note that if $d_1 > 1/2 + 6\varepsilon$, then the fault line formed by the bottom sides of Q_7 and Q_8 is tangent to Q_1 and extends 5ε to either side.)

Now consider the point p_9 which is above T_1 and to the left of Q_0 with $d(p_9, T_1) = d(p_9, Q_0) = \delta$ and choose $Q_9 \in \mathcal{T}$ with $p_9 \in Q_9$. It follows that Q_9 touches Q_1 and since Q_0 has side length $d_1 > \frac{1}{2}$, we conclude that Q_9 has side length $d_1 - \varepsilon$.

However, this is a contradiction since it implies that the bottom side of Q_8 and the top side of Q_9 violate the Tunnel Property since each extends at least $\frac{1}{2} - 7\varepsilon$ to the right of Q_1. With this observation, the proof of the theorem is complete. ▮

We proceed immediately to the proof of THEOREM 6.

THEOREM 6. *For every $n \geq 3$, there exists a homeomorph S of the open ball $\{(x_1, x_2, \ldots, x_n) \in \mathbb{R}^n : x_1^2 + x_2^2 + \cdots + x_n^2 < 1\}$ which cannot be tiled with n-cubes each of which touches the boundary of S.*

Proof. We first establish the result when $n = 3$. The extension to the case $n > 3$ will follow easily. Let S_0 denote the open subset of $\mathbb{R}^2$ constructed in the proof of THEOREM 3. Then let $P = 1000 = 10M$.

Next let A be the closed subset of $\mathbb{R}^3$ defined by $A = (\bar{S}_0 \times [0, P]) \cup ([-M, M] \times [-M, M] \times [P, P + 1])$. We may view A as being obtained from the rectangular solid (3-box) $R = [-M, M] \times [-M, M] \times [0, P + 1]$ by drilling two square holes into the square face of R to a depth of P. Since the holes do not go all the way through R, it follows that A is homeomorphic to the closed ball $\{(x_1, x_2, x_3) \in \mathbb{R}^3 : x_1^2 + x_2^2 + x_3^2 \leq 1\}$. Thus the open set $S_1 = \text{Int}(A)$ is homeomorphic to the open ball in $\mathbb{R}^3$. (See FIGURE 13.)

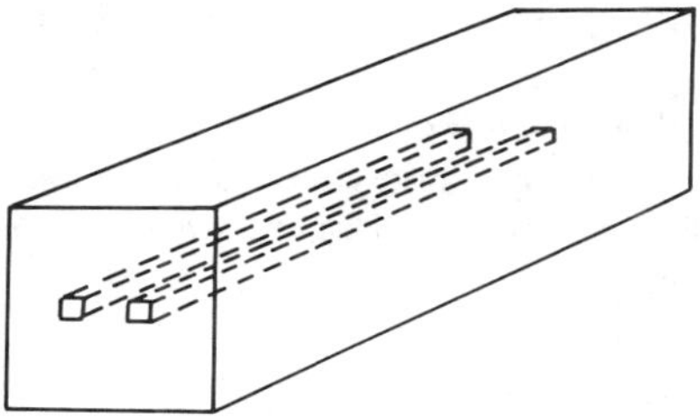

FIGURE 13

Now let $I = (P/2, 1 + P/2)$ and for every $z \in I$, let $p(z)$ denote the plane $x_3 = z$; i.e., $p(z)$ is parallel to the plane determined by the x_1-axis and x_2-axis and passes through the point $(x_1, x_2, x_3) = (0, 0, z)$. It follows that $p(z) \cap S_1$ is a copy of S_0 for every $z \in I$.

Now suppose that $\mathscr{T}$ is a tiling of S_1 by cubes so that each cube in $\mathscr{T}$ touches the boundary of S_1. Then $\mathscr{T}$ is countable so we may choose a point $z_0 \in I$ so that no cube in $\mathscr{T}$ has a face contained in the plane $p(z_0)$. Now let $\mathscr{T}(z_0)$ be the subcollection of all cubes in $\mathscr{T}$ that intersect $p(z_0)$. Then $\mathscr{T}' = p(z_0) \cap Q: Q \in \mathscr{T}(z_0)$ is a tiling of $p(z_0) \cap S_1$ by squares. However, since $P/2 < z_0 < P/2 + 1$ and $P/2 > 2M$, no cube in $\mathscr{T}(z_0)$ can have a face in either of the planes $x_3 = 0$ or $x_3 = P + 1$. It follows that each square in $\mathscr{T}'$ touches the boundary of $p(z_0) \cap S_1$ which contradicts THEOREM 3.

As promised, the extension to the case $n > 3$ is immediate since we need only consider the Cartesian product $S_1 \times (0, P + 1) \times (0, P + 1) \times \cdots \times (0, P + 1)$ of S_1 with $n - 3$ copies of $(0, P + 1)$. ▮

The reader may conjecture that convex sets in $\mathbb{R}^n$ can be tiled with n-cubes but this is not true. Although we do not include the details here, the reader may verify that the convex set S which is the interior of the convex hull determined by $\{(1, 0, 0), (-1, 0, 0), (0, 1, 0), (0, -1, 0), (0, 0, 1), (0, 0, -1)\}$ in $\mathbb{R}^3$ cannot be tiled with cubes each of which touches the boundary of S.

Before proving THEOREM 4, we pause to introduce some notation and terminology. We will refer to a rectangle $R = [a, b] \times [c, d]$ as a *strip* and call a strip $R' = [a, b] \times [e, f]$ a *substrip* of R when $c \leq e < f \leq d$.

We then let $\mathscr{S}(R, S)$ denote the set of all substrips of R which satisfy the requirement that Int $(R) \subseteq S$. Partial order $\mathscr{S}(R, S)$ by set inclusion and let $\mathscr{M}(R, S)$ denote the set of all maximal elements of $\mathscr{S}(R, S)$. Observe that every strip in $\mathscr{M}(R, S)$ touches the boundary of S except possibly for R itself. However, we can ensure that every strip in $\mathscr{M}(R, S)$ touches the boundary of S by further requiring that Int (R) contain a point which does not belong to S. Next, we let $\mathscr{H}_1(R, S)$ consist of those strips in $\mathscr{M}(R, S)$ which satisfy the additional condition that their height is at least as large as their width. Note that $\mathscr{H}_1(R, S)$ is always finite.

Since $\mathscr{H}_1(R, S)$ is finite it follows that there is a finite collection $\mathscr{H}_2(R, S)$ of substrips of R which forms a tiling of minimum cardinality of $R - \bigcup \mathscr{H}_1(R, S)$. Finally, in a strip $R = [a, b] \times [c, d]$, we let $l(R) = [a, (a + b)/2] \times [c, d]$ and $r(R) = [(a + b)/2, b] \times [c, d]$. We consider $l(R)$ and $r(R)$ as the left half and right half of R, respectively.

THEOREM 4. *If S is a bounded open subset of $\mathbb{R}^2$, then S can be tiled with a family of rectangles each of which touches the boundary of S.*

Proof. Without loss of generality, we may assume that $S \subseteq (0, 1) \times (0, 1)$. Then let $R_0 = [0, 2] \times [0, 2]$, $\mathscr{F}_0 = \varnothing$, and $\mathscr{G}_0 = \{R_0\}$. Then suppose that for some k with $k \geq 0$, we have determined $\mathscr{F}_k$ and $\mathscr{G}_k$ so that:

1. $\mathscr{F}_k \cup \mathscr{G}_k$ is a tiling of R_0 by a finite collection of rectangles.
2. $\text{Int}(R) \subseteq S$ and $R \cap \text{Bdry}(S) \neq \varnothing$ for every $R \in \mathscr{F}_k$.
3. $\text{Int}(R) \not\subseteq S$, for every $R \in \mathscr{G}_k$.
4. $\text{Width}(\mathscr{R}) \leq 2^{1-k}$ for every $R \in \mathscr{G}_k$.
5. If $0 < i \leq k$, then $\mathscr{F}_{i-1} \subseteq \mathscr{F}_i$.

We observe that each of these five properties is satisfied when $k = 0$. Note that R_0 was chosen sufficiently large to ensure that the third property would hold. We now describe a procedure for determining $\mathscr{F}_{k+1}$ and $\mathscr{G}_{k+1}$.

Let $R' \in \mathscr{G}_k$ and consider the collections $\mathscr{H}_1(R', S)$ and $\mathscr{H}_2(R', S)$. Since $\text{Int}(R') \not\subseteq S$, we know that $R' \notin \mathscr{H}_1(R', S)$ and therefore, every strip in $\mathscr{H}_1(R', S)$ touches the boundary of S.

Then let $\mathscr{H}_3(R', S) = \{R'' \in \mathscr{H}_2(R', S)\text{: Int}(R'') \subseteq S\}$ and $\mathscr{H}_4(R', S) = \mathscr{H}_2(R', S) - \mathscr{H}_3(R', S)$. Note that each strip in $\mathscr{H}_3(R', S)$ touches the boundary of S, and that $\text{Int}(R) \not\subseteq S$ for every $R \in \mathscr{H}_4(R', S)$.

Now let $\mathscr{H}_4(R', S) = \mathscr{H}_5(R', S) \cup \mathscr{H}_6(R', S) \cup \mathscr{H}_7(R', S)$ be the partition defined by $\mathscr{H}_5(R', S) = \{R'' \in \mathscr{H}_4(R', S)\text{: Int}(l(R'')) \subseteq S\}$, $\mathscr{H}_6(R', S) = \{R'' \in \mathscr{H}_4(R', S) - \mathscr{H}_5(R', S)\text{: Int}(r(R'')) \subseteq S\}$, and $\mathscr{H}_7(R', S) = \mathscr{H}_4(R', S) - \mathscr{H}_5(R', S) - \mathscr{H}_6(R', S)$.

Next, let $R'' = [a, b] \times [c, d] \in \mathscr{H}_5(R', S)$ and then define $f(R'') = \text{l.u.b. } \{t\text{: Int}([a, t] \times [c, d]) \subseteq S\}$. Then $(a + b)/2 \leq f(R'') < b$. Then let $\mathscr{H}_8(R', S) = \{[a, f(R'')] \times [c, d]\text{: } R'' = [a, b] \times [c, d] \in \mathscr{H}_5(R', S)\}$. Note that each strip in $\mathscr{H}_8(R', S)$ touches the boundary of S on its right side.

Next define $\mathscr{H}_9(R', S) = \{[f(R''), b] \times [c, d]\text{: } R'' = [a, b] \times [c, d] \in \mathscr{H}_5(R', S)\}$ and let $\mathscr{H}_{10}(R', S) = \{R'' \in \mathscr{H}_9(R', S)\text{: Int}(R'') \subseteq S\}$. Note that each strip in $\mathscr{H}_{10}(R', S)$ touches the boundary of S on its left side. Then let $\mathscr{H}_{11}(R', S) = \mathscr{H}_9(R', S) - \mathscr{H}_{10}(R', S)$. Note that $\text{Int}(R) \not\subseteq S$ for every $R \in \mathscr{H}_{11}(R', S)$.

Next, let $R'' = [a, b] \times [c, d] \in \mathscr{H}_6(R', S)$ and define $g(R'') = \text{g.l.b. } \{t\text{: Int}([t, b] \times [c, d]) \subseteq S\}$. Then $a < t \leq (a + b)/2$. Then define $\mathscr{H}_{12}(R', S) = \{[g(R''), b] \times [c, d]\text{: } r'' = [a, b] \times [c, d] \in \mathscr{H}_6(R', S)\}$ and note that each strip in $\mathscr{H}_{12}(R', S)$ touches the boundary of S on its left side. Then let $\mathscr{H}_{13}(R', S) = \{[a, g(R'')] \times [c, d]\text{: } R' = [a, b] \times [c, d] \in \mathscr{H}_6(R', S)\}$ and $\mathscr{H}_{14}(R', S) = \{R'' \in \mathscr{H}_{13}(R', S)\text{: Int}(R'') \subseteq S\}$. Note that each strip in $\mathscr{H}_{14}(R', S)$ touches the boundary of S on its right side. Then let $\mathscr{H}_{15}(R', D) = \mathscr{H}_{13}(R', S) - \mathscr{H}_{14}(R', S)$. Note that $\text{Int}(R) \not\subseteq S$ for every $R \in \mathscr{H}_{15}(R', S)$.

Next, let $\mathscr{H}_{16}(R', S) = \{l(R'')\text{: } R'' \in \mathscr{H}_7(R', S)\} \cup \{r(R')\text{: } R'' \in \mathscr{H}_7(R', S)\}$. Note that $\text{Int}(R) \not\subseteq S$ for every $R \in \mathscr{H}_{16}(R', S)$. Now let $\mathscr{H}_{17}(R', S) = \mathscr{H}_1(R', S) \cup \mathscr{H}_3(R', S) \cup \mathscr{H}_8(R', S) \cup \mathscr{H}_{10}(R', S) \cup \mathscr{H}_{12}(R', S) \cup \mathscr{H}_{14}(R', S)$. Then it follows that $\mathscr{H}_{17}(R', S)$ is a finite collection of rectangles so that $\text{Int}(R) \subseteq S$ and $R \cap \text{Bdry}(S) \neq \varnothing$ for every $R \in \mathscr{H}_{17}(R', S)$. Next, let $\mathscr{H}_{18}(R', S) = \mathscr{H}_{11}(R', S) \cup \mathscr{H}_{15}(R', S) \cup \mathscr{H}_{16}(R', S)$. Note that $\mathscr{H}_{18}(R', S)$ is finite, and $\text{Int}(R) \not\subseteq S$ and $\text{width}(R) \leq \text{width}(R')/2 \leq 2^{1-k}/2 = 2^{1-(k+1)}$ for every $R \in \mathscr{H}_{18}(R', S)$. Furthermore, note that $\mathscr{H}_{17}(R', S) \cup \mathscr{H}_{18}(R', S)$ is a tiling of the rectangle R'.

Finally, we set $\mathscr{F}_{k+1} = \mathscr{F}_k \cup (\bigcup\{\mathscr{H}_{17}(R', S)\text{: } R' \in \mathscr{G}_k\})$ and $\mathscr{G}_{k+1} = \bigcup\{\mathscr{H}_{18}(R', S)\text{: } R' \in \mathscr{G}_k\}$. We then define $\mathscr{F} = \bigcup_{k=0}^{\infty} \mathscr{F}_k$. In the same manner as used in the proof of THEOREM 1, we refer to this recursive procedure as Algorithm 2 and refer to $\mathscr{F}$ as the family of rectangles obtained by applying Algorithm 2 to S. As was the case in the proof of THEOREM 1, we have an algorithm which produces a family of rectangles which is easily seen to satisfy P_1, P_2, and P_4. The question is whether $\mathscr{F}$ also satisfies P_3; i.e., Is it true that $S \subseteq \bigcup \mathscr{F}$? We now proceed to show that the answer is yes.

Choose an arbitrary point $x_0 \in S$. Since S is open, there exists $\varepsilon > 0$ so that the square $Q = [\pi_1(x_0) - \varepsilon,\ \pi_1(x_0) + \varepsilon] \times [\pi_2(x_0) - \varepsilon,\ \pi_2(x_0) + \varepsilon] \subseteq S$. Now choose $k \geq 0$ so that $2^{1-k} \leq \varepsilon$. Since $\mathscr{F}_k \cup \mathscr{G}_k$ is a tiling of R_0 and $x_0 \in R_0$, there exists a strip $R' \in \mathscr{G}_k$ with $x_0 \in R'$. Since Int $(R') \nsubseteq S$, it follows that Int $(R') \nsubseteq Q$. Thus the strip R' contains the center x_0 of the square Q and must either extend out the top of Q or out the bottom of Q. However, this implies that there exists a substrip $R'' \in \mathscr{H}_1(R', S)$ so that $x_0 \in R''$. Since $\mathscr{H}_1(R'', S) \subseteq \mathscr{F}$, we conclude that $x_0 \in \bigcup \mathscr{F}$. Since x_0 is arbitrary, we conclude that $S \subseteq \bigcup \mathscr{F}$ and the proof of the theorem is complete. ∎

In the interest of brevity, we do not include the details of the proof of THEOREM 5 since it is a straightforward extension of the previous result.

4. CONCLUDING REMARKS

The problems discussed in this paper were proposed by Bennett and Sharpley [1] who tried to extend the elementary result that every bounded open subset S of $\mathbb{R}^1$ can be tiled by closed intervals each of which touches the boundary of S. It is interesting to note that the theorems in this paper were obtained in the following order. The first result established was the result for rectangles—THEOREM 4. THEOREM 5 came as an immediate corollary. The next result was THEOREM 1 which was extended to the case where S is a simply connected open set, but we do not include the argument for the result here. The next result was to produce an open set in the plane which could not be suitably tiled with squares. The first counterexample had a "large—but finite" number of holes. Later, an example with six holes was found, and then one with four. Bernhard Ganter and Bill Sands (as communicated by George McNulty) produced an example with two holes; the proof given here for THEOREM 3 is a simplification of their argument. Only after this result was obtained was the problem involving the annulus investigated. Of course, the negative result for tiling homeomorphs of the unit ball in $\mathbb{R}^n$ with n-cubes was obtained concurrently with the first counterexample for squares.

It would be of interest to investigate the following tiling problems:

1. For $n \geq 3$, which bounded open sets can be tiled with n-cubes each of which touches the boundary?

2. For $n \geq 2$, which bounded open sets are tiled by squares each of which touches the boundary by applying the "Greedy Algorithm" of always choosing a square of maximum size in the region not already tiled? The reader may note that this algorithm fails for certain convex sets in $\mathbb{R}^2$.

ACKNOWLEDGMENTS

The author gratefully acknowledges many stimulating conversations with his colleagues Ron DeVore, George McNulty, and Jim Roberts on these problems.

References

1. Bennett, C. & R. Sharpley. 1979. Weak-type inequalities for H^p and BMO. *In* Harmonic Analysis in Euclidean Spaces, Proceedings of Symposia in Pure Mathematics **35:** 201–229.
2. Gardner, M. 1979. Mathematical games. Sci. Amer. **241:** 18–26.
3. Kazarinoff, N. D. & R. Weitzenkamp. 1973. Squaring rectangles and squares. Amer. Math. Monthly **80:** 877–888.
4. Tutte, W. T. 1965. The quest of the perfect square. Amer. Math. Monthly **72:** 29–35.

REMARKS ON THE SPHERE OF INFLUENCE GRAPH

David Avis[a]

School of Computer Science
McGill University
Montreal, Quebec, Canada

Joe Horton

School of Computer Science
University of New Brunswick
Fredericton, New Brunswick, Canada

1. INTRODUCTION

Let S be a finite set of points in the plane. For each point $x \in S$, let r_x be the closest distance to any other point in the set, and let C_x be the circle of radius r_x centered at x. Toussaint has defined the sphere of influence graph as a graph on S with an edge between points x and y if and only if the circles C_x and C_y intersect in at least two places. It is shown that:

(i) The sphere of influence graph has at most $29n$ edges ($n = |S|$).

(ii) Every decision tree algorithm for computing the sphere of influence graph requires at least $\Omega(n \log n)$ steps in the worst case.

As an application of (i), ElGindy observed that an algorithm of Bentley and Ottman can be used to find the sphere of influence graph in $O(n \log n)$ time.

Several recent papers in the pattern recognition literature have investigated various graph structures that can be defined on a set of points in the plane. The goal is to find a graph that preserves the perceptual relevance of the pattern. Such graphs include the minimum spanning tree, gabriel graph, relative neighborhood graph, and Delaunay triangulation. For a general discussion of these structures and their applications to pattern recognition, the reader is referred to the excellent survey papers by Toussaint [5, 6].

Let $S = \{x_1, \ldots, x_n\}$ be a set of n points in the plane and let

$$r_i = \min_{k \neq i} d(x_i, x_k)$$

where d represents Euclidean distance. Finally, let C_i be the circle of radius r_i centered at x_i. Then the *closed sphere of influence graph* $\overline{G}(S)$ can be defined on S by joining points x_i and x_j of S by an edge whenever $C_i \cap C_j \neq \varnothing$.

The open sphere of influence graph $G(S)$ is a subgraph of $\overline{G}(S)$: Points x_i and x_j are joined by an edge whenever C_i and C_j intersect in precisely two points. FIGURE 1 gives examples of these graphs.

The sphere of influence graph is not simply related to any of the above-mentioned graphs derived from a set of points in the plane. As Toussaint has shown [5], the minimum spanning tree is a subgraph of the relative neighborhood graph,

[a] Research supported by Natural Science and Engineering Research Council Grant A3013 and Le Programme de Formation de Chercheurs et d'Action Concertée Grant EQ-1678.

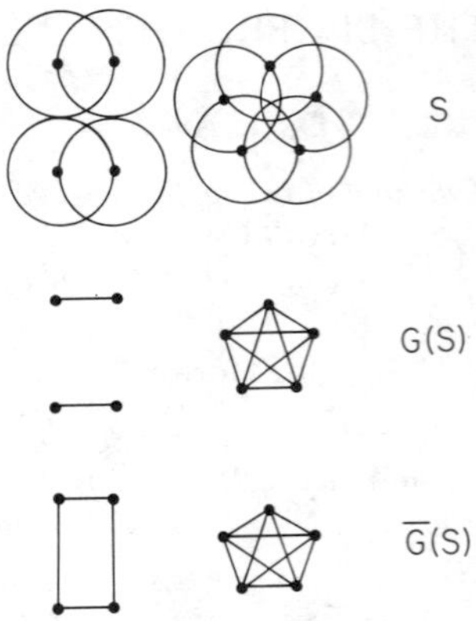

FIGURE 1. A set of points S and its sphere of influence graphs.

which is a subgraph of the gabriel graph, which is a subgraph of the Delaunay triangulation. However, the sphere of influence graph does not necessarily include the (necessarily connected) minimum spanning tree, since $\overline{G}(S)$ need not be connected. Nor is the sphere of influence graph necessarily a subgraph of the (necessarily planar) Delaunay triangulation, since $G(S)$ need not be a planar graph. Indeed, $G(S)$ can include K_6, and $\overline{G}(S)$ can include K_7. Thus the sphere of influence graph is neither a subgraph, nor a supergraph, of any of the above graphs.

The main result of this paper is to settle affirmatively the following conjecture of Toussaint: Does there exist a constant c such that $\overline{G}(S)$ [and hence $G(S)$] has at most cn edges? In fact, we prove the following theorem.

THEOREM. *$\overline{G}(S)$ has at most 29n edges.*

ElGindy has found an interesting application of this result. He observed that the set of circles C_i can be computed in $O(n \log n)$ time by using the Voronoi diagram of the point set [4]. Then he noted that $\overline{G}(S)$ can be computed using an algorithm of Bentley and Ottman [2; see also Reference 3] for reporting all intersections of a set of circular arcs. This algorithm runs in $O(n \log n + k \log n)$ time, where k is the number of intersections. Hence, applying the above theorem, $\overline{G}(S)$ can be computed in $O(n \log n)$ time. We conclude this paper by showing that this algorithm is optimal to within a constant factor, by exhibiting an $\Omega(n \log n)$ lower bound[b] for decision tree algorithms that compute $\overline{G}(S)$.

2. MAIN RESULTS

Before proving the main theorem, we will require the following lemma. Let S, x_i, r_i, and C_i be defined as in Section 1.

LEMMA. *No point in the plane but not in S, lies in more than six of the circles C_i. All but a finite number of points in the plane lie in at most five of the circles.*

Proof. Let $y \notin S$ be any point on or inside circles C_i and C_j, centered at x_i and x_j, respectively. Then

$$d(y, x_i) \leq d(x_i, x_j) \quad \text{and} \quad d(y, x_j) \leq d(x_i, x_j).$$

[b] $f(n)$ is $\Omega(g(n))$ if there exists a constant $c > 0$ such that $f(n) \geq cg(n)$ for all n sufficiently large.

From elementary geometry it follows that the angle $x_i y x_j$ equals or is greater than $\pi/3$ radians. Equality holds if and only if

$$d(y, x_i) = d(y, x_j) = d(x_i, x_j),$$

i.e., y, x_i, and x_j are the points of an equilateral triangle. Therefore y can be in at most six circles C_i, and exactly six circles only if the six circles all have the same radius r and the six centers all lie on a circle of radius r with center y. Thus a given set of six circles can contain at most one common point not in S. Hence, there are no more than $n + \binom{n}{6}$ points lying in six circles. ▮

Before proving the main result, we state the following simplifying assumption:

$$r_1 = \min_{1 \leq i \leq n} r_i = 1.$$

Further, let C be the circle centered at x_1 with radius $\sqrt{3}$. Corresponding to each circle C_i, we define a *disk* D_i which is the circle C_i together with its interior. We now state and prove:

THEOREM. *x has degree at most* 29 *in* $\overline{G}(S)$.

Proof. Suppose circle C_j intersects C_1. Let S_j denote the segment of C defined by

$$S_j = C \cap D_j.$$

We claim that the segment S_j subtends an angle of at least $\pi/3$ radians. The situation is depicted in FIGURE 2. Let x, y, and θ be as shown in FIGURE 2. $1 \leq x \leq y \leq 1 + x$ and

$$\cos \theta = \frac{3 + y^2 - x^2}{2\sqrt{3y}}.$$

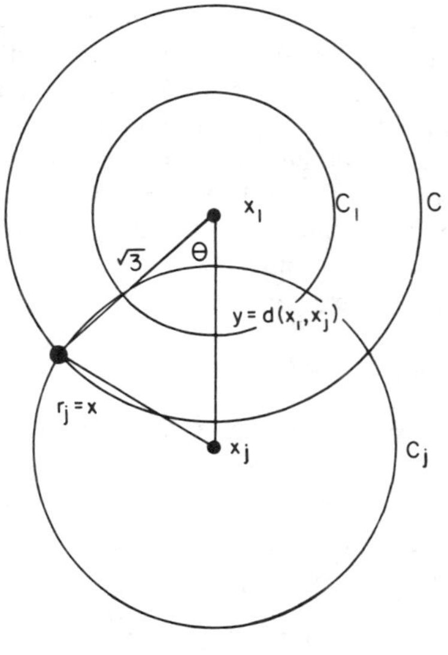

FIGURE 2. Finding a bound on θ.

An elementary calculation shows that

$$\cos\theta \le \frac{\sqrt{3}}{2};$$

hence $\theta \ge \pi/6$ radians. This proves the claim. In addition, $\cos\theta = \sqrt{3}/2$ and $\theta = \pi/6$ only if $x = 1$ and $y = 1$ or 2.

By the lemma, all but a finite number of points on C can lie in at most five circles. Thus the number of circles intersecting C_1 is at most

$$\frac{5.2\pi}{\pi/3} = 30.$$

We can improve this bound slightly by observing that at most 18 points of S can be at a distance 1 or 2 from x_1. At most 6 can lie on C_1, and at most 12 can lie on a circle of radius 2 centered at x_1. (Recall that all radii are greater than or equal to one.) Thus there is at least one circle C_i that intersects C_1 and yields a segment S_i of C that subtends an angle greater than $\pi/3$ radians. Thus 30 is not achievable and C_1 intersects at most 29 circles. ▮

COROLLARY. *$\bar{G}(S)$ contains at most 29n edges.*

Proof. Let $x_1 \in S$ be defined as in the theorem. Then note that any edge of $\bar{G}(S)$ not touching x_1 is an edge of $\bar{G}(S - \{x_1\})$, since removing x_1 can only increase the radii of the circles. The corollary now follows by mathematical induction on the cardinality of S. ▮

The best lower bound on the maximum degree of x_1 known to the authors is 18. An example is provided by the hexagonal lattice.

3. Lower Bounds

We conclude by showing a $\Omega(n \log n)$ lower bound for decision tree algorithms that compute $\bar{G}(S)$ or $G(S)$. Let $D(n)$ be the number of different possible labeled graphs on n vertices that are sphere of influence graphs. The information theoretic lower bound is $\log_2(D(n))$. Since the log of the number of labeled graphs with $29n$ or fewer edges is $O(n \log n)$, and taking into consideration our upper bound on the required number of operations, the best lower bound we can expect is $\Omega(n \log n)$. We obtain this bound by exhibiting $n!/2$ different sphere of influence graphs, since $\log(n!) = O(n \log n)$ by Stirling's formula.

Let $\Pi = (\Pi_1, \Pi_2, \ldots, \Pi_n)$ be a permutation of the first n integers. Let

$$S = \{x_i \mid x_i = (\Pi_i, 0),\ i = 1, 2, \ldots, n\},$$

a set of points in the plane. For $n = 5$, $\bar{G}(S)$ is illustrated in FIGURE 3.

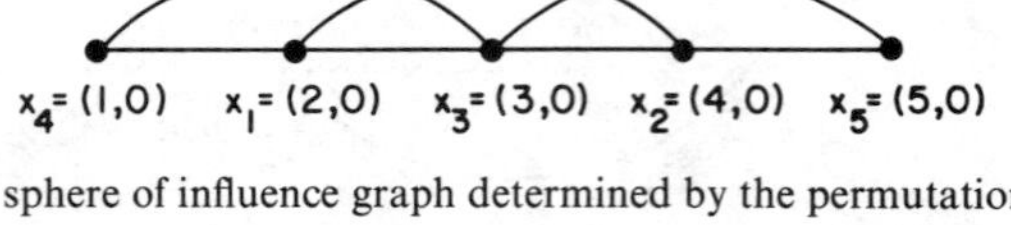

FIGURE 3. The sphere of influence graph determined by the permutation (2, 4, 3, 1, 5).

Each Π gives rise to a different sphere of influence graph, $n \geq 5$, except that Π and $\Pi' = (\Pi_n, \Pi_{n-1}, \ldots, \Pi_1)$ give rise to the same graph. Since there are $n!$ permutations, the result follows. This is sufficient to prove a $\Omega(n \log n)$ lower bound for computing $\overline{G}(S)$ under a comparison-based model. However, as shown in [1], comparison trees, or even the more powerful linear decision trees, are not powerful enough to solve geometric problems in finite time.

We can obtain a $\Omega(n \log n)$ bound under a stronger model of computation by observing that the sphere of influence graph can be used to solve the nearest neighbor problem for a set of points in the plane. All that needs to be done is to find the shortest edge, which can be done in linear time. As this problem has a lower bound of $\Omega(n \log n)$ [4], it also shows that the sphere of influence graph has a lower bound that is $\Omega(n \log n)$. The model of computation used in these lower bounds is somewhat stronger than the linear decision tree model, and the algorithm of [2, 3] has $O(n \log n)$ complexity under this model [1].

Acknowledgments

The authors gratefully acknowledge useful discussions with Godfried Toussaint and Hossam ElGindy. Also, the authors acknowledge the work of the Simon Fraser Combinatorics Seminar in the spring of 1981, where the problem was introduced by Geňa Hahn.

References

1. Avis, D. 1980. Lower bounds for geometric problems. *In* Proceedings, 18th Allerton Conference, Urbana, Illinois. Pp. 35–40.
2. Bentley, J. & T. Ottman. 1979. Algorithms for reporting and counting geometric intersections. IEEE Trans. Comput. **C-28**: 643–647.
3. Brown, K. Q. 1981. Comments on "Algorithms for reporting and counting geometric intersections." IEEE Trans. Comput. **C-30**: 147–148.
4. Shamos, M. & D. Hoey. 1975. Closest point problems. *In* Proceedings, 16th IEEE Symposium on Foundations of Computer Science. Pp. 151–162. Publ. IEEE. New York, N.Y.
5. Toussaint, G. T. 1980. Pattern recognition and geometric complexity. *In* Proceedings, 5th International Conference on Pattern Recognition, Miami Beach (December 1980). Pp. 1–24.
6. Toussaint, G. T. 1981. Computational geometric problems in pattern recognition. *In* Pattern Recognition Theory and Applications. J. Kittler, Ed. NATO Advanced Study Institute, Oxford University. Oxford, England.

A NEW BOUND FOR EUCLIDEAN STEINER MINIMAL TREES

F. R. K. Chung and R. L. Graham

Bell Laboratories
Murray Hill, New Jersey 07974

Introduction

Let X denote a finite set of points in the (Euclidean) plane. A *minimum spanning tree* for X, denoted by $M(X)$, is a connected network of line segments joining various pairs of points of X so that the sum of the lengths of the line segments, called the *length* of $M(X)$, is as small as possible. We denote this minimum possible total length by $L_M(X)$.

It may happen for a set X that by embedding X in a suitable larger set Y, $L_M(Y)$ is actually less than $L_M(X)$. For example, if X consists of the three vertices of an equilateral triangle and Y consists of X together with the centroid of the triangle, then an easy calculation shows that

$$\frac{L_M(Y)}{L_M(X)} = \frac{\sqrt{3}}{2} = 0.866025\ldots.$$

For any set X we define a *Steiner minimal tree* $S(X)$ for X to be a minimum spanning tree $M(Y)$ having the least possible length over *all* sets $Y \supseteq X$. We denote this minimum possible length by $L_S(X)$. Although a set X may have many different Steiner minimal trees, the length $L_S(X)$ is uniquely determined. Points in $Y - X$ are called *Steiner points* of $S(X)$. Points in X are called regular points. Note that $L_S(X) \leq L_S(Y)$ for any $X \subseteq Y$.

The problem of determining $S(X)$ and $L_S(X)$ for a given set X has an old and venerable history, dating back to Steinhaus, Maxwell, Steiner, and, in a primitive form, to Fermat. As is the case with many combinatorial optimization problems, the (Euclidean) Steiner minimal tree problem is known to be NP-complete [4]. (For a complete discussion of this concept, see [5].) This has reinforced recent efforts to find good approximations for $S(X)$ that can be computed efficiently. One of the most natural of these is just the minimum spanning tree $M(X)$. A common measure of how well $M(X)$ approximates $S(X)$ in length is given by the worst-case ratio

$$\rho := \inf_X \frac{L_S(X)}{L_M(X)},$$

where X ranges over all finite sets in the plane.

A long-standing conjecture, due to Gilbert and Pollak [6], is the following:

Conjecture. $\rho = \sqrt{3}/2 = 0.866025\ldots$.

If true, this would be best possible, as the previously noted example of an equilateral triangle shows. This conjecture has been verified in the case where $|X| \leq 5$ (see [3, 10]) and also if $S(X)$ contains at most one Steiner point (see [9]).

Over the past few years, lower bounds for ρ have improved from the observation of E. F. Moore (see [6]) that $\rho \geq 1/2$ (in fact, for any metric space), to $\rho \geq 1/\sqrt{3} = 0.57735\ldots$ by Graham and Hwang [7] (for any Euclidean space), and then to

$$\rho \geq (2 + 2\sqrt{3} - \sqrt{7 + 2\sqrt{3}})/3 = 0.74309\ldots$$

by Chung and Hwang [1]. Very recently, Du and Hwang [11] proved that $\rho \geq 0.8$. The purpose of this paper is to tighten the bound on ρ even further with the following result.

THEOREM.

$$\rho \geq \rho_0 = 0.82416874\ldots, \tag{1}$$

where ρ_0 is the unique real root of the polynomial

$$P(x) = x^{12} - 4x^{11} - 2x^{10} + 40x^9 - 31x^8 - 72x^7 + 116x^6$$
$$+ 16x^5 - 151x^4 + 80x^3 + 56x^2 - 64x + 16$$

satisfying $0.8 < \rho_0 < 1$.

We point out that if X_n denotes a regular simplex in Euclidean n-space, then it is known [2] that the ratio $L_S(X_n)/L_M(X_n)$ can be arbitrarily close to $\sqrt{3}/(4 - \sqrt{2}) = 0.66984\ldots$ for n sufficiently large so that the previously mentioned lower bound of $1/\sqrt{3}$ is not as bad as it looks. In the case of the *rectilinear* plane, with the (l_1) distance between (x, y) and (x', y') given by $|x - x'| + |y - y'|$, and where $L_{RS}(X)$ and $L_{RM}(X)$ denote the lengths of the rectilinear Steiner minimal tree for X and the rectilinear minimum spanning tree for X, respectively, Hwang [8] proved

$$\frac{L_{RS}(X)}{L_{MS}(X)} \geq \frac{2}{3}, \qquad \text{for all } X,$$

and that further, this is best possible [as the four points $(\pm 1, \pm 1)$ show, for example]. Thus, the Euclidean plane seems to be a somewhat more stubborn case.

Before proceeding to the main results of the paper, we should comment on its general style. Even a casual perusal shows that a fair amount of algebraic computation is displayed. We felt that this was unavoidable for two reasons. First, in fairness to the reader who might really wish to verify the various claims, we tried to include enough detail so that it would actually be possible. In fact, we omitted most of the actual computations and, of course, all of the computer symbolic computations we needed. Second, on more than one occasion in the past, assertions made by various authors in proofs of bounds for this problem have proven to be incorrect (usually because of sketchy and incomplete arguments). We hope to avoid this pitfall by being more explicit than usual.

PROPERTIES OF STEINER MINIMAL TREES

In this preliminary section we will mention several properties of Steiner minimal trees that will be needed later. Some of these can be found in the excellent (but now somewhat out-of-date) survey of Gilbert and Pollak [6].

LEMMA 1. (i) *In any Steiner minimal tree $S(X)$ for X we can assume without loss of generality that every Steiner point is incident to exactly three edges of $S(X)$, each of which meets the other two at* 120°.

(ii) *If $|X| = n$, then $S(X)$ has at most $n - 2$ Steiner points. If $S(X)$ has $n - 2$ Steiner points it is called a full Steiner tree.*

(iii) *Any Steiner minimal tree $S(X)$ for X can be decomposed into full Steiner trees in the following sense. There always exist unique subsets $X_i \subseteq X$ such that $|X_i \cap X_j| \leq 1$ for $i \neq j$, $S(X)$ restricted to X_i is a full Steiner tree on X_i, and these full trees partition the set of all edges of $S(X)$.*

LEMMA 2 [10]. *Let four points A, B, C, D be given so that both possible full Steiner trees exist. Then A and B are adjacent to a common Steiner point in the unique Steiner minimal tree if and only if the angle AEB is acute where E is the intersection of lines AD and BC* (see FIGURE 1).

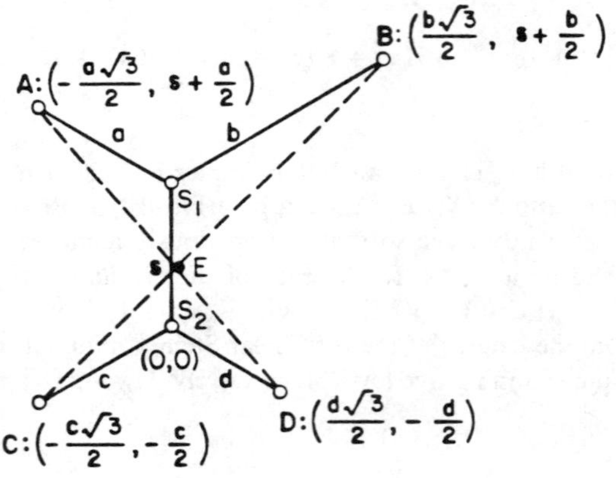

FIGURE 1

With coordinates assigned as shown in FIGURE 1 we have

$$\text{slope } AD = -\frac{s + (1/2)(a + d)}{(\sqrt{3}/2)(a + d)},$$

$$\text{slope } BC = \frac{s + (1/2)(b + c)}{(\sqrt{3}/2)(b + c)}.$$

Thus, by LEMMA 2, if A and B have a common Steiner point S_1 and the angle AEB is acute, then we have

$$\frac{s + (1/2)(a + d)}{(\sqrt{3}/2)(a + d)} \cdot \frac{s + (1/2)(b + c)}{(\sqrt{3}/2)(b + c)} \geq 1,$$

i.e.,

$$(2s + a + d)(2s + b + c) \geq 3(a + d)(b + c). \quad \textbf{(2)}$$

In fact, for $a' \leq a$, $b' \leq b$, $c' \leq c$, $d' \leq d$, we have

$$(2s + a' + d')(2s + b' + c') \geq 3(a' + d')(b' + c'). \quad \textbf{(2')}$$

THE PROOF OF THE THEOREM

We now proceed to the proof of (1). It will be enough to show that for any finite set X in the Euclidean plane,

$$\frac{L_S(X)}{L_M(X)} \geq \rho_0, \tag{3}$$

where ρ_0 is defined as in (1). The proof is by induction on $|X|$. Inequality (3) certainly holds if $|X| \leq 4$ as we previously noted. Assume that (3) holds for all $\bar{X}$ with $|\bar{X}| < n$ for a fixed integer $n \geq 5$ and suppose $|X| = n$ violates (3), i.e.,

$$\frac{L_S(X)}{L_M(X)} < \rho_0.$$

We will eventually derive a contradiction through a sequence of claims. The occasionally rather formidable algebraic manipulations needed in the proofs of the claims have been relegated to the Appendix.

CLAIM 1. *Any Steiner minimal tree $S(X)$ on X must be a full Steiner tree. For if $S(X)$ were not full, then by* LEMMA 1, *we could partition the edges of $S(X)$ into (full) Steiner minimal trees on smaller sets $X_i \subseteq X$, and*

$$\frac{L_S(X)}{L_M(X)} \geq \frac{\sum_i L_S(X_i)}{\sum_i L_M(X_i)} \geq \rho_0$$

since by induction, for each i

$$\frac{L_S(X_i)}{L_M(X_i)} \geq \rho_0.$$

Thus, we can also assume that some Steiner point S_1 of $S(X)$ is adjacent to two points A and B of X and another Steiner point S_2.

Let $\mathbf{d}(U, V)$ denote the (Euclidean) distance between U and V. Referring to FIGURE 2, we can assume (without loss of generality) that $\mathbf{d}(A, S_1) = a \leq b = \mathbf{d}(B, S_1)$ and $\mathbf{d}(A, B) = 1$.

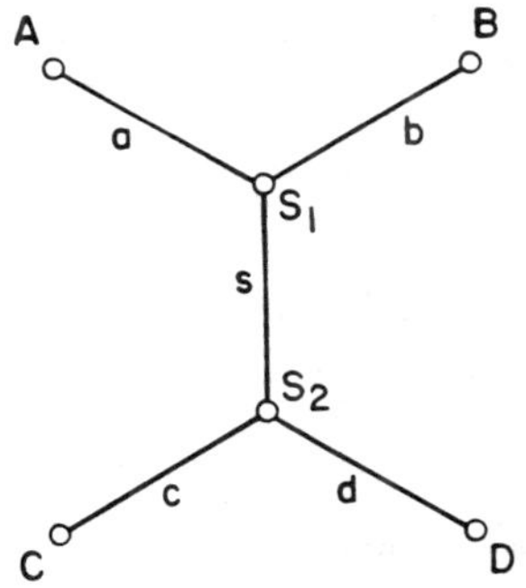

FIGURE 2

Note that $a^2 + ab + b^2 = 1$. Set

$$\mathbf{d}(S_1, S_2) = s, \quad \mathbf{d}(C, S_2) = c, \quad \mathbf{d}(D, S_2) = d.$$

Note that at least one of C and D must be a Steiner point.

CLAIM 2. $s + c + d < ((\rho_0 - b)(2a + b - \rho_0))/(a + 2b - 2\rho_0)$.

Proof. Let s' denote $L_S(\{A, C, D\})$ and let X' denote $X - \{B\}$. Suppose

$$s + a + b + c + d - s' \geq \rho_0.$$

Now,

$$L_S(X') \leq L_S(X) - (s + a + b + c + d) + s'$$

since the right-hand side is the length of *some* tree connecting X' together. Since $|X'| < n$, then by induction

$$L_S(X') \geq \rho_0 L_M(X').$$

Combining these we obtain

$$\frac{L_S(X)}{L_M(X)} \geq \frac{L_S(X) - (s + a + b + c + d) + s' + \rho_0}{L_M(X') + 1}$$

$$\geq \frac{\rho_0 L_M(X') + \rho_0}{L_M(X') + 1} = \rho_0$$

which is a contradiction. Thus, we can assume

$$m + a + b - s' < \rho_0, \qquad \text{where } m = s + c + d.$$

A straightforward computation gives

$$s' = (a^2 + am + m^2)^{1/2}.$$

Therefore,

$$(m + a + b - \rho_0)^2 < a^2 + am + m^2$$

which implies

$$m(a + 2b - 2\rho_0) < (\rho_0 - b)(2a + b - \rho_0).$$

Suppose we have $2\rho_0 > a + 2b$. Then by taking $a^2 + ab + b^2 = 1$, we obtain $b \geq 1.15$ or $b < 0.49$ which is impossible. Therefore $b < \rho_0$ and

$$m = s + c + d < \frac{(\rho_0 - b)(2a + b - \rho_0)}{(a + 2b - 2\rho_0)} := f(b).$$

Note that since $f(b)$ is a decreasing function in increasing b,

$$m \leq f(1/\sqrt{3}) = 2.677\ldots. \ \blacksquare$$

CLAIM 3. *The angle AEB is acute* (*see* FIGURE 1).

CLAIM 3 is proved in the Appendix.

CLAIM 4. $b \leq 0.7574$ *and* $a \geq 0.3728$.

Proof. Choose S_2' on the extension of S_1S_2 such that $\mathbf{d}(S_1, S_2') = f(b)$. It is easily checked that the angle $AS_2'B$ is smaller than the angle AS_2B which is smaller than the acute angle AEB. Therefore by (**2′**) we have

$$(2f(b) + a)(2f(b) + b) \geq 3ab.$$

Together with $a^2 + ab + b^2 = 1$, we obtain

$$b \leq 0.7574.$$

This also implies $a \geq 0.3728$. ▮

CLAIM 5. (i) *If* C *is a Steiner point, then there is a point* $Y \in X$ *with* $\mathbf{d}(A, Y) \leq \mathbf{d}(A, C)$ *or* $\mathbf{d}(A, Y) \leq \mathbf{d}(A, S_2)$.

(ii) *If* D *is a Steiner point, then there is a point* $Z \in X$ *with* $\mathbf{d}(B, Z) \leq \mathbf{d}(B, D)$ *or* $\mathbf{d}(B, Z) \leq \mathbf{d}(B, S_2)$.

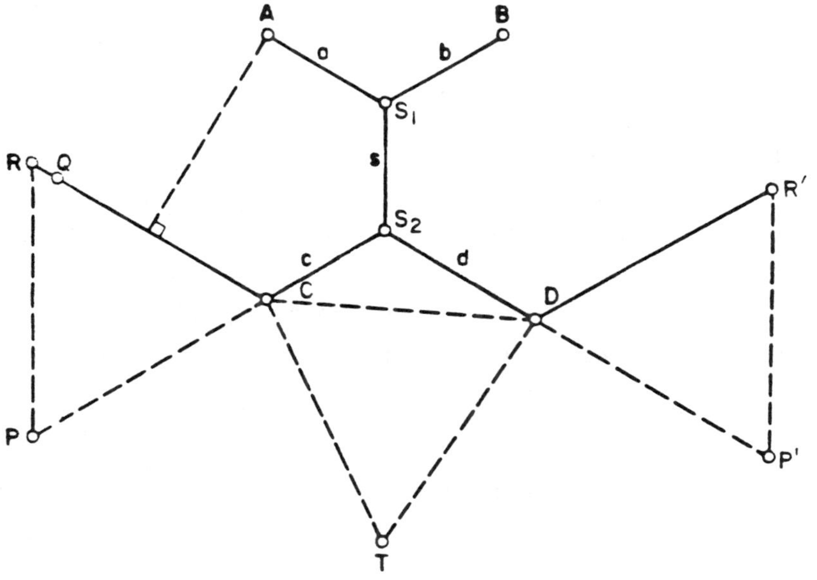

FIGURE 3

Proof of (i). Suppose C is a Steiner point but no such point Y exists. Referring to FIGURE 3, locate the point R so that RC is parallel to AS_1 and $\mathbf{d}(R, C) = s + 2a - c$ so that $\mathbf{d}(A, R) = \mathbf{d}(A, C)$. If some point of X lies on the line segment RC, then we are done.

Suppose some Steiner point Q lies on RC. Let us move along the edges of $S(X)$ on $AS_1S_2CQQ_1Q_2 \ldots Q_i \ldots Y$ by always turning to the right at the Steiner points Q_i. If the angle ACQ is no more than 60°, we have $\mathbf{d}(Q, C) \leq \mathbf{d}(A, Q) \leq \mathbf{d}(A, C)$ because of the minimality of $S(X)$. Also, we note that the angle Q_1QA is no more than 60°. Therefore we have $\mathbf{d}(A, Q_i) \leq \mathbf{d}(A, C)$ and $\mathbf{d}(A, Y) \leq \mathbf{d}(A, C)$ if Q_i is not in the region AS_1S_2C. If the angle ACQ is greater than 60°, it can easily be seen that the angle AQC is less than 60°. By a similar argument we have

$$\mathbf{d}(A, Y) \leq \max\{\mathbf{d}(A, C), \mathbf{d}(A, S_2)\}.$$

We may assume some Steiner point Q' of $S(X)$ connected to C lies on the extension of the line segment CR.

We next need the following inequality which is proved in the Appendix:

$$a + b + c + d + s + \mathbf{d}(R, C) \geq L_S(A, C, D, R) + \rho_0. \tag{4}$$

By deleting the point B and the five edges incident to S_1 or S_2 and reconnecting the points (excluding B) by $S(A, C, D, Q')$ we have

$$L_S(A, C, D, Q') \leq L_S(A, C, D, R) + \mathbf{d}(R, Q')$$

and so, by induction and **(5)**,

$$\frac{L_S(X)}{L_M(X)} \geq \frac{(L_S(X) - (a + b + c + d + s + \mathbf{d}(C, R)) + L_S(A, C, D, R)) + \rho_0}{L_M(X - \{B\}) + 1}$$

$$\geq \frac{L_S(X - \{B\}) + \rho_0}{L_M(X - \{B\}) + 1} \geq \rho_0$$

which is a contradiction. ▮

Proof of (ii). The argument used to prove (ii) is analogous to that used to prove (i) except that in place of **(4)** we need the inequality **(4′)** (also proved in the Appendix):

$$a + b + c + d + s + \mathbf{d}(D, R') \geq L_S(B, C, D, R') + \rho_0. \tag{4′}$$

To complete the proof of **(1)** we need a final inequality (proved in the Appendix):

$$a + b + c + d + s \geq \mathbf{d}(C, D) + \rho_0\,\omega, \tag{5}$$

where

$$\omega = \min\{1 + \omega_1, 1 + \omega_2, \omega_1 + \omega_2\}$$

and

$$\omega_1 = \max\{\mathbf{d}(A, C), \mathbf{d}(A, S_2)\},$$
$$\omega_2 = \max\{\mathbf{d}(B, D), \mathbf{d}(B, S_2)\}.$$

Granting these, we then have

$$\frac{L_S(X)}{L_M(X)} \geq \frac{(L_S(X) - (a + b + c + d + s)) + \mathbf{d}(C, D) + \rho_0\,\omega}{L_M(X - \{A, B\}) + \omega}$$

$$\geq \frac{L_S(X - \{A, B\}) + \rho_0\,\omega}{L_M(X - \{A, B\}) + \omega}$$

$$\geq \rho_0$$

which again is a contradiction.

In the last sequence we used the induction hypothesis on the set $X - \{A, B\}$, using the fact (by CLAIM 5) that there are points of $X - \{A, B\}$ relatively close to A and B.

This completes the proof of the theorem. ▮

APPENDIX

Proof of CLAIM 3. Suppose the angle AEB is not acute. LEMMA 2 indicates that the full Steiner tree on $ABCD$ in the "other" direction (i.e., the full Steiner tree in which A and C are joined to the same Steiner point) does not exist. This implies (as shown in [10]) that

$$\angle ACS_2 + \angle BDS_2 < 60^\circ \qquad \text{or} \qquad \angle CAS_1 + \angle DBS_1 < 60^\circ.$$

Case 1. $\angle ACS_2 + \angle BDS_2 < 60^\circ$.

We then have

$$\frac{a+s}{c+s} = \frac{\sin\theta_1}{\sin(60^\circ + \theta_1)},$$

$$\frac{b+s}{d+s} = \frac{\sin\theta_2}{\sin(60^\circ + \theta_2)},$$

where $\theta_1 = \angle ACS_2$ and $\theta_2 = \angle BDS_2$.

Since $\theta_1 + \theta_2 < 60^\circ$, we have

$$\begin{aligned}(c+s) + (d+s) &\geq (a+s)\frac{\sin(60^\circ + \theta_1)}{\sin\theta_1} + (b+s)\frac{\sin(60^\circ + \theta_2)}{\sin\theta_2}\\ &\geq (a+s)\frac{\sin(60^\circ + \theta_1)}{\sin\theta_1} + (b+s)\frac{\sin(60^\circ + \theta_1)}{\sin(60^\circ - \theta_1)}\\ &:= \omega(\theta_1).\end{aligned}$$

Now suppose

$$\frac{d\omega}{d\theta_1} = -\frac{(a+s)\sin 60^\circ}{(\sin\theta_1)^2} + \frac{(b+s)\sin 60^\circ}{(\sin(60^\circ - \theta_1))^2}$$

has a zero at $\bar{\theta}$. It can be shown that

$$\frac{d^2\omega}{d\theta^2}(\bar{\theta}) \geq 0.$$

Therefore ω has a minimum at $\bar{\theta}$ and

$$\omega(\bar{\theta}) = (\sqrt{a+s} + \sqrt{b+s})^2.$$

Thus

$$c + d > a + b + 2\sqrt{(a+s)(b+s)}$$

and $f(b) \geq a + b + s + 2\sqrt{(a+s)(b+s)}$. Since $\theta_1, \theta_2 < 60^\circ$, we have $c \geq a$ and $d \geq b$.

Choose B_0, C_0, D_0 on BS_1, CS_2, DS_2, respectively, such that $\mathbf{d}(B_0, S_1) = \mathbf{d}(C_0, S_2) = \mathbf{d}(D_0, S_2) = a$. Then the subtree of $S(x)$ connecting A, B_0, C_0, D_0 is minimal. Therefore, $s \geq (\sqrt{3} - 1)a$.

We have

$$f(b) \geq \sqrt{3}a + b + 2\sqrt{\sqrt{3}a(b + (\sqrt{3} - 1)a)}.$$

Combining this with the fact that $a^2 + ab + b^2 = 1$ and $b \geq a$, we get

$$b \leq 0.561 \qquad \text{or} \qquad b \geq 0.853$$

which is impossible. Therefore Case 1 cannot occur.

Case 2. $\angle CAS_1 + \angle DBS_1 < 60°$.

Then by a method similar to that in Case 1, we have

$$a + b > c + d + 2\sqrt{(c + s)(d + s)}.$$

We consider the following subcases:

Subcase (i). $c \leq a - s$ and $d \leq b - s$.

Since $\angle ACR$ and $\angle BDR'$ (see FIGURE 3) are not more than 60°, by an argument similar to that used in CLAIM 5, it can be shown that there are regular points $\mathbf{d}(A, Y) \leq \mathbf{d}(A, S_2)$ and $\mathbf{d}(B, Z) \leq \mathbf{d}(B, S_2)$. Since $s \leq a \leq b$, the minimal spanning tree on A, B, S_2 consists of AS_2 and BS_2. Therefore we have

$$a + b + s \geq \frac{\sqrt{3}}{2}(\mathbf{d}(A, S_2) + \mathbf{d}(B, S_2))$$

and

$$\begin{aligned}\frac{L_S(X)}{L_M(X)} &\geq \frac{L_S(X) - a - b - s + (\sqrt{3}/2)(\mathbf{d}(A, S_2) + \mathbf{d}(B, S_2))}{L_M(X - \{A, B\}) + \mathbf{d}(A, Y) + \mathbf{d}(B, Z)}\\ &\geq \frac{L_S(X - \{A, B\}) + (\sqrt{3}/2)(\mathbf{d}(A, S_2) + \mathbf{d}(B, S_2))}{L_M(X - \{A, B\}) + \mathbf{d}(A, S_2) + \mathbf{d}(B, S_2)}\\ &\geq \frac{\rho_0 L_M(X - \{A, B\}) + \rho_0(\mathbf{d}(A, S_2) + \mathbf{d}(B, S_2))}{L_M(X - \{A, B\}) + \mathbf{d}(A, S_2) + \mathbf{d}(B, S_2)}\end{aligned}$$

which is a contradiction.

Subcase (ii). $c > a - s$ and $b \leq b - s$.

We know that

$$2s + c + d \leq a + b.$$

Choose C' on the extension of CS_2 such that $\mathbf{d}(C', S_2) = a + b - 2s$. Let E' denote the intersection of AS_2 and BC'. Then the angle $AE'B$ is greater than the angle AEB. Therefore from **(2)** we know that

$$(2s + a)(2s + b + c') \leq 3a(b + c'),$$

where $c' = a + b - 2s$, i.e.,

$$(2s + a)(a + 2b) \leq 3a(a + 2b - 2s)$$

which implies

$$s \leq \frac{(2b + a)a}{2(b + 2a)} = s_0 < a.$$

We want to show that

$$(2 - \sqrt{3})a + b > \rho_0\sqrt{b^2 + s_0^2 + bs_0}.$$

Using $a^2 + ab + b^2 = 1$, this can be derived by straightforward manipulation (using the VAXIMA symbolic system, for example). In fact, this holds for $1 > b \geq 0.2$.

Now choose C'' on CS_2 so that $\mathbf{d}(C'', S_2) = a - s$. Then $a + b + s + \mathbf{d}(C'', S_2) - L_S(\{A, S_2, C\}) - \rho_0 \mathbf{d}(B, S_2) \geq 0$ since $L_S(\{A, S_2, C''\}) = \sqrt{3}a$. We consider

$$\frac{L_S(X)}{L_M(X)} \geq \frac{L_S(X) - a - b - s - \mathbf{d}(C'', S_2) + L_S(\{A, S_2, C\}) + \rho_0 \mathbf{d}(B, S_2)}{L_M(X - \{B\}) + \mathbf{d}(B, Z)}$$

$$\geq \frac{L_S(X - \{B\}) + \rho_0 \mathbf{d}(B, S_2)}{L_M(X - \{B\}) + \mathbf{d}(B, S_2)}$$

$$\geq \rho_0.$$

Again this is a contradiction.

Subcase (iii). $c \leq a - s$ and $d > b - s$.
As in (ii) we can get

$$s \leq \frac{(2a + b)b}{2(a + 2b)} := s_1 < b,$$

and

$$a + (2 - \sqrt{3})b > \rho_0\sqrt{a^2 + as_1 + s_1^2}$$

(which holds for $1 > b > 0.43$). We can also derive a contradiction by the same method as in Subcase (ii). ∎

Proof of (**4**). Referring to FIGURE 3, we first note that $c \geq a$ in this case. It is easily seen that

$$a + b + c + d + s + \mathbf{d}(C, R) \leq L_S(A, B, C, D, R).$$

Choose P so that PCR forms an equilateral triangle (see FIGURE 3). Thus, PCS_2 forms a straight line and

$$L_S(A, B, D, P) = a + b + c + d + s + \mathbf{d}(C, R)$$

(by applying LEMMA 2 and using the fact that PCR is equilateral).

Thus, by (**2**) we have

$$(2s + a + d)(2s + b + c') \geq 3(a + d)(b + c'), \tag{6}$$

where

$$c' = c + \mathbf{d}(C, R) = s + 2a.$$

In fact, for $d' < d$ we have

$$(2s + a + d')(2s + b + c') \geq 3(a + d')(b + d'). \tag{6'}$$

If $f(b) \leq s + c' + d$, then by a calculation similar to that used in CLAIM 2 we can prove

$$a + b + c' + d + s - L_S(A, P, D) \geq \rho_0$$

which is equivalent to (**4**).

We may assume

$$f(b) \geq s + c' + d = 2(s + a) + d. \tag{7}$$

Suppose $d \geq b$. By (**6′**) we have

$$(2s + a + b)(2s + b + c') \geq 3(a + b)(b + c'),$$

i.e.,

$$s \geq \frac{\sqrt{13b^2 + 40ab + 28a^2} - b - 2a}{6} \geq a.$$

From (**7**) we have

$$4a + b \leq f(b).$$

This implies

$$g(a, b) := \rho_0^2 + (-4b - 10a)\rho_0 + 3b^2 + 11ab + 4a^2 \leq 0.$$

On the other hand, we note that $g(a, b)$ can be viewed as a function of a single variable b since $a^2 + ab + b^2 = 1$ and the derivative of $g_1(b) := g(a, b)$ with respect to b is

$$\frac{d}{db} g_1(b) = -4\rho_0 + 10\left(\frac{2b + a}{2a + b}\right)\rho_0 + \frac{14a^2 - 16b^2 - 4ab}{2a + b}$$

$$\geq \frac{14a^2 - 4b^2 - 2ab}{2a + b} \geq 0$$

since $da/db = -(2b + a)/(2a + b)$, and $\rho_0 \geq 0.76 \geq b \geq a \geq 0.37$. By a straightforward calculation we have $g(a, b) \geq g_1(1/\sqrt{3}) > 0$, which is a contradiction. Therefore we may assume $d < b$. It follows from earlier arguments that there is a regular point Z such that

$$\mathbf{d}(B, Z) \leq \mathbf{d}(B, S_2) \qquad \text{or} \qquad \mathbf{d}(B, Z) \leq \mathbf{d}(B, D).$$

We consider the following two possibilities.

Case 1. $\mathbf{d}(B, Z) \leq \mathbf{d}(B, S_2)$.

Suppose the following is true:

$$a + b + c' + s - \mathbf{d}(A, P) - \rho_0 \mathbf{d}(B, S_2) \geq 0. \tag{8}$$

Then we have

$$\frac{L_S(X)}{L_M(X)} \geq \frac{(L_S(X) - (a + b + c' + s + \mathbf{d}(C, R))) + L_S(A, R, C) + \rho_0 \mathbf{d}(B, Z)}{L_M(X - \{B\}) + \mathbf{d}(B, Z)}$$

$$\geq \frac{L_S(X - \{B\}) + \rho_0 \mathbf{d}(B, Z)}{L_M(X - \{B\}) + \mathbf{d}(B, Z)}$$

$$\geq \rho_0$$

which is a contradiction. Therefore we must have

$$a + b + c' + s - \sqrt{3}(a + s) - \rho_0(b + s) < 0,$$

i.e.,

$$s \geq \frac{(3 - \sqrt{3})a + (1 - \rho_0)b}{\rho_0 - 2 + \sqrt{3}} \geq 1.5b,$$

since $a \geq 0.3728$ and $b \leq 0.7574$.

From **(7)** we have

$$2a + 3b \leq f(b).$$

However, this implies $f(b) \geq 4a + b$, which has been shown to be impossible.

Case 2. $\mathbf{d}(B, S_2) \leq \mathbf{d}(B, Z) \leq \mathbf{d}(B, D)$.

In this case we have $d \geq b - s$. Suppose the following is true:

$$a + b + c + d + s - \mathbf{d}(C, D) - \rho_0(1 + \mathbf{d}(B, D)) \geq 0.$$

Then we have

$$\begin{aligned}\frac{L_S(X)}{L_M(X)} &\geq \frac{L_S(X) - a - b - c - d - s + \mathbf{d}(C, D) + \rho_0(1 + \mathbf{d}(B, Z))}{L_M(X - \{A, B\}) + 1 + \mathbf{d}(B, Z)} \\ &\geq \frac{L_S(X - \{A, B\}) + \rho_0(1 + \mathbf{d}(B, Z))}{L_M(X - \{A, B\}) + 1 + \mathbf{d}(B, Z)} \\ &\geq \rho_0,\end{aligned}$$

which is a contradiction.

Thus we may assume

$$g_2(a, b, c, d, s) := a + b + c + d + s - \mathbf{d}(C, D) - \rho_0(1 + \mathbf{d}(B, D)) \leq 0.$$

Since g_2 is increasing in c and decreasing in d we then have

$$\begin{aligned}g_3(a, b, s) &:= g_2(a, b, 0, b, s) \\ &= 2a + 2b + s - 1 - \rho_0(1 + b + s) \leq 0.\end{aligned}$$

By choosing $d' = b - s$ in **(6′)** we have

$$s \geq \frac{\sqrt{4b^2 + 13ab + 10a^2} - b - 2a}{3} \geq \frac{b + a}{3}.$$

Therefore

$$g_3(a, b, s) \geq 2a + b - 1 - \rho_0 + 7(1 - \rho_0)(a + b)/3 \geq 0.02 > 0$$

since $a > 0.37$ and $a + b \geq 1.12$. This again yields a contradiction. This completes the proof of **(4)**. ∎

The proof of **(4′)** is basically the same as that of **(4)** so the details will be omitted.

Proof of **(5)**. The one remaining inequality to prove is **(5)**. We consider several cases (refer to FIGURE 3).

Case 1. $\mathbf{d}(A, C) \leq \mathbf{d}(A, S_2)$, $\mathbf{d}(B, D) \leq \mathbf{d}(B, S_2)$.

In this case (**5**) holds if it holds for $c = d = 0$. However, since it does hold for $c = d = 0$, Case 1 is completed.

Case 2. $\mathbf{d}(A, C) \leq \mathbf{d}(A, S_2)$, $\mathbf{d}(B, D) \geq \mathbf{d}(B, S_2)$.

It is easily checked that $d \geq b - s = d'$ and $s \leq a$.

Suppose we have

$$a + (2 - \sqrt{3})b \geq \rho_0 \mathbf{d}(A, S_2).$$

Then we can choose D' on $S_2 D$ with $\mathbf{d}(S_2, D') = b - s$. Suppose $a + b + d' + s - L_S(B, S_2, D') \geq \rho_0 \mathbf{d}(A, S_2)$. Then

$$\frac{L_S(X)}{L_M(X)} \geq \frac{(L_S(X) - (a + b + d' + s) + L_S(B, S_2, D')) + \rho_0 \mathbf{d}(A, S_2)}{L_M(X - \{A\}) + \mathbf{d}(A, S_2)}$$

$$\geq \frac{L_S(X - \{A\}) + \rho_0 \mathbf{d}(A, S_2)}{L_M(X - \{A\}) + \mathbf{d}(A, S_2)}$$

$$\geq \rho_0,$$

which is a contradiction.

Thus we must have

$$a + (2 - \sqrt{3})b \leq \rho_0 \sqrt{a^2 + as + s^2}.$$

This implies

$$s \geq \frac{\sqrt{a^2(4 - 3\rho_0^2) + 28b^2 + \sqrt{3}(-16b^2 - 8ab) + 16ab} - a\rho_0}{2\rho_0} := s_0.$$

We want to prove

$$g_4(a, b, s) := a + b + s - \rho_0(\sqrt{a^2 + as + s^2} + 1) \geq 0.$$

Since g_4 is increasing in s, by straightforward calculation we have

$$g_4(a, b, s_0) \geq (\sqrt{3} - 1)b + s_0 - \rho_0 \geq 0.16 > 0.$$

Recalling that $b \leq 0.7574$ and $a \geq 0.3728$, we therefore have

$$a + b + c + d + s \geq g_4(a, b, s) + \mathbf{d}(C, D) + \rho_0(1 + \omega_1) \geq \mathbf{d}(C, D) + \rho_0(1 + \omega_1)$$
$$\geq \mathbf{d}(C, D) + \rho_0 \omega$$

and (**5**) is proved.

Case 3. $\mathbf{d}(A, C) \geq \mathbf{d}(A, S_2)$, $\mathbf{d}(B, D) \leq \mathbf{d}(B, S_2)$.

Since any point on $L_S(X) - L_S(A, B, S_2)$ cannot be within distance b of B, there is a regular point Z' in the shaded area of FIGURE 4.

We know that

$$\frac{\sqrt{3}}{2} L_M(A, B, Z') \leq L_S(A, B, Z')$$
$$\leq \sqrt{(a + b + s)^2 + d^2 + (a + b + s)d}. \qquad (\mathbf{9})$$

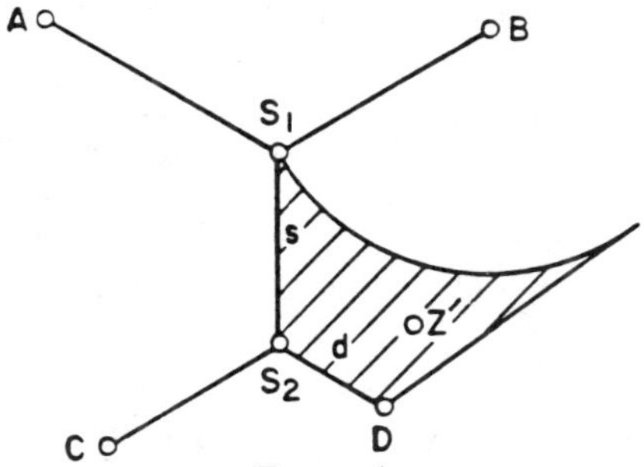

FIGURE 4

We now consider

$$g_5(a, b, d, s) := a + b + s - \rho_0 \frac{2}{\sqrt{3}} \sqrt{(a+b+s)^2 + d^2 + (a+b+s)d}.$$

Note that

$$\frac{a+b+s}{\sqrt{(a+b+s)^2 + d^2 + (a+b+s)d}} = \frac{\sin\theta}{\sin 60^\circ},$$

where θ is the angle S_2DW (see FIGURE 5).

If $\sin\theta \geq \rho_0$, we have $g_5(a, b, d, s) \geq 0$ and

$$\frac{L_S(X)}{L_M(X)} = \frac{L_S(X) - a - b - s + \rho_0 L_M(A, B, Z')}{L_M(X - \{A, B\}) + L_M(A, B, Z')}$$

$$= \frac{L_S(X - \{A, B\}) + \rho_0 L_M(A, B, Z')}{L_M(X - \{A, B\}) + L_M(A, B, Z')}$$

$$\geq \rho_0$$

which is impossible.

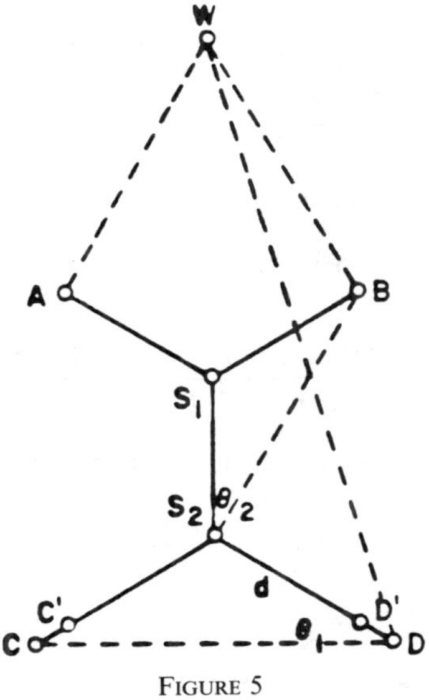

FIGURE 5

We may assume $\sin\theta < \rho_0 := \sin\theta_0$. Hence

$$d \geq (a + b + s)\frac{\sin(60° - \theta_0)}{\sin\theta_0} := d_0. \tag{10}$$

Suppose we have

$$b + (2 - \sqrt{3})a \geq \rho_0\,\mathbf{d}(B, S_2),$$

where $\mathbf{d}(C', S_2) = a - s$. Then

$$\frac{L_S(X)}{L_M(X)} \geq \frac{L_S(X - \{B\}) + \rho_0\,\mathbf{d}(B, S_2)}{L_M(X - \{B\}) + \mathbf{d}(B, S_2)} \geq \rho_0$$

which is impossible. We may therefore assume

$$b + a\sqrt{3} \leq \rho_0\sqrt{b^2 + bs + s^2}. \tag{11}$$

Consider the function

$$g_6(a, b, s) := 2a + b + d_0 - \mathbf{d}(C', D') - \rho_0(1 + \sqrt{b^2 + bs + b^2}),$$

where $\mathbf{d}(D', S_2) := d_0$.

Note that g_6 is decreasing in s since

$$g_6(a, b, s + \varepsilon) - g_6(a, b, s) = \varepsilon(t - t\cos\theta_1 - \rho_0\cos\theta_2) < 0$$

when $t = \sin(60° - \theta_0)/\sin\theta_0 \leq 0.1$ and $\theta_1 = \angle S_2D'C'$, $\theta_2 = \angle S_1S_2B \leq 60°$. Thus

$$g_6(a, b, s) \geq g_6(a, b, a) = 2a + b - 2\rho_0.$$

If $b \leq 0.6539$, then $g_6(a, b, s) \geq 0$ and

$$\begin{aligned}\frac{L_S(X)}{L_M(X)} &\geq \frac{L_S(X) - (a + b + s + c' + d') + L_S(A, C', D') + \rho_0\,\mathbf{d}(B, Z')}{L_M(X - \{B\}) + \mathbf{d}(B, Z')}\\ &\geq \frac{L_S(X - \{B\}) + \rho_0\,\mathbf{d}(B, Z')}{L_M(X - \{B\}) + \mathbf{d}(B, Z')}\\ &\geq \rho_0, \qquad \text{a contradiction.}\end{aligned}$$

Thus, we may assume $b \geq 0.6539$. Also, if $a + b + s - \rho_0(1 + \mathbf{d}(A, C)) \geq 0$, then we have

$$\frac{L_S(X)}{L_M(X)} \geq \frac{L_S(X) - a - b - s_0 + \rho_0(1 + \mathbf{d}(A, C))}{L_M(X - \{A, B\}) + 1 + \mathbf{d}(A, C)} \geq \rho_0$$

which is impossible. Consequently, we may assume

$$\mathbf{d}(A, C) \geq \frac{a + b + s - \rho_0}{\rho_0}. \tag{12}$$

We now consider

$$g_7(a, b, c, d, s) := c + d + s - f(b)$$

for a, b, c, d and s satisfying

$$a^2 + ab + b^2 = 1 \tag{13}$$

together with **(10)**, **(11)**, **(12)** and $0.6539 \le b \le 0.7574$.

Let $(a_0, b_0, c_0, d_0, s_0)$ denote the 5-tuple for which g_6 achieves its minimum value m_0. It is easily seen that a_0, b_0, c_0, d_0, and s_0 must satisfy **(10)**, **(11)**, and **(12)** with equality. Thus, g_7 can be considered a function of the single variable b. We now outline the (brute-force) method used to prove that $g_7(b) > 0$ for $0.6539 \le b \le 0.7574$. The actual mechanism by which this was done was through the symbolic computation system VAXIMA, a modified form of the Macsyma system at the Massachusetts Institute of Technology.

Step 1. Determine the derivatives da/db, dc/db, dd/db, ds/db in terms of a, b, c, d, s.

Step 2. Show that in the ranges

$$0.37 \le a \le 0.5, \quad 0.65 \le b \le 0.76, \quad 0 \le s, c, d \le 1$$

we have $|da/db| \le 2$, $|ds/db| \le 2$, $|dc/db| \le 10$, $|dd/dc| \le 1$, and $df(b)/db \le 0$. It then follows that $dg_7/db \ge -15$.

Step 3. Compute the values of $g_7(b_i)$ for $b_i = 0.65 + i/200$, $0 \le i \le t$, where $b_{t-1} < 0.76 < b_t$. It turns out that $g_7(f_i) > 0.1$ for $0 \le i \le t$.

Step 4. Estimate $g_7(b)$ for $0.65 \le b \le 0.76$ using the results from Steps 2 and 3. In particular, since g_7 is continuous, then for any $b \in [0.65, 0.76]$ we can choose the largest b_i satisfying $b > b_i$ and conclude

$$\begin{aligned} g_7(b) &\ge g_7(b_i) - 15(b - b_i) \\ &\ge 0.1 - 0.075 > 0. \end{aligned}$$

By this method we can prove that $g_7(b) > 0$. However, this contradicts Claim 2 so we conclude that Case 3 cannot occur.

Case 4. $\mathbf{d}(A, C) \ge \mathbf{d}(A, S_2)$, $\mathbf{d}(B, D) \ge \mathbf{d}(B, S_2)$.

We want to prove that

$$g_8(a, b, c, d, s) := a + b + c + d + s - d(C, D) - \rho_0(1 + \omega_3) \ge 0,$$

where $\omega_3 = \min\{\mathbf{d}(A, C), \mathbf{d}(B, D)\}$ for a, b, c, d and s satisfying **(13)** and

$$s + c + d \le \frac{(\rho_0 - b)(2a + b - \rho_0)}{(a + 2b - 2\rho_0)}, \tag{14}$$

$$(2s + a + d)(2s + b + c) \ge 3(a + d)(b + c), \tag{15}$$

$$0.6539 \le b \le 0.7574. \tag{16}$$

Note that **(14)** and **(15)** follow from Claims 1 and 2.

Let $(a_0, b_0, c_0, d_0, s_0)$ denote the 5-tuple for which g_7 achieves its minimum value m. Suppose $\mathbf{d}(A, C) < \mathbf{d}(B, D)$. For a small positive value ε we note that

$$g_8(a_0, b_0, c_0, d_0 - \varepsilon, s_0) < g_8(a_0, b_0, c_0, d_0, s_0)$$

and $(a_0, b_0, c_0, d_0 - \varepsilon, s)$ satisfies **(13)**–**(15)**, and $d_0 - \varepsilon > 0$ since $\mathbf{d}(B, D) > \mathbf{d}(B, S_2)$. This contradicts the minimality of $(a_0, b_0, c_0, d_0, s_0)$. Similarly, if $\mathbf{d}(A, C) > \mathbf{d}(B, D)$, we derive a contradiction. Thus we may assume $\mathbf{d}(A, C) = \mathbf{d}(B, D)$, i.e.,

$$(a_0 + c_0)s_0 + a_0^2 + c_0^2 - a_0 c_0 = (b_0 + d_0)s_0 + b_0^2 + d_0^2 - b_0 d_0. \qquad \textbf{(17)}$$

Suppose $(a_0, b_0, c_0, d_0, s_0)$ does not satisfy equality in **(15)**. We consider the 5-tuple $(a_0 - \varepsilon_a, b_0 + \varepsilon_b, c_0 + \varepsilon_c, d_0 - \varepsilon_d, s_0)$, where $\varepsilon_a, \varepsilon_b, \varepsilon_c, \varepsilon_d$ are small positive values chosen such that equality holds in **(13)** and **(14)**, and $\mathbf{d}(A', C') = \mathbf{d}(A, C) = \mathbf{d}(B', D')$ (see FIGURE 6). Looking at first-order approximations, we conclude:

(i) $(2a_0 + b_0)\varepsilon_a = (2b_0 + a_0)\varepsilon_b$,
(ii) $\varepsilon_a(s_0 + 2a_0 - c) = \varepsilon_c(s_0 + 2c - a)$,
(iii) $\varepsilon_b(s_0 + 2b_0 - d) = \varepsilon_d(s_0 + 2d_0 - b)$.

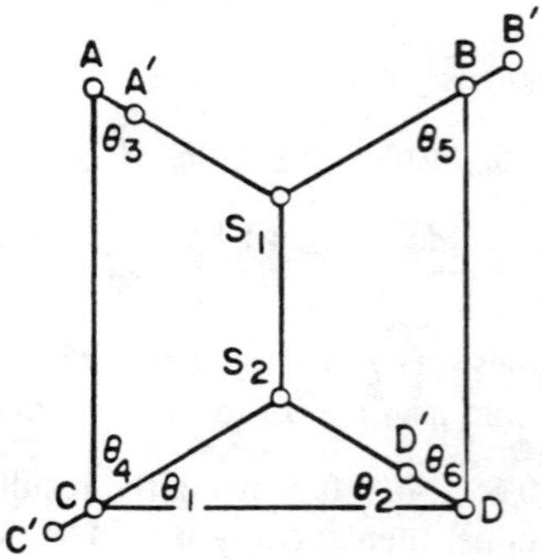

FIGURE 6

Thus

$$\varepsilon_a + \varepsilon_d - \varepsilon_b - \varepsilon_c \geq 3\varepsilon_b\left(\frac{b_0 - d_0}{s_0 + 2d_0 - b_0} + \left(\frac{2b_0 + a_0}{b_0 + 2a_0}\right)\left(\frac{c_0 - a_0}{s_0 + 2c_0 - a_0}\right)\right)$$

$$:= 3\varepsilon_b\, g_9(a_0, b_0, c_0, d_0, s_0).$$

Since $\mathbf{d}(A, C) = \mathbf{d}(B, D)$, we have $b_0 \geq a_0$, $c_0 \geq d_0$. It suffices to consider the following three cases:

Subcase 1. $b_0 \geq a_0 \geq c_0 \geq d_0$.
Here, it is immediate that $g_9(a_0, b_0, c_0, d_0, s_0) \geq 0$.

Subcase 2. $c_0 \geq d_0 \geq b_0 \geq a_0$.
Then

$$\begin{aligned} g_9(a_0, b_0, c_0, d_0, s_0) &\geq \frac{b_0 - d_0}{s_0 + 2d_0 - b_0} + \frac{c_0 - a_0}{s_0 + 2c_0 - a_0} \\ &\geq \frac{(b_0 + c_0 - a_0 - d_0)s_0 + b_0 c_0 - a_0 d_0}{(s_0 + 2d_0 - b_0)(s_0 + 2c_0 - a_0)} \\ &\geq 0. \end{aligned}$$

Subcase 3. $b_0 \geq d_0$, $c_0 \geq a_0$.

Then

$$g_9(a_0, b_0, c_0, d_0, s_0) \geq \frac{(b_0 + c_0 - a_0 - d_0)s_0 + b_0 c_0 - a_0 d_0}{(s_0 + 2d_0 - b_0)(s_0 + 2c_0 - a_0)} \geq 0.$$

Thus we have $\varepsilon_a + \varepsilon_d \geq \varepsilon_b + \varepsilon_c$. It follows that $(a_0 - \varepsilon_a, b_0 + \varepsilon_b, c_0 + \varepsilon_c, d_0 - \varepsilon_d, s)$ satisfies **(15)** since the angle determined by $A'C'$ and $B'D'$ containing S_1 is no more than 90°. Therefore,

$$\begin{aligned} g_8(a_0, b_0, c_0, d_0, s_0) &- g_8(a_0 - \varepsilon_a, b_0 + \varepsilon_b, c_0 + \varepsilon_c, d_0 - \varepsilon_d, s_0) \\ &= \varepsilon_a - \varepsilon_b - \varepsilon_c(1 - \cos\theta_1) + \varepsilon_d(1 - \cos\theta_2) \\ &> (1 - \cos\theta_1)(\varepsilon_a - \varepsilon_b - \varepsilon_c + \varepsilon_d) \geq 0, \end{aligned}$$

where $\theta_1 = \angle S_2CD$ and $\theta_2 = \angle S_2DC$. However, this contradicts the minimality of $(a_0, b_0, c_0, d_0, s_0)$. Thus $(a_0, b_0, c_0, d_0, s_0)$ satisfies the inequalities **(13)**, **(14)**, **(15)**, and **(17)**, with equality. Since we have five variables and four equations, a_0, c_0, d_0, s_0, and g_8 can all be viewed as functions of the single variable b_0. Again we use a brute-force technique to prove $g_8(b) \geq 0$ for $1/\sqrt{3} \leq b \leq 0.76$ as follows:

Step 1. Determine da/db, dc/db, dd/db, ds/db in terms of a, b, c, d, s using **(13)**–**(17)**.

Step 2. Determine dg_8/db in terms of a, b, c, d, s. It turns out that

$$|dg_8/db| \leq 1500.$$

Step 3. Determine d^2g_8/db^2 in terms of a, b, c, d, s. It turns out for this computation that

$$|d^2g_8/db^2| \leq 6 \times 10^8.$$

Step 4. Sample $(dg_8/db)(b)$ at the points $b_i \in (1/\sqrt{3}, 0.58]$ where

$$b_i = 1/\sqrt{3} + i(2.2 \times 10^{-8}), \qquad i = 1, 2, \ldots, 1.3 \times 10^5.$$

All these values of $(dg_8/db)(b)$ happen to be greater than 60.

Step 5. Sample $g_8(b)$ at points b_j in $[0.58, 0.76]$ where

$$b_j = 0.58 + j(6 \times 10^{-5}).$$

All these values of g_8 are greater than 0.1.

Now from Steps 3 and 5 we know that

$$\frac{df_8}{db}(b) \geq \frac{df_8}{db}(b_i) - 6 \times 10^8 \cdot 2.2 \times 10^{-8} > 0$$

for $b \in [1/\sqrt{3}, 0.58]$. Since $g_8(1/\sqrt{3}) = 0$, we have $g_8(b) \geq 0$ for $b \in [1/\sqrt{3}, 0.58]$. From Steps 2 and 5, we have, for b in $[0.58, 0.76]$,

$$g_8(b) \geq g_8(b_i) - 1500 \cdot 6 \times 10^{-5} > 0.$$

This completes the proof of **(5)**. ▮

We point out that the actual value of ρ_0 is obtained by solving the system of equations formed by taking equality in **(13)**, **(14)**, **(15)**, and **(17)** together with the equation $g_8(1/\sqrt{3}) = 0$. We remark in conclusion that it would appear that new ideas will be needed to prove that $\rho \geq \sqrt{3}/2$, although this does not diminish our belief in its truth.

REFERENCES

1. CHUNG, F. R. K. & F. K. HWANG. 1978. A lower bound for the Steiner tree problem. SIAM J. Appl. Math. **34:** 27–36.
2. CHUNG, F. R. K. & E. N. GILBERT. 1976. Steiner trees for the regular simplex. Bull. Inst. Math. Acad. Sinica **4:** 313–325.
3. DU, D. Z., F. K. HWANG & E. Y. YAO. The Steiner ratio conjecture is true for five points. J. Combin. Theory Ser A. In press.
4. GAREY, M. R., R. L. GRAHAM & D. S. JOHNSON. 1977. The complexity of computing Steiner minimal trees. SIAM J. Appl. Math. **32:** 835–859.
5. GAREY, M. R. & D. S. JOHNSON. 1979. Computers and Intractability, a Guide to the Theory of *NP*-Completeness. W. H. Freeman and Co. San Francisco, Calif.
6. GILBERT, E. N. & H. O. POLLAK. 1968. Steiner minimal trees. SIAM J. Appl. Math. **16:** 1–29.
7. GRAHAM, R. L. & F. K. HWANG. 1976. Remarks on Steiner minimal trees, I. Bull. Inst. Math. Acad. Sinica **4:** 177–182.
8. HWANG, F. K. 1976. On Steiner minimal trees with rectilinear distance. SIAM J. Appl. Math. **30:** 104–114.
9. KALLMAN, R. 1973. On a conjecture of Gilbert and Pollak on minimal trees. Stud. Appl. Math. **52:** 141–151.
10. POLLAK, H. O. 1978. Some remarks on the Steiner problem. J. Combin. Theory Ser. A **24**(3): 278–295.
11. DU, D. Z. & F. K. HWANG. 1983. A new bound for the Steiner ratio. Trans. Amer. Math. Soc. **278:** 137–148.

GEOMETRIC SORTING THEORY

Jacob E. Goodman[a]

City College
City University of New York
New York, New York 10031

Richard Pollack[b]

Courant Institute
New York University
New York, New York 10012

Suppose the points $P_1, \ldots, P_n$ are arranged on a straight line, and we are told—for each index i—the cardinality $\lambda(i)$ of the set $\Lambda(i)$ of points which lie to the left of P_i:

$$\lambda(i) = |\Lambda(i)| = |\{j | P_j < P_i\}|.$$

In FIGURE 1a, for example, we have $\lambda(1) = 3$, $\lambda(2) = 0$, etc. Then of course we can immediately deduce what the sets $\Lambda(i)$ themselves are: just take all the points P_j for which $\lambda(j) < \lambda(i)$.

Now suppose $P_1, \ldots, P_n$ no longer lie on a line, but in a plane, and suppose—for each *pair* i, j now—we know how *many* points $\lambda(i, j)$ lie to the left of the directed line P_iP_j, and we would like to know *which* ones ($\Lambda(i, j)$) do. For the configuration in FIGURE 1b, for example, the value of $\lambda(i, j)$ is the i, jth entry of the matrix

$$(\lambda(i, j)) = \begin{pmatrix} 5 & 0 & 3 & 1 & 2 \\ 3 & 5 & 1 & 2 & 0 \\ 0 & 2 & 5 & 1 & 3 \\ 2 & 1 & 2 & 5 & 1 \\ 1 & 3 & 0 & 2 & 5 \end{pmatrix};$$

here the "undefined" values of λ are listed as n (in this case $n = 5$) on the diagonal of the λ-matrix. Now it is not so simple to deduce $\Lambda(i, j)$. But surprisingly enough, it turns out that we can again determine the sets $\Lambda(i, j)$ just from knowing their cardinalities. And in fact we can do this in any dimension d: if we know—for each d-tuple $i_1, \ldots, i_d$—the *number* of points on the positive side of the oriented hyperplane spanned by $P_{i_1}, \ldots, P_{i_d}$, then we can determine the points themselves. This works even if the configuration $\{P_1, \ldots, P_n\}$ is not simple—in other words, if more than two points are sometimes collinear, or more than three sometimes coplanar, and so on—and in fact even if the points $P_1, \ldots, P_n$ are not all distinct.

[a] Research partially supported by National Science Foundation Grant MCS 82-01831.
[b] Research partially supported by National Science Foundation Grant MCS 82-01342.

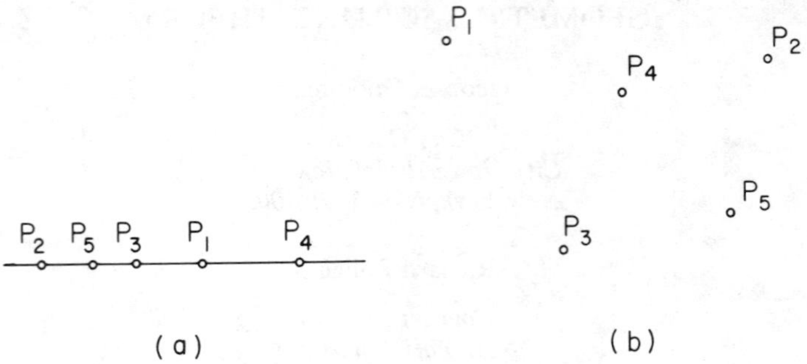

FIGURE 1

THEOREM. *If* $\{P_1, \ldots, P_n\}$ *is a configuration in* $\mathbb{R}^d$ *and, for every ordered d-tuple* $(i_1, \ldots, i_d)$, *we have*

$$\Lambda(i_1, \ldots, i_d) = \{j \mid P_j \text{ is on the positive side of the oriented hyperplane } \langle P_{i_1}, \ldots, P_{i_d} \rangle\}$$

and

$$\lambda(i_1, \ldots, i_d) = |\Lambda(i_1, \ldots, i_d)|,$$

then λ *determines* Λ.

A proof appears in [2] and—in a generalization of the two-dimensional case to "generalized configurations" or oriented matroids of rank 3—in [3]. Here we present the proof for a planar configuration, where the argument is fairly straightforward, at least if we restrict ourselves to the simple case (no collinear triples).

Suppose then that $P_1, \ldots, P_n$ are distinct points, no three collinear, in the plane, and suppose for each i, j we know the number $\lambda(i, j)$ of points to the left of the directed line $P_i P_j$. The first step is to realize that we can immediately tell which points P_i are *extreme* points of the convex hull of the set $\{P_1, \ldots, P_n\}$: they are precisely the ones for which $\lambda(i, j) = n - 2$ for some j.

Now if P_n, say, is an extreme point, and if we swing a ray counterclockwise around P_n, we can tell the order in which it passes through the points $P_1, \ldots, P_{n-1}$ by listing them in the order in which $\lambda(n, i)$ decreases. This means we can tell just *which* points P_j lie to the left of $P_n P_i$, and this means we can tell whether P_n is to the right or left of $P_i P_j$ (see FIGURE 2). That, in turn, means that if we let λ_n be the corresponding λ-function for the set $\{P_1, \ldots, P_{n-1}\}$, we can determine λ_n from λ:

$$\begin{aligned} \lambda_n(i, j) &= \lambda(i, j) - 1 && \text{if } P_n \text{ is to the left of } P_i P_j, \\ &= \lambda(i, j) && \text{otherwise.} \end{aligned}$$

This suggests using induction on n, the theorem being trivial if $n = 3$ (there are only two possible configuration types), so by induction hypothesis we can determine Λ_n,

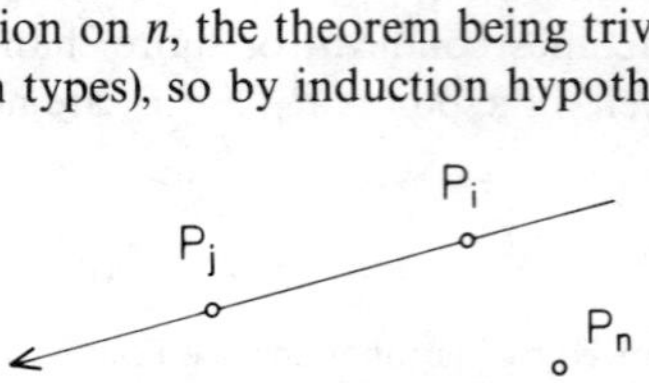

FIGURE 2

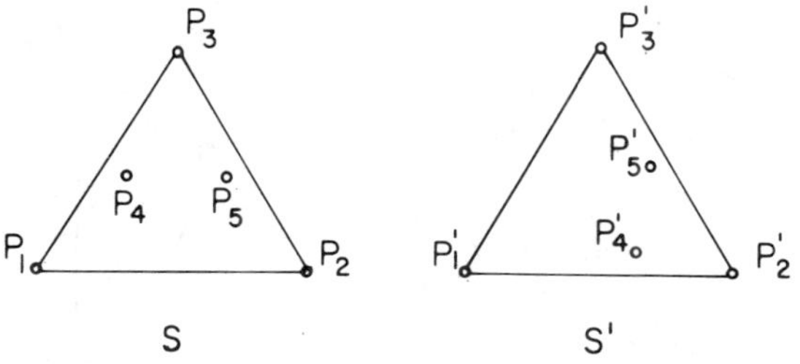

FIGURE 3

the Λ-function for $\{P_1, \ldots, P_{n-1}\}$. But since we know where P_n itself lies with respect to each $P_i P_j$, we can reconstruct the Λ-function for the full configuration:

$$\begin{aligned}\Lambda(i, j) &= \Lambda_n(i, j) \cup \{n\} && \text{if } P_n \text{ is to the left of } P_i P_j, \\ &= \Lambda_n(i, j) && \text{otherwise,}\end{aligned}$$

and that finishes the proof.

Note, by the way, that if we do not distinguish left from right, the theorem becomes false: FIGURE 3 shows two sets of points, S and S', such that each line $P_i P_j$ cuts S into two subsets of the same pair of cardinalities as $P_i' P_j'$ cuts S', but not into corresponding pairs of sets.

Now let us see what this theorem has to do with sorting.

Suppose we have two numbered sets of distinct points on the line: $\{P_1, \ldots, P_n\}$ and $\{Q_1, \ldots, Q_n\}$. When are they in the same order? Clearly when—for every i and j—

$$P_i < P_j \Leftrightarrow Q_i < Q_j.$$

This takes n^2 comparisons to verify, or actually just $n(n-1)/2$; but of course we can do it in much less time by "sorting" the two sets first, i.e., finding the permutations π and π' for which

$$P_{\pi(1)} < P_{\pi(2)} < \cdots < P_{\pi(n)} \qquad \text{and} \qquad Q_{\pi'(1)} < Q_{\pi'(2)} < \cdots < Q_{\pi'(n)},$$

and then just checking whether $\pi(i) = \pi'(i)$ for every i. Once the numbers are sorted, which can be done in time $O(n \log n)$ [4], to see whether they are in the same order then just takes n comparisons instead of n^2. The moral is that sorting a set of numbers encodes all the n^2 pieces of information "for every i, j, $P_i < P_j$ or $>$" into a more compact form of size n, which carries all the information within it.

Now suppose we have two numbered sets of distinct points in the *plane*, $\{P_1, \ldots, P_n\}$ and $\{Q_1, \ldots, Q_n\}$, as in FIGURE 4, and let us assume for simplicity that no three points are collinear in either set. When are *they* in the same order, in some sense? The question is, of course, what do we mean by that? Well, just as the basic unit of order on the line is a pair, $P_i P_j$, which can be oriented either positively or negatively, let us use as our basic unit of order in the plane a *triple*, $P_i P_j P_k$, which—likewise—can be oriented either positively or negatively; here, of course, that would mean either counterclockwise or clockwise. Then for the two sets to have the same "order type" means that for each triple i, j, k,

$$P_i P_j P_k \text{ is counterclockwise} \Leftrightarrow Q_i Q_j Q_k \text{ is counterclockwise.}$$

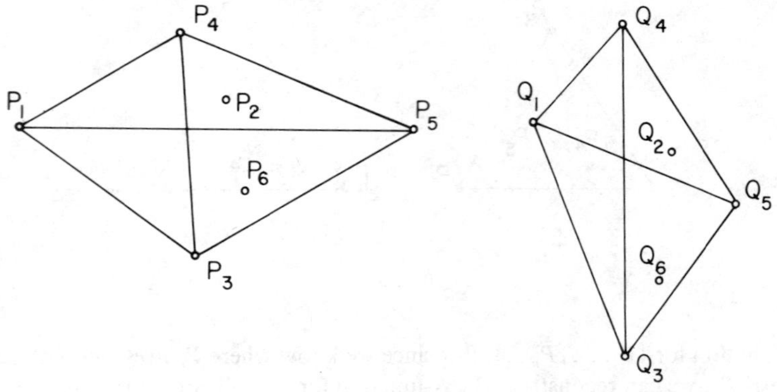

FIGURE 4

Now this would take n^3 [or actually $n(n-1)(n-2)/6$] comparisons to verify; but it follows immediately from the theorem above that it can be done in just n^2 comparisons, as soon as the points are "sorted" in the sense that—for each i, j—we determine just how many lie to the left of the directed line P_iP_j, i.e. determine the λ-matrix, since that in turn determines *which* ones do, i.e. determines the orientation of every triple $P_iP_jP_k$. So this process of *sorting geometrically*, which is the name we give to calculating the λ-matrix, once again compresses all the n^3 pieces of information "$P_iP_jP_k > 0$" or "< 0" into size n^2 (namely, the λ-matrix, which occupies that storage, carries all the n^3 pieces of information within it, by the theorem above). So just as in the linear case, we can save essentially an order of magnitude by sorting first. And the sorting procedure itself—which on the line takes time $O(n \log n)$—can be accomplished in the plane in time $O(n^2 \log n)$. Here is how this is done:

Suppose we have a configuration $S = \{P_1, \ldots, P_n\}$ in the plane. Fix a point P_i. For each $j \neq i$, let $\bar{P}_j$ be the reflection of P_j in P_i, as in FIGURE 5. Let a ray R sweep counterclockwise around P_i. Then R hits the points $\{P_j, \bar{P}_j\}_{j \neq i}$ in some circular order, and we can sort these points according to their polar angles with respect to P_i; for P_3 in FIGURE 5, for example, we get the circular sequence

$$P_3\colon \ldots, \bar{P}_2, P_5, P_6, \bar{P}_4, \bar{P}_1, P_2, \bar{P}_5, \bar{P}_6, P_4, P_1, \bar{P}_2, \ldots.$$

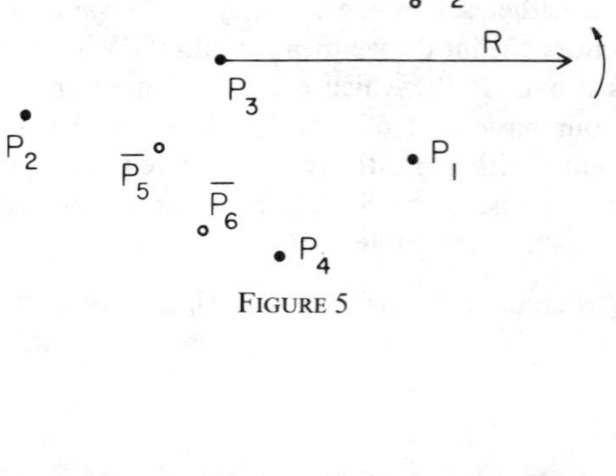

FIGURE 5

This sorting can be done in time $O(n \log n)$, by various methods [4]. We can then determine, for each j, how many original points P_k lie in the counterclockwise direction from P_j around to $\bar{P}_j$; this number will be precisely the number of original points lying to the left of the directed line $P_i P_j$, in other words $\lambda(i, j)$. [In the P_3-sequence, for example, we see that between P_5 and $\bar{P}_5$ there are precisely *two* of the original points, namely, P_6, P_2; hence $\lambda(3, 5) = 2$.] Now do this for each i from 1 to n. Then the resulting time is $O(n \cdot n \log n) = O(n^2 \log n)$.

In higher dimensions, for points in $\mathbb{R}^d$, say, we can do something similar, by rotating a hyperplane around the $d - 2$-flat determined by each set of $d - 1$ points in general position; see [2] for details.

Note, by the way, that if we just wanted the vertices of the convex hull of the set, in counterclockwise order, say, we could find them—by various algorithms [5]—in time $O(n \log n)$; but when we sort the points geometrically, we determine not only the convex hull (which can be read off immediately from the λ-matrix, as in the proof earlier), but the positions of *all* the points in relation to each other. So it is not surprising that it should take an order of magnitude more time. Note also that just as the problem of finding the minimum and maximum of a set S of points on a line becomes, in higher dimensions, the problem of finding the convex hull of a configuration C, so the problem of sorting *all* the points of S becomes the problem of sorting C geometrically, i.e., finding its λ-matrix.

Thus the result is that in the plane we can sort and compare, together, in time $O(n^2 \log n)$; in higher dimensions the exponent of n goes up, just as it does in the convex hull algorithms.

But suppose, now, that we are given two configurations to compare which are *unnumbered*, or only *randomly* numbered, and that we want to determine whether there is some renumbering in which their order types match. (This problem is more likely to come up in "real life" than the problem of matching numbered configurations, for example if we are looking at an application to pattern recognition or classification—see below.) On the face of it, it looks like we would have to try all the $n!$ different renumberings of one of them and change the λ-function correspondingly in order to find a renumbering in which their order types might agree. But it turns out that we can do much better than this—in fact, we can get away with at most k renumberings, where k is the number of extreme points, and so certainly with at most n renumberings. Here is an example:

FIGURE 6 shows two configurations, S and S', which obviously have the same order type, but not with the numbering shown. Sorting each gives the λ-matrices

$$\lambda = \begin{pmatrix} 6 & 4 & 1 & 0 & 3 & 2 \\ 0 & 6 & 4 & 2 & 1 & 3 \\ 3 & 0 & 6 & 4 & 2 & 1 \\ 4 & 2 & 0 & 6 & 3 & 1 \\ 1 & 3 & 2 & 1 & 6 & 3 \\ 2 & 1 & 3 & 3 & 1 & 6 \end{pmatrix}, \qquad \lambda' = \begin{pmatrix} 6 & 3 & 1 & 1 & 3 & 2 \\ 1 & 6 & 2 & 0 & 4 & 3 \\ 3 & 2 & 6 & 3 & 1 & 1 \\ 3 & 4 & 1 & 6 & 2 & 0 \\ 1 & 0 & 3 & 2 & 6 & 4 \\ 2 & 1 & 3 & 4 & 0 & 6 \end{pmatrix}.$$

Choose an extreme point i of S by finding an entry $\lambda(i, j) = 0$; say $i = 1$. Similarly, find *all* the extreme points of S' from λ': they are 2, 5, 6, 4. (Note that if our configurations are not simple, this method of locating extreme points must be modi-

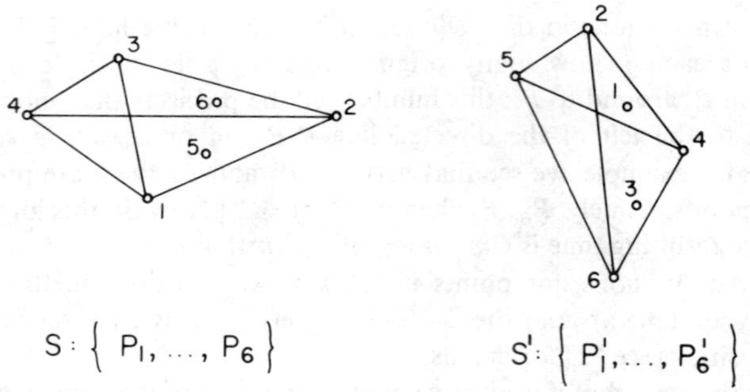

FIGURE 6

fied; see [2].) Next, read off the 1-sequence from λ by reading the indices j such that $\lambda(1, j)$ are in decreasing order: 125634. Do the same for each extreme point of S':

$$256314, \quad 563412, \quad 643125, \quad 421536.$$

Then we have only to try renumbering S according to each permutation

$$\begin{pmatrix}125634\\256314\end{pmatrix}, \begin{pmatrix}125634\\563412\end{pmatrix}, \begin{pmatrix}125634\\643125\end{pmatrix}, \begin{pmatrix}125634\\421536\end{pmatrix},$$

and if there is *any* renumbering which will make the λ'-matrix agree with the λ-matrix, it must be one of these, since the extreme points must agree, in cyclic order, and the sequences for the corresponding extreme points must also agree. We see that only the third of these, call it π, satisfies

$$\lambda(i, j) = \lambda'(\pi(i), \pi(j))$$

for every i, j, which means that if S is renumbered according to π it will have the same order type as S'.

This matrix of $\leq n$ renumberings that must be checked means that the time for comparing two unnumbered (or randomly numbered) configurations is $O(n^3)$, which is considerably better than the $O(n! \cdot n^2)$ we would need if we had to try all the possible renumberings!

[Note also that we can get all the orientation-preserving symmetries (up to order type) in this way—just take $S' = S$.]

In higher dimensions this renumbering procedure generalizes in an interesting way: it turns out to be necessary to consider all the flags of faces of the convex hull of the configuration; each one gives rise to an ordering of the points in a canonical way, and we have to try all of these "canonical orderings" when we renumber. There are roughly $n^{[d/2]}$ in all, in dimension d; for details, see [2].

Now all of this leads to a number of possible applications. For one thing, we now have a powerful tool that we can use in pattern recognition problems, which can easily be implemented on a computer in order to compare and identify point patterns (and therefore many other images, since these can be reduced to point patterns by various techniques). And, in fact, one does not necessarily have to insist that the λ-matrices in question be *identical*, but only that they *correlate* highly, in order to have two point patterns that resemble each other. This might be useful in machine recognition of hand-printed characters, for example, where one would not expect to get a perfect match.

Another application is to what is called stereochemistry, which deals with the structure of molecules in three-dimensional space; it turns out that the right-handedness or left-handedness of certain configurations of atoms within a molecule plays a role in determining the chemical properties of the molecule, and the sorting procedure we have been discussing gives a compact way of encoding all this information. In fact, recently a chemist, A. Dreiding, together with a mathematician, K. Wirth, came up with a scheme for encoding precisely the same information [1]: their idea was to list all the ordered quadruples of points in a three-dimensional configuration in lexicographic order, and write a "1" for a right-handed one and a "0" for a left-handed one, thereby getting the binary representation of an integer. The result, which they call the "signature" of the configuration, has $\Omega(n^4)$ bits and hence requires a calculation time of $\Omega(n^4)$, while geometric sorting in 3-space, as a result of the theorem discussed earlier, carries the same information but takes only $O(n^3 \log n)$; so geometric sorting constitutes an improvement by a factor of $n/\log n$.

One more application we can mention is to cluster analysis, a branch of statistics which deals with the identification of "clusters" of points in d-dimensional space. There are many ways of getting hold of these clusters, but one technique consists of repeatedly splitting off what are called semispaces, that is, subconfigurations which are separated from the remaining points by hyperplanes. It turns out [3] that two numbered configurations have the same semispaces if and only if they have precisely the same order type in our sense; this means that our sorting methods give an efficient way of classifying point sets from the point of view of cluster analysis. In fact, we can easily read off all the semispaces of a configuration during the process of sorting it; see [2] for further remarks.

Let us conclude by mentioning an interesting problem that comes up in relation to this work; it has to do with the question of the optimality of our sorting procedure. The problem is: how many distinct configurations of n points are there in $\mathbb{R}^d$? Here, "distinct" means having distinct order types, in other words distinct λ-functions. This problem is unsolved even in the plane. An upper bound of $\exp(cn^2 \log n)$ follows immediately from our sorting theory for what are called information-theoretic reasons—the storage needed for a λ-matrix being $O(n^2 \log n)$ bits, there cannot be more than $\exp(cn^2 \log n)$ distinct λ-matrices. On the other hand, there are certainly more than $n!$ [$\sim \exp(cn \log n)$ by Stirling's formula] distinct configurations of n points in $\mathbb{R}^2$; for $n = 5$, for example, there are already 264. So the answer lies somewhere between $\exp(cn \log n)$ and $\exp(cn^2 \log n)$. We can show, however, that there are at least $\exp(cn^2)$ *generalized* configurations in the plane [2], a generalized configuration being a configuration in which each pair of points is joined not by a straight line, but by a "pseudoline," or topological line, any two of which meet just once and cross there. (Generalized configurations in the plane are essentially equivalent to oriented matroids of rank 3.) This therefore narrows the gap between the lower and upper bounds in the case of generalized configurations, since λ-matrices also encode their order types, but the gap remains wide for ordinary configurations.

[**Note added in proof:** Since the writing of this article, H. Edelsbrunner, J. O'Rourke, and R. Seidel have found an $O(n^2)$ algorithm for geometric sorting in the plane, which is clearly optimal; their method extends to higher dimensions. (Constructing arrangements of lines and hyperplanes, with applications. *In* Twenty-Fourth Annual FOCS Symposium (1983), 83–91.)]

References

1. Dreiding, A. S. & K. Wirth. 1980. The multiplex—a classification of finite ordered point sets in oriented *d*-dimensional spaces. Match **8:** 341–352.
2. Goodman, J. E. & R. Pollack. 1983. Multidimensional sorting. SIAM J. Comput. **12:** 484–507.
3. Goodman, J. E. & R. Pollack. 1984. Semispaces of configurations, cell complexes of arrangements. J. Combin. Theory Ser. A **37:** 257–293.
4. Knuth, D. E. 1973. The Art of Computer Programming, Vol. 3, Sorting and Searching. Addison–Wesley. Reading, Mass.
5. Shamos, M. I. 1978. Computational Geometry. Ph.D. Thesis. Yale University. New Haven, Conn.

LEAST SQUARES FITTING PROCEDURES FOR PROBLEMS INVOLVING MORE THAN ONE HYPERPLANE

A. C. Hayes[a] and D. G. Larman

Department of Mathematics
University College London
London WC1E 6BT, England

1. Introduction

Variants of the well-known procedure for the fitting of a hyperplane to a set of points in Euclidean space by minimizing the sum of vertical least squares distances are discussed. In each case, the problem requires the fitting of two or more hyperplanes simultaneously and methods are developed having, for fixed dimension, polynomial complexity in the number of data points considered.

We shall call a finite set P of points in $(d + 1)$-dimensional real Euclidean space $\mathbb{R}^{d+1}$ $(d \geq 1)$ a data set. Regarding $\mathbb{R}^d$ as being embedded in $\mathbb{R}^{d+1}$ in the natural way, we shall frequently write elements of P in the form (x, y), where $x \in \mathbb{R}^d$ and $y \in \mathbb{R}$.

It is common in regression analysis to have a linear model of dependence of a variable y on variables $x_1, x_2, \ldots, x_d$. Given a data set P it is then required to find a hyperplane modeling the dependence of y on the independent variables $x_1, x_2, \ldots, x_d$ which is, in some sense, the "best" possible. The usual method employed is that of vertical least squares. Then, given a data set P, the one hyperplane least squares (1HLS) problem is to find a hyperplane $h(x) = a_0 + a_1x_1 + \cdots + a_dx_d$ so that the sum $\sum_{(x, y) \in P} (h(x) - y)^2$ is minimized. This is a standard problem and an optimal solution can be found by solving a set of linear simultaneous equations (the so-called normal equations). The reader is referred to [1] for details. For a fixed dimension d, the complexity of such a procedure is $O(n)$, where throughout this paper n shall denote the number of points of P.

A variation of this fitting procedure is the two hyperplane least squares (2HLS) problem where after a suitable assignment of points to hyperplanes we find two hyperplanes minimizing the sum of squared vertical distances. One similarly envisages a k hyperplane least squares (kHLS) problem.

Another variant is the bent hyperplane least squares (BHLS) problem. Here a d-hyperplane (in $\mathbb{R}^{d+1}$) is "hinged" along some $(d - 1)$-plane it contains and the sum of squared vertical distances to P is minimized. We formally define the kHLS $(k \geq 2)$ and BHLS problems in the sections to follow.

In the case of the plane $\mathbb{R}^2$, both the 2HLS and BHLS problems were studied in [2]. Algorithms of polynomial time complexity in n were discussed and the authors conjectured that their results could be extended to higher dimensions and to arbitrary numbers of hyperplanes. We carry out such extensions in this paper. We refer

[a] The research by this author was carried out while he was in receipt of an S.E.R.C. research studentship.

the reader to [2] for the motivations and applications of the fitting procedures described and also for a short bibliography of work in this field.

In the final section we mention an interesting extension of the BHLS problem.

2. The Two Hyperplane Least Squares Problem

We shall suppose that P is a data set in $\mathbb{R}^{d+1}$ having n data points. We begin with a formal statement of the problem:

Find a partition (Q_1, Q_2) of P into subsets Q_1, Q_2 (one possibly empty) and find hyperplanes h_1, h_2 where

$$h_j(x) = a_{0j} + \sum_{i=1}^{d} a_{ij} x_i, \qquad j = 1, 2,$$

such that the sum

$$\sum_{j=1}^{2} \sum_{(x,\, y) \in Q_j} (h_j(x) - y)^2 \tag{1}$$

is minimized over all possible choices of h_1, h_2, Q_1, Q_2.

We shall extend the ideas developed in [2] in order to give an algorithm of polynomial time complexity. Two observations can be made concerning an optimal solution $(h_1, Q_1; h_2, Q_2)$ for (1). These are:

(i) h_1, h_2 are the ordinary 1HLS solutions for Q_1, Q_2, respectively, and

(ii) the set $Q_1 \subseteq P$ can be chosen to satisfy

$$Q_1 = \{(x, y) \in P: |h_1(x) - y| \leq |h_2(x) - y|\}. \tag{2}$$

The naive approach to the problem, suggested by (i), is simply to compute all possible partitions (Q_1, Q_2). Then for each we solve the 1HLS problem for Q_1 and Q_2 and select an optimal solution. However, this procedure has exponential time complexity with respect to the number of data points.

We therefore require a method that avoids consideration of a large number of feasible solutions. The idea for such a method is found in [2].

For an optimal solution $(h_1, Q_1; h_2, Q_2)$ let Q_0 denote the set of points in $\mathbb{R}^{d+1}$ for which equality holds in (2). After rearrangement we find that

$$\begin{aligned} Q_0 &= \{(x, y): |h_1(x) - y| = |h_2(x) - y|\} \\ &= (L(h_1, h_2) \times \mathbb{R}) \cup \{(x, h_0(x))\}, \end{aligned} \tag{3}$$

where $h_0(c) = \frac{1}{2}(h_1(x) + h_2(x))$ and, in general, $L(h_1, h_2)$ is a $(d-1)$-plane in $\mathbb{R}^d \subset \mathbb{R}^{d+1}$ so that $L(h_1, h_2) \times \mathbb{R}$ is a "vertical" hyperplane with normal in $\mathbb{R}^d$. More precisely, $L(h_1, h_2)$ is the orthogonal projection into $\mathbb{R}^d$ of $h_1 \cap h_2$; that is, $L(h_1, h_2) = \{x \mid h_1(x) = h_2(x)\}$. A special case arises when h_1, h_2 are parallel, so that $L(h_1, h_2)$ is empty. However, in the algorithm to come, an appropriate choice of the degenerate partition $(\varnothing, P)$ of P allows the possibility of a parallel optimal solution, if one exists, to be found. Thus, in what follows, we shall suppose h_1, h_2 are not parallel; that is, $L(h_1, h_2) \neq \varnothing$.

Let S be an s-dimensional affine plane containing an $(s - 1)$-dimensional affine plane T. We say $T \subset S$ is orientated in S if the closed half spaces generated by T relative to S are unambiguously defined. We denote them by T^+ and T^-. Then from (**2**), (**3**) we rewrite (**2**) as

$$Q_1 = \{(x, y) \in P: x \in L^-, y \leq h_0(x)\} \cup \{(x, y) \in P: x \in L^+, y \geq h_0(x)\} \qquad \textbf{(4)}$$

where $L = L(h_1, h_2)$ is appropriately orientated in $\mathbb{R}^d$.

From the above discussion we see that to find an optimal solution minimizing (**1**), we need only consider feasible solutions in which the partition (Q_1, Q_2) is defined by a "vertical" hyperplane and another, arbitrary hyperplane. In general, this divides $\mathbb{R}^{d+1}$ into four sections and Q_1 lies in two sections meeting only in the intersection of the defining hyperplanes. In $\mathbb{R}^2$, the situation is shown in FIGURE 1.

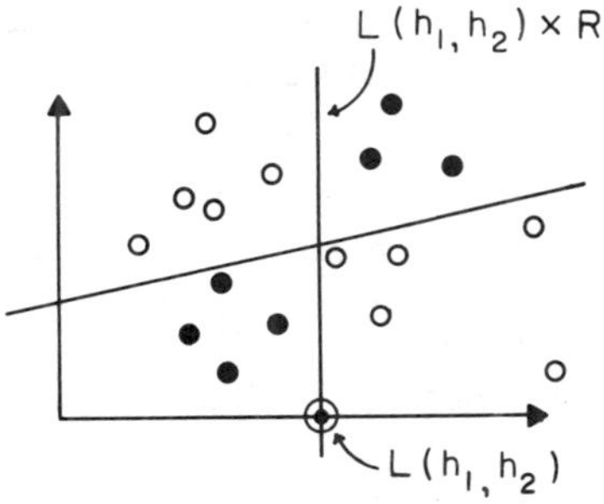

FIGURE 1

For each of the partitions (Q_1, Q_2) generated, we solve the 1HLS problem for Q_1 and Q_2 to obtain h_1, h_2, respectively. An optimal solution is then found by selecting from these feasible solutions one which minimizes (**1**).

In order to show that this method of generating an optimal solution has polynomial time complexity with respect to n, we need to show that the number of such partitions is polynomially bounded, and we also need to show how to find them.

To do this, we shall make the following three assumptions. First, we shall suppose that $n > d + 1$. Second, we shall suppose that no subset of $d + 2$ elements of P lies in any d-hyperplane. From this we may deduce that, for $0 \leq s \leq d$, no subset of $s + 2$ points of P lies in any s-plane. Third, we shall suppose that no two points of P both orthogonally project into the same point in $\mathbb{R}^d$. That is, if $(x_1, y_1) \neq (x_2, y_2)$, then $x_1 \neq x_2$.

Let m denote the greatest integer less than, or equal to, $\frac{1}{2}(d + 1)$. For $0 \leq s \leq m$, let $\omega(n, s)$ denote the set of all 3-tuples of the form $(\alpha_1, \alpha_2, \tau)$, where α_1, α_2 are subsets of P and such that

(i) α_1 has s elements,
(ii) $\alpha_1 \cap \alpha_2 = \varnothing$,
(iii) $\alpha_1 \cup \alpha_2$ has $d + 1$ elements, and
(iv) $\tau \in \{+, -\}$.

Define

$$\Omega(n, d + 1) = \bigcup_{s=0}^{m} \omega(n, s).$$

We now show how elements of $\Omega(n, d+1)$ give rise to partitions of P, separated by a hyperplane. First, for each hyperplane generated by $d+1$ elements of P, we assign an orientation. Let $(\alpha_1, \alpha_2, \tau) \in \Omega(n, d+1)$. Define a hyperplane $H = \text{aff}(\alpha_1 \cup \alpha_2)$. We generate a partition (Q_1, Q_2) of P by putting

$$\begin{aligned} Q_1 &= \alpha_1 \cup (\text{int } H^{\tau} \cap P), \\ Q_2 &= \alpha_2 \cup (\text{int } H^{-\tau} \cap P) \\ &= P \backslash Q_1, \end{aligned}$$

where int denotes interior.

We note that it is quite possible for two distinct elements of $\Omega(n, d+1)$ to give rise to the same partition of P. Let M denote this method of generating partitions from $\Omega(n, d+1)$. We show:

THEOREM. *There is an injection of the set of partitions (Q_1, Q_2) of P by an arbitrary hyperplane into the set of partitions generated by M.*

Proof. Let (Q_1, Q_2) be the partition of P, separated by a hyperplane. Let CQ_i denote the convex hull of Q_i. It is clear that there is a hyperplane H separating Q_1 and Q_2 and containing at least one point of each. In particular, CQ_1, CQ_2 are in opposite half spaces of H. Since CQ_1, CQ_2 are polytopes, any support hyperplane of CQ_i meets CQ_i in a face of dimension $t \geq 0$. We recall that a t-dimensional face is the convex hull of its vertices. Hence, any t-face of CQ_i is the convex hull of $t+1$ points of Q_i.

Suppose that H supports CQ_1 and CQ_2 in faces f_1, f_2, respectively, each of dimension at least zero. Suppose that $\dim \text{aff}(f_1 \cup f_2) < d$. Then, by a suitable rotation of H which keeps fixed $\text{aff}(f_1 \cup f_2) \subset H$ we can find a hyperplane H' containing $\text{aff}(f_1 \cup f_2)$ and supporting CQ_1 and CQ_2 in faces f'_1, f'_2 and with the property that $f'_1 \supseteq f_1, f'_1 \supseteq f_2$ with at least one of the inclusions holding strictly. Hence $\dim \text{aff}(f'_1 \cup f'_2) > \dim \text{aff}(f_1 \cup f_2)$. After at most d repetitions of the process we will have produced a hyperplane $H = \text{aff}(f_1 \cup f_2)$ supporting CQ_1, CQ_2 in faces f_1, f_2 each of dimension at least zero. By our remarks concerning points of Q_i in faces of CQ_i, we conclude that there is a subset $\alpha_0 \subset P$ containing $d+1$ points such that $H = \text{aff}(\alpha_0)$ and that there are (nonempty) subsets α_1, α_2 of α_0 with $f_i = \text{conv}(\alpha_i)$.

Recalling that H has an orientation, then for a suitable choice of $\tau \in \{+, -\}$ we have

$$\begin{aligned} Q_1 &= \alpha_1 \cup (\text{int } H^{\tau} \cap P), \\ Q_2 &= \alpha_2 \cup (\text{int } H^{-\tau} \cap P) \\ &= P \backslash Q_1. \end{aligned}$$

Thus every partition (Q_1, Q_2) of P, given by a hyperplane, can be seen to arise as a partition generated by an element of $\Omega(n, d+1)$; that is, by M. ▌

We note that since each $\omega(n, s)$ has $2\binom{n}{d+1}\binom{d+1}{s}$ elements, $\Omega(n, d+1)$ has

$$2 \sum_{s=0}^{m} \binom{n}{d+1}\binom{d+1}{s} = O(n^{d+1})$$

elements.

We now describe an algorithm to solve the 2HLS problem in $O(n^{2d+2})$ time. We recall that we need to consider partitions of P by an arbitrary hyperplane and a "vertical" hyperplane. We note that separation by a vertical hyperplane is equivalent to separating $P' = \{x: (x, y) \in P\}$ by a $(d-1)$-plane in $\mathbb{R}^d$. By the above theorem, we know that there are at most $O(n^d)$ such partitions. Also, there are at most $O(n^{d+1})$ partitions arising from the arbitrary hyperplane. Thus the total number of feasible partitions of P is $O(n^{2d+1})$.

For each such partition, we must solve the 1HLS problem twice. As remarked in the Introduction, this can be accomplished in $O(n)$ time. Combining these facts, we see that the entire algorithm runs in $O(n^{2d+2})$ time.

It is interesting to note that the algorithm in [2] for the case of the plane $\mathbb{R}^2$ has a running time of $O(n^3)$, whereas ours above is $O(n^4)$. This is because it is possible to utilize the total ordering of the real line to enable the solution of the 1HLS problems in constant time. The reader is referred to [2] for the details.

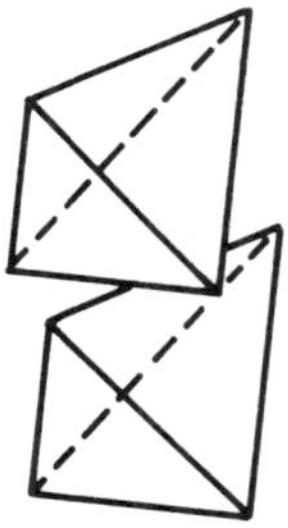

FIGURE 2

Finally, it may be of interest to see why such an elaborate method for generating the partitions is necessary. The natural method of extracting all subsets of $d+1$ elements of P and generating the hyperplanes they span does not give all possible partitions. An example is shown in FIGURE 2. Taking Q_1 as the vertex set of the upper simplex and Q_2 as the vertex set of the lower simplex, we see that the partition (Q_1, Q_2) is not separated by any hyperplane which is the affine hull of three points of $Q_1 \cup Q_2$. Here, and elsewhere, we shall say two sets A and B are separated by a hyperplane H if A and B are contained in opposite closed half spaces of H.

3. THE k HYPERPLANE LEAST SQUARES PROBLEM

The kHLS problem is a generalization of the 2HLS problem. It is stated thus:

Find a partition $(Q_1, Q_2, \ldots, Q_k)$ of P into subsets $Q_1, Q_2, \ldots, Q_k$ $(k \geq 3)$ and find hyperplanes $h_1, h_2, \ldots, h_k$ where

$$h_j(x) = a_{0j} + \sum_{i=1}^{d} a_{ij} x_i, \qquad j = 1, 2, \ldots, k,$$

such that the sum

$$\sum_{j=1}^{k} \sum_{(x, y) \in Q_j} (h_j(x) - y)^2 \tag{5}$$

is minimized over all possible choices of $h_i, Q_i, 1 \leq i \leq k$.

The method employed to provide an algorithm is a development of that in the previous section. Consequently, we shall only outline the principal points. As in Section 2, we can make two observations about an optimal solution $(h_1, Q_1; \ldots; h_k, Q_k)$:

(i) h_1 is the 1HLS solution for Q_1 and $(h_2, Q_2; \ldots; h_k, Q_k)$ is the $(k-1)$HLS solution for $P - Q_1$,

(ii) $Q_1 \subseteq P$ can be chosen to satisfy

$$Q_1 = \left\{(x, y) \in P \colon |h_1(x) - y| \le \min_{2 \le j \le k} |h_j(x) - y|\right\}$$

$$= \bigcap_{j=2}^{k} \{(x, y) \in P \colon |h_1(x) - y| \le |h_j(x) - y|\}.$$

Let Q_{0j} denote the set of $(x, y) \in \mathbb{R}^{d+1}$ for which $|h_1(x) - y| = |h_j(x) - y|$. By exact analogy with (**3**) we can write Q_{0j} in the form

$$Q_{0j} = (L_j \times \mathbb{R}) \cup \{(x, h_{0j}(x))\},$$

where L_j is the $(d-1)$-plane in $\mathbb{R}^d$ which is the orthogonal projection into $\mathbb{R}^d$ of $h_1 \cap h_j$, and $h_{0j}(x) = \frac{1}{2}(h_1(x) + h_j(x))$, $2 \le j \le k$.

We can therefore see that Q_1 will be the intersection of sets of the form of (**4**). That is, if, for $j = 2, \ldots, k$,

$$B_j = \{(x, y) \in P \colon x \in (L_j)^-, y \le h_{0j}(x)\} \cup \{(x, y) \in P \colon x \in (L_j)^+, y \ge h_{0j}(x)\},$$

then

$$Q_1 = \bigcap_{j=2}^{k} B_j. \tag{6}$$

From the analysis of the 2HLS problem we know that there are at most $O(n^{2d+1})$ possibilities for each B_j and hence there are, from (**6**) and induction, at most $O(n^{(2d+1)(k-1)})$ possibilities for Q_1. For each Q_1 so generated we solve the $(k-1)$HLS problem for $P - Q_1$ and the 1HLS problem for Q_1. By using the 2HLS result as a base for the induction, we conclude that we can solve the kHLS problem in $O(n^\beta)$ time, where $\beta = \frac{1}{2}(2d+1)k(k-1) + (k-1)$.

4. The Bent Hyperplane Least Squares Problem

Given a data set P, the BHLS problem is formulated thus:

Find hyperplanes h_1, h_2 with $h_j(x) = a_{0j} + a_{1j}x_1 + \cdots + a_{dj}x_d$ and a partition (Q_1, Q_2) of P so that the sum

$$S(h_1, h_2) = \sum_{(x,y) \in Q_1} (h_1(x) - y)^2 + \sum_{(x,y) \in Q_2} (h_2(x) - y)^2 \tag{7}$$

is minimized over all possible choices of h_1, h_2, Q_1, Q_2 and subject to the condition that if $Q'_j = \{x \mid (x, y) \in Q_j\}$, $j = 1, 2$, then $L(h_1, h_2)$ separates Q'_1, Q'_2.

For some such pair (h_1, h_2) an optimal bent hyperplane $h^*(h_1, h_2)$ is then defined by

$$h^*(x) = h_j(x), \qquad j = 1, 2,$$

according as x is on the same side of $L(h_1, h_2)$ as Q_j. (h^* has the appearance of a hyperplane hinged along $h_1 \cap h_2$.)

In [2] an algorithm is given for the solution of the problem in the case of the plane $\mathbb{R}^2$. We extend the method of [2] to higher dimensions.

We begin by noting that since $L(h_1, h_2)$ is a hyperplane in $\mathbb{R}^d$ separating (Q'_1, Q'_2), we may use the method M of Section 2 to generate all possible partitions of $P' = \{x: (x, y) \in P\}$ to which we can fit a bent hyperplane. We recall that there are at most $O(n^d)$ such partitions. We also recall that we can generate these partitions by means of elements of a set Ω' of 3-tuples of the form $(\alpha'_1, \alpha'_2, \tau)$, where $\alpha'_1, \alpha'_2 \subset P'$ and $\tau \in \{+, -\}$. We note that if $x \in P'$ and (Q'_1, Q'_2) is a separation of P' by a hyperplane in $\mathbb{R}^d$, then there is a separation of P' into (Q'_1, Q'_2) by a hyperplane in $\mathbb{R}^d$ through x if, and only if, x appears as an element of $\alpha'_1 \cup \alpha'_2$ for some $(\alpha'_1, \alpha'_2, \tau) \in \Omega'$. Similarly, if $\{x_1, x_2, \ldots, x_s\} \subseteq P'$, $1 \leq s \leq d$, then there is a hyperplane in $\mathbb{R}^d$ containing $\{x_1, x_2, \ldots, x_s\}$ separating Q'_1, Q'_2 if, and only if, $\{x_1, x_2, \ldots, x_s\} \subset \alpha'_1 \cup \alpha'_2$ for some $(\alpha'_1, \alpha'_2, \tau) \in \Omega'$. We shall use these facts to solve the BHLS problem as we generate separations of P'.

Suppose we have generated a partition (Q'_1, Q'_2) of P'. If h_1, h_2 are the 1HLS solutions for (Q_1, Q_2) and $L(h_1, h_2)$ preserves (weakly) the separation (Q'_1, Q'_2), then $h^*(h_1, h_2)$ is an optimal solution for the partition (Q_1, Q_2). If $L(h_1, h_2)$ does not preserve (Q'_1, Q'_2), then we can find a pair of hyperplanes h'_1, h'_2 so that $L(h'_1, h'_2)$ preserves the separation (Q'_1, Q'_2), contains some element x of P', and which is optimal for (Q_1, Q_2). To see this, suppose (h''_1, h''_2) is any feasible pair with $L(h''_1, h''_2)$ preserving (Q'_1, Q'_2) and not containing any $x \in P'$. Since (7) is a convex function in (h_1, h_2) for a fixed (Q_1, Q_2), then for a suitable choice of λ with $0 < \lambda < 1$ and with $h'_j = \lambda h_j + (1 - \lambda)h''_j$, we have that $L(h'_2, h'_1)$ will contain a point $x \in P'$ and $S(h'_1, h'_2) \leq S(h''_1, h''_2)$. For as λ increases from 0 to 1, $L(h''_1, h''_2)$ will transform to $L(h_1, h_2)$. Although the transformation may not be continuous (for example, if the hyperplanes become parallel), it is straightforward to show that in all cases for a suitable λ with $0 < \lambda < 1$ the hyperplane $L(h'_1, h'_2)$ will meet a point $x \in P'$ while still preserving the partition (Q'_1, Q'_2).

Suppose we now fix some point $p \in P'$ through which we know there passes a hyperplane in $\mathbb{R}^d$ separating (Q'_1, Q'_2). Then we may minimize (7) subject to the additional constraint that $p \in L(h_1, h_2)$ [that is, $h_1(p) = h_2(p)$]. If $L(h_1, h_2)$ preserves (Q'_1, Q'_2) we are done. If not, we may repeat the preceding argument to show that there is an optimal solution for (Q_1, Q_2) containing two points of P'. Proceeding inductively at each stage we see that if (h_1, h_2) is a solution minimizing S subject to the condition that $L(h_1, h_2)$ contains $\{p_1, \ldots, p_s\} \subset P'$ $(0 \leq s \leq n - 1)$ and such that $L(h_1, h_2)$ does not preserve (Q'_1, Q'_2), then there is an optimal solution for (Q_1, Q_2) containing an additional point $p_{s+1} \in P'$, $p_{s+1} \neq p_i$, $1 \leq i \leq s$.

It is clear that if we find a solution (h_1, h_2) subject to the constraint that $L(h_1, h_2)$ contains a fixed subset of d elements of P' with $L(h_1, h_2)$ preserving (Q_1, Q_2), then, since $L(h_1, h_2)$ is totally determined by these d points, the solution (h_1, h_2) minimizing S so obtained will automatically preserve (Q'_1, Q'_2).

We are now in a position to describe the algorithm. For each element of Ω' we shall generate a feasible solution for the problem which is optimal for the given element of Ω'. We shall then take the best of these solutions as the optimal solution.

Let $z = (\alpha'_1, \alpha'_2, \tau) \in \Omega'$. We calculate the partition (Q'_1, Q'_2) generated by z. We now solve the 1HLS problem for (Q_1, Q_2) to obtain a pair (h_1, h_2). If $L(h_2, h_2)$

preserves (Q'_1, Q'_2) we are done. If not, then for each $p \in \alpha'_1 \cup \alpha'_2$ we minimize S subject to $p \in L(h_1, h_2)$ for the minimal pair (h_1, h_2). We discard those which do not preserve (Q'_1, Q'_2). We next minimize S subject to $L(h_1, h_2)$ containing two elements of $\alpha'_1 \cup \alpha'_2$. We may proceed inductively for subsets of $\alpha'_1 \cup \alpha'_2$ having s elements, $1 \leq s \leq d$. Since, when $s = d$, we must obtain solutions which preserve (Q'_1, Q'_2), we will have at least one solution for z which preserves (Q'_1, Q'_2). We pick a solution h^*_z among those preserving (Q'_1, Q'_2) which is minimal. We may repeat this for each $z \in \Omega'$.

It is now a consequence of the previous discussion concerning subsets of P' being contained in a hyperplane separating Q'_1, Q'_2 if, and only if, they arise in an element of Ω', that we see an optimal solution for the BHLS problem will be a member of $\{h^*_z\} z \in \Omega'$ which minimizes S.

With regard to the complexity of these calculations, we first note that Ω' has $O(n^d)$ elements each of which is generated in constant time. For each member $z \in \Omega'$ we solve $\sum_{s=0}^{d} \binom{d}{s}$ subproblems, which, for fixed d, involves a constant number of operations. Each subproblem involves minimizing (7) subject to $L(h_1, h_2)$ containing some subset of P. This is a convex quadratic programming problem and for fixed d involves a constant number of operations. Each coefficient can be calculated in $O(n)$ time. Thus for each $z \in \Omega'$ we can find a minimal solution h^*_z in $O(n)$ time. Combining these facts we see the entire procedure can be carried out in $O(n^{d+1})$ steps.

Comparing with [2], where the algorithm requires $O(n \log n)$ steps in $\mathbb{R}^2$, we see our algorithm will require $O(n^2)$ steps. This improvement in the case of the plane is attributable to the ordering of the real line which implies the generation of the partitions is equivalent to the ordering of the data set which is of complexity $O(n \log n)$.

We would like to thank the referees for the following observation. Since a solution in which $L(h_1, h_2)$ is forced to go through some subset T of $0 \leq t \leq d$ points of P' cannot be better than a solution in which $L(h_1, h_2)$ is forced to go through any subset of these points, it is only required to solve the problem with $L(h_1, h_2)$ forced to contain T if no subset of T has yielded a separation preserving solution. In a practical implementation of the algorithm, this may enable a considerable savings in the number of calculations required.

5. A Segmented Hyperplane Problem

In this section we mention briefly a fitting problem which seems to be a fairly natural extension of the BHLS problem. For the sake of simplicity we define the problem only in the case of $\mathbb{R}^3$. We require two preliminary definitions.

Definition 1. We say that (R_1, R_2, R_3, q) is a feasible splitting of $\mathbb{R}^2$ if

(i) R_1, R_2 are closed convex cones with common apex $q \in \mathbb{R}^2$ and having intersection precisely a common boundary ray,

(ii) $R_3 = \text{closure}\,[\mathbb{R}^2 - (R_1 \cup R_2)]$.

Definition 2. A segmented hyperplane ρ in $\mathbb{R}^3$ is a set of hyperplanes h_1, h_2, h_3 and a feasible splitting (R_1, R_2, R_3, q) of $\mathbb{R}^2 \subset \mathbb{R}^3$ so that

$$\rho(x) = h_j(x) \qquad \text{if, and only if, } x \in R_j,\ 1 \leq j \leq 3.$$

DEFINITIONS 1 and 2 taken together imply that ρ is continuous.

The segmented hyperplane least squares (SHLS) problem may now be stated thus:

Given $P \subset \mathbb{R}^3$, find a partition (Q_1, Q_2, Q_3) of P and a segmented hyperplane $\rho = \rho(h_1, h_2, h_3, R_1, R_2, R_3, q)$ so that the sum

$$\sum_{j=1}^{3} \sum_{(x, y) \in Q_j} (h_j(x) - y)^2$$

is minimized over all choices of $h_1, h_2, h_3, Q_1, Q_2, Q_3$ subject to the condition that $Q'_j \subset R_j$, $1 \leq j \leq 3$. (Recall that Q'_j is the projection into $\mathbb{R}^2$ of Q_j.)

Despite a seemingly complicated statement, the geometric idea behind the SHLS problem is simple. Given P, we are fitting a continuous surface comprising (in general) an apex and three hyperplane segments to minimize the sum of vertical least squares distances. The situation is illustrated in FIGURE 3.

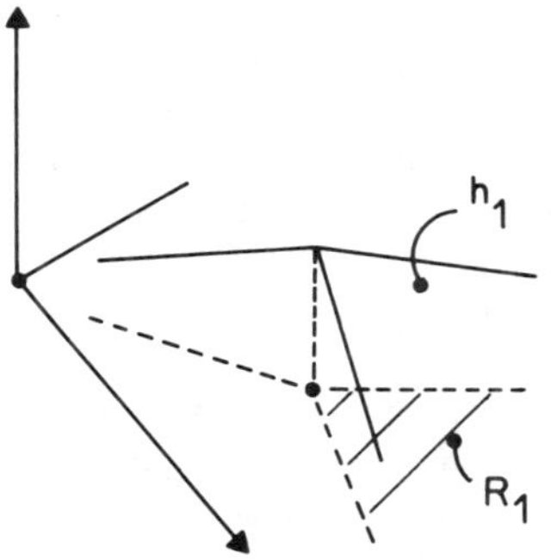

FIGURE 3

It is easy to generalize DEFINITIONS 1 and 2 to higher dimensions, and state the SHLS problem in this case. Certainly in the case of $\mathbb{R}^3$ methods similar to those developed for solving the BHLS problem yield an algorithm for the solution of the SHLS problem which has a complexity polynomial in n. The arguments involved are more intricate than those for the BHLS problem, although essentially the same. Details of the algorithm in $\mathbb{R}^3$ will appear in the thesis of the first author (A.C.H.). It seems reasonable to expect that these methods will extend to higher dimensions.

REFERENCES

1. BEAUMONT, G. P. Elementary Mathematical Statistics. Chapter 16. McGraw–Hill Book Co., Inc. New York, N.Y.
2. LENSTRA, A. K., J. K. LENSTRA, A. H. G. RINNOOY KAN & T. J. WANSBEEK. Two lines least squares. Math. Centrum. Preprint, 1981.

AN UPDATE ON THE ELLIPSOID ALGORITHM

P. F. Pickel

Polytechnic Institute of New York
Farmingdale, New York 11735

1. Introduction

The basic idea underlying the ellipsoid algorithm was formulated by Shor [28] (not in terms of ellipsoids) as an approach to convex nonlinear programming problems. Later, Iudin and Nemirovskii [18] expressed the method in terms of ellipsoids for their study of the computational complexity of convex optimization. Khachian's work on the computational complexity of linear programming [20] and the ensuing publicity [23, 30] focused attention, somewhat inappropriately, on the ellipsoid method as an algorithm for solving linear programming problems. When it became clear that the ellipsoid method could not compete with the simplex algorithm on linear programming problems, interest in the ellipsoid method waned. With the focus on linear programming applications, two aspects of the method received little notice. First, Khachian used the algorithm as a theoretical tool to obtain a result in computational complexity and, second, the algorithm was originally developed for nonlinear optimization.

In this paper, we survey some recent work illustrating the uses of the ellipsoid algorithm in nonlinear programming and computational complexity. After a brief review of the algorithm from two viewpoints, we discuss reports of Ecker and Kupferschmid [5, 6] comparing the ellipsoid algorithm with existing methods for nonlinear programming. We then give an overview of some work of Grotschel *et al.* [15] who have applied the ellipsoid algorithm in an ingenious way to obtain computational complexity results for some problems in graph theory. For a more comprehensive survey, oriented toward linear programming, see Bland *et al.* [2].

2. The Algorithm

From the viewpoint of computational complexity, the basic problem involves a convex body (closed bounded convex set with nonempty interior) K in R^n and a convex function f. The goal is to find a point x^* where f assumes its (absolute) minimum on K. Briefly, the ellipsoid algorithm constructs a sequence of ellipsoids each of which contains the optimum point x^* with volumes decreasing in geometric ratio. As the ellipsoids shrink, the centers of the ellipsoids become better approximations to the desired solution.

For the purposes of the algorithm, each ellipsoid, E, may be represented in terms of its center, x, and a matrix, B, as

$$E = \{y + x \mid y^t B^{-1} y \leq 1\}. \tag{1}$$

(The superscript t denotes transpose.) We begin with an ellipsoid which contains the feasible set K. At each step, we choose a half space $\bar{H}$ in one of two ways. If the center of the current ellipsoid, x, is not in K, we choose H to be a separating

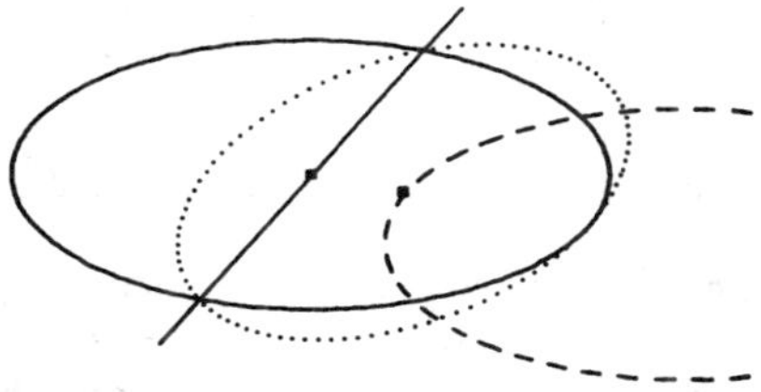

FIGURE 1a. Ellipsoid step at a nonfeasible point. Feasible region indicated by dashed line. New ellipse indicated by dotted line.

hyperplane between x and K and $\bar{H}$ to be the half space, determined by H, that contains K (see FIGURE 1a). On the other hand, if x is in K and is not optimal, we choose a hyperplane through x, so that all points y with $f(y) < f(x)$ lie in the half space $\bar{H}$ on one side of H (see FIGURE 1b). In the latter case, if $f(x)$ were differentiable at x, then H would be determined by the gradient of f at x. In either case, $E \cap \bar{H}$ contains all possible solutions. We then construct a new ellipsoid, E_H, of minimum volume containing $E \cap \bar{H}$ (see FIGURES 1a and b). The new ellipsoid E_H contains all possible solutions. In addition, the ratio of volumes of the ellipsoids satisfies

$$V(E_H)/V(E) \leq \exp(-1/2n) < 1. \tag{2}$$

Let D be the diameter of the initial ellipsoid. If K contains a ball of radius r and if ε is the required accuracy in the objective function value, then the number of steps required to solve the problem is $O(n^2 \log(D/r\varepsilon))$ [14].

To summarize: Implementation of the ellipsoid algorithm for computational complexity requires

(1) A way of determining an initial ellipsoid containing the feasible set K.

(2) A procedure for determining the separating hyperplane H.

(3) Update formulas for x and B. Here it is necessary to maintain sufficient accuracy in x and B so the new ellipsoid using rounded values will contain $E \cap \bar{H}$.

(4) A stopping criterion which determines when the center of the current ellipsoid is sufficiently close to the optimum point.

While this view of the ellipsoid algorithm is appropriate for studies of computational complexity, it is too restrictive for practical problems. We now give a view more appropriate to practical optimization problems and relate the ellipsoid approach to some other nonlinear optimization methods.

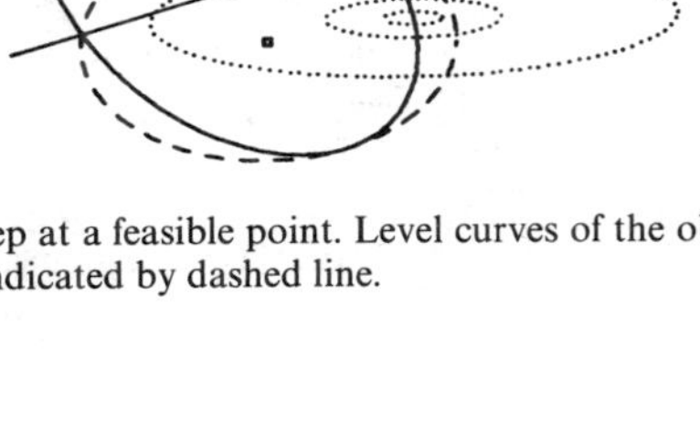

FIGURE 1b. Ellipsoid step at a feasible point. Level curves of the objective function shown as dotted lines. New ellipse indicated by dashed line.

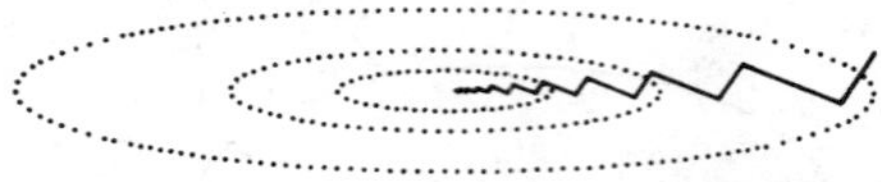

FIGURE 2a. Search path of simple gradient descent.

Consider the problem of minimizing a convex function f (with no constraints). A common iterative approach is to choose a search direction at each step, then search for a (one-dimensional) optimum in the chosen direction. The minimum point for f along the search line is the next approximate solution. A classical method, steepest descent, uses the negative gradient of f as the search direction. If the level surfaces of f are highly eccentric, the iteration sequence follows a zigzag path and the procedure converges slowly (FIGURE 2a). Quasi-Newton or variable metric algorithms maintain an approximation to the Hessian matrix $H = (\partial^2 f/\partial x_i\, \partial x_j)$ of f which is updated at each step. This approximate Hessian is used to modify the search direction given by the gradient. The modification can be interpreted as a change in the metric of the space of variables so that, in the new metric, level surfaces are approximately spherical. As a result, the modified gradient search direction leads more directly to the optimum (FIGURE 2b). The ellipsoid algorithm can be viewed as a kind of variable metric gradient search. As can be seen in FIGURES 1a and b, the direction to the center of the next ellipsoid is not the same as the gradient direction. The search direction is modified using the ellipsoid matrix B in the same way that the inverse Hessian H^{-1} is used in quasi-Newton methods. (See Goffin [11] for some results on the relation of B to H^{-1}.) Instead of searching in the modified gradient direction for a minimum of f, the ellipsoid method takes a uniform step in that direction. The step size is controlled by the size of the ellipsoid at each step. Methods employing deep cuts [5], in which the hyperplane H cuts off less than half of the ellipsoid E making E_H smaller, may be viewed as including line search in the ellipsoid algorithm. The view of the ellipsoid algorithm as a variable metric algorithm is essentially that expressed in Shor's original paper [28].

The size and shape of the new ellipsoid, E_H, are determined by several parameters [13]. For computational complexity applications, these parameters are chosen so that E_H is the ellipsoid of minimum volume containing $E \cap \bar{H}$. Goffin [12] has pointed out that this condition on E_H may be too restrictive for practical optimization. If the ellipsoid algorithm is viewed merely as a variable metric search algorithm, we can choose the parameters of E_H so that the ellipsoids shrink more rapidly. Goffin has experimented with many choices of parameters on three convex, nondifferentiable optimization problems. He has reported that the method converges even when the volume of the ellipsoid is reduced by a factor of 2 at each step. This drastic improvement in convergence rates certainly warrants further investigation.

FIGURES 3a and b illustrate two sequences of steps taken by the ellipsoid algorithm on a two-dimensional problem with a nonconvex objective function.

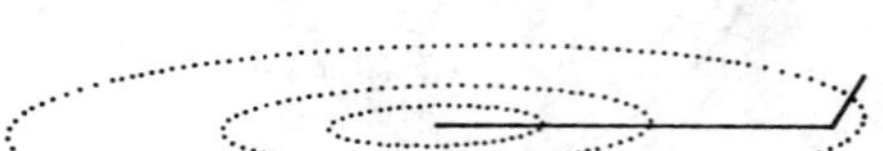

FIGURE 2b. Search path of variable metric gradient descent.

FIGURE 3a shows the initial steps while FIGURE 3b shows a typical sequence once the ellipsoid center is in the neighborhood of the solution. The feasible region is bounded by the dashed line and level curves of the objective function are given by dotted lines. The ellipsoid centers move fairly rapidly to the neighborhood of the optimum. Once the center of the ellipsoid is near the solution, feasible points tend to be separated by at most n (=number of variables) steps.

3. PRACTICAL APPLICATIONS

Ecker and Kupferschmid [5] have reported on comparisons of the ellipsoid method with a commercial code in solving inequality constrained nonlinear programming problems:

$$\begin{aligned} &\text{minimize:} && f(x), \\ &\text{subject to:} && g_i(x) \leq 0,\ i = 1, \ldots, m. \end{aligned} \tag{3}$$

In a more recent paper [6], they reported on more extensive comparisons of the ellipsoid method (EA) with an augmented Lagrangian code (NAG [9]), two quadratic subproblem codes (IQP, RQP), and a generalized reduced gradient code (GRG). Before discussing their results, we give a brief overview of these other methods.

The augmented Lagrangian method [8] makes use of an augmented Lagrangian function

$$F(x, \lambda) = f(x) + \sum \tfrac{1}{2}c_i[(g_i(x) + \lambda_i/c_i)_+]^2, \tag{4}$$

where $(y)_+$ denotes max $(y, 0)$, and $\lambda_i \geq 0$ are approximate Lagrange multipliers. If $g_i(x) + \lambda_i/c_i \geq 0$, the corresponding summand term is

$$\lambda_i g_i(x) + \tfrac{1}{2}c_i(g_i(x))^2 + \tfrac{1}{2}\lambda_i^2/c_i. \tag{4a}$$

The first term contributes to a Lagrangian function while the second (penalty) term is intended to force solutions to feasible points and ensure that the Hessian of F is positive definite near the minimum. The method involves an inner and an outer loop. In the inner loop, we search for a minimum of F by a quasi-Newton method keeping λ_i fixed. In the outer loop, each time F is minimized, the multipliers λ_i are adjusted to move them closer to the true Lagrange multipliers. The current minimum of F is accepted if the necessary (Kuhn–Tucker) conditions

$$\begin{aligned} &g_i \leq 0, \quad \lambda_i \geq 0, \quad \lambda_i g_i = 0, \\ &\nabla f = -\sum y_i \nabla g_i \end{aligned} \tag{5}$$

are satisfied by the current x and λ to within a preset error tolerance.

In the quadratic subproblem approach [26] (IQP), we again have an inner and an outer loop. At each step of the outer loop, we have some approximate solution y. We approximate the given problem (3) by a quadratic programming problem:

$$\begin{aligned} &\text{minimize:} && c^t x + 1/2x^t Gx, \\ &\text{subject to:} && Ax \leq b. \end{aligned} \tag{6}$$

The linear constraints are determined by the gradients of the constraint functions g at y. The objective function is quadratic with the linear part determined by the

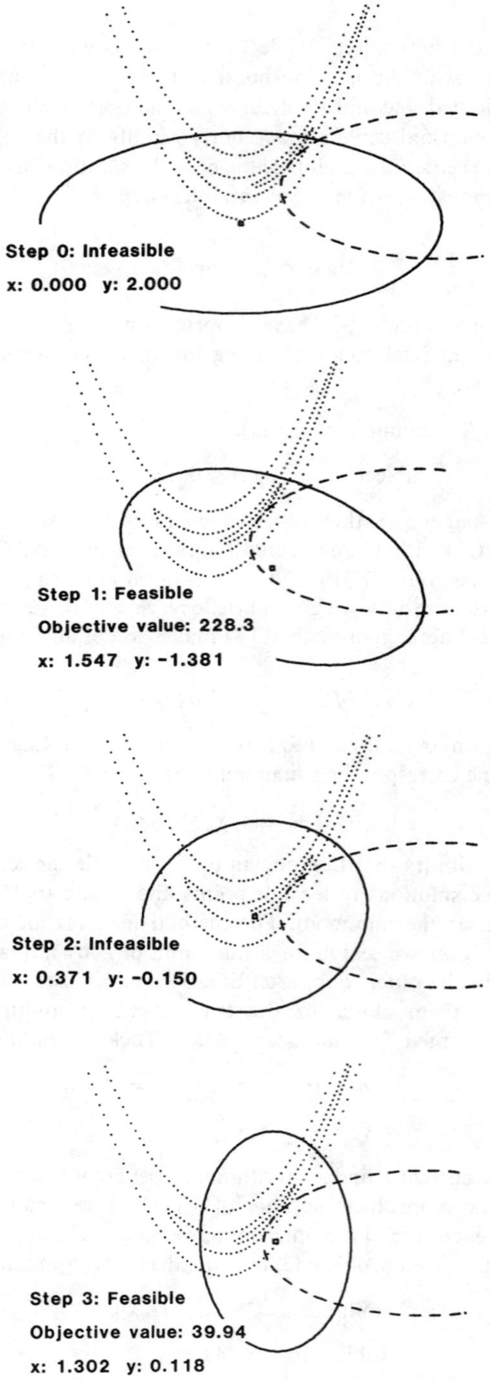

FIGURE 3a. Initial sequence of steps for the ellipsoid method applied to a nonconvex problem. Feasible region outlined by dashed line. Level curves of the objective function shown as dotted lines.

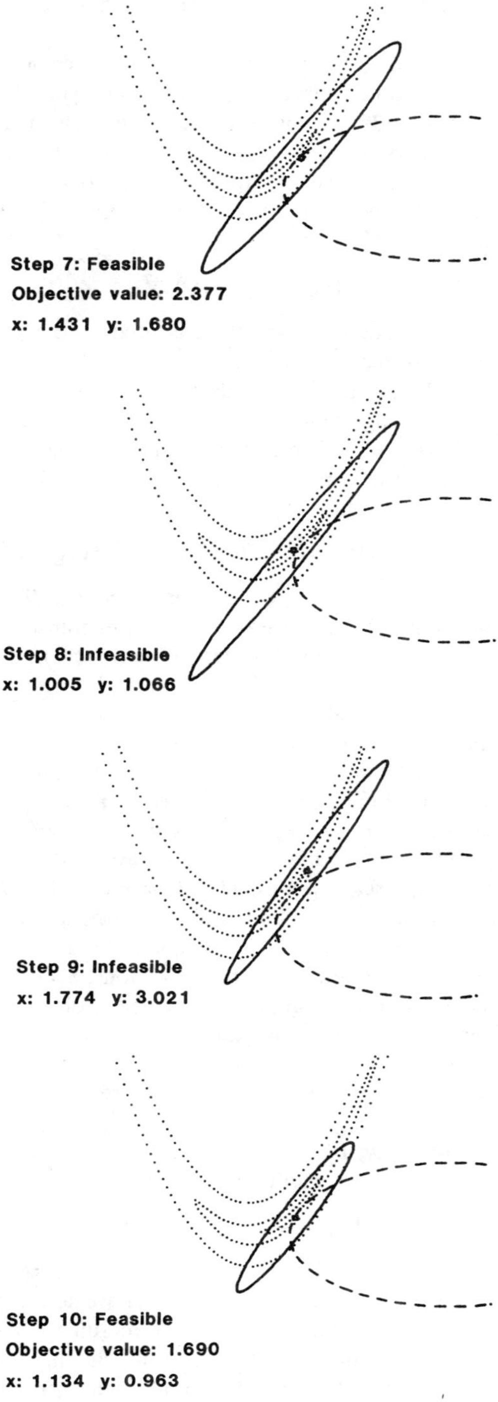

FIGURE 3b. Intermediate sequence of steps for the ellipsoid method applied to a nonconvex problem.

gradient of f at y and matrix G an approximation to the Hessian of the Lagrangian. If the solution to the quadratic subproblem is sufficiently close to y, we accept it as the solution to the problem (3). Otherwise, the solution to the quadratic subproblem determines a search direction to find a new approximate solution y. Information at the new approximate solution is used to update the approximate Hessian G and the outer loop continues with a new subproblem (6). In RQP [1], we form a quadratic approximation to the penalty function

$$f(\bar{x}) + r \sum ((g_i(\bar{x}))_+)^2. \quad \textbf{(6a)}$$

The solution to the quadratic approximation is used as the next iterate and the penalty parameter r is adjusted.

The generalized reduced gradient method [22] is somewhat similar to the simplex method of linear programming. At each point $\bar{x}$, we have a set of nb active constraints [for which $g_i(\bar{x}) = 0$]. These nb equalities are used to solve for nb of the variables, $\bar{x}_1$, in terms of the remaining variables $\bar{x}_2$. We then form the reduced objective function

$$f(\bar{x}) = f(\bar{x}_1, \bar{x}_2) = f(\bar{x}_1(\bar{x}_2), \bar{x}_2) = F(\bar{x}_2).$$

We use the reduced gradient $\nabla F(\bar{x}_2)$ to determine a search direction $\bar{d}$ within the manifold of solutions of $g_i(\bar{x}) = 0$. We search for a minimum of $F(\bar{x}_2 + \alpha\bar{d})$ projecting onto the manifold of solutions at each step of the search. If, during this search, a previously nonbinding constraint is violated, a point $\bar{x}_2 + \alpha\bar{d}$ is found where the constraint is satisfied as an equality, the constraint is added to the list of binding constraints, and a new F is formed. If we find a minimum of $F(\bar{x}_2 + \alpha\bar{d})$, we test the Kuhn–Tucker conditions (5) and either accept the current point as the minimizer or choose a new descent direction. In the latter case, we may drop a constraint from the list of binding constraints before repeating the iterative step.

Hock and Schittkowski [17] have compared several different nonlinear programming codes on a large set of published test problems. In general, codes based on the augmented Lagrangian method and the generalized reduced gradient approach required less CPU time, while the quadratic subproblem code almost always required fewer function and gradient evaluations. Schittkowski [27] carried out a more extensive study using computer generated problems. In that study, the quadratic subproblem approach and the generalized reduced gradient methods were highly ranked while the augmented Lagrangian codes were ranked below average. In addition, augmented Lagrangian codes were reported as more likely to fail to converge to the correct solution.

In their most recent study, Ecker and Kupferschmid defined a measure of the error at a particular approximate solution x by

$$p(x) = |f(x) - f(x^*)| + \sum \lambda_i^* |g_i(x)|, \quad (7)$$

where λ_i^* are the Lagrange multipliers at the optimum. Thus, $p(x)$ measures error in the objective function and errors in constraints which are active at the optimum. In their experimental design, they allowed each code to run to completion (or error return). After each iteration, statistics were recorded including relative error $E(\bar{x}) = p(\bar{x})/p(\bar{x}_0)$ and problem state CPU time. The test problems included 25 geometric programming problems and 25 general nonlinear programming problems all of which had only inequality constraints. Of the 50 problems, 13 were convex and 37

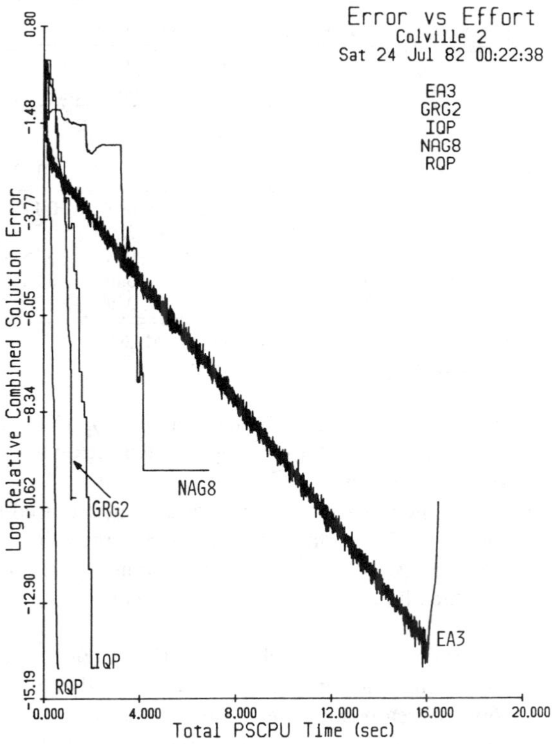

FIGURE 4. Plot of log error vs. effort (problem state CPU time). Reprinted with permission from Ecker and Kupferschmid [6].

nonconvex. In each problem, bounds were set on the variables (usually as part of the problem data), the initial point was chosen as the center of the region defined by the bounds, and the initial ellipsoid for EA was chosen to contain this region.

For several of the problems, graphs were given of log $E(\bar{x})$ vs. problem state CPU time. (FIGURE 4 reproduces one such graph.) These plots show EA with a sharp initial drop in the error followed by linear convergence to quite small error levels. The other methods show errors decreasing more slowly at first, then, due to superlinear convergence, decreasing rapidly to a level beyond which there is no significant improvement. The CPU time elapsed before superlinear behavior begins and the final error level obtained vary greatly from problem to problem and among the methods. To give an overview of the performances of the methods, Ecker and Kupferschmid included several tables comparing the methods on whole groups of problems. TABLES 1 and 2 are reproductions of two of these tables. TABLE 1 gives, for each of several relative error levels and each algorithm, the fraction of the test problems on which the given method was the first to reach the particular error level. This table shows EA to be superior for the initial error levels while GRG and RQP are better for higher error levels. TABLE 2 gives (as a measure of the robustness of the methods) the fraction of problems that were eventually solved to each error level by each of the codes. This table shows EA to be the most robust at all error levels.

TABLE 1
EFFICIENCY AT VARIOUS ERROR LEVELS FOR ALL 50 TEST PROBLEMS[a]

Method	Solution Error Level Required:	Fraction of Test Problems Solved First							
		10^{-1}	10^{-2}	10^{-3}	10^{-4}	10^{-5}	10^{-6}	10^{-7}	10^{-8}
EA		0.66*	0.44*	0.20	0.24	0.16	0.12	0.14	0.14
GRG		0.14	0.12	0.24	0.24	0.32*	0.36*	0.28*	0.18
IQP		0.00	0.08	0.14	0.16	0.16	0.18	0.22	0.28
NAG		0.08	0.10	0.10	0.06	0.08	0.08	0.08	0.10
RQP		0.12	0.26	0.30*	0.28*	0.26	0.24	0.24	0.30*

[a] Reprinted with permission from Ecker and Kupferschmid [6].
* Denotes best performance at this error level.

The combination of linear convergence, better initial error reduction, and almost certain convergence on this class of problems has led Ecker to compare EA to interval halving methods for root finding.

When the number of function and gradient evaluations required to reach a given error level are compared, IQP is by far the best. Among the others, EA is competitive but it does take an average of 10 times the evaluations of IQP (as do the others). Thus, for problems in which function and gradient evaluations are expensive, IQP is preferred.

Ecker and Kupferschmid tested other characteristics of the codes as well. To test dependence on starting point (and initial ellipsoid for EA), they ran some problems with some of the bounds greatly increased. This modification moves the starting value and for EA results in a much larger starting ellipsoid. They found that the qualitative behavior of EA was unaffected while the other methods often failed to converge. In fact, EA decreased the (initially larger) error more rapidly for the larger ellipsoids. The codes were also tested on a problem with a large number of stationary points (gradient = 0) only two of which were global minima. The codes were started from several different points. On this problem, EA consistently found a global minimum while the other methods often converged to a nonglobal local minimum or a saddle point.

The good initial behavior of EA reported by Ecker and Kupferschmid and similar experience with the ellipsoid method on linear programming problems [25]

TABLE 2
ROBUSTNESS AT VARIOUS ERROR LEVELS FOR ALL 50 TEST PROBLEMS[a]

Method	Solution Error Level Required:	Fraction of Test Problems Solved							
		10^{-1}	10^{-2}	10^{-3}	10^{-4}	10^{-5}	10^{-6}	10^{-7}	10^{-8}
EA		0.98	0.98	0.96	0.96	0.96	0.96	0.94	0.92
GRG		0.90	0.80	0.76	0.76	0.76	0.76	0.60	0.30
IQP		0.80	0.76	0.76	0.76	0.76	0.76	0.76	0.76
NAG		0.82	0.74	0.66	0.56	0.54	0.52	0.50	0.42
RQP		0.74	0.68	0.66	0.60	0.60	0.60	0.58	0.56

[a] Reprinted with permission from Ecker and Kupferschmid [6].

suggest that EA be used to find a starting point for another method with superlinear local convergence. Alternatively, Goffin's idea (see Section 2) of changing the parameters of EA to make the ellipsoids shrink more rapidly could be used when iterates are close enough to the solution that the problem is approximately convex.

In either case, some criterion must be developed to determine an appropriate point to change methods. In addition, most superlinearly convergent methods develop an approximation to the Hessian, H, of the Lagrangian function to obtain superlinear convergence. Since the ellipsoid matrix, B, approximates a multiple of H^{-1}, it would be useful to have a way of using the information in B to form an initial approximation to H rather than starting with the identity as is usually done.

In this context, it should also be noted that Ecker and Kupferschmid reported that the code for EA is much smaller than that for any other method. Thus, inclusion of EA as a preprocessor for another code would lead to a small increase in the size and complexity of the code, with most of the increase coming from the interface.

On the negative side, it should be noted that EA does poorly on problems with nonlinear equality constraints or "gully type" objective or constraint functions. For instance, consider the objective function of FIGURES 3a and b (Rosenbrock's function):

$$f(x_1, x_2) = (x_1 - 1)^2 + 100(x_2 - x_1)^2.$$

This function defines a surface which is a steepsided, parabolic gully, in which the bottom slopes relatively slowly to the minimum. If we try to minimize f using EA, starting in the second quadrant, we find that the ellipsoid aligns with the gully and cannot "turn the corner." Similar problems arise with equality constraints since the ellipsoid tends to become aligned along the constraint surface. Attempting to include equality constraints as penalties in the objective function

$$F(\bar{x}) = f(\bar{x}) + r \sum (g_i(\bar{x}))^2$$

(sum over equality constraints) leads to the kind of gully functions described above. One possible approach to equality constraints could be to apply EA in a reduced space obtained by projecting onto the manifold determined by the equality constraints. Thus, EA might be included in a generalized reduced gradient type procedure in which EA was used in the manifold determined by the equality constraints to find a good starting point and identify the (initial) set of active inequality constraints. Since the iterates of EA tend to be widely separated initially, more work would probably be involved in projecting onto the constraint surface at each step than in GRG.

The results of Ecker and Kupferschmid show that an ellipsoid method for relatively small, inequality constrained nonlinear optimization problems can be competitive with existing codes and indicate that combination with another algorithm could be an approach superior to these existing methods. Application to equality constrained problems is not recommended without changes to the algorithm and considerable testing.

4. Complexity Applications

The ellipsoid algorithm has been used by several authors to show that various problems can be solved in polynomial time and space. In order to apply the ellipsoid method, we must formulate the problem as a convex optimization problem with feasible set K, containing a ball of radius r, and objective function f in such a way that [14]:

(1) An initial ellipsoid with diameter D, containing K, can be found in polynomial time such that $\log D$ is a polynomial in the input data. Also, r can be found in polynomial time with $-\log r$ a polynomial in the input data such that K contains a ball of radius r.

(2) A procedure can be found for determining in polynomial time whether x is in K. In addition, if x is not in K, the procedure must determine a separating hyperplane in polynomial time as well.

(3) The time required to update x and B (which depends on the accuracy with which x and B are stored) must be polynomial in the input data.

(4) A stopping criterion is determined so that $-\log \varepsilon$ is a polynomial in the input data, where ε is the required accuracy in f.

Given this information, the problem is solved in $O(n^2 \log (D/r\varepsilon))$ steps so the number of steps is polynomial in the input data. By conditions 2 and 3, each of these steps takes polynomial time.

In Khachian's proof that linear programming problems can be solved in polynomial time, the formulation as a convex optimization problem and the separation problem (2) were trivial. The major work was in showing conditions 1, 3, and 4.

Here we survey some work of Grotschel, Lovasz, and Schrijver on perfect graphs. In this work, the formulation of the problem as a convex optimization problem for which the separation problem can be solved in polynomial time is highly nontrivial. We begin with some definitions:

Recall that an algorithm for a problem is said to be polynomial if the time and storage required to solve any instance of the problem are bounded by a fixed polynomial in the size of the input for the instance. A problem is polynomial (in P) if there is a polynomial algorithm for the problem. A problem is in NP if there is a polynomial algorithm for guessing and then checking a solution for each instance of the problem. It is presently unknown whether $P = NP$. A class of problems is said to be NP-complete if the existence of a polynomial time algorithm for the class would imply that $P = NP$. For a more detailed and precise discussion of these concepts, see Garey and Johnson [10].

Since it will not affect the question at hand and simplifies exposition, we assume all graphs are simple (no loops or multiple edges).

A set of vertices in a graph is called *stable* if no two of the vertices of the set are adjacent.

A set of vertices is called a *clique* if each vertex of the set is adjacent to all other vertices in the set.

Note that at most one element of a stable set can be in any clique.

The *stable set number*, $\alpha(G)$, of a graph G is the maximum number of elements in a stable set of G.

The *clique cover number*, $\rho(G)$, of G is the minimum number of cliques required to cover the vertices of G.

Since each element of a stable set must lie in a different clique, $\alpha(G) \leq \rho(G)$ for any graph G.

A subgraph H of G is called a *vertex induced subgraph* if the edges of H are exactly all the edges of G which join two vertices of H.

A graph G is said to be *perfect* if $\alpha(H) = \rho(H)$ for each vertex induced subgraph H of G.

For general graphs G, the problems of computing $\alpha(G)$ and $\rho(G)$ have been shown to be NP-complete [10]. Using the ellipsoid algorithm, Grotschel, Lovasz, and Schrijver showed that, if G is a perfect graph, $\alpha(G)$ and thus $\rho(G)$ can be computed in polynomial time.

There are several ways in which this discrete problem might be formulated as a convex optimization problem. I' $V = \{v_1, \ldots, v_n\}$ is the vertex set of graph G and if W is any subset of V, define the incidence vector $x^W = \{x_1^W, \ldots, x_n^W\} \in R^n$ by

$$x_i^W \begin{cases} = 1 & \text{if vertex } v_i \in W \\ = 0 & \text{otherwise.} \end{cases} \tag{8}$$

Let $P(G)$ denote the convex hull of $\{x^W \mid W \text{ is a stable set of } G\}$. Then $\alpha(G) = \max \{\sum x_i \mid x \text{ in } P(G)\}$ expresses $\alpha(G)$ as the solution to a convex (linear) optimization problem. Unfortunately, the constraints of $P(G)$ are not given explicitly and the general separation problem cannot be solved in polynomial time.

Since the separation problem for the set $P(G)$ is difficult, we look for another formulation in which we have constraints that are given explicitly. If C is a clique in G and x is the incidence vector of a stable set W in G, then

$$\sum \{x_i \mid v_i \text{ is in } C\} \leq 1. \tag{9}$$

Since the inequalities (**9**) are linear and $P(G)$ is the convex hull of the incidence vectors of stable sets in G, for any clique C, the inequalities (**9**) are satisfied by all points of $P(G)$. If we define $P^*(G)$ to be the set of x in R^n which satisfy (**9**) for all cliques C and

$$\alpha^*(G) = \max \{\sum x_i \mid x \text{ in } P^*(G)\}, \tag{10}$$

then

$$P(G) \subseteq P^*(G) \qquad \text{and} \qquad \alpha(G) \leq \alpha^*(G).$$

By adding together the constraints (**9**) for all C in a minimal clique cover of G, it is easy to see that $\alpha^*(G) \leq \rho(G)$. Thus, for perfect graphs, $\alpha(G) = \alpha^*(G)$. For general graphs, however, the complexity of the separation problem for $P^*(G)$ is closely related to the complexity of a problem known to be NP-complete (the weighted clique problem) [15]. Thus, one cannot expect to find a polynomial time separation algorithm for $P^*(G)$ in general.

Grotschel, Lovasz, and Schrijver used another parameter $\Theta(G)$ introduced by Lovasz [24] in his study of the Shannon capacity of a graph. For any graph G, $\Theta(G)$ is the solution to a convex, nonlinear optimization problem to which the ellipsoid algorithm may be applied while $\alpha(G) \leq \Theta(G) \leq \alpha^*(G)$. Thus, in the case of perfect graphs, since $\alpha(G)$ and $\rho(G)$ are equal to $\Theta(G)$, they can be computed in polynomial time. $\Theta(G)$ can be defined in a variety of equivalent ways. For our purpose, it is convenient to proceed as follows. Let $\mathscr{B}(G)$ be the set of $n \times n$ symmetric matrices that are positive semidefinite, have trace equal to 1, and have entry b_{ij} equal to zero if vertex v_i is adjacent to vertex v_j. Then

$$\Theta(G) = \max \{\sum b_{ij} \mid (b_{ij}) \in \mathscr{B}(G)\}. \tag{11}$$

While $\mathscr{B}(G)$ is a convex set, $\mathscr{B}(G)$ is not fully dimensional since many equality constraints appear in the definition of $\mathscr{B}(G)$. In order to apply the ellipsoid algorithm, we project $\mathscr{B}(G)$ to a lower dimensional space. We project from n^2 dimensional space to m dimensional space where $m = (n(n-1)/2 + n - 1 -$ (number of edges of G)). The projection map merely discards certain matrix entries which are redundant or zero. We keep those entries below the diagonal which are not required to be 0 and the first $n - 1$ diagonal entries. From these entries, the original matrix may be reconstructed. $\mathscr{B}(G)$ contains the scalar matrix, $1/n$ times the identity matrix. The projection of $\mathscr{B}(G)$, $\mathscr{B}'(G)$, contains a ball about the image of this scalar matrix. To solve the separation problem for $\mathscr{B}'(G)$, we need only check if the reconstructed matrix is positive semidefinite since all other conditions are automatically satisfied. This can be done in polynomial time using Gaussian elimination and some determinant calculations [15]. Since the matrices in $\mathscr{B}(G)$ are positive semidefinite with trace one, all entries in the matrices must have absolute value less than one. Thus an initial ellipsoid can be easily determined. Also, $\Theta(G) = \alpha(G)$ must be an integer so it is sufficient to solve the optimization problem to an accuracy in the objective function of less than one-half. Thus the conditions for applying the ellipsoid method are satisfied.

A major drawback of all ellipsoid based algorithms is the required accuracy for storage and computation of intermediate results. In order to prove that the ellipsoid method yields the correct solution, the number of bits of accuracy required tends to be about the same as the number of ellipsoid steps required [14]. While this is a polynomial in the size of the problem, it is far beyond the capacity of most hardware. Thus implementation of ellipsoid methods involves either attaining this computational accuracy in software or using only the hardware accuracy. The first alternative drastically slows down computations while use of the second alternative results in an algorithm that is no longer provably correct. For example, Grotschel, Lovasz, and Schrijver reported some computational experiments with the separation algorithm for $\mathscr{B}'(G)$ described above. They stated that, for problems with 10 to 20 vertices using standard computer accuracy, the separation algorithm does not give consistently correct decisions as to whether $\bar{x}$ is in $\mathscr{B}'(G)$. They recommended that the method not be used as a practical procedure for determining the stability number of perfect graphs.

In practice, the ellipsoid method seems to be relatively insensitive to errors in the separation routine which occur when $\bar{x}$ is close to the boundary of the feasible region. Also, it has not been necessary to carry the ellipsoid data to greater than machine accuracy. As pointed out above, while the algorithm works in practice in many cases, it is no longer provably correct. Because of these accuracy considerations, an efficient implementation based on the ellipsoid algorithm that is also provably correct seems quite unlikely.

Since Edmonds [7] suggested identifying polynomial time complexity algorithms with efficient or "good" algorithms, the idea has been a useful heuristic for the development of algorithms. While most exponential time methods tend to be variants of exhaustive search, polynomial time methods seem to reflect a deeper understanding of the underlying structure of the problem [10]. There are certainly efficient algorithms such as the simplex method in linear programming [21] and certain shortest path algorithms [19] that have exponential time complexity. These efficient algorithms with exponential time complexity are rare. On the other hand, before Khachian's result, most polynomial time algorithms arising in practice had time complexity a very low degree polynomial [10]. Thus, while some efficient algorithms might not be polynomial, most polynomial algorithms are efficient.

Since the appearance of Khachian's result, a large number of polynomial time algorithms based on the ellipsoid method have appeared. For instance, in another paper [14], Grotschel *et al.* have shown how many combinatorial problems, previously known to be solvable in polynomial time, can be solved using the ellipsoid method. Also, Grotschel and Pulleyblank [16] have used the ellipsoid algorithm to show that the max-cut problem for certain graphs is polynomial. Because of the slow shrinkage rate of the ellipsoids and the high accuracy required, all of these algorithms are certain to be inefficient. While the ellipsoid algorithm has provided a powerful new tool for proving polynomial time complexity, the resulting algorithms have eroded the heuristic value of the polynomial time criterion. As a consequence, a polynomial time complexity result can only be regarded as a strong indication that an efficient exact algorithm exists just as an NP-completeness result indicates that such an algorithm is unlikely. In fact, Grotschel *et al.* [14] view results obtained by the ellipsoid method as pointing out directions for the search for practical algorithms. As further indications of this view, Khachian's result seems to have stimulated interest in the average case complexity of the simplex method (Dantzig [4], Borgwardt [3], and Smale [29]) and the result described in this section has encouraged rather than discouraged efforts to find a good algorithm for stable sets in perfect graphs.

Note added in proof:

5. A New Polynomial Time Algorithm

Recently, N. Karmarkar [31] has developed and implemented a new polynomial time algorithm for the linear programming problem:

$$\text{Find } x \text{ to minimize } c^T x,$$

$$\text{subject to: } x \in \Omega \cap S, \tag{12}$$

$$\text{where } \Omega = \{x \mid Ax = 0\} \text{ and } S = \{x \mid x \geq 0, \textstyle\sum x_i = 1\},$$

where the minimum value of $c^T x$ is known to be zero and the initial point, $x^{(0)}$, is known to be feasible.

At each step, a projective transformation, T, maps the simplex S onto itself, taking the current iterate, $x^{(k)}$, to the center, a_0, of S. The transformed objective function, c', is projected onto c_P in $\Omega' \cap \{x \mid \sum x = 0\}$ where Ω' is the image of Ω under T. Thus $a_0 - \alpha c_P$ is in $\Omega \cap S$ for sufficiently small α. For an appropriate choice of α, we choose $b' = a_0 - \alpha c_P$ and set $x^{(k+1)} = T^{-1}(b')$.

Karmarkar shows that the algorithm completes in $O(nL)$ steps and $O(n^{3.5}L^2)$ time, where n is the number of variables and L is the number of bits required to encode the problem. These results compare favorably with $O(n^2L)$ steps and $O(n^6L^2)$ time for the ellipsoid method. Each iterate of the Karmarkar algorithm is an improved feasible point, whereas, in the ellipsoid method, each feasible point is followed by several nonfeasible points and any particular feasible point need not be better than its predecessor. Also, in the Karmarkar algorithm, each step depends only on the current point so that small errors at any step can slow convergence but not prevent it. Thus regular floating point arithmetic may be used in contrast to the accuracy required by the ellipsoid theory.

Note that, at any point, we may guess that the variables that are currently the largest will be basic at the optimum and test feasibility and optimality as in the revised simplex algorithm. Karmarkar has used this idea in his implementation. He has reported [33] that, in practice, the algorithm seems to take $O(\log n)$ steps to identify the optimal basis. Some impressive comparisons [34] with a commercial simplex-based code have been reported. It should be mentioned that problems in the form (**12**) will have an artificial starting basis that is almost totally degenerate. Thus the simplex method would start with a severe handicap.

Several other questions require further investigation:

(1) Most of the time on each iteration is spent in projecting c' onto c_P. Currently, each step of the Karmarkar algorithm takes many times what a simplex iteration would take. An efficient projection method that exploits the sparsity of the coefficient matrix A is needed.

(2) The algorithm requires that the minimum objective value be zero. If the actual minimum, c_m, is known, we may subtract c_m from each objective coefficient to achieve this (since $\sum x_i = 1$). If c_m is not known, it must be estimated and the efficiency of the method appears to depend on the accuracy of the estimate.

(3) A more general system of constraints, $Ax = b$, may be rewritten in the required form $Ax - b\sum x_i = 0$ using $\sum x_i = 1$. However, even if A were sparse, it is now dense, although the change is only of rank one. Thus more general linear programming problems can be expected to take more time per iteration.

(4) A thorough theoretical study of the algorithm might explain the observed average case behavior and suggest ways of making a good choice for c_m and the stepsize parameter α.

Karmarkar has also suggested that the algorithm would apply fairly directly to quadratic programming problems and might be applicable to more general convex programming problems. He suggests several other directions for further research in [32].

Acknowledgments

I would like to thank J. Ecker, E. Lutwak, D. Shmoys, and P. Wolfe for their comments on previous versions of this paper which have helped improve its accuracy and readability.

References

1. Biggs, M. C. 1975. Constrained minimization using recursive quadratic programming. *In* Toward Global Optimization. L. C. W. Dixon & G. P. Szego, Eds.: 341–349. North-Holland. Amsterdam, the Netherlands.
2. Bland, R. G., D. Goldfarb & M. J. Todd. 1981. The ellipsoid method: a survey. Oper. Res. **29:** 1039–1091.
3. Borgwardt, K.-H. The average number of pivot steps required by the simplex method is polynomial. Preprint.

4. DANTZIG, G. B. 1980. Expected Number of Steps of the Simplex Method for a Linear Program with a Convexity Constraint. Technical Report SOL 80-3R. Department of Operations Research, Stanford University. Stanford, Calif.
5. ECKER, J. G. & M. KUPFERSCHMID. 1981. An Ellipsoid Algorithm for Nonlinear Programming. Research Report. Rensselaer Polytechnic Institute. Troy, N.Y.
6. ECKER, J. G. & M. KUPFERSCHMID. 1982. A Computational Comparison of Several Nonlinear Programming Algorithms. Report OR & S-82-2. Rensselaer Polytechnic Institute. Troy, N.Y.
7. EDMONDS, J. 1965. Paths, trees and flowers. Canad. J. Math. **17:** 449–467.
8. FLETCHER, R. 1974. Methods related to Lagrangian functions. *In* Numerical Methods for Constrained Minimization. P. E. Gill & W. Murray, Eds. Academic Press, Inc. New York, N.Y.
9. FOX, L. & J. H. WILKINSON. 1978. NAG Fortran Library Manual. Numerical Algorithms Group Ltd. Oxford, England.
10. GAREY, M. R. & D. S. JOHNSON. 1979. Computers and Intractability, a Guide to the Theory of NP-Completeness. Freeman. San Francisco, Calif.
11. GOFFIN, J. L. 1981. Variable Metric Relaxation Methods. I. A Conceptual Algorithm. Technical Report SOL 81-16. Systems Optimization Laboratory, Department of Operations Research, Stanford University. Stanford, Calif.
12. GOFFIN, J. L. 1981. Convergence results in a class of variable metric subgradient methods. *In* Nonlinear Programming 4. O. Mangasarian *et al.*, Eds. Academic Press, Inc. New York, N.Y.
13. GOLDFARB, D. & M. J. TODD. 1982. Modifications and implementation of the ellipsoid algorithm for linear programming. Math. Programming **23:** 1–19.
14. GROTSCHEL, M., L. LOVASZ & A. SCHRIJVER. 1981. The ellipsoid method and its consequences in combinatorial optimization. Combinatorica **1:** 169–197.
15. GROTSCHEL, M., L. LOVASZ & A. SCHRIJVER. Polynomial algorithms for perfect graphs. Preprint, 1981.
16. GROTSCHEL, M. & W. R. PULLEYBLANK. 1981. Weakly bipartite graphs and the max-cut problem. Oper. Res. Lett. **1:** 23–27.
17. HOCK, W. & K. SCHITTKOWSKI. 1981. Test Examples for Nonlinear Programming Codes. Lecture Notes in Economics and Mathematical Systems Vol. 187. Springer-Verlag. West Berlin, FRG.
18. IUDIN, D. B. & A. S. NEMIROVSKII. 1976. Informational complexity and effective methods of solution for convex extremal problems. Ekonom. i Mat. Metody **12:** 357–369. [Translated in Matecon **13:** 25–45 (1977).]
19. KERSCHENBAUM, A. 1981. A note on finding shortest path trees. Networks **11:** 399-400.
20. KHACHIAN, L. G. 1979. A polynomial algorithm in linear programming. Dokl. Akad. Nauk SSSR **244:** 1093–1096. [Translated in Soviet Math. Dokl. **20:** 191–194 (1979).]
21. KLEE, V. & G. L. MINTY. 1972. How good is the simplex algorithm. *In* Inequalities III. O. Shisha, Ed.: 159–175. Academic Press, Inc. New York, N.Y.
22. LASDON, L. S., A. D. WAREN, A. JAIN & M. RATNER. 1978. Design and testing of a generalized reduced gradient code for nonlinear programming. ACM Trans. Math. Software **4:** 34–50.
23. LAWLER, E. L. 1980. The great mathematical sputnik of 1979. The Sciences (September): 12–15, 34–35.
24. LOVASZ, L. 1979. On the Shannon capacity of a graph. IEEE Trans. Inform. Theory **IT–25:** 1–7.
25. PICKEL, P. F. A comparison of some improvements to the ellipsoid method in linear programming. Polytechnic Institute of New York. Farmingdale, N.Y. Preprint, 1980.
26. POWELL, M. J. D. 1978. A fast algorithm for nonlinearly constrained optimization calculations. *In* Numerical Analysis (Dundee, 1977). G. A. Watson, Ed. Lecture Notes in Mathematics No. 630. Springer-Verlag. West Berlin, FRG.
27. SCHITTKOWSKI, K. 1980. Nonlinear Programming Codes. Lecture Notes in Economics and Mathematical Systems Vol. 183. Springer-Verlag. West Berlin, FRG.
28. SHOR, N. Z. 1970. Convergence rate of the gradient descent method with dilation of the space. Kibernetika **6:** 80–85. [Translated in Cybernetics **6:** 102–108.]
29. SMALE, S. On the average speed of the simplex method of linear programming. Preprint, 1982.

30. WOLFE, P. 1980. A Bibliography for the Ellipsoid Algorithm. IBM Research Center Report. Yorktown Heights, N.Y.
31. KARMARKAR, N. A new polynomial-time algorithm for linear programming. AT&T Bell Laboratories. Murray Hill, N.J. Preprint, 1984.
32. KARMARKAR, N. Some comments on the significance of the new polynomial-time algorithm for linear programming. AT&T Bell Laboratories. Murray Hill, N.J. Preprint, 1984.
33. KARMARKAR, N. Talk at NYU-Courant Institute. October 19, 1984.
34. KOLATA, G. 1984. A fast way to solve hard problems. Science **225:** 1379–1380.

Index of Contributors

Author Index

Subject Index